W9-BES-473

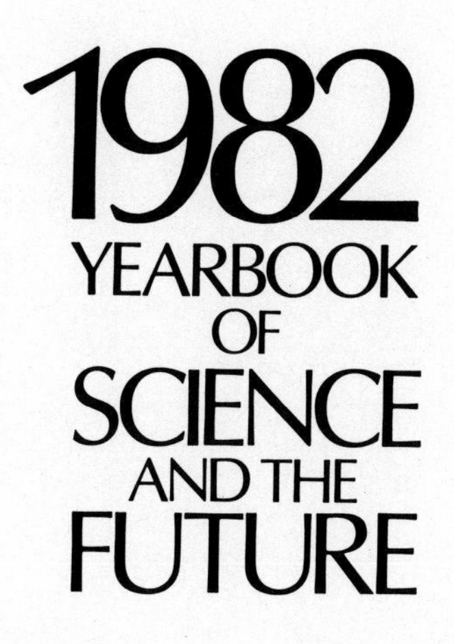
1982
YEARBOOK
OF
SCIENCE
AND THE
FUTURE

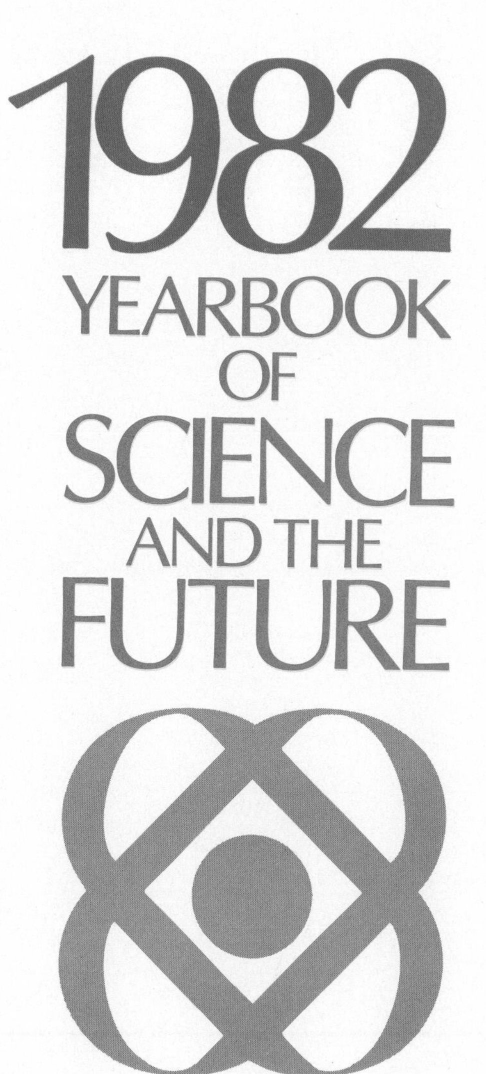

Encyclopædia Britannica, Inc.

Chicago Toronto London Geneva Sydney Tokyo Manila Seoul

Library of Congress Catalog Card Number: 69-12349
International Standard Book Number: 0-85229-384-4
International Standard Serial Number: 0096-3291

(Trademark Reg. U.S. Pat. Off.) Printed in U.S.A.

The University of Chicago
The Yearbook of Science and the Future
is published with the editorial advice of the faculties of
the University of Chicago.

Encyclopædia Britannica, Inc.

Chairman of the Board
Robert P. Gwinn

President
Charles E. Swanson

Vice-President, Editorial
Charles Van Doren

1982
YEARBOOK OF SCIENCE AND THE FUTURE

CONTENTS

Feature Articles

HUMAN-POWERED TRANSPORTATION

As fuel costs continue to rise, the use of human muscles for transportation becomes an increasingly attractive alternative.

by David Gordon Wilson

ZB
HPV ONLY
EXPRESS LANE
LANE1-60mph
LANE2-40mph
LANE3-30mph

Throughout much of the world more people are interested in physical exercise than ever before. Jogging, tennis, racquetball, roller-skating, and cycling all have enjoyed increased popularity in the last few years. But it is also true that the use of fossil fuels is still rising and, simultaneously, the employment of human muscles for useful tasks is still declining, despite the high prices and shortages of fuel.

Within this paradoxical situation the bicycle is enjoying its third period of development. Whether or not this era of bicycle enthusiasm, which began in the 1960s, will last longer than its predecessors is one of the questions that this review will attempt to answer. Human-powered propulsion on the water may also be due for a revival, with the principal impact being in recreation and sports. Muscle-powered aircraft have already achieved notable successes, but their future use appears to be limited, for reasons discussed below.

Rise of the automobile

The principal impact of future human-powered transportation will undoubtedly be in commuting vehicles. Bicycles have been almost excluded from widespread use in Western countries through a two-stage process, which now seems likely to be reversed. First, the prices of automobiles and the cost of operating them were steadily reduced in relative terms through the first seventy years of the century until those costs together constituted a comparatively minor part of an average U.S. family's annual budget. Almost every family owned and used at least one car. The roads became clogged and were progressively redesigned to be suitable almost exclusively for cars. Bicycles were "engineered" out of the urban and interurban picture. Then the average family, having the mobility to escape and not liking the congestion which the automobile itself had brought not only to the cities but also to the suburbs, moved farther and farther away from town and work. Thus for most people it became impractical to ride a bicycle to work even if roads that could be safely used existed.

But even at the peak of automobile use in the U.S., more than half the trips made by automobile were of fewer than eight kilometers (five miles), a distance easily accomplished by bicycle by most people in most areas and in most weather conditions. Furthermore, the average speed of bicyclists in urban areas over such short distances has been estimated to be somewhat higher than the average speed of automobiles on comparable trips. During the New York City transit strike of 1980 at least 70,000 people found that they could travel to their work by bicycle, and most of them seem to have enjoyed the experience.

In the future the use of human-powered vehicles seems bound to increase. It appears inevitable that fuel prices and the other costs of automobile ownership will continue to increase relatively and absolutely. This will lead an increasing proportion of people to choose to live nearer their places of work. More will, therefore, be able to commute by bicycle. When the concentration of bicycles reaches some critical density, highways will be modified again to facilitate their use, riding bicycles will become even more attractive, and so on.

__DAVID GORDON WILSON__ is a Professor of Mechanical Engineering at the Massachusetts Institute of Technology, Cambridge.

(Overleaf) Illustration by John Youssi

Improving the bicycle

But there is another factor in this process. Will the bicycle itself be improved? Or has it already reached something near perfection?

The principal justification for believing that the bicycle cannot be improved much further is that it reached approximately its present configuration in the 1890s. Since then improvements and modifications have been minor. The model that evolved by the 1890s was such a vast improvement over the bicycles of the early nineteenth century that some leveling off in its rate of development was to be expected. And in comparison with other forms of transport its characteristics are remarkable. Its mass is one-tenth of its payload, while the automobile weighs of the order of ten times the mass of its average load. The bicycle's speed and range are much greater than that of the horse, and yet its appetite is tiny (even if that of the rider is included). It has indeed been brought to a refined state of development.

And yet that development has ceased, and in order to understand why one must look at the history of the bicycle. Its origins are unknown, but by the early 1800s the bicycle existed as a crude beam carrying two heavy nonsteering wheels. There were no pedals, and the operator ran or walked along the ground. This model was not taken seriously as a means of transportation. And then in about 1816 Baron Karl de Drais de Sauerbrun made the front wheel steerable (though still without pedals), discovered that one could balance with the feet up when coasting, showed that he could travel long distances faster than the horse-drawn "posts" of the period, patented his machine in Paris, and started a craze. This was the first bicycle revival.

But it died away. It may be that the railroads had more appeal and more apparent promise. For whatever reason no thread of development came from the "Draisiennes" or at least not until the 1860s, when Pierre Michaux put pedals on the front wheel. Undoubtedly others had done so earlier, just as others had developed and sold advanced "velocipedes" of various types, but none was accepted and used in sufficient numbers to influence either technology or transportation. Michaux was simultaneously a craftsman, a businessman, and a showman, and he started the second revival. Paris was delirious with his velocipedes, and the craze spread in time to much of the rest of Europe and to the U.S.

The velocipede eventually evolved into what came to be called the "ordinary," a bicycle with a very large front wheel and a very small back one. During the period from about 1870 to 1885 these vehicles engaged the enthusiasms and clannishness of well-to-do young men, much as sports cars did later. They required a degree of athletic skill to mount and ride, and the dangers from a forward fall were great enough that family men, however fit, were reluctant to use them. Designers then began to satisfy the desires of those who could or would not ride ordinaries. First they introduced tricycles, which had a short period of popularity in the 1880s, and then chain-driven "safety" bicycles with wheels of equal size. After the reintroduction of the pneumatic tire in 1888, the safety bicycle reached something very close to its present form by 1895. Gears were introduced early in the 20th century but were not included on most bicycles in the United States until the 1960s.

Bicycles of the 19th century included (right) the "Draisienne" of about 1816, the first to have a steerable front wheel, and (below) the Michaux "velocipede" of the 1860s, one of the first with pedals. At the bottom right is a high-wheel "ordinary" of 1879, a model that enjoyed great popularity from about 1870 to 1885. Because the ordinary required some degree of athletic ability to mount and ride, tricycles such as that at the bottom left were introduced.

The "American Star" version of the ordinary (top left), introduced in 1880, featured a small front wheel and large rear one. Among the first of the chain-driven "safety" bicycles, with wheels of equal size, was the "Rover" of 1880 (top center). A variation that failed was a two-wheeled German velocipede of 1864 (top right). By 1890, when the "Cleveland" (above) was introduced, bicycles had achieved the basic form they were to retain to the present.

Cyclists finish a race in an illustration from the late 19th century. The bicycles of today differ little from these models.

Up to that time advances in design and performance had been made as a result of competition, not only in the marketplace but in road and track races. The devotees of the high-wheelers scorned the "safeties" until they were trounced in races. Users of the new pneumatic tires were derided until they overcame all opposition. And immediately after they were successful in racing, the new ideas were accepted in everyday use.

But then something unfortunate for future development occurred. A body known as the Union Cycliste Internationale (UCI; International Cyclists' Union) was set up to control the rules for international bicycle racing. The rules stated, in effect, that racing bicycles must conform to the type and style existing at the start of the 20th century.

These rules have had a stifling effect on bicycle development. One example concerns the recumbent bicycle, on which the rider sits back on a chairlike support rather than perching on a saddle. Such bicycles have been made at intervals at least since the turn of the century. In 1934 a relatively unknown French rider on a recumbent bicycle (known misleadingly as the Velocar) was allowed to ride against the then world bicycling champion. The Frenchman won and also set several world records. The UCI met in some confusion but resolved the situation swiftly. The recumbent, it ruled, was not a bicycle. The message to would-be developers and entrepreneurs was clear. No improvements to the 1900s bicycle would be allowed.

Recently there have been some startling improvements in human-powered vehicle performance. These were brought about because a few people in California, led by Prof. Chester Kyle of California State University at Long Beach, set up an organization, the International Human Powered Vehicle Association (IHPVA), that had almost no restrictions on the design of the competing vehicles. (The only confining rule is that there be no energy storage other than that in the rider.) In May 1980 the 200-meter flying-start speed championships were won at over 90 kilometers per hour (one kilometer equals 0.621 miles) in the single-rider category and at almost 101 kilometers per hour in the multiple-rider machines, the winner in the latter having two riders. On conventional bicycles the comparable speed has been under 69 kilometers per hour. This difference in speeds between the bicycles bound by turn-of-the-century rules and those unfettered by such restrictions is increasing. The speeds attained in the annual IHPVA speed trials are still, after six years, showing an almost linear rise with time. Calculations indicate that the peak speed that could be reached by a single rider on such a vehicle should be over 112 kilometers per hour.

If history were to repeat itself, one should expect an almost immediate transference of the technology used in the winning vehicles to those employed by the everyday commuting or recreational rider. But so far this has not happened. There are several small firms that make recumbent bicycles and tricycles, not derived from any of the IHPVA vehicles but influenced by the public acceptance brought about by news stories of the races. These machines have so far offered rather small advances in performance. The question of importance for the future of human-powered commuting vehicles is whether or not the full technology of race winners will be used in the way it was in the 1880s and 1890s.

Al Voigt

New bicycle and tricycle designs feature streamlined shapes, enclosed bodies, and recumbent seats for one or two riders. The tricycle below won a 1980 international championship race with a maximum speed of 101.3 kph (62.9 mph).

George J. Naoum

Two men lie prone to pedal a long vehicle covered with clear plastic (above). A view from below reveals the pedals and chain mechanisms (right). The Avatar 2000 (opposite, top left and right) features a reclining seat that allows efficient power delivery to the pedals and also has a low center of gravity to prevent over-the-handlebar spills and to improve the weight distribution of the rider. Opposite, bottom, is a vehicle in which the rider lies prone and steers, while cranking, by twisting his or her wrists in handgrip gimbals; the steering motion is transmitted to the fork by cables.

(Top, left and right) Michelle Forrestall; (bottom) Paul Van Valkenburgh

There are several significant differences between the present time and that of the second bicycle revival. In those days almost the only powered vehicles competing for the use of the streets were electric trolley cars, very limited in extent and predictable in behavior. Bicyclists also encountered years of hostility from horse users, but by the 1890s they had won their right of passage and were campaigning for better roads. New styles of human-powered vehicles did not, therefore, have to pay great regard to the impact on or from other road users. Now, however, the highways are completely dominated by motor vehicles, from high-speed motorcycles to huge trucks and buses. The winning IHPVA machines are generally very low to the ground, with the rider(s) almost totally recumbent or supine, a configuration that would be unsafe for use in traffic.

Another difference is that the technology of aerodynamics had not been used to permit increased speeds in racing machines in the 1880s and 1890s. When the rules for racing were frozen, aerodynamic enclosures were disallowed. The principal reason for the increases in the speeds of the IHPVA vehicles up to the present has been the reduction in aerodynamic drag, which dominates all other losses at speeds above about 19 kilometers per hour. Some of the drag reduction has been achieved by lowering the frontal area through having riders lie on their backs or stomachs. But most reduction has resulted from the streamlined bodies or enclosures which have encased the riders. These have aerodynamic drag coefficients as low as 0.1, one tenth that of a rider on a conventional bicycle. (The drag coefficient when multiplied by the frontal area and the pressure of the relative wind gives the drag force.)

In the early IHPVA races the enclosures used were quite crude in form and construction. Some of those employed in 1980 reached the sophistication of computer-calculated shapes and refined manufacture. Usually the riders had to be fitted into their vehicles as did late medieval jousting knights into their armor. Clearly this technology is not yet ready for transfer to commuting vehicles.

(Top, left) Don Monroe; (others) George J. Naoum

The vehicles shown above all have competed in international speed championships. The quadracycle (center left) won first place in a 1977 race with a speed of 79.5 kph (49.4 mph).

Experimental cycles come in a variety of shapes and sizes. A major aim of the designers is to reduce aerodynamic drag, achieved in part by lowering the frontal area and enclosing the rider.

If it were, would commuters accept it? The advantages of traveling at 32.4 kilometers per hour on a level road for the same energy expenditure (75 watts) as one would need for 18 kilometers per hour on a conventional machine are substantial, and they go beyond merely accomplishing given journeys in a shorter time. Much of the danger in highway travel comes from the relative speeds between different road users. By being able to travel near the average speed of automobile traffic in most urban and suburban areas, the rider of a streamlined vehicle would avoid a large proportion of the conflicts which the user of slower bicycles continually encounters.

An aerodynamic enclosure would also protect the rider from rain and road dirt. It could give some protection from injury in falls and minor accidents, although this aspect needs research. (Bicycles in which riders sit in a semi-recumbent position with the handlebars beneath the seat seem to give the riders greatly enhanced protection from the often traumatic forward spills, partly because they land first on their feet and partly because they are able to ward off the later impact with the pavement with their hands and arms, rather than their skulls. An enclosure must not, presumably, trap the arms.)

Some form of enclosure could also provide more convenient and safer storage for packages. And it would allow the rider the pleasure of listening to a radio, even a stereo, without having to use earphones.

All these characteristics must not be obtained, however, at the expense of convenience. A conventional bicycle can be mounted, ridden, and left against a store window before a car can be driven out of its garage, if the journey is short. If climbing inside an enclosed vehicle is a lengthy procedure, it will find favor only for long trips.

On balance it seems likely that the advantages of a streamlined enclosure are so great that for highway use some will be developed, particularly for recumbent bicycles, in a compromise length (that is, shorter than would give minimum drag). The riding position will be sitting rather than lying, because the need to see and be seen must be paramount. The resulting increased height of the center of gravity will rule out tricycles and four-wheeled vehicles, because the track width required to enable short-radius turns to be negotiated without overturning would be large, greatly reducing the usefulness of the machine. However, it is possible that small outrigger or "training" wheels could be deployed at stops and for very-low-speed maneuvering.

Would such a vehicle become popular? It should, even on its merits alone. There would be the added effect of status, the desire of some people to be seen using the latest development, and also the discovery that much of the "sweat" of bicycling—especially the belief among many that one always has a headwind—had been greatly reduced. If bicycles were to become generally accepted, the reduction in the use of petroleum fuels could be quite large.

Water vehicles

Oars and paddles have been the favored means for the application of human power to boat propulsion since prehistoric times. Roman galleys had hundreds of oars in up to three banks, one above the other. A drawing of a large 17th-century galley in the British Museum shows 54 oars, or sweeps, with five men on each. The men were likely to be criminals, chained to their

(Top) The Bettmann Archive; (center) Heinz Kluetmeier—Sports Illustrated ©Time Inc.; (bottom) adapted from a drawing by David Gordon Wilson; illustration by Ramon Goas

A "water velocipede" of the late 19th century (top) delivered power through the pedals to the screw propeller at the rear. At the left the Norwegian double sculls team wins the 1976 Olympic Games gold medal. These lightweight shells are the fastest human-powered boats in general use, but their speed could be exceeded by a pedal boat conforming to the design below. The outrigger pontoons provide stability.

benches. A central gangplank was patrolled by overseers, equipped to provide persuasion for anyone considered to be taking life too easy.

War canoes of the Maori of New Zealand were typical of others of the South Pacific in being propelled by sometimes more than 100 paddlers, usually in a kneeling position. Ordinary rowboats employed to move around harbors and waterways had (and still have) two oars pulled by one person, sitting on a central bench. Occasionally boats are propelled by a single oar resting in a notch in the stern and given an oscillating motion by a standing oarsman in some similarity to the movement of a fish's tail.

Another form of propulsion in which the human motive power stands is poling, used on gondolas in Venice and on punts on the River Cam in England, to give two well-known examples. This system is suitable only when the water is smooth, sheltered from high winds, and of constant moderate depth.

In all these older forms of propulsion the muscle actions are typical of those considered appropriate in the ancient world. The hand, arm, and back muscles are those principally employed. The largest muscles in the body, those in the legs, are used merely to provide reaction forces. Often the motion is one of straining mightily against a slowly yielding resistance. It seems obvious now that to get maximum useful power from the human body or, equally important, a given amount of power with maximum comfort, the leg muscles must be primarily used. It was not until the last century that this principle was applied successfully to boat propulsion with the development of sliding-seat lightweight skiffs or shells, which have formed the basis for racing ever since. Improvements in this century have resulted mainly from the use of high-strength materials.

Though the skiffs were the fastest generally used human-powered boats, their speeds were exceeded in the 1890s by pedal-driven boats in which the force is transmitted to screw propellers. Such a craft, powered by three bicyclists, covered the 162.5 kilometers (100.9 miles) on the River Thames in England from Oxford to Putney in 19 hours 28 minutes, while a triple-sculls boat needed just over 22 hours to cover the same course. And it may be assumed from illustrations from the period that the screw-driven boat was far from the optimal. The reason it was not adopted for racing was once again, as for bicycling, restrictions in racing rules.

Now new races are being established by several groups, including the IHPVA, with few restrictions on the design of the boats. Consequently, a number of alternative designs are being considered. In all of them oars are completely abandoned. They lose energy in wind resistance, in useless side thrust, and in kinetic energy that is not conserved in either forward or return strokes. They are difficult to manage in waves, and a light boat can capsize if an oar digs into the water too deeply or comes out of the water during the power stroke.

Until more is discovered about the wonderful and highly efficient propulsion system of fishes, humans will use pedals and propellers. These have been designed to give hydraulic efficiencies (ratios of the useful energy delivered to that supplied) of more than 80%. One problem with them, however, is that if they are designed to give a high propulsive efficiency, which is a function of the area of the propeller disk, slow-speed "utility" boats will require screws of the order of 300 millimeters in diameter; this will limit their movement considerably in shallow water. The ducted propeller or water-jet form of propulsion may then be advantageous, or, as an alternative, the use of several propellers of smaller diameter. (The ducted propeller is an axial-flow pump that operates inside a circular pipe or duct and discharges a jet of water at the rear of the boat.) The size of the screw required for high-speed racing boats is in the range of one-half the above figure, or 150 millimeters, which poses fewer problems.

The continuously turning screw is more easily harnessed to the large muscles of the legs than are oars, but oars provide lateral stability in narrow, high-speed boats that screws do not. Most fast screw-propelled boats have, therefore, been catamarans. Utility boats have sufficient stability in a single hull. One type of ship's lifeboat once was fitted with a screw that was turned by a crankshaft pedaled by six men, three on each side, at the stern of the boat. This arrangement has advantages over oars in efficiency, saving space, and reducing the danger of being swamped.

For light, high-speed recreational or racing boats an alternative to a catamaran is to use a main hull similar to a shell, and outrigger pontoons. A third intriguing possibility for a recreational boat is to have only a single hull and a bow rudder, thus requiring the rider to balance the craft by steering into the direction of fall, as on a bicycle.

The fastest form of human-powered boat may, however, be a hydrofoil. According to a recent study it was calculated that a speed of 18 kilometers per hour may be possible with a single-rider craft, enabling it to challenge an eight-rider shell. Such a vehicle would need years of rather expensive development and is perhaps more likely to be arrived at through efforts of enthusiastic amateurs in racing than by industrial research.

Human-powered flight

For centuries men and women dreamed of being able to soar like birds. Hopeful but impractical inventors tried innumerable times to adapt human arms to flap wings, with apparently inevitable and often tragic failure. True soaring became possible in this century with sailplanes or gliders. These are

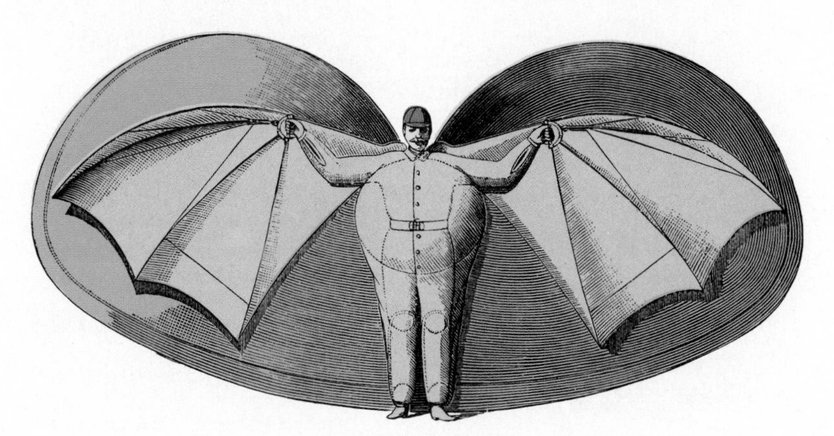

Mary Evans Picture Library

An early concept of human self-propelled flight dates from Germany in 1886.

carried by updrafts but are mechanically launched by catapults, motor vehicles, or airplanes. A later design, the hang-glider, enabled people to launch themselves from hills and subsequently to glide and soar. This is the nearest approach to the "coasting" flight of birds that human beings have achieved, and at least some of the energy comes from human muscles because altitude is first gained by climbing a hill.

During the last few decades there have been many attempts to achieve human-powered flight with airplanes driven by pedals and propellers. Recently these attempts were stimulated by the offer of British industrialist Henry Kremer of a prize for covering a figure-eight course around two pylons a half mile apart at above a minimum altitude and, later, a larger prize for a human-powered flight across the English Channel. The difficulty of even the first task was emphasized by the failures of the many teams from several nations. But both prizes were eventually won with apparent ease by a team organized by Paul MacCready with aircraft named the "Gossamer Condor" (for the figure-eight) and the "Gossamer Albatross," each piloted and pedaled by bicyclist Bryan Allen.

Do these successes mean that a new sport of human-powered flight will become popular? It seems doubtful. Flight is obviously a power-limited phenomenon. Human power is a function of body weight, but the power-weight ratio tends to fall as weight increases. An athletic male of 75–80 kilograms (165–175 pounds) can produce a kilowatt for less than a minute, a half kilowatt for almost half an hour, and 250 watts for 12 hours.

Views from head-on and above reveal the design and dimensions of the "Gossamer Albatross" (below), the first human-powered aircraft to cross the English Channel. Pedaled/piloted by Bryan Allen from his seat in the narrow cockpit (opposite top and bottom), the "Gossamer Albatross" encountered headwinds but still managed to cross the 35-kilometer- (22-mile)-wide channel in 2 hours and 49 minutes on June 12, 1979.

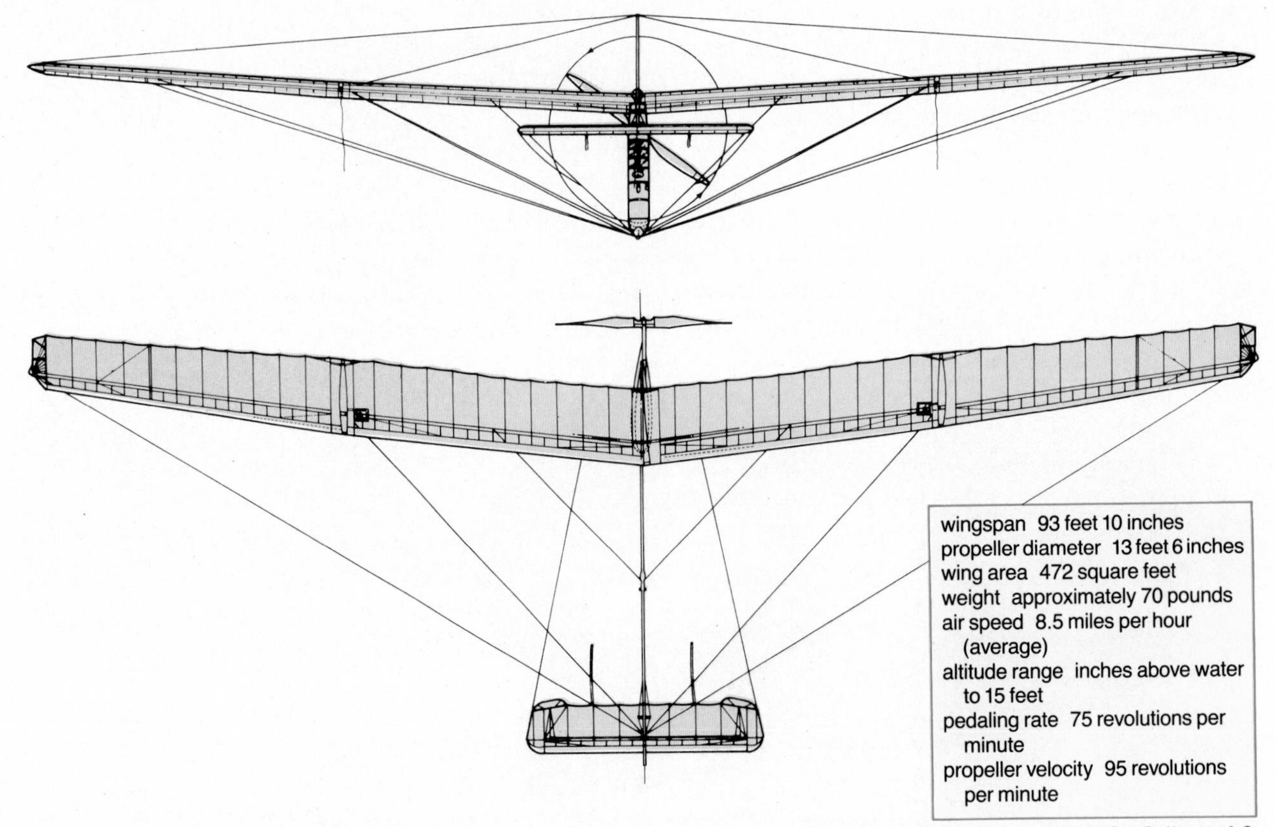

Adapted from information obtained from E. I. Du Pont De Nemours & Co.

James Joseph

The power required in level flight is the product of the drag and the speed. If the lift-drag ratio of the airfoils is only slightly affected by speed, then minimum power is required by flight at the lowest practicable speed. A principal reason why the MacCready team was successful in attaining the Kremer goals was its choice of about 10 kilometers per hour as the cruising speed instead of approximately 19 kilometers per hour, which was used by most other teams. But the wingspan of such a craft had to be about 30 meters, and flight was impossible or impracticable when the wind rose much above one meter per second. Such a plane wants to become a kite in even a mild zephyr and requires many loving hands to hold it gently down and to take it to a large (and therefore expensive) hangar.

Human-powered flight as a sport or recreation will therefore be an occupation only of the rich. It seems unlikely that there could be any developments that could outwit the laws of physics to change this view. The wings and rudimentary fuselages of human-powered aircraft could be filled with hydrogen to add lift and thereby reduce the required wing size, but hydrogen would diffuse fast through the thin membranes used as wing covering and flights would be short. It seems more likely that human-powered flight will be reserved for special-occasion public displays.

Human-powered travel should, then, become faster, more comfortable, and safer on the highways; faster and much more fun on the water; and principally an enjoyable and occasional spectacle in the air.

Don Monroe

SCREENPLAY BY MOTHER NATURE

Dedicated natural history filmmakers are pioneering technical advances in photography to better communicate their sense of wonder and excitement about the natural world.

by Peter Parks

For more than 140 years photographers and scientists have been tackling the problems inherent in visually recording what goes on about them. For the first hundred of those years many of the problems were concerned with the chemistry of the photographic process. During the last 40 years many of these problems have been solved, and more and more attention has been given to enriching the scope of the art. The film industry, the commercial advertising industry, and human vanity have contributed enormously to the vigor with which these new approaches have been made. Yet, for years nature motion-picture photography was left in relative stagnation, a poor cousin to the more commercial fields of the art.

With the advent of high-fidelity color and no doubt assisted by the explosive assault of television and mass communications, the great and largely untapped resource of the natural world as a supply of information and entertainment quite quickly became apparent. The remarkable innovation and perfection in both still and motion-picture cameras of the reflex viewing system, which allows the photographer to see through the picture-taking lens, was perhaps the final key that opened wide the natural world to human challenge. Because of the sophistication of modern wide-screen and commercial photography and the extraordinary appetite of television entertainment and documentary, that challenge today is very appreciable.

The dauntless professional

Most wildlife filming is conducted by single dedicated individuals. As such, the financial resources available with which to assault mountainous challenges are seldom adequate. Natural history photographers, however, seem to exude ingenuity and stubbornness, two very effective substitutes for extravagant outlays of cash. With modern audiences becoming intoxicated with the wizardry of presentation, the wildlife cameraman has to think as elaborately as the crew responsible for filming the special effects of *Superman* or as lavishly as the director of *Doctor Zhivago* or *Jaws*. Moreover, if one considers natural history to include not only plants and animals but also geophysical and climatic subjects, then the array of ingenious endeavor demanded of the nature photographer ranks second to none.

Ascent of a lacewing from the tip of a seed pod is frozen by three successive bursts from a high-speed flash unit. This kind of technical challenge taxes the inventiveness and patience of today's nature photographer.

Stephen Dalton—The Natural History Photographic Agency

Everyday shots of animals are no longer enough to satisfy either audience or cameraman. In addition, the extra effort to capture on film what the eye ordinarily misses can give scientifically revealing results. Triple exposure of a little owl as it slowed for a landing (facing page, top) was taken with a high-speed strobe at intervals of $^{1}/_{30}$ of a second and made use of a photoelectric eye to trigger lights and camera. Record of a swallow scooping up water in flight (bottom) required the same stop-action technique as well as strategies to encourage the bird to swoop in front of the camera. The photo answered the question of exactly how swallows skim across the water.

It is perhaps the total lack of predictability that lends to nature filming an air of opportunism that in turn makes every production a kind of exploratory expedition. Could one cameraman, for instance, have guessed that his wife would be caught by the leading edge of a right whale's fluke as it plunged into the depths? Or another that an adult gorilla would sit on his chest while being filmed for a television series? A third, photographing an underwater hippopotamus, must have been surprised to end up six feet in the air with a tooth of the animal in his calf. Yet a fourth, on a surface snorkel dive to photograph a large manta ray, would not have believed that the three-quarter-ton fish would suddenly slam full tilt into his chest and shove him along for several yards. These examples of the unpredictable are the spectacular ones that thankfully happen only rarely. For every one like them, however, a thousand much more subtle quirks of nature crop up to torment the wildlife cameraman.

In years past just capturing the animal on film was enough to enthrall both the cameraman and the audience. Today not only must photographers still contend with the vagaries of the wild but they also must match the technical sophistications of the wide screen and must better their rivals' attentions on the same subjects. Convention was first shattered in the late 1950s and early 1960s with the work of such innovators as West German forest-life specialist Heinz Sielmann, who filmed from within a woodpecker's nest hole; British biologist Gerald Thompson, later a co-founder of Oxford Scientific Films (OSF), Ltd., in the U.K., who photographed from within the timber bored by a wood-eating insect; and BBC photographer Ron Eastman, who recorded a diving kingfisher from below the water's surface. Following those achievements the gauntlet was on the ground, and the great wave of brilliant camera technique that presently is regarded as commonplace began to build.

Today no wildlife cameraman thinks in terms of obvious shots. Australian small-animal specialist Jim Frazier was not content to film a fast-retreating frilled lizard of Australia with an immobile camera. He insisted on tracking the creature, at its own speed, with a wide angle lens and so produced a memorable record. Stephen Dalton, a Briton with a keen interest in bird filming, was satisfied only when he had captured a swallow drinking on the wing, the water actually flowing up the lower bill and a reflection of the same event clearly visible in the water's surface beneath the bird. Des Bartlett, an Australian and a pioneer large-animal photographer, spared no effort to capture a close-up, slow-motion portrait of snow geese in flight from a perspective in front of and traveling with the birds. Sean Morris, another co-founder of OSF, was determined to film, in time lapse, the entire process of decomposition of a dead mouse. As a result of painstaking work to develop both the required equipment and the conditions necessary to the biological process, he succeeded in compressing the nine-day event into a most spectacular two-minute film in which 200 fly maggots consumed the entire carcass to the bones. In response to a bet from a television network John Paling and others of the OSF production team agreed to film a bird called a dipper moving underwater in a fast-flowing trout stream in its search for trout eggs. This study, ultimately achieved in a studio set and in slow motion,

***PETER PARKS,** a former marine zoologist, is a co-founder and co-director of Oxford Scientific Films, England, and a specialist in aquatic photography, photographic engineering design, and microphotography.*

(Top) Stephen Dalton—The Natural History Photographic Agency; (bottom) Stephen Dalton—Oxford Scientific Films

John Cooke—Oxford Scientific Films

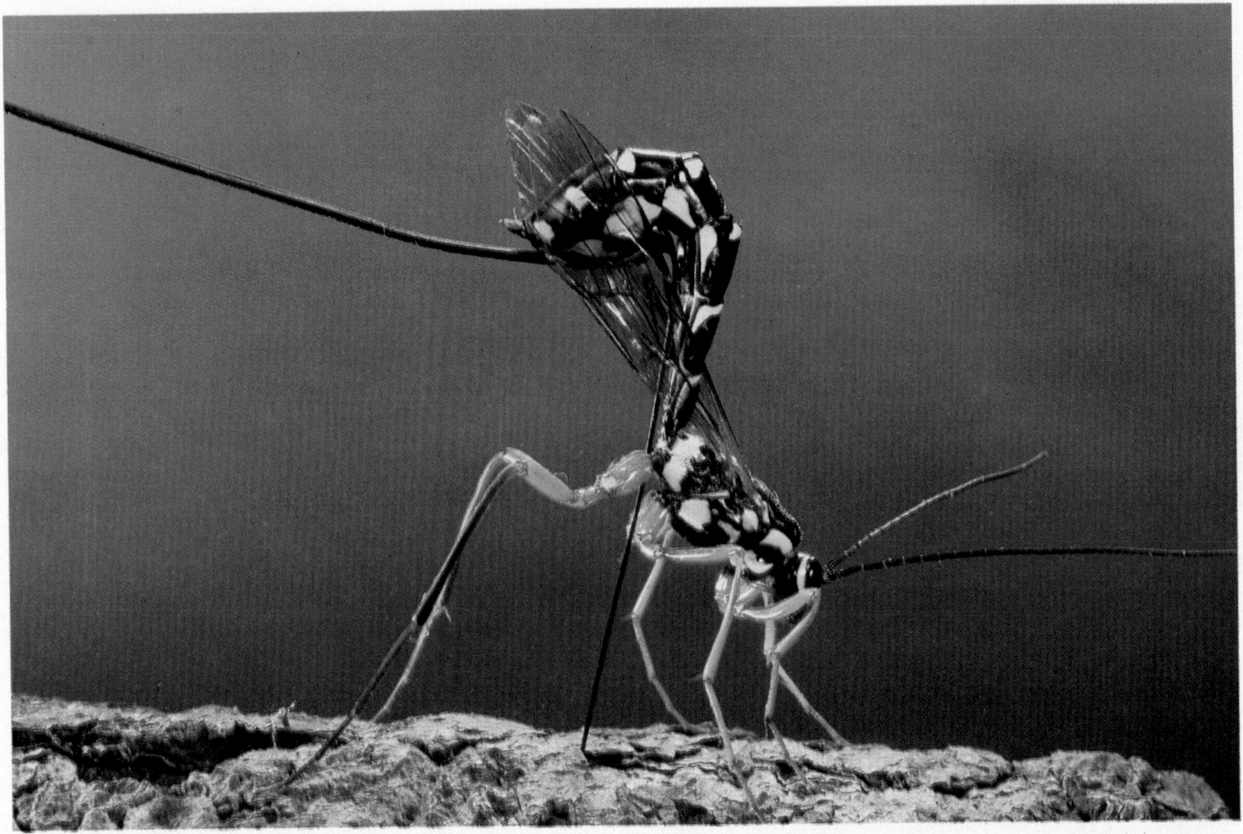

A parasitic ichneumon fly penetrates tree bark with its long needlelike ovipositor in order to lay eggs on its host, the wood-boring larva of a wasp. Zoologist Gerald Thompson's documentary on The Alder Woodwasp and Its Insect Enemies, *completed in 1960 and featuring footage taken from within the timber bored by the wasp, helped usher in a new era of brilliant nature filmmaking.*

offered a truly privileged view since the dipper is one of the few birds to actually fly underwater, rather than paddle. Among the film's beauties was the revelation that a layer of air remains trapped beneath and within the bird's feathers, enveloping it in a case of silver.

Small subjects

Though in extent much reduced, but in effect no less spectacular than these examples, are the photographic opportunities offered by the world of the small. One such instance was well exploited in OSF team member George Bernard's simple but effective photography of a froghopper insect from within its protective globule of spittle. Although reduction in the size of a wildlife subject sometimes lessens the work load of recording from an unusual vantage point, seldom does it reduce the intrigue and subterfuge necessary to crack the problems. Moreover, if one tackles the extremely small, then even the work load or necessary hardware may not be reduced. Often it is substantially increased. For example, in order to film, with a three-dimensional tracking capability, marine or freshwater organisms that are no bigger than one-tenth the size of a period on this page, OSF photographers employ an optical bench system that weighs 254 kilograms (560 pounds) and that took ten years to develop. Nevertheless, this hardware has permitted OSF both to bring minute creatures to the wide screen and to film tiny events—both biological and nonbiological—as special effects for major motion pictures.

Equipment size or weight in itself is not essential for filming small creatures. At times, both can produce problems of their own. Yet, if weight is a by-product of rigidity, then it may be important for equipment designed to film at extremely high magnifications. If animals invisible to the naked eye are to be filmed, then magnifications of 200 times or greater must be used. At such extremes even breathing on the body of the camera can cause it to move in relation to the subject and so produce a blurred image. Consider, then, the vibration caused by a heavy camera motor jolting the film-transport system into action. One technique for overcoming this problem is to link camera and subject rigidly so that both vibrate together.

A good example of just such a vibration problem confronted OSF in 1979 while filming on the Great Barrier Reef of Australia. Crew members were using the well-tested optical bench system mentioned above, which mounts a camera and a subject holder, or stage, together on a very solid support, when a surprising vibration problem appeared. After a frustrating three weeks, they finally found the cause. One of the horizontal platforms in the stage was clamped to the base plate only along one side. A piece of grit had become lodged between the platform and the base plate and had turned the platform into a very efficient springboard. Each time the camera was run, the subject became a soft blur. Once the problem was corrected the vibration disappeared, allowing the crew to film microscopic single-celled creatures such as radiolarians and dinoflagellates at magnifications of 400 times.

The equipment used for such filming enables OSF to record over a wide range of microscopic magnifications with a constantly focused five-to-one zoom capability and to track in three dimensions of space with great steadiness and at almost any speed. Furthermore the camera view can be made to roll (rotate around its center), pitch and yaw, circumnavigate, or spiral and to do it horizontally, down, or up. The same optical bench has been used extensively to produce special effects for the motion pictures *Superman*, *Flash Gordon*, *Dracula*, and *Altered States*. By contrast, 12 years earlier the prototype system consisted of nothing more than two wooden survey tripods —one each for camera and subject—held together with clamps.

Studio magic

There is no rule or understanding among nature photographers that requires wildlife behavior to be filmed in the field. A recreated natural environment housed in a studio can be quite authentic to both the subject of a film and its audience. At the same time it allows the photographer to work in a convenient enclosure and to use equipment that would be virtually impossible to operate at an outdoor site.

For better or worse the design and construction of large studio equipment has been a key element in the evolution of OSF, and by 1980 it had completed two $1^3/_4$-ton remote-controlled camera-and-suspension systems. This equipment permits filming on a wide range of scales anywhere in and over a natural environment that has been set up in the studios. Initially the systems were intended for use with relatively small sets, about the size of a large tabletop. In the process of designing the necessary camera head and suspension it was realized that, if one was going to cover areas ten feet square, one

(Top and bottom, left) Peter Parks—Oxford Scientific Films; (bottom, right) S. R. Morris—Oxford Scientific Films

On location in Bermuda, Peter Parks films tiny marine organisms on OSF's optical bench (above). Under bench illumination marine diatoms (top) shimmer prismatically; close-up of deep-sea jellyfish (left) reveals fine details.

Photos, Peter Parks—Oxford Scientific Films

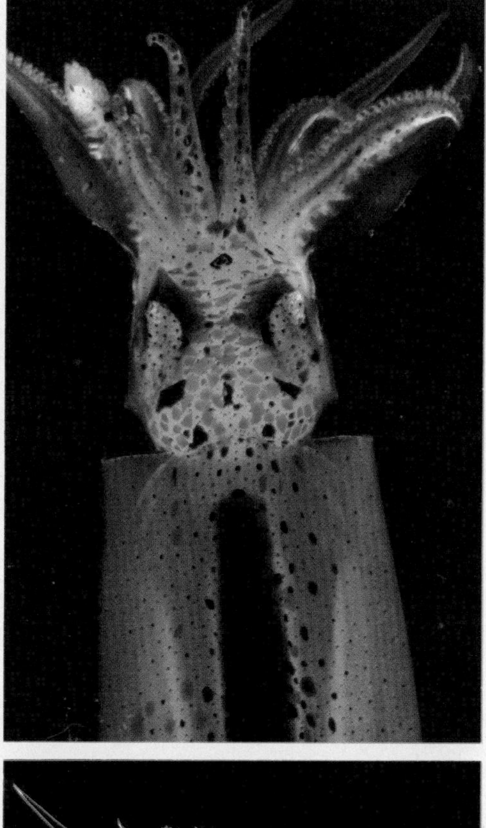

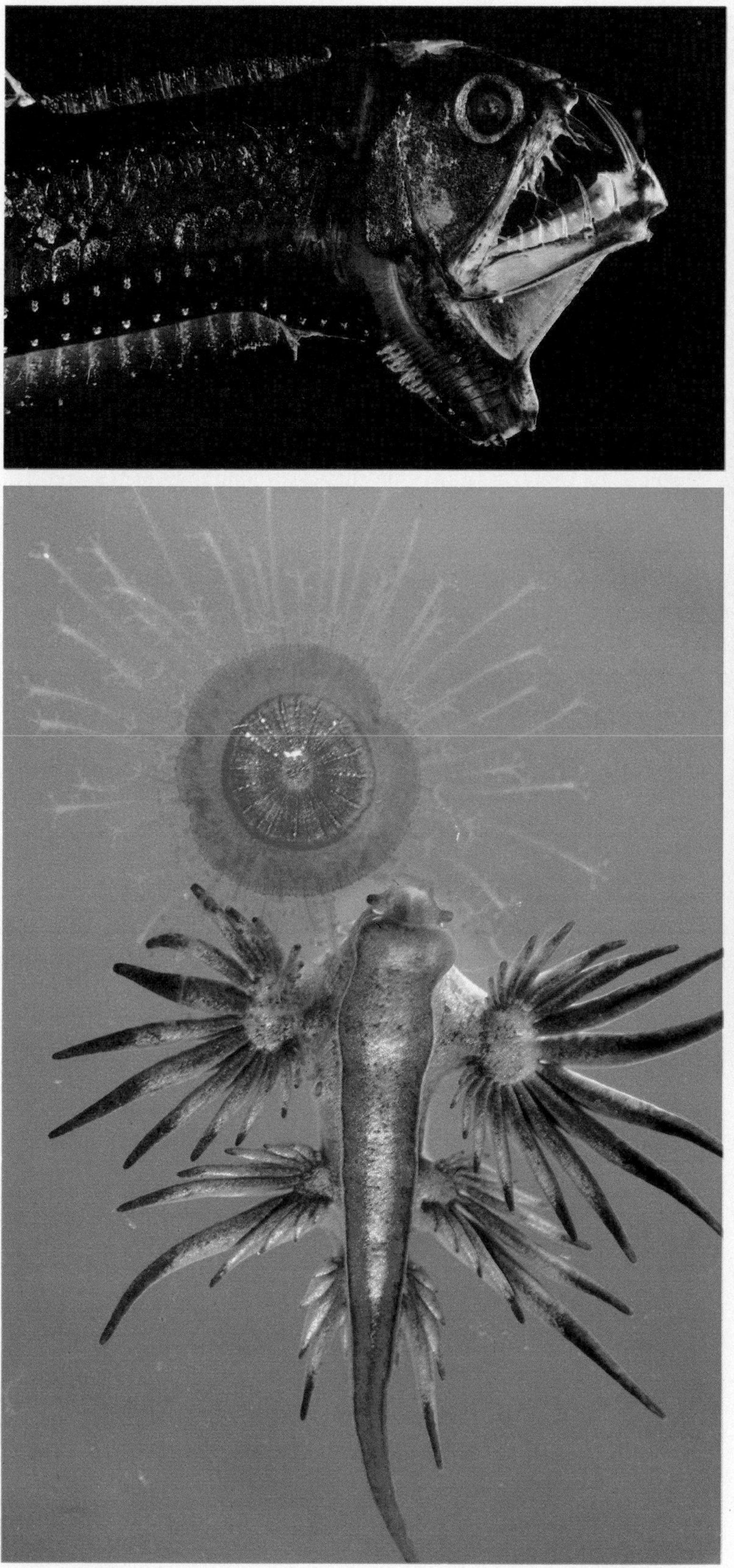

Colorful subjects for OSF*'s close-up camera work include (clockwise, from top left) a deep-sea viperfish, a squid, and a decapod crustacean from the ocean bottom. (Left) Round, tentacled siphonophore and feathery nudibranch, or sea slug, are dwellers of the tropical ocean surface. Tracking some of the smaller of these creatures, and those on the facing page, while filming on the optical bench can give the impression that the camera is drifting on a journey through space.*

might as well consider covering entire floors. At that time OSF's "big studio" was 60 feet by 25 feet, (one foot equals about 0.3 meters). When it had originally been erected three years earlier, the builders had been asked to leave exposed the reinforced steel joists in the roof just in case the photographers ever wanted to hang anything from them. Such foresight was fortunate, because by 1980 those steel beams supported seven tons of equipment.

These camera-and-suspension systems presently permit remote tracking in the "big studio," which has been expanded since its erection, for 90 feet and anywhere over its floor area. The camera view can crane, or elevate, as high as eight feet above the floor. It can pan, tilt, roll, pitch, yaw, roll over (tilt through 180° or more), spiral, tumble (do a partial or complete loop-the-loop), and circumnavigate at almost any desired speed. The photographic axis can be as low as one-eighth of an inch off the ground (one inch equals 2.54 centimeters). The cameras can retain a horizontal lens angle of 60°–65°, which is a very pleasantly wide-angle view akin to human vision. The operator can shift the focus from infinity to the head and shoulders of a fly by simply turning a control knob, all while seeing and recording the results through a closed-circuit television system built into the movie camera eyepiece. Furthermore, this entire array of ability is possible with the lens submerged in water. As many as 11 people may be required to control the equipment when the most sophisticated moves are being conducted. At such times the whole operation becomes very much a team effort.

Hanging from the exposed beams of OSF's "big studio," two remote-controlled camera suspension systems track along a specially built trough during the filming of space-effects footage for the feature film Saturn 3. *The Cosmoglide, on the left, is controlled by stationary operators, whereas the Pathfinder, on the right, accommodates operators who ride along with the camera head on the system's 90-foot track.*

G. I. Bernard—Oxford Scientific Films

The two remote-controlled suspensions that accomplish these wide-ranging and potentially far-reaching abilities are termed the Pathfinder and the Cosmoglide. A third suspension used at OSF, the Stellaglide, is a manually operated precision crane. Onto all of these may be slung either of two specially designed camera heads, dubbed the Galactoscope and Cosmoscope. The Cosmoscope is a refinement of the Galactoscope; it incorporates more sophisticated optics to reorient the image to the cameraman's advantage and possesses a completely different focusing system. Both camera heads give great accessibility to the sets over which they operate because both end in lens tubes only an inch or so in diameter. In addition, both can operate underwater. On the Pathfinder suspension these camera heads are actually accompanied by the operators, who ride along on a 90-foot track. When using the Cosmoglide the operators are positioned remotely; electrical cables pass from their control units to the camera head.

In due course OSF intends to make it possible to attach yet another camera head, the Astroprobe, which has the rather unique ability to make the camera view fly through pinholes. For instance, with this device on hand, the photographer can cut a minute hole into the back of a flower and then, without actually passing either camera or lens through or into that hole, can gain a wide-angle view looking out from the base of the flower as an approaching insect enters the flower to pollinate it. The Astroprobe was used effectively by Sean Morris during a four-year effort, completed in the summer of 1980, to produce a one-hour television special on the incredible pollination mechanisms used by plants in their reproduction. Part of the film reveals that some flowers mimic female insects so well that males try to mate with them. In the process the insects become carriers for adhering sacs of pollen, which they then transfer to other flowers of the same species.

A step beyond nature

The sophistication of animal and plant behavior demands, or at least deserves, equivalent sophistication and elaboration of camera technique to do it justice. Such was the intention when OSF initiated design and construction of much of its novel equipment. It soon became apparent, however, that this equipment had uses other than the recording of natural history. First one feature film company and then several others asked OSF to film a few special effects for use in space scenes or other imaginative situations. When these first efforts proved both successful and effective, requests became more elaborate and more demanding of OSF's special equipment and techniques and photographic ingenuity. Some efforts called for filming at frame rates ranging from 10,000 frames per second to one frame every 4,000 seconds; others called for black holes, spaceships, alien planets and their surfaces, hallucinatory effects, and—possibly most demanding of all—a journey through the debris that makes up the rings of Saturn.

Although these efforts seem far removed from natural history, filming imaginative recreations of what exists, or may well exist, on Earth or in any other part of the universe can be considered extensions of nature photography. In order to create realistic and believable effects it is essential that many of the rules gleaned from studying the natural world be obeyed.

A trip through the studio—by submersible

In filming a major sequence for the television special "The Seas Must Live," a production of Anglia Television Group in the U.K., OSF resorted entirely to special effects—but with some justification because it had been asked to depict a journey in a deep-diving submersible traveling across the floor of the Atlantic Ocean. The course took the audience down part of the length of the Mid-Atlantic Ridge, out along one of the side fissures, west across the the seafloor toward the subterranean mountain upon which Bermuda sits, and finally up the Hudson Canyon to emerge alongside Long Island and Manhattan, with the Appalachian chain of low mountains visible beyond and the snow-capped peaks of the Rockies far in the distance. Intercut with this sequence were to be sequences of marine life shot by OSF and other wildlife cameramen. At an early stage in the planning of its production Anglia was able to warn of the possible requirement for this complex sequence. A description of the various stages of production that occurred between that initial warning and the finished result offers a good example of how a technically difficult film sequence is made.

Once Anglia had conceived of a sequence concerned with the Mid-Atlantic Ridge, it asked OSF's technical director Peter Parks if the Cosmoglide remote-tracking equipment could be used, with model work, to film a sustained sequence and if OSF could describe the sort of sequence that might result. Agreeing to study the proposal committed OSF to about three or four weeks of research, outlining, and finally a storyboard. A storyboard is a kind of visual script made up of a layout of drawings, photographs, or both in cartoon-strip style with corresponding verbal descriptions of the action taking place. Previous experience with feature and commercial work had given OSF an appreciation of the value of carefully designed storyboards in indicating to directors and producers just what could be done and how much it would cost. Most film sequences comprise a series of shots, each shot representing a separate camera position, lens focal length, frame rate, or key point in a camera move or lens zoom. In OSF's plan, however, there was only one long sustained shot covering the first two minutes of the sequence, so the storyboard concentrated on full-color detail with amplifying notes.

In OSF's original proposal the camera is first fixed on a sunrise, then draws back through the space between the twin towers of the World Trade Center in New York City. It continues the retreat over water back across the Hudson basin and the western Atlantic, past Bermuda, and then out over the mid-Atlantic. Shortly thereafter, the view submerges beneath the waves and is overlaid first with footage of microscopic marine organisms previously shot in Bermuda and then later with unusual footage shot on board the R.R.S. "Discovery," a British deep-sea research vessel, of amazing creatures trawled alive from two miles down in the Atlantic. Descending still further to the deepest realms of the Atlantic, the camera tracks forward to examine the abyssal features of the ocean until it again reaches the Hudson Canyon. Finally it surfaces to show its proximity to New York City in time to witness, symbolically and factually, the offshore dumping of nuclear and chemical wastes. Anglia carefully considered this proposal and, perhaps most pertinently, its cost. Considering that it included construction and materials for

Filming a simulated journey by submersible for the British television special "The Seas Must Live" involved the construction of a 50-foot-long, timber-supported model of the Atlantic basin and part of the North American coastline (facing page, top left). Detail (top right) depicts in polystyrene and plaster the Hudson River basin as it flows into the Atlantic and a caricature of New York City sculpted from Plasticine. Completed set viewed from sea level off the Atlantic coast (bottom) shows submarine peaks in the foreground with the Appalachian chain and the Rockies in the far distance. Galactoscope camera head, in the left of the photo, was used to film the final sequence.

(Top, left) Peter Parks—Oxford Scientific Films; (top, right and bottom) G. I. Bernard—Oxford Scientific Films

Photographing in the wild often demands the same elaborate preparations, if not the sophisticated equipment, of a studio setup. Numerous flash units, a camera, a blind for concealment, and a nature photographer's vast stubbornness were needed (bottom) for this stop-action record of a coal tit leaving its nest hole (right).

Photos, Stephen Dalton—The Natural History Photographic Agency a 50-foot-long model as well as video monitor rental, rental of extra lighting, and utilization of the studio, its facilities, and the Cosmoglide and Galactoscope, a budget of £15,000 (about $35,000) seemed reasonable. Anglia accepted it, requesting that shooting be completed within an eight-week period, if possible.

After further discussions it was mutually decided to shoot the main scene dry but carefully lit to look watery. This scene would also be preceded with a totally separate sequence to be shot, in time lapse or stop frame, over a 12-foot-square model of the complete North Atlantic as viewed from a satellite, from the Norwegian coast to the Gulf of Mexico, to show how that part of the globe would appear if the "plug were pulled out" and the Atlantic drained of every last drop of water.

Six weeks of model construction followed, although not all was continuous hard work. Some time was spent waiting for plaster to dry and some for the technical director to recover from a foot injury that resulted from dropping one of the 16-foot-long timber supports for the model. One might imagine that a 50-foot-long model would also incorporate some large vertical contours. Such is the unusual ability of the Galactoscope, however, that not even the loftiest submarine mountains, some of which are as high as Mt. Everest from base to peak, had to be represented at any stage by polystyrene and plaster models that were more than five inches above the base height. It should be mentioned that, because of the extremely small ratio of vertical height to horizontal extent for natural features on the real Earth, the vertical scale of the model (as with any geographic contour model) was purposely exaggerated.

For Parks and Dave Thompson, who did all this work, the most rewarding phase was the painting and lighting, during which the drab model foundation underwent an exciting transformation to the more familiar shapes and colors of nature. A caricature of New York City was sculpted from Plasticine, a claylike material long a favorite of model makers. Achieving the desired depth of field, or range of sharp focus, required small lens apertures and consequently a considerable lighting load, which drained nearly all the available 125-kilowatt power supply. Although the builders experienced exhausting days of walking from a glowing brazier, set outside the studio, with red-hot irons with which to cut the base of expanded polystyrene prior to plaster dressing—courting a fire hazard and the risk of poisoning from plastic fumes in the process—they found the results satisfyingly realistic. So did representatives from Anglia, who viewed it with great enthusiasm and gave their approval to commence shooting. During the following two weeks, work with the Galactoscope and Cosmoglide secured all the film Anglia required.

Following the total demolition of the model, two other distinct sections of the final sequence were prepared. The first was the satellite-view scale model of the North Atlantic basin, which had to be not only accurate but also functional, as it had to hold water and satisfactorily permit several filling and emptying operations. This requirement in fact did create many problems: plaster softened and disintegrated, buoyant polystyrene land masses tore free of their weighted moorings and drifted across the Atlantic, and coast-

One call for OSF's brand of special effects involved a close-up view within the living body (right), for the feature film Altered States. *Aerial image relay, a technique with great special-effects potential, permits bringing background and close-up foreground scenes into sharp focus simultaneously; unusual juxtapositions (bottom) can be photographed at one time without resorting to double exposure or multiple printings. Equipment developed at OSF takes the technique a step further, into the realm of motion pictures.*

(Top) G. I. Bernard—Oxford Scientific Films; (bottom) Stephen Dalton and Peter Parks—Oxford Scientific Films lines needed careful working and reworking to ensure that all continents assumed the correct shape when the Atlantic was filled. The first time-lapse runs, over a period of two hours, picked up an annoying stray reflection in the water. Eventually, however, it all worked well.

The second section of shooting was conducted on the optical bench at very high magnification. Here was filmed some of the more beautiful microscopic plant life in all its glory. As a group, the diatoms have skeletons of silica in their outer cell walls with intricate striations upon them. In water these structures catch the sunlight and split it prismatically to produce beautiful structural colors. Careful use of the three-dimensional tracking ability of the optical bench produced a spacelike feel to the watery environment of these minute plants such that the audience could well believe that it was drifting through them in a continuous journey.

This material was then linked with footage taken in the field on marine expedition four years earlier to an isolated island off the coast of Bermuda. The animals and plants filmed there were all collected from the water with fine meshed nets. They were then filmed upon the optical bench back on the island at night with light powered by a portable generator. Some of the creatures filmed on that expedition and two subsequent trips have yet to be identified. Rare or common, all were singularly beautiful.

It was only two years after the entire sequence was assembled that Parks got to see the final film version. By then it had been edited in a way that differed from the original script, but on a 50-foot screen it looked impressive and those five-inch mountains loomed.

New directions

As a by-product of research to produce the Cosmoscope and Galactoscope, OSF recently developed a novel photographic system that is opening up amazing realms of wildlife filming for exploration. Moreover, like much of OSF's other equipment, it has enormous special-effects potential, and once again the two applications are being pursued side by side. The new system is derived in part from aerial image relay, a technique used in the animation and printing divisions of the film business. Aerial image relay uses ordinary optics to project the image of a subject not onto a screen but into space, the way a simple magnifying glass held at arm's length seems to produce the inverted image of a distant object just behind the glass. The technique is employed to project, or relay, an in-focus image of a background into the plane in space occupied by a foreground subject. An OSF member, John Cooke, had used it for some work with still photographs to relay backgrounds into combination with close-up foregrounds, usually insects. Parks attempted to take the idea a stage further, into the realm of motion pictures, an effort that eventually produced three pieces of equipment known to OSF as space-image-relay systems.

Although the details of the systems are presently one of OSF's precious secrets, their inherent ability can be summarized in the following example. Because of the depth-of-field limitations of the ordinary camera system, a standard frame-filling close-up shot of an insect will render the background out of focus, or soft. Usually it is so soft as to be unrecognizable. If it were

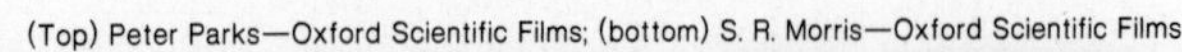
(Top) Peter Parks—Oxford Scientific Films; (bottom) S. R. Morris—Oxford Scientific Films

possible to increase depth of field, these backgrounds could be sharpened. With the space-image-relay systems that now are operational at OSF, it is possible to bring into focus any background scenes or objects one desires. For instance, one can film a foreground locust being approached from beyond by a desert tarantula. The entire approach of that spider as well as the panoramic desert scene that it inhabits can remain in sharp focus while the locust continues to occupy a large fraction of the screen. Only as the tarantula took its final steps would one lose a little of the crispness of the distant horizons.

When Parks and OSF colleague Ian Moar were first designing and constructing this equipment it was their intention only to bring the immediate background beyond the foreground subject into sharp focus. While setting up the system one day, however, Moar moved into the field of view of the equipment and placed his face where formerly had been background set and subject. The result was astonishing: not only could his face be sharply focused, but as he stepped back first his head and shoulders and then his entire body could be brought into focus at the same time as an insect in foreground. But then far more than increased focusing depth became apparent. By careful adjustment Moar's body could be made to appear half the size of the insect. In other words the new system could actually miniaturize objects—in camera—at one take. Traditionally such magic is worked only by complex post-shooting optical combinations. The advantages of the method

Peter Parks—Oxford Scientific Films

OSF had stumbled upon became obvious, and the potential was staggering. In 1980 OSF was in the process of mounting a major project based upon techniques of this nature, a full-length feature film integrating an imaginative fantasy with spectacular wildlife and natural history photography.

So, what in several instances began as wildlife-filming research programs ended by providing not only far-reaching advantages for natural history photography but also exciting possibilities for imaginative filming in other fields. Since the original desire at OSF—one that first brought together the five biologists who founded it—was to enrich the presentation to their audience of the biological stories that so intrigued themselves, this outcome is not inappropriate.

Recorded on a slide titled "The First Step," the optical-bench forerunner of OSF's complex modern camera systems consisted of tripods held together with C-clamps (facing page, bottom). More than a decade later OSF's image relay equipment (facing page, top) is promising solutions for imaginative photographic projects once thought as difficult to accomplish as the task suggested above.

OUR CROWDED AIRWAYS

The heavy flow of planes into and out of major U.S. airports is posing a challenge to the nation's air traffic control system.

by Richard L. Collins

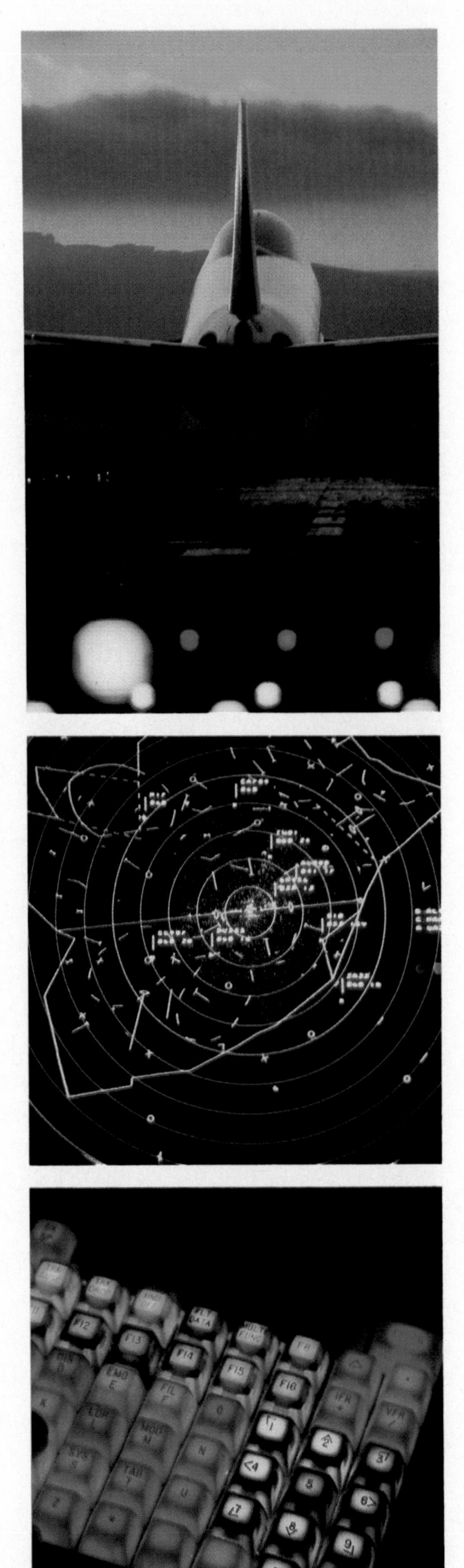

The adequacy and safety of the air traffic control system in the United States has been a subject of considerable controversy during recent years. Controllers throughout the nation maintain that their computerized radar systems are breaking down with increasing frequency, posing a "serious menace" to air safety. During a recent six-month period, according to a leader of the controllers' organization, 143 computer outages occurred at the Leesburg, Virginia, Air Route Traffic Control Center; similar malfunctions were claimed to have taken place at other installations throughout the nation.

The controllers also argue that at certain exceptionally busy airports traffic is often so heavy that they must work long hours at great speed to keep it flowing. To dramatize this charge they have occasionally "worked to the rule book" at some of these airports; the results have been long lines of planes on taxiways and waiting in holding patterns.

On the other side of the dispute is the Federal Aviation Administration (FAA). To the charges about computer breakdowns an FAA official commented that the present equipment offered "an adequate level of safety"; the agency is, however, working to develop improved models. In the following discussion an experienced flier describes the evolution of the U.S. air traffic control system to the present day. He points out its strengths and weaknesses and offers possible remedies for the future.

Early efforts

The need for control of air traffic from the ground was first experienced in the 1930s, when the fledgling airline industry was working to provide reliable service regardless of weather. Meeting the schedule often required flying in clouds. Pilots could not see and avoid one another, and as traffic increased it became apparent that a system of separation was needed. The first control of traffic in the United States was provided by the airlines. When the federal government took over on July 6, 1936, the air traffic control system consisted of facilities at Newark, New Jersey, Chicago, and Cleveland, Ohio. It soon spread to the rest of the country.

(Opposite page, top) © Dirck Halstead—Gamma/Liaison; (center) courtesy, Philadelphia International Airport; (bottom) Fred Ward—Black Star; (above) © Michael J. Pettypool—Uniphoto

(Left) Courtesy of the Reference-Room of the Crawford Auto-Aviation Museum, Western Reserve Historical Society, Cleveland, Ohio; (right) UPI

The first airways radio station at the Cleveland (Ohio) Municipal Airport (above) was constructed in 1925. Also at the Cleveland airport the managing traffic controller demonstrates the new airway traffic control system in 1936 (above right). The locations of planes flying from Cleveland are shown on the blackboard. At that time the Cleveland airport was one of four in the U.S. providing traffic control facilities.

Before radar was developed and installed, air traffic controllers separated aircraft based on knowledge of position and the pilot's radioed estimate of arrival time over various radio fixes. Navigational aids were four-course radio ranges, on which the pilot could navigate precisely in any one of four directions from the station. Traffic control was quite simple: do not let two airplanes arrive at the same place at the same time. Assigning different altitudes to different flights would take care of passing and crossing traffic. This worked well, given the relatively light air traffic.

This "system" was one that pilots used only when bad weather made it absolutely necessary. When visibility was adequate, flying was by visual flight rules with pilots navigating by landmarks and watching for other traffic out the window. The Douglas DC-3 was the airliner of the time; the weather information available to pilots was sketchy at best; and while some airports had control towers to direct arriving and departing traffic, many did not. The system was rudimentary but adequate.

World War II taught many lessons on air traffic control, but the system in use during the immediate postwar period was much the same as before the war: there were some control towers, en route traffic was handled by estimated and reported knowledge of position, and in good weather there was a total dependence on seeing and avoiding other air traffic.

The introduction of higher performance aircraft made change inevitable. On an airport bulletin board in the early 1950s was a white poster with a tiny speck in the middle. Close examination of the speck showed it to be a front view of an F-80 jet fighter, and below was the comment that the jet would be zipping past just a few seconds after first being seen the size of the speck on the poster. The message was clear: in some situations see-and-be-seen was a questionable means of separating air traffic.

***RICHARD L. COLLINS** is Editor of Flying magazine.*

Dramatic as it was, that poster did not do much to change the air traffic control system. It took tragedy to spark action.

On June 30, 1956, TWA Flight 2, a Lockheed 1049 Super Constellation, left Los Angeles at 9:01 AM bound for Kansas City. Three minutes later United Air Lines Flight 718, a Douglas DC-7, left Los Angeles bound for Chicago. TWA 2 was cleared to fly at 19,000 feet, and United 718 was assigned to 21,000 feet. At 9:21 AM TWA 2 requested a change in altitude to 21,000 feet. This was denied because the projected flight path crossed that of United 718. The TWA crew was so advised; they then elected to fly under visual flight rules, uncontrolled, at 21,000 feet. The crew of United 718 was not informed of the potentially conflicting traffic. The weather was good, but there were some clouds building through the 21,000-foot level.

At 10:31 AM an unidentified radio transmission was heard at Salt Lake City and San Francisco. The message was not understood when received, but a later playback of the recording of the transmission revealed it as United 718. It was interpreted as: "Salt Lake, United 718 . . . ah . . . we're going in." The two airplanes had collided over the Grand Canyon, probably at 21,000 feet, and both crashed in the canyon with no survivors among the total 128 on board.

Evolution of the present system

This dramatic and tragic collision gave impetus to the development of the present air traffic control system. Congress appropriated money, and radar was soon added. Air traffic controllers could then *see* traffic moving on the radarscope.

Hardware was, however, to be only part of the system. New rules and operating procedures were needed, and in 1961 U.S. Pres. John F. Kennedy accepted the report of a blue-ribbon panel and instructed the Federal Aviation Agency (FAA) to implement its recommendations. All traffic above 24,000 feet over mountainous areas and over 14,500 feet elsewhere would be put under "positive" air traffic control; that is, it would be separated at all times by controllers on the ground. Additionally, such control would extend down to 8,000 feet on some busy routes. Transponders, devices that electrically identify aircraft on radarscopes, would be required for operations in busy areas. Virtually all commercial jets, just then coming into widespread use, and other airplanes using high altitudes and busy airports would spend most flight time with separation from other airplanes provided from the ground. The reliance on pilots seeing and avoiding other traffic would be applicable only in areas of light traffic and to smaller airplanes flying at low altitudes.

The FAA did not move very quickly in implementing this program. In fact, 20 years later, in 1981, the 5-year plan had yet to be fully implemented. During those years the air traffic control system did not evolve smoothly. There were big traffic jams at major metropolitan area airports in the 1960s, and late in that decade three major midair collisions between airline jets and general aviation (primarily personal and business-use) airplanes emphasized that the system was still in need of much work.

Disaster makes things move; the Grand Canyon collision started the U.S.

Courtesy, Lockheed-California Company

The array of instruments in the cockpit of a Lockheed L-1011 TriStar reveals the complexity of today's large jet airliners.

on the road to the present air traffic control system, and the three collisions of the late 1960s gave the program further direction. Grand Canyon emphasized the need for separation at high altitudes; the later collisions pointed to a need for work at the lower altitudes, where airplanes flying by visual flight rules and airplanes flying instrument flight rules are mixed in good weather.

All air traffic is now under positive control at and above 18,000 feet (referred to as Flight Level 180). Minimum vertical separation is 1,000 feet up to 29,000 feet (Flight Level 290); at and above that level 2,000 feet of vertical separation is required. Aircraft at the same level are kept a minimum of five miles apart except in areas near terminals, where, within 40 miles of the radar antenna, minimum separation is three miles.

Positive control of all air traffic below 18,000 feet is provided in airport traffic areas (the airspace up to 3,000 feet within five miles of the airport) and in terminal control areas, called TCAS by most pilots. By 1980 in the U.S. 23 airports had or would soon have TCAS, and 27 more were proposed by the FAA. Most TCAS extend up to 7,000 or 8,000 feet and include the airspace used by an aircraft as it maneuvers for landing at an airport. (The FAA is in the process of extending TCAS up to 12,500 feet.)

In airspace that is not under positive control (that below 18,000 feet and outside airport traffic and terminal control areas) there is a mixture of aircraft on visual and instrument flight rules in good weather. The instrument aircraft are separated from one another but not from the visual aircraft. FAA regulations require that all airplanes flying above 12,500 feet carry transponders that automatically report position and altitude on the traffic controller's scope; also, many aircraft that operate below 12,500 feet carry this equipment. Thus a controller can give a controlled aircraft both the position and altitude of an uncontrolled aircraft, enhancing the see-and-be-seen concept.

Because see-and-be-seen is compromised by increased speed, the FAA imposes a 250-knot speed limit on aircraft flying below 10,000 feet, where there is the greatest mixing of fast and slow, controlled and uncontrolled traffic. Additionally, air traffic control procedures are designed to keep jets at high altitudes for as long as possible before they start their descents. Where there is a TCA and where the TCAS are extended up to 12,500 feet, jets will have a minimal exposure to airspace that is not positively controlled at all times. Where there is not a TCA, exposure is substantially greater but traffic is lighter and all aircraft are in contact with air traffic control when flying to or from an airport with a control tower.

Heavy flow of air traffic at major airports can be seen in the lineup of planes waiting to take off at Chicago's O'Hare International Airport.

Courtesy, Federal Aviation Administration, Great Lakes Region Public Relations Office

(Left) Mike Malyszko—Black Star; (right, top and bottom) Philip Jon Bailey—Stock, Boston

One of the busiest airports in the U.S. is Logan International, located across the harbor from downtown Boston. A plane lands at the airport (above right) guided by controllers in the tower (above and top right).

Air traffic control is provided from three places. The familiar control tower handles traffic on the ground and in the takeoff, final approach, and landing phases of flight. It is, in effect, the runway referee. At busy airports the tower has radar both for tracking airplanes on the ground (useful when the visibility is low) and for monitoring the sequence of airplanes during their approach.

The control of aircraft in the area of the airport—within 30 or 40 miles and generally below 7,000 to 10,000 feet, depending on the airport—is handled by a terminal radar control facility. It provides both departure and approach control services. Where there are a number of airports in the area, this facility provides service to all of them.

Away from terminal areas air route traffic control centers handle traffic. There are 20 of these in the contiguous U.S.; each handles en route traffic over a wide area.

Flight 40RC

These three basic air traffic control facilities communicate with one another through phone lines, and close coordination is required for a flight. For an example of how the control system works, one can follow a flight from Trenton, New Jersey, to Asheville, North Carolina. This flight was in a general aviation airplane and was flown on instrument flight rules.

At 6:00 AM a telephone call was made to the FAA flight service station, a facility that provides weather information and accepts flight plans for transmission to the appropriate air traffic control facilities. The plan included a proposed routing and altitude, a proposed time of takeoff, and an estimated time en route. This information goes into a computer and is distributed to controllers who will handle the flight. The computer provides no air traffic control function, though. That is, it does not examine flight plans for possible conflicts. It performs a bookkeeping function, triggered by the pilot's flight plan.

At the airport, with the airplane loaded and engine running, the first radio call is to ground control. The pilot tells the controller his location on the field, that he is ready to taxi for takeoff, and that the flight is IFR (instrument flight rules) to Asheville, North Carolina. (Busier airports have recorded information on weather and airport conditions as well as a separate radio frequency and controller to call for clearances.) In this case the ground controller already had the air traffic control clearance: "Cessna 40RC is cleared to the Asheville Airport via a Trenton Two departure to Cedar Lake, as filed. Maintain 3,000 feet. Expect 14,000 feet at Cedar Lake. Contact departure control on 128.4, squawk 2771."

This clearance okayed the route filed by the pilot. The Trenton Two departure is published in charts used for IFR flying and prescribes a path that minimizes conflict with traffic at the busy Philadelphia International Airport, 32 miles to the southwest. The airplane was first cleared to climb to 3,000 feet (above sea level) and to expect clearance on up to 14,000 feet after passing Cedar Lake, New Jersey. The radio frequency to be used after takeoff was 128.4 MHz. The aircraft's transponder was to be set on 2771, a code issued to this airplane only (in its area of flight). The computerized air traffic control radar would read this code and identify the airplane on the scope with its registration number, N40RC (or flight number in the case of an airliner), its speed across the ground, and its altitude.

Ground-based facilities form an important part of the U.S. air traffic control system. Below is a long-range radar installation, and at the left is an antenna used to aid pilots in making instrument landings.

Photos, courtesy, Federal Aviation Administration, Great Lakes Region Public Relations Office

Courtesy, Federal Aviation Administration, Great Lakes Region Public Relations Office

Controllers in the radar room monitor arrivals and departures at Chicago's O'Hare International Airport.

While the tower at Trenton issues the en route as well as takeoff clearance, it cannot let an IFR airplane go until getting permission from the terminal radar control facility at Philadelphia, which controls IFR traffic in the area. On this morning there was no conflicting traffic, and so Philadelphia issued the clearance promptly.

There was not much traffic, and the controller at Philadelphia approved a shortcut off the Trenton Two as well as a climb to an altitude higher than 3,000 feet before the aircraft reached Cedar Lake. As the aircraft was climbing, though, the controller called and told the pilot to maintain 6,000 feet. He called again and affirmed the order when it did not appear that the pilot was stopping the climb. A jet airliner was nearby at 7,000 feet, and the paths of the airplanes intersected; the controller saw that the climb of N40RC would not get it above the airliner's altitude quickly enough. If the controller had not noticed this, an automated system would have alerted him to the possible conflict.

As Cessna 40RC moved away from the terminal area, control of the flight was transferred to the New York air route traffic control center and the pilot was cleared to climb to 14,000 feet, the requested cruising altitude. Later, control was passed on to the Washington Center at Leesburg, Virginia, and then to the Atlanta Center. Within each of those areas control was achieved by different controllers handling specific sectors of airspace.

Each time control of a flight is transferred, the pilot changes radio frequency. The computerized radar system and airborne transponder and altitude reporting system automatically identify the airplane to each controller as it comes onto the radar depiction of the sector being controlled by that

Courtesy, Federal Aviation Administration, Great Lakes Region Public Relations Office

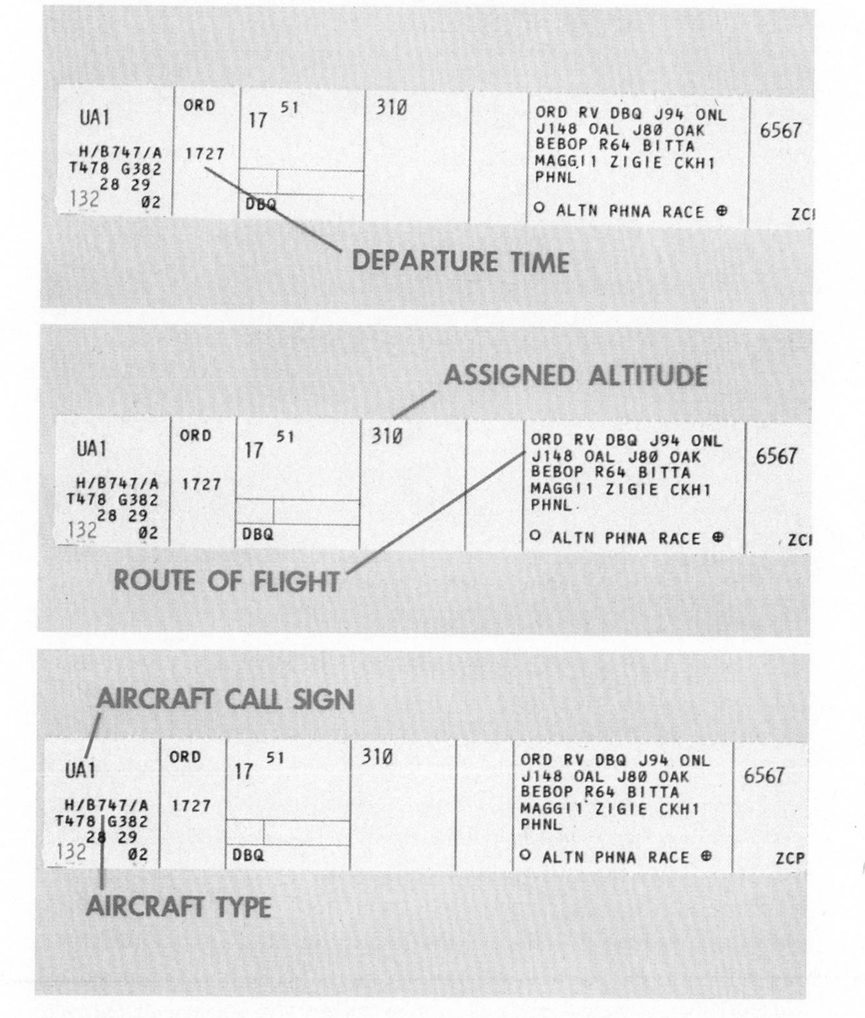

Flight plan for an airliner taking off from O'Hare International Airport (ORD) indicates departure time, assigned altitude (flight level: 31,000 feet) and route of flight, and identification of aircraft.

person. If the computerized radar system is inoperative or if the controller has some doubt, voice communication and the transponder can be used to verify the identity of a target. Transponders have an "ident" feature: when the pilot presses a button, it enhances the return of the target on the controller's radarscope.

On that day the computerized radar was working perfectly. When control of the flight was handed from one controller to another, the conversation was routine and there was no necessity to use the ident function of the transponder.

Controllers monitor the progress of traffic and watch for other airplanes in the vicinity. The weather was good enough that day for flying by means of visual flight rules (VFR). Positive separation was not provided from other planes sighted by the controller, nor was he obligated to report them to the pilot; unidentified traffic is called to pilots on a workload-permitting basis. Most of these planes were undoubtedly flying at a lower altitude than 40RC —almost all VFR flying is done below 12,500 feet because either oxygen or a pressurized cabin is required above that level. Positive separation from all aircraft flying IFR was provided.

Photos, George W. Gardner

Instruments in the cockpits of modern airplanes include (top) an attitude horizon indicator, which reveals that the plane is climbing at a 20° attitude with the wings level; the airplane is designated by the orange pointer at the top and the orange bars and dot in the center. A transponder (above) shows the speed, in knots, over the ground and the distance, in nautical miles, to the nearest navigation station. When the pilot pushes an identification button on his transponder, a ground controller is provided with a display of locations and altitudes (above, right).

Courtesy, Federal Aviation Administration, Great Lakes Region Public Relations Office

This flight could have flown in the VFR mode to the Asheville terminal area, where there was cloud cover. The tops of the clouds were at about 8,000 feet, and the bases were 800 feet above the ground. The visibility on the surface was five miles, and conditions were good for an approach to the airport using the instrument landing system. (At most airports a half-mile minimum visibility is specified for instrument landing.)

There were three airplanes in the Asheville area maneuvering for landing, and the controller assigned flight paths that sequenced the airplanes properly and provided the required three-mile spacing on final approach. About five miles from the airport, control of the aircraft was tranferred to the tower, and 40RC was cleared to land.

This flight was typical. It was cleared for takeoff when the pilot was ready, or within a few minutes of that time. It flew the route selected by the pilot and at the altitude preferred by the pilot. On arrival there was a minimum of maneuvering, and the landing took place without delay. There were no incidents; no other airplane came close enough to be seen, much less to be concerned about.

Shortcomings

An air traffic control system that allows most flights to be as routine as that of 40RC is basically good. Some slight element of risk will always be present, though. There is no way to eliminate all of it from any form of transportation, and no envisioned system, regardless of cost, could be considered as an absolute solution to collisions between airplanes. As long as more than one airplane is allowed to fly at a time, some risk will remain.

The key to the successful future of air traffic control is in limiting collisions to extremely random one-in-a-million occurrences that are not likely to recur. The system has come a long way in accomplishing this, but the current controversy over the adequacy of the air traffic control system, as well as its ability to absorb future growth, indicates that there is more to be done. Busy areas, where the system already operates at a level some think is greater than full capacity, are causing the most concern.

Near-misses receive considerable publicity as do controller grievances about overwork and inadequacies of current radar systems. There are times when traffic delays around major airports result in long lines of airplanes waiting to take off. Holding patterns fill with airplanes waiting to land. And the collision between an airliner and a Cessna on a training flight at San Diego, California, in September 1978 highlighted problems inherent when airline and general aviation traffic are mixed at busy airports.

The San Diego collision is an interesting case because it occurred with both airplanes operating in air space that was under positive control from the ground and with the conflict alert feature of the automated radar system warning controllers of the possibility of trouble. The crew of the airliner, though, had reported that they had the Cessna in sight; when that occurs, the controller can instruct the crew to keep the other airplane in sight and continue the approach to the airport. This tranfers the responsibility for separation of the two airplanes from the ground to the air. The airline crew at San Diego subsequently lost sight of the airplane and did not regain visual contact before hitting the airplane from behind.

The accident at San Diego might be used as an argument against visual separation, but millions of successful uses of this procedure argue for it. It certainly helps expedite traffic at busy airports, and if the procedure were outlawed the traffic jams of bad weather days might also occur at some airports on days of sunshine and clear skies.

If an attempt were made to eliminate all use of visual separation and visual flight rules and to control all air traffic all the time from the ground, even in good weather, the system would be in serious trouble. The required expansion would cost a prohibitive amount of money, and the practical capability of general aviation airplanes would be severely compromised. And, in the end, there would not likely be any demonstrable safety benefit.

Boeing 727 airliner plunges toward the ground at San Diego, California, in September 1978, moments after colliding with a small Cessna on a training flight. The airliner crew, placed on visual separation by the San Diego controllers, lost sight of the Cessna and struck it from behind.

Adapted from an illustration by Ray Pioch, "Automatic Alert System Will Help Prevent Midair Collisions," Doug Garr, POPULAR SCIENCE, vol. 215, no. 3, pp. 82–85, 162, September 1979; © 1979, Times Mirror Magazines, Inc.

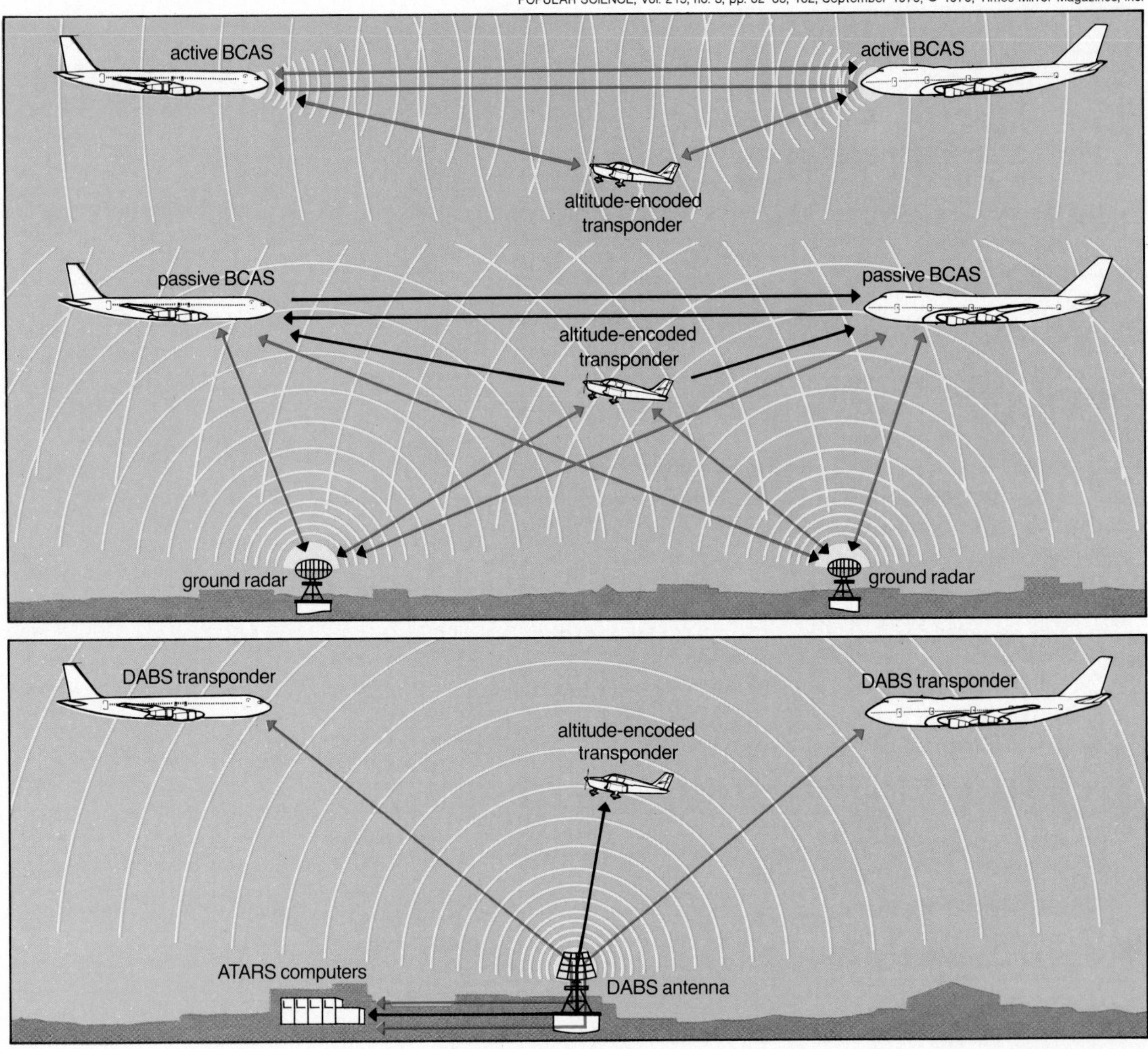

Beacon Collision Avoidance System (BCAS) uses the beacon reply signal that a plane's altitude-reporting transponder sends to ground radar. In an active BCAS (top) planes can interrogate and reply to one another; planes with altitude transponders can only reply. With passive BCAS (center) a plane can only listen as ground radar interrogates. ATARS, the Automated Traffic Advisory and Resolution Service, tracks planes and determines evasive maneuvers to avoid collisions (bottom). The ATARS computers obtain their data from a Discrete Address Beacon System (DABS).

The risk of collision due to human error would remain. Whether or not visual separation in positive control air space, as used at San Diego, is considered a continuing hazard depends on how that particular accident is regarded. Was it a one-in-a-million that is not likely to recur? Or was it an accident that was bound to happen and that will be repeated? Unfortunately, the answer is somewhere in between because of the way traffic is mixed.

The U.S. air traffic control system falls short on the procedural separation of fast and slow traffic. At San Diego the Boeing 727 overtook and collided with a Cessna 172 that was flying approximately 100 knots slower. It would have been one thing to instruct a 727 crew to maintain visual separation with another 727; it is quite another matter to tell a pilot to maintain visual separation with a much smaller airplane traveling at a much slower speed. At any airport the maintenance of separate traffic flow patterns to accommodate fast and slow airplanes could do more for airport safety than could any expansion of the man/machine aspects of air traffic control.

The FAA has done some work in the area of procedural as opposed to direct control of air traffic. For example, at the annual Experimental Aircraft Association conference in Oshkosh, Wisconsin, the world's largest fly-in meeting of general aviation aircraft, a published approach is followed by VFR traffic to the airport. Use of this approach and the fact that most traffic operates at roughly the same speed makes it possible to handle more daily traffic at Oshkosh than at the year-round busiest airport, Chicago O'Hare, which has more runways as well as complete radar service.

Some reports of malfunctioning radar, overworked air traffic controllers, and near-misses are caused by personnel problems. The relationship between the FAA and the air controllers' union is not the best, and as in any labor/management dispute each side dramatizes its points. Even when that factor is dismissed, however, some question exists about the adequacy of the system and the equipment in busy areas such as New York City and Chicago.

When the controller's computerized radar system fails, the backup is the older radar system that shows targets without identification and altitude. The controller has to ascertain which airplane is which (by having pilots "squawk ident" on their transponders) and must get information on the altitude by voice contact. If a controller is working ten airplanes and the computer mulfunctions, he or she must positively identify all ten and get a verification of altitude on each. It is like having one's electronic calculator fail halfway through a series of square-root problems. They can be done with a brain, a pencil, and paper, but it takes longer and the chance of error is greater.

The FAA has plans for more advanced radar, but the controllers want the equipment sooner rather than later. And while a computer breakdown might not be critical matter for low-altitude en route traffic and around terminals with moderate traffic, it is serious on heavily traveled high-altitude airways and around busy airports.

Controller workload increases as airplanes get closer together on their approaches to an airport. They come from random directions, all bound for the same place (the runway), and the most complex part of the control job is getting them properly sequenced for landing. Some airports operate at a level that might be considered beyond their practical capacity. Traffic is continuous, and the slightest glitch causes a problem. At airports such as National near Washington, D.C., O'Hare in Chicago, and LaGuardia in New York City the level of traffic is such that unless controllers work very quickly backups can easily develop. When controllers at such places "work to the rule book," as they have occasionally to protest their working conditions, long lines develop on taxiways and the holding patterns fill.

The answer to the question of capacity is elusive. The present controller force has been moving traffic at the busiest airports relatively safely for years. Some think that more controllers and equipment will increase capacity for future growth. But the problem is not that of running out of air space. There is plenty of that. Rather, the shortage is of runway capacity. One runway can handle just so many airplanes per hour, and the limit has been reached at some airports. Any plan to increase the use of a runway beyond

Experimental Aircraft Association conference, held annually at Oshkosh, Wisconsin, is the world's largest fly-in meeting of general aviation aircraft. During the conference the Oshkosh airport handles more daily traffic than Chicago's O'Hare.

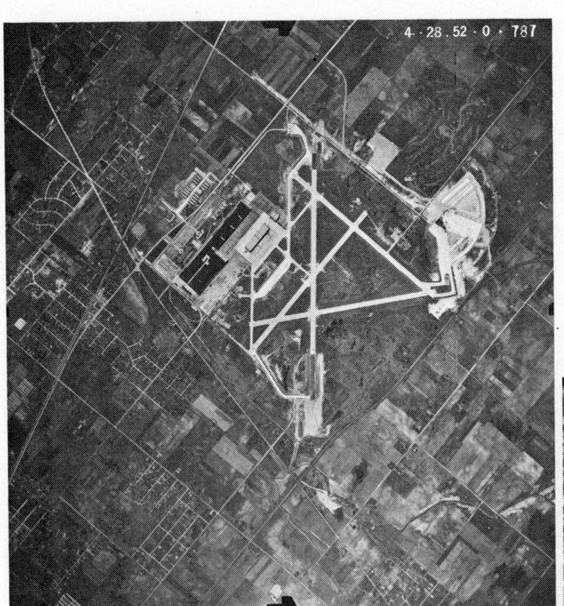

Aerial views of Chicago's O'Hare International Airport reveal expansion of runways between 1952 (left) and 1979 (below). To achieve the desired amount of runway space most new major airports must be built far from city centers.

Photos, National Oceanic and Atmospheric Administration

what the busiest ones are now handling would be questionable from a safety standpoint.

The solution to the problem of lack of capacity is to build more runways. But that can be difficult. The public is more disturbed about a new or expanded airport than about air traffic delays, and new airport construction is extremely expensive. Development of a new airport in any major urban area is highly improbable. The new airports built in recent years to increase capacity in given areas are inconveniently far from the city centers (Dulles International for Washington, D.C., Dallas/Fort Worth, and Kansas City International are examples); and they were possible only because large tracts of relatively undeveloped real estate were available. Any increase in capacity in other urban areas by means of new airport construction will come at the price of inconvenience to air travelers and will simply not be available in areas of solid urban sprawl.

Remedies

There are technological steps to take that will reduce controller workload. Data-link equipment to eliminate verbal transmission of instructions will simplify procedures, and airborne collision avoidance systems will someday provide an important backup for the ground-based traffic control system. When looking for technological solutions, though, the government must heed the lessons learned by the military in procuring weapons. It is self-defeating to develop systems so costly that no one can afford them and so technically complex that they are unreliable and difficult to maintain.

From a regulatory standpoint there will no doubt be new rules that will restrict the use of certain air space to certain types of aircraft so that capacity, at least for those planes, has been increased. But as more and more airports reach the ultimate limit of their runway capacity we will learn and relearn that the adequacy, capacity, and safety of our air traffic control system is determined as much by plain old concrete as it is by highly trained manpower and electronic marvels.

Courtesy, Federal Aviation Administration, Great Lakes Region Public Relations Office

A stuffed owl is placed near a runway to scare off birds. Flocks of birds on runways have been sucked into jet engines, causing planes to lose power and crash.

MAN AT HIGH ALTITUDES

Man's acclimatization to the reduced supply of oxygen at high altitudes is being studied because of its implications for people who suffer from a lack of oxygen because of diseases of the heart, lungs, or blood.

by Drummond Rennie

On May 8, 1978, Reinhold Messner and his colleague Peter Habeler, part of an Austrian expedition, became the 62nd and 63rd climbers to reach the summit of Mt. Everest (8,848 meters; 29,028 feet) and the first to do so without supplemental oxygen. In the succeeding 12 months the three highest mountains of the world were all climbed without bottled oxygen: Everest (for the second time); K2 (8,611 meters; 28,250 feet); and Kangchenjunga (8,598 meters, 28,208 feet). No fewer than 13 people in the space of that single year exceeded the previous altitude record for climbing without oxygen, which had been set in 1924 by two British climbers who had finally come to a halt at 8,565 meters (28,100 feet), only 283 meters (928 feet) short of Everest's summit.

This discussion shall argue that these long, lonely, dangerous exploits by the toughest of the tough have clinical relevance to every patient who ever needed oxygen or lapsed into heart failure and that when the physician looks at man at high altitudes, he does so to benefit us all.

The explosive increase in the popularity of skiing and trekking has brought hundreds of thousands of people who dwell around sea level to high altitudes. Thus, to the few members of mountain-climbing expeditions have been added a constant flood of tourists, plodding through once remote Himalayan villages, clambering up volcanoes in Mexico or Kenya, and overfilling ski resorts in the Rocky Mountains and the Alps. First the railways in the Andes, and then roads, ski lifts, planes, and helicopters have made the world's higher elevations ever more accessible. This expansion, due itself to advances in technology, has had results that are physiologically and clinically fascinating.

Wide World

Gamma/Liaison

Edmund Hillary (top left) and Tenzing Norgay (top right) in 1953 became the first to climb to the top of Mount Everest, at 8,848 meters (29,028 feet) the highest mountain in the world. A French team (bottom) in 1979 climbs the south-southwest ridge of K2, the first to attempt that route up the world's second-highest peak. Bad weather forced them to give up less than 200 meters below the summit.

DRUMMOND RENNIE *is Deputy Editor of The New England Journal of Medicine and Associate Professor of Medicine at the Harvard University Medical School.*

(Overleaf) Keith Gunnar—Bruce Coleman Inc.

It is easy for the well-equipped modern tourist to imagine that he or she must be venturing into unknown fields and to forget man has lived at high altitudes for millennia. Paleontologists have found human artifacts in Peru at about 4,000 meters (13,120 feet) above sea level dating back more than 10,000 years, and the Inca Empire once extended 4,000 kilometers ($\simeq$2,500 miles) along the Andes Mountains. It was largely a high-altitude society that was highly organized and efficient, creating a stark contrast to the tiny warring and primitive tribes of the nearby low-lying basin of the Amazon River. This contrast suggests that the altitude itself might in the long run have had survival value. The windswept tundra of the Andes may have been far less hostile than the insect- and disease-ridden swamps of the jungle far below.

Some 2.5% of the world's population now live at altitudes above 3,000 meters (about 10,000 feet), and so their sheer numbers make it important to understand how their bodies function (or go wrong) at such levels. The principal fact, however, is that 97.5% live at lower altitudes. The lower lying plains and the coastal regions offer man an easier living, and the higher altitudes are rightly considered hostile and dangerous to the newcomer.

Fred Ward—Black Star

Railroad train in the Andes Mountains of South America demonstrates the coming of industrial civilization to the high altitudes.

The high-altitude environment

Lack of oxygen (hypoxia) is the most important feature of life at high altitudes and is due to the progressive fall in barometric pressure with increases in elevation. The invention of the barometer is attributed to the Italian Evangelista Torricelli (1608–47), who followed a suggestion by Galileo, but it was the extraordinary Frenchman Blaise Pascal (1623–62), theologian, inventor of perhaps the earliest calculating machine, and founder both of the science of hydraulics and of the theory of chance, who in 1646 proposed his "Great Experiment." Pascal viewed the world as a sphere surrounded by a thin membrane, the atmosphere. He proposed that as the height of mercury in a barometer was a measure of the weight of the atmosphere pressing down upon it, the higher above sea level one measured the barometric pressure the less atmosphere would press down upon the mercury and so the lower the pressure would be. He also noted that all his scientific colleagues opposed this reasoning.

Pascal was proved triumphantly correct when a group of witnesses measured the barometric pressure at about 500 meters (1,640 feet) above sea level and then at the top of the nearest mountain, the Puy de Dôme (1,465 meters). Pascal's brother-in-law Périer recorded that the drop in pressure (of 3.09 inches) "overcame us all with wonder and astonishment." Scientists now know that the relation between altitude and barometric pressure is a complex one, depending not only on elevation but also on such factors as temperature and latitude. Broadly speaking, however, barometric pressure declines in a logarithmic fashion with ascent. If the sea-level atmospheric pressure is 760 millimeters of mercury (abbreviated mm Hg or "torr"—named after Torricelli—or, in the units of the Système Internationale d'Unités, 101 kPa or kilopascals), then it will be 380 mm Hg or half an atmosphere at 5,740 meters (18,832 feet) and 250 mm Hg or one-third of an atmosphere on top of Mount Everest (8,848 meters; 29,028 feet).

Culver Pictures

French scientist Blaise Pascal in Paris in 1648 demonstrates that barometric pressure drops with increasing altitude.

It had long been recognized that there was a substance in the atmosphere necessary for the respiration and life of animals as well as for combustion. During the late 18th century three scientists, Carl Wilhelm Scheele in Sweden (1772), Joseph Priestley in England (1774), and, most importantly, Antoine Lavoisier in France (1775), all more or less independently isolated this unknown gas in the air; Lavoisier named it oxygen. Later measurements showed that oxygen constituted 20.946% of air and that this proportion of the atmosphere was the same at all altitudes. The total pressure of the air being the sum of the partial pressures of its constituents (almost entirely nitrogen and oxygen, with a little carbon dioxide and some inert gases), the partial pressure of oxygen at sea level was 20.946% of 760 mm Hg, or 159 mm Hg. This, too, would be halved (79.6 mm Hg) at 5,740 meters (18,872 feet) and, of course, reduced to one-third (53 mm Hg) at the summit of Mount Everest—a figure that could have been calculated well before Mount

Everest was identified by a governmental survey of India as the highest point on Earth's surface in 1852 but which, in fact, had not as of 1981 been verified by actual experiment.

A dwindling partial pressure of atmospheric oxygen (or, as it is often called, "oxygen tension"), producing an inevitably lowered partial pressure of oxygen in a person's blood, is only one of the consequences of ascent. All other things being equal, the temperature drops about 1.95° C for every 305 meters (1,000 feet) of ascent. Thus at the same time that a temperature of −29° C (−20° F) was observed at the summit of Everest the nearest sea-level temperature was +27° C (+81° F). Clearly this temperature difference depends to a great extent on the season and upon the latitude. At levels above 4,572 meters (15,000 feet) temperatures will inevitably be some 29° C (84 ° F) lower than at sea level, and so it is not surprising to find that the high-altitude climate is an arctic one and that the Altiplano (High Plains) of Peru are habitable only because they are so near the Equator.

Strong winds are another characteristic of high altitudes, though only at the highest elevations. The jet stream is a constant and powerful west-to-east flow of air that zigzags around the Northern Hemisphere, usually at about 60 knots (111 kilometers or 69 miles per hour) but at its vigorous core sometimes at more than 200 knots. It is most forceful at an altitude of about ten kilometers (six miles), but its vagaries and its opposition by the summer monsoon are major factors that must be taken into account by those climbing the highest peaks of the Himalayan range. Other attributes of the high-altitude environment are decreased humidity of the air and increased irradiation from the Sun, since there is less atmosphere to filter the Sun's rays and also because there is considerable reflection from snow and ice.

It is not surprising, given the prevailing conditions, that frostbite is common among climbers and skiers. Also, the low humidity combined with the cold air and the increased breathing rate tends to lower resistance to laryngeal and bronchial infections, and the radiation may cause severe sunburn of the skin or of the cornea and conjunctiva of the eye, the latter resulting in the temporary but painful condition known as snow blindness.

Most of these problems, however, are typical of any arctic environment at whatever altitude. The only one that is unique, and which usually turns out to be the major difficulty for sea-level man taken to high altitude, is lack of oxygen. On ascent, then, what happens to the body as a consequence of hypoxia and how does the body respond? Here it should be stressed that this discussion is not dealing with sudden exposure to air low in oxygen, such as would be found in an unpressurized balloon, going up in a few minutes to 8,000 meters (26,240 feet) with no oxygen cylinders on board. Such exposure would produce coma and probable death in a few minutes. Instead the focus is on relatively gradual ascent to moderate heights, 3,000–4,000 meters (approximately 10,000–13,000 feet). An understanding of how the body responds normally to a gradual ascent (by beginning the infinitely complex process called acclimatization) and abnormally (by developing "acute mountain sickness") is relevant not merely to sea-level dwellers taken to high altitudes but to hundreds of thousands of low-altitude patients with heart and lung conditions.

Courtesy, Drummond Rennie

High lenticulate clouds, made up of ice crystals that are continually forming, hover over the top and north face of King Peak (5,173 meters; 16,967 feet) in the Yukon territory of Canada.

Physiology of ascent and acclimatization

Those who have driven in a few hours from Lima on the coast of Peru to one of the mining towns situated 4,000–5,000 meters (13,100–16,400 feet) up in the Andes will be struck on arrival by their own total incapacity to engage in any physical activity. They will certainly contrast it with the exuberance with which the local miners, having just finished a hard eight-hour shift, will plunge into an energetic game of soccer, and they will rightly ascribe the difference to "acclimatization." The long-term effects of exposure to a relative lack of oxygen are shown by every organ in the body, and they are demonstrated best by those who have been born at and lived longest at the highest altitudes. Though some effects, such as an increase in the number of oxygen-carrying red blood cells, begin to be detectable within minutes, most are not complete for months or even years, and it seems impossible ever fully to acclimatize to altitudes that are above 5,400 meters (17,700 feet).

Acclimatized to life at high altitudes, residents of such villages as San Pedro de Casta (top), situated at 3,192 meters (10,533 feet) in the Peruvian Andes, can engage in such vigorous sports as soccer (bottom).

(Top) Eric Simmons—Stock, Boston; (bottom) Owen Franken—Stock, Boston

Courtesy, Drummond Rennie

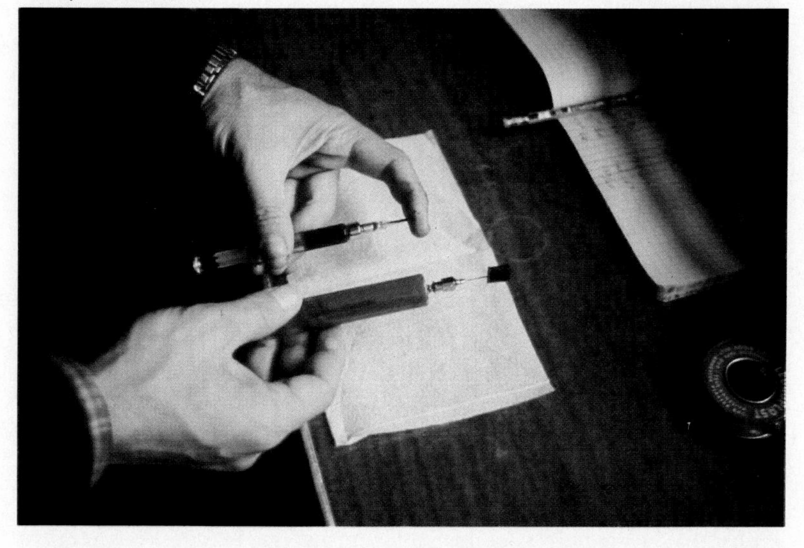

Arterial blood taken at sea level, held in the right hand, is compared with arterial blood taken from a miner at Cerro de Pasco in Peru, located at an altitude of 4,259 meters (13,973 feet). This change in the blood results from a lack of oxygen in the circulating hemoglobin and causes people at high altitudes to look blue.

Large and accessible populations in whom one may measure such effects live only in the towns of the Altiplano, and Drummond Rennie has made six visits to Peru to work with Peruvian researchers both at sea level and in the mining towns above 4,200 meters (13,800 feet). Of particular interest to him was the function of the kidneys. One reason for this was that Rennie had observed protein in the urine (proteinuria) in young adults who had been born in London and had lived there all their lives; these people were suffering from congenital abnormalities of their hearts that shunted blood away from their lungs so that their bodies were never properly oxygenated and they looked literally "blue." This blueness is due to the fact that more than a third of their circulating hemoglobin, the substance in red blood cells that carries oxygen, is unoxygenated and blue-black in color. Physicians call this condition "cyanosis." Rennie wished to see how the very fit people of the Altiplano, who, because of their life at a high altitude, had also been cyanosed continuously since birth, managed to preserve the function of an organ—the kidney—that is normally "bathed" in oxygen.

Hundreds of experiments were performed on people in the Altiplano. It became clear that they, like the Londoners, had excessive protein in their urine—almost always a sign that the delicate sieve through which blood flows and from which urine begins to be formed has become more leaky. The experiments showed further that the Peruvians had normal kidney blood flow and that though each milliliter of blood contained far less oxygen per gram of hemoglobin than did the blood of people living at low altitudes, there was far more hemoglobin per milliliter of blood. The blood of the Altiplano Peruvians was not composed of the normal sea-level 45% of red cells and 55% plasma. but instead consisted of 65% red blood cells and 35% plasma. This adjustment allowed the delivery of oxygen to such tissues and organs as the kidney to be the same as for healthy people at sea level.

Such experiments and those by workers in Colorado on residents of Leadville and by others on natives in the remote regions of the Pamirs in the Soviet Union demonstrated clearly that as the partial pressure of oxygen is inescapably decreased at any high altitude, the body tries to compensate ("acclimatizes") by arranging an easier passage of what oxygen there is in the air to where it is needed: the tissues. For example, the respiration rate is increased, the oxygen-carrying capacity of the blood is vastly improved, and there is huge expansion in the number of open capillaries, or tiny blood vessels, so that oxygen diffusing from the red cells in the capillaries to the mitochondria in the cells has less far to go. There is a brisk increase both in the number of mitochondria (which is where oxygen is actually used) and in the chemical compounds and enzymes that handle all these reactions as well as many other physiological changes.

The adjustments, collectively called acclimatization, are so dramatically successful that miners can work at over 5,800 meters and climbers carry on heavy sustained exertion at well over 8,000 meters (26,240 feet): altitudes at which the unacclimatized newcomer would die in minutes. No amount of acclimatization, however, can bring the body's physical capacity back to what it was at sea level because it cannot increase the maximal amount of oxygen that anyone can take up and use in any one minute. This figure at sea level is an indicator of an individual's state of training, but even in trained people it declines steadily with increasing altitude until at about 7,500–8,500 meters (24,600–27,880 feet) it has fallen to the level necessary to preserve bodily functions at rest. On the very slightest exertion at such an altitude a climber is breathing as if he were in the final stretch of a 10,000-meter race. To be able actually to climb, therefore, at such altitudes indicates a phenomenal capacity, both physical and mental, to exist on the outer edge of what is theoretically possible. A further intriguing aspect about acclimatization is that all adjustments made by natives such as the Quechua Indians of Peru, whose ancestors have lived at high altitudes for 10,000 years or more, have been made just as effectively by people born at those altitudes but who are of low-altitude stock.

After it was shown that in people fully acclimatized to hypoxia since birth there is a change in their kidney sieving mechanism which allows some blood proteins to leak into the urine, the next questions were whether this leakiness also occurs in sea-level dwellers taken to high altitudes, whether

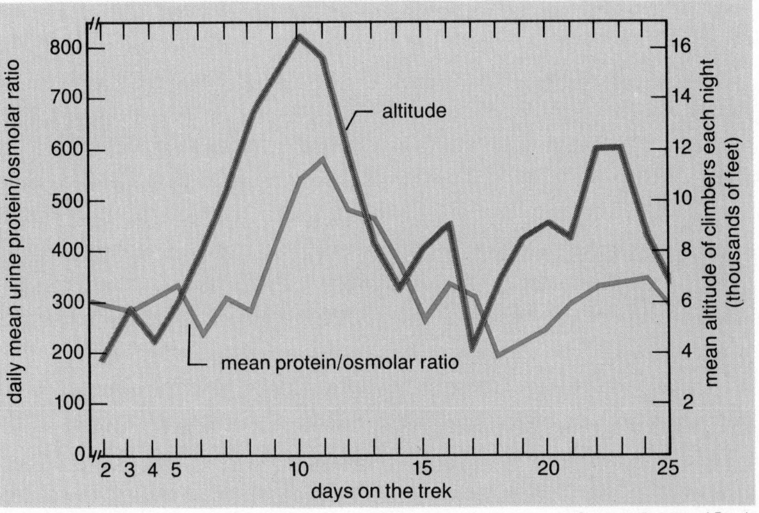

Courtesy, Drummond Rennie

Urine secretions of mountain climbers increase with altitude as part of the process of acclimatization.

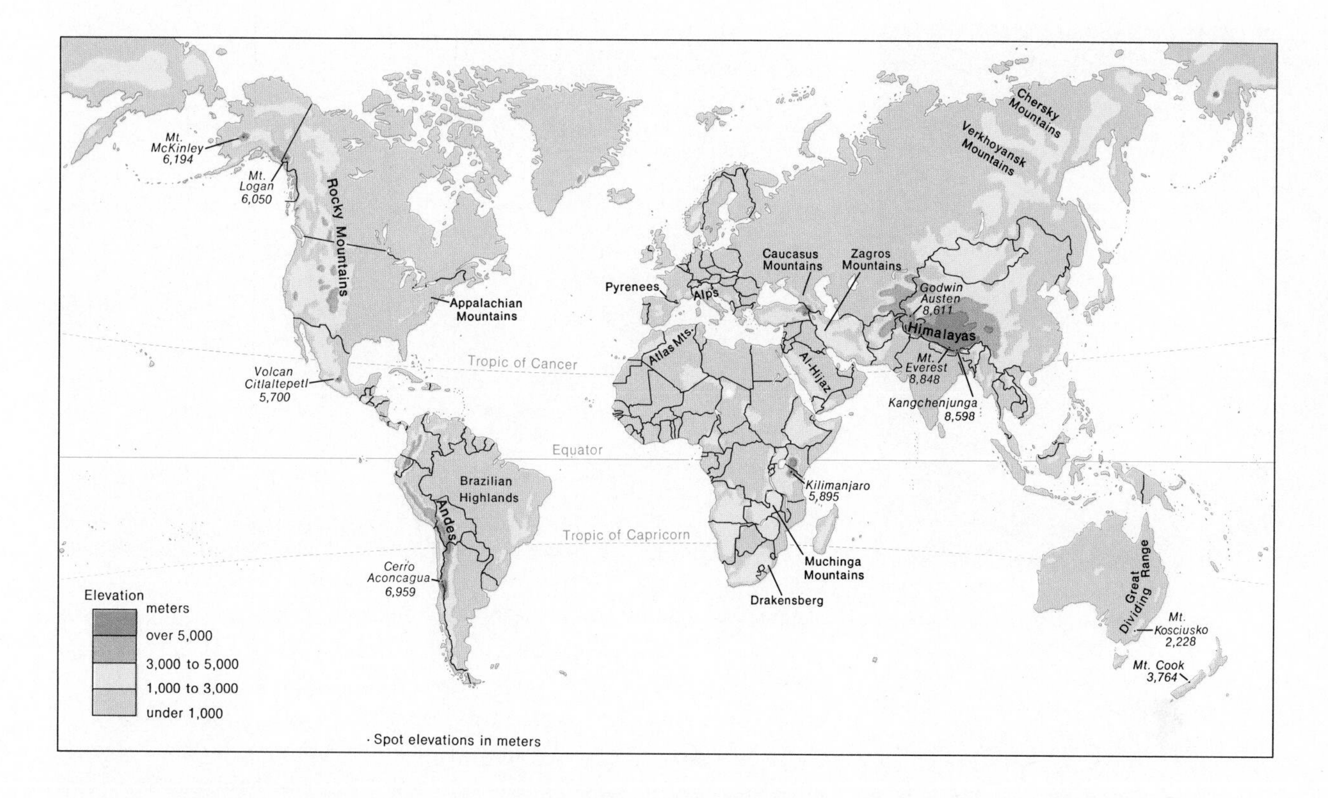

High mountains are found in many parts of the world, and the people living among them have become acclimatized to life at such altitudes. The highest peaks on the Earth are the Himalayas in Asia.

its extent corresponds to the altitude, and whether it disappears on descent. In a group of unacclimatized U.S. climbers in the Himalayas Rennie found that indeed all these questions should be answered yes: the higher the climber, the more his kidneys leaked proteins.

The results suggested that slight hypoxia has a striking effect on the way kidney capillaries filter large molecules such as proteins. To check that this was not due to some alteration in the blood proteins or even due to the extra exertion, it was necessary to go to a place where one could study lowlanders taken effortlessly to relatively high altitudes and kept at rest while injected with harmless molecules of various sizes. The appearance of these molecules in the urine was then measured. Rennie accomplished this in the far northwest of Canada on Mount Logan (6,050 meters; 19,849 feet). There, each year a team led by Charles Houston of the University of Vermont and of the Arctic Institute of North America sets up a cluster of tents on a glacier at 5,360 meters (17,585 feet). As part of a much larger series of studies, Rennie was able to show that in both unacclimatized and fairly well acclimatized people who had been flown up to this high camp and who were resting the kidney capillaries filtered and allowed into the urine larger molecules than they allowed at sea level.

Acute mountain sickness

While engrossed with these studies many researchers as well as their subjects suffered variously but frequently from bouts of "acute mountain sickness." Like others before him, Rennie soon became involved in investigating this troublesome and sometimes fatal condition.

Photo and diagram, courtesy, Drummond Rennie

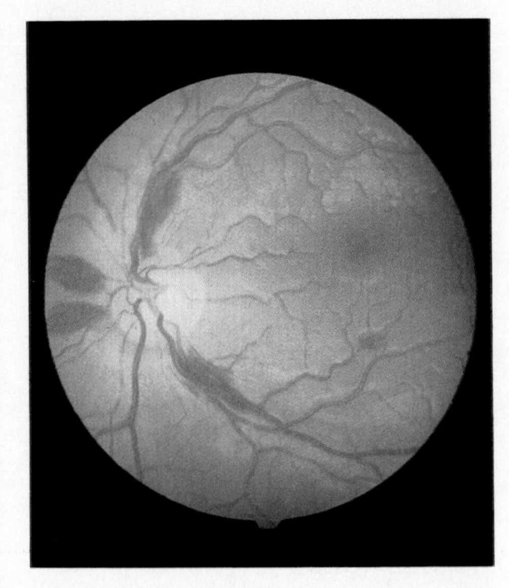

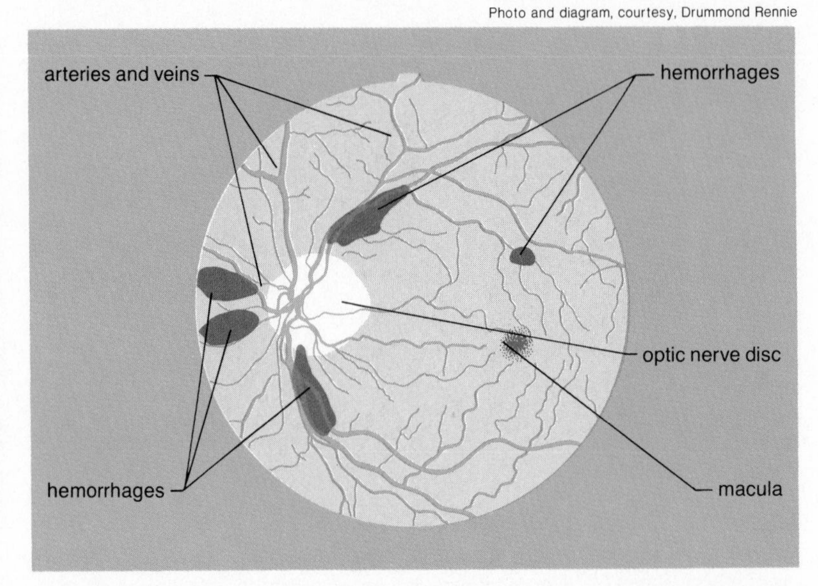

Photograph of the retina of an eye (above) was taken at 5,913 meters (19,400 feet) of a fit climber who had descended from 7,620 meters (25,000 feet). As shown in the drawing at the right, the dark patches indicate minute hemorrhages. They are caused by a massive increase in retinal blood flow, which probably occurs in order to help the eye deal with the lack of oxygen at high altitudes.

Acute mountain sickness (AMS) was well described by the Jesuit priest José de Acosta in his account of his travels with the Spanish viceroy through the Peruvian Andes in the 1570s. This malady, usually a transient nuisance rather than an illness, begins from 12–72 hours after ascent to higher altitudes, usually above 3,000 meters (about 10,000 feet). Sufferers characteristically lose all appetite and may vomit. They feel lethargic, dizzy, and may even lose their balance completely. Their sleep is disturbed by nightmares and attacks of breathlessness, and a progressively severe headache increases the misery. But after about one or two days at rest, all the symptoms disappear. Individuals vary widely in their susceptibility to AMS.

De Acosta reasoned that the illness affecting him and his companions was due to the very high altitude of the Andes, and he stated presciently—for this was well before the time of Pascal—that the real cause was that at such altitudes the air was excessively thin. Present-day researchers have spent much time investigating AMS partly because it is now commonly seen wherever skiers and mountain tourists congregate but chiefly because it has two potentially fatal forms, high-altitude pulmonary edema and high-altitude cerebral edema. A better understanding of these could lead to improved treatment for patients at low altitudes who are suffering from the common conditions of cerebral and pulmonary edema.

Working originally on Mount Logan, Rennie was one of a group headed by Regina Frayser, who first showed that some people taken abruptly to high elevations developed minute hemorrhages from the blood vessels supplying the retina of the back of the eye. By injecting a fluorescent dye and photographing it as it passed through the retinal blood vessels, they demonstrated that these hemorrhages were associated with a massive increase in retinal blood flow, presumably to combat the hypoxia (though the blood contained less oxygen, an increased blood flow might make up for it). There was some suggestion that the hemorrhages occurred more commonly with AMS, but to investigate that the researchers needed far greater numbers than the dozen or so people they could airlift each year.

Peter Hackett, a young physician from Chicago, had staffed a tiny clinic at the village of Pheriche in Nepal, situated at 4,243 meters (13,920 feet) on the approach route to Everest. Pheriche was an ideal place for investigating lowlanders ascending to high altitudes because large numbers of tourist-trekkers had to pass through it on their way around the vast, projecting ridge of Nuptse (7,879 meters; 28,850 feet) before they could see Everest itself from the base camp at the foot (5,300 meters) of the Southwest Icefall, where climbing really begins. The tourists arrived from Nepal's capital, Kathmandu, via the village of Lukla (2,800 meters; 9,186 feet), which they had reached either by a ten-day hike over two passes or by an hour's flight to the landing strip. From Lukla to the Everest base camp it was an easy hike along a well-trodden path with no climbing of any sort.

The investigators tried to see every tourist, questioning each closely about symptoms and giving each a brief physical examination. After examining some 500 such people Hackett and Rennie were able to show that mild AMS was common and harmless; that it affected the fit and the unfit and females and males equally; and that its frequency was related to the speed of ascent (it was commoner in those who had flown to Lukla) and amount of exertion. Not only was it an illness of those who went too high too fast, independent of their sea-level athletic prowess, but also its incidence steadily dropped off with increasing age: the young were more susceptible. The research confirmed that it was an illness that was almost entirely preventable by ascending slowly.

On the 1973 U.S. expedition to Dhaulagiri, Rennie had previously shown when photographing the backs of climbers' eyes at 5,900 meters (19,400 feet) that fit, exceedingly strong climbers with no trace of AMS developed retinal hemorrhages (which cleared up on descent) in roughly the same proportion (one-third) as people on Mount Logan at 5,364 meters (17,600 feet). Hackett and Rennie then showed that lower down at Pheriche (4,243 meters; 13,920 feet) the incidence of retinal hemorrhage was far lower (4%)

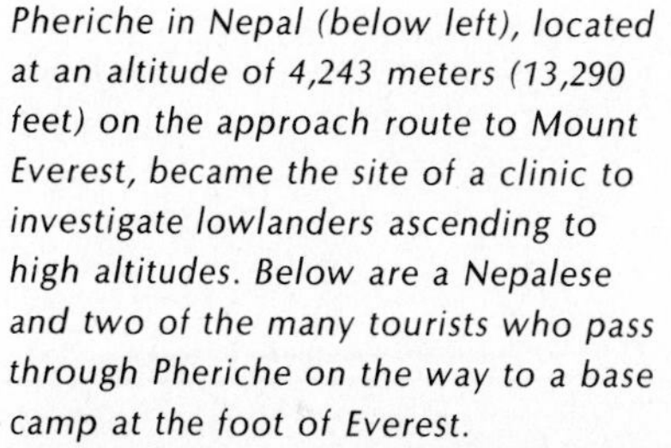

Pheriche in Nepal (below left), located at an altitude of 4,243 meters (13,290 feet) on the approach route to Mount Everest, became the site of a clinic to investigate lowlanders ascending to high altitudes. Below are a Nepalese and two of the many tourists who pass through Pheriche on the way to a base camp at the foot of Everest.

(Left) Diane M. Lowe—Stock, Boston; (right) © Stuart Cohen—Stock, Boston

Courtesy, Drummond Rennie

Drummond Rennie (left) treats a climber at a base camp at 5,878 meters (19,400 feet) after the latter had returned from the summit of Dhaulagiri (8,177 meters; 26,811 feet) in the Himalayas. The patient is taking oxygen for therapy, and from that point he had to be carried to a lower altitude to receive further treatment.

but that there it was definitely higher in those with AMS. Since both on Logan and at Pheriche those with AMS tended to have lower levels of oxygen in their blood than would be expected for any given altitude, this seemed to suggest that the retinal hemorrhages were related to severe hypoxia.

What causes AMS and how can it be fatal? Although the disease is usually no more than a transient nuisance, some 2.5% of the tourists studied by Hackett and Rennie developed high-altitude pulmonary edema (HAPE). This is a condition in which the lungs suddenly flood with fluid exuded into the air sacs from the blood. Victims rapidly become breathless, deeply cyanosed, and helpless, and may die if not quickly taken to a lower altitude. Furthermore, some 1.8% developed cerebral edema, a swelling of the brain. This resulted in mental changes, weakness, disorientation, and, occasionally, rapid coma and death. Descent to a lower altitude reversed the symptoms, usually even in those who were very ill, though some deaths still occur.

The cause of HAPE remains unknown. It is not failure of the heart, causing a backup of fluid in the lungs, and it is not a direct effect of hypoxia on the lung capillaries, causing them to leak into the air sacs. What probably happens is that hypoxia causes some lung arteries to shut, and the whole circulation is forced through the remaining, open capillaries whose overstrained walls then leak. The cause of cerebral edema is equally obscure. Perhaps hypoxia causes sensitive cells in the brain to lose their power to pump out salts and fluid; indeed, when this occurs in a very minor degree, it may merely cause the usual symptoms of AMS. It is only when severe disorientation and coma occur that the diagnosis of cerebral edema is made, but minor edema of parts of the lower brain may be the cause of all AMS.

Hackett and Rennie showed that AMS may be completely prevented by slow ascent and cured by immediate descent in the event of trouble. Notices containing such information were placed along mountain trails, and these measures seemed to be dealing with the problem satisfactorily.

Recently, however, Hackett and Rennie returned to Nepal to study the control of respiration and the body's handling of fluids in Kathmandu and

(Top, left and right) Gamma/Liaison; (bottom, left) Jonathan T. Wright—Bruce Coleman Inc.

Pierre Mazeaud (above), who in 1978 climbed to the top of Mount Everest at the age of 49, was the oldest person to conquer the world's highest peak. At the top left are Sherpas in Nepal, long acclimatized to high altitudes. A powder avalanche on Everest (bottom left) is one of the hazards faced by mountain climbers.

in Pheriche. For the following 72 hours the two physicians never went to bed. They were first caught up in a lengthy night-time rescue of a 28-year-old Swiss, a climber who became unconscious from cerebral edema at 6,000 meters (19,700 feet) and whose coma was so deep that his breathing stopped, with the result that for 60 hours oxygen had to be pumped into his lungs by continually squeezing a bellows by hand. Meanwhile, treatment became necessary for a French woman and a West German man who had mild pulmonary edema but who were too breathless and confused to help themselves; for one elderly Austrian man who had had a hemorrhage from a duodenal ulcer; and for a middle-aged West German man with cerebral edema who was wildly hallucinated and antagonistic before he, too, lapsed into stupor. The final case involved a French man with cerebral and pulmonary edema and massive retinal hemorrhages, who had no idea who or where he was; he was given an escort and sent down the trail to safety.

After three harried days and nights a helicopter appeared (by chance), and Hackett and Rennie used it to evacuate the three sickest patients to Kathmandu (where all except the Swiss climber, who died, recovered); the others were sent down on foot. The physician-physiologists then returned to get some sleep and after that to get on with the work that had brought them to that part of the world, reflecting that no matter how much they knew about hypoxia or AMS there would always be many cases to remind them of mankind's eagerness to explore and, ultimately, to teach them something about all those hypoxic patients at sea level.

THE AMAZING NEWBORN

Recent research has shown that human newborns are more competent than anyone heretofore supposed. In some respects they are even superior to the adults they will become.

by T. G. R. Bower

An infant literally is someone who cannot speak or understand language. The study of infancy is thus the study of that period of life between birth and the onset of language use. It is a short segment of a human life, about 18–20 months on the average. Despite its brevity infancy has fascinated parents and scientists for centuries. How well do infants perceive the world? To what extent do they understand what is going on around them? What underlies the myriad changes one sees during infancy? These are typical of the questions that have prompted thousands of articles and books.

Until relatively recently the answers given to such questions suffered from a severe shortage of empirical data. Since infants do not talk, they cannot be asked directly about their views of the world. Because no investigator would wish to starve or shock a baby, certain techniques used with nonhuman animals can hardly be applied. While charming, infants make difficult experimental subjects; they are not terribly active and are liable to fall asleep or demand food, cleaning, or other services at frequent intervals, so that simply watching babies, without doing anything with them, is rather unprofitable as well. Despite these problems recent work has allowed scientists to estimate with some accuracy the abilities of infants of various ages. It now appears that infants, even newborn infants, are much more competent creatures than had been suspected.

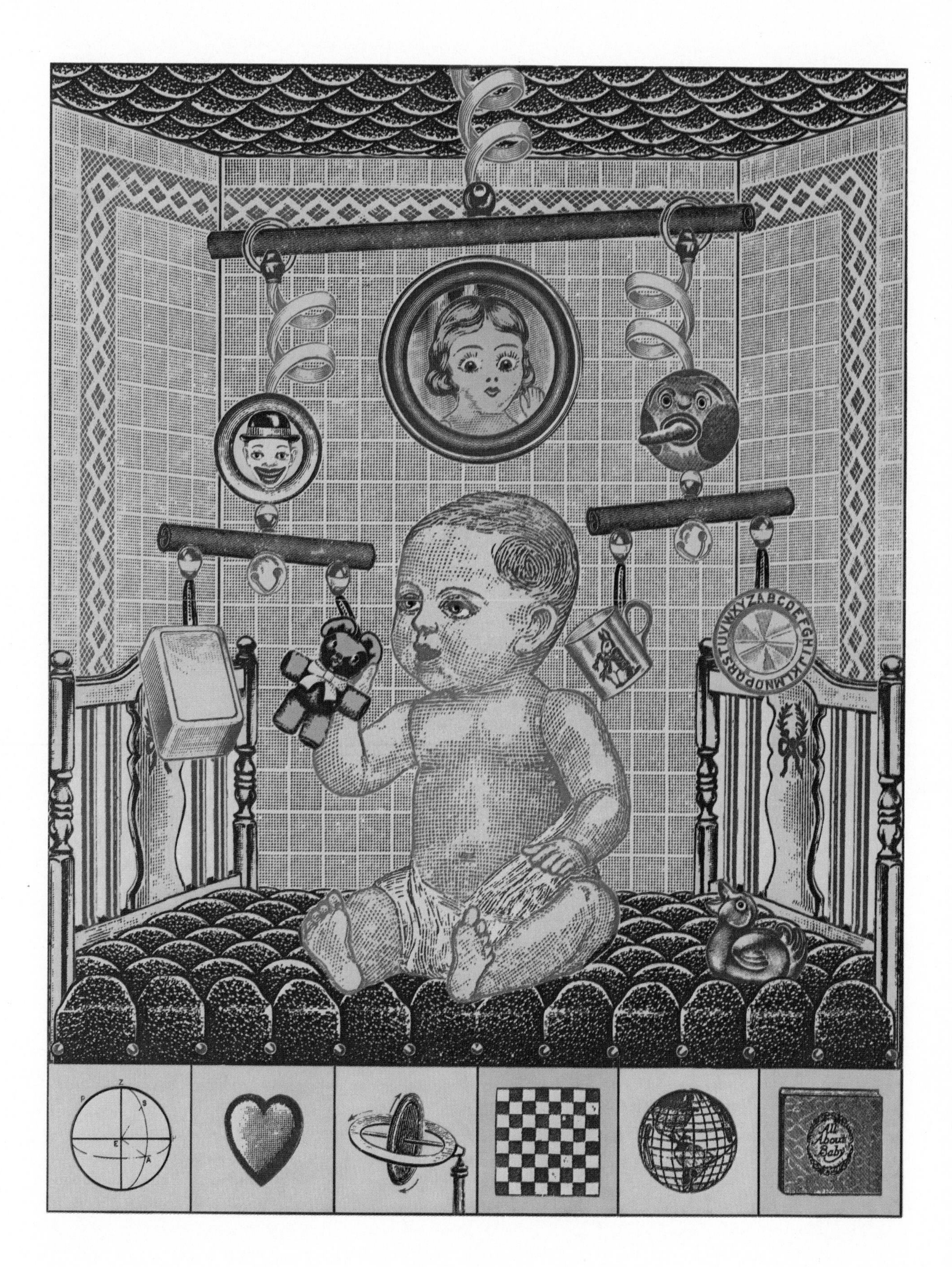
All About Baby

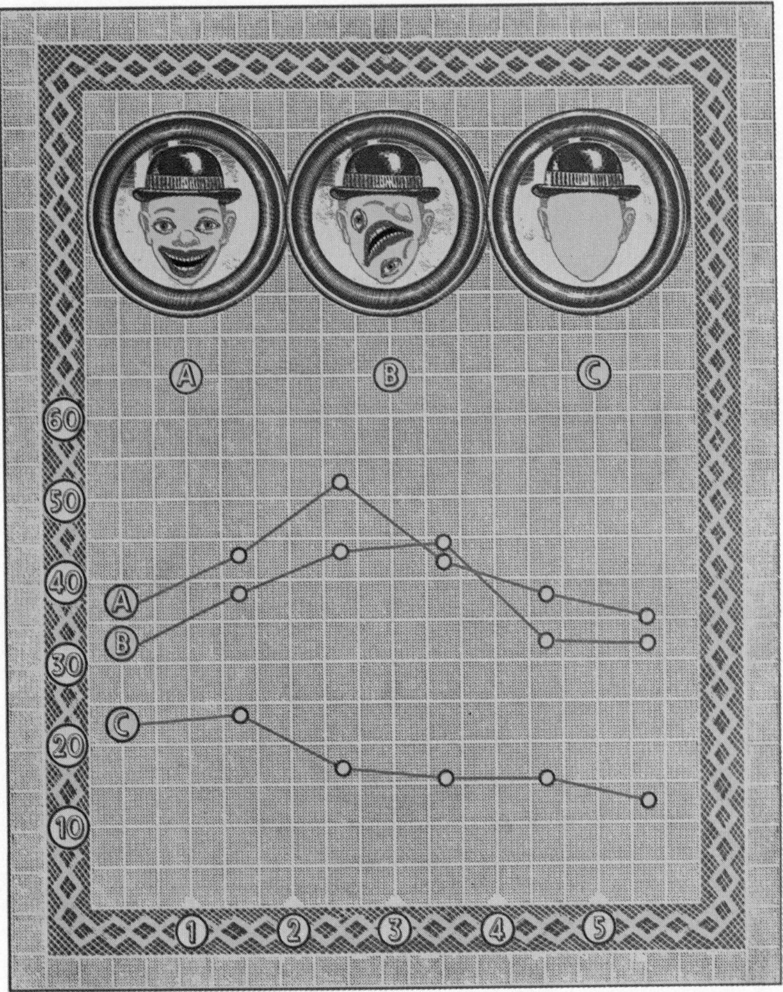

What newborns see and hear

The results of a few simple experiments, taken together, begin to answer some of the more fundamental questions about infants and indeed about human development. The first question often asked by new parents—and by scientists as well—is "Can the baby see?" In light of recent research the answer appears to be resoundingly affirmative. In a study done in 1970 by C. Jirari at the University of Chicago newborns were shown pictures of a lifelike face and a scrambled face, the latter with the features reassembled such that, for example, the mouth was above the nose and the eyes were no longer parallel. When the face was moved back and forth, the babies, averaging ten minutes of age, often followed the lifelike face with their eyes but were far less likely to follow the scrambled face. This experiment shows that these babies could tell the faces apart, a discrimination based on pattern, since the components in the faces were identical. It also shows that babies only minutes old already prefer objects that are like human faces.

Even more impressive perceptual abilities have been demonstrated in studies of imitation done in the mid-1970s by Jane Dunkeld at the University of Edinburgh and Andrew N. Meltzoff and M. Keith Moore at the Univer-

T. G. R. BOWER is a Reader in Psychology at the University of Edinburgh, Scotland, and an expert in infant development.

Illustrations by John Craig

Courtesy, Andrew N. Meltzoff, Child Development & Mental Retardation Center, University of Washington

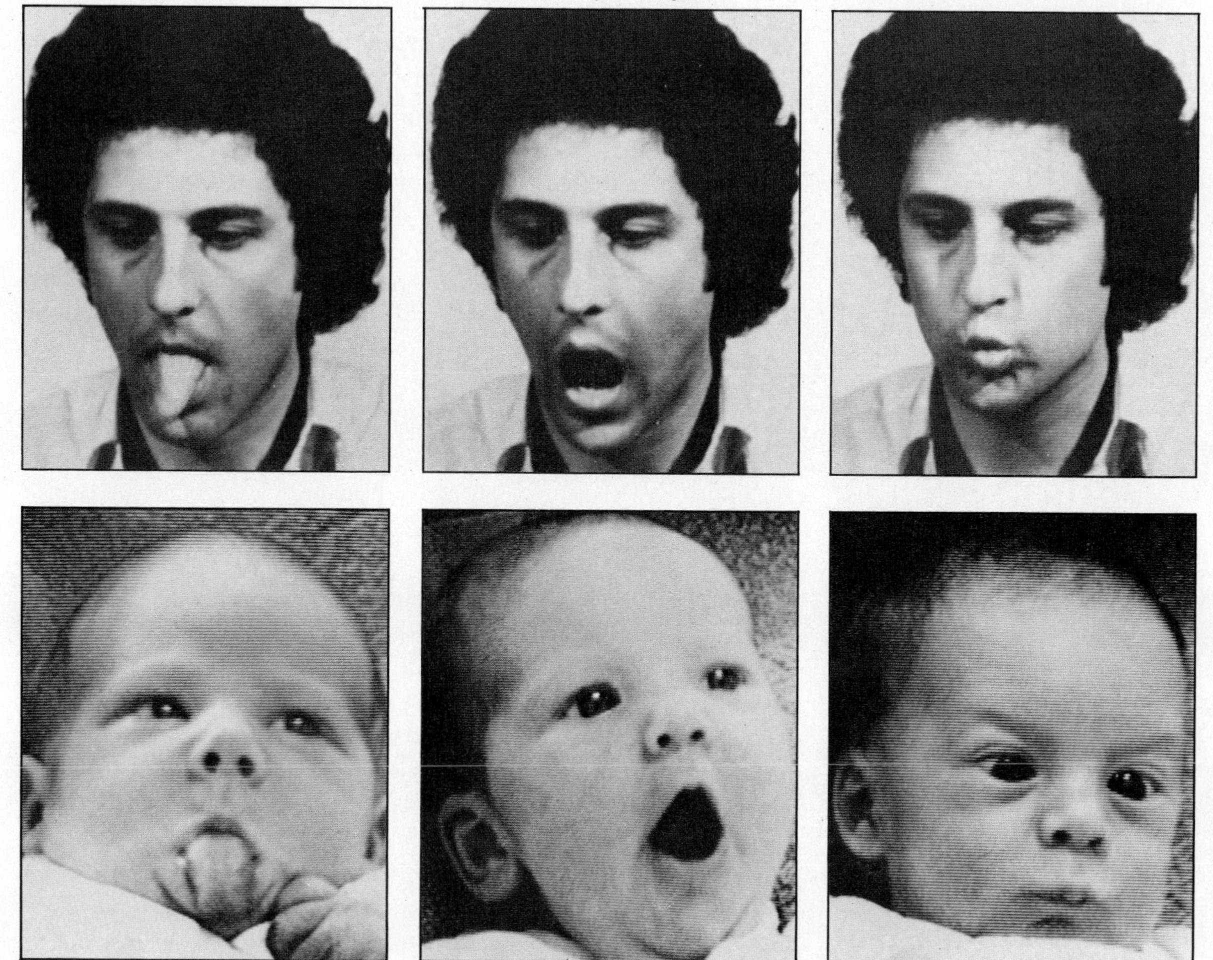

sity of Washington. These studies showed that if an adult, for instance, a mother, sticks out her tongue at her newborn, within a short time the baby will begin to stick his tongue out at her. Suppose she then stops sticking her tongue out and begins to flutter her eyelashes; the baby will flutter back. If she next starts to open and close her mouth, the baby will begin to open his mouth in synchrony. Of course, the baby will also stick out his tongue, flutter his eyelashes, and open his mouth spontaneously, but he does so to a far greater extent if an adult model is present. Furthermore, the newborn actually seems to enjoy engaging in this mutual imitation game.

One must give serious consideration to the level of organization required by such behavior. How does a newborn baby know he has a mouth? How does he know he has a tongue? And how does he know that his mouth and tongue are like the mouth and tongue he sees before him? It appears that there must be an astonishing amount of built-in intersensory communication for the baby to watch an adult sticking out her tongue and transform this information such that he knows, given the social situation, that he should return the compliment. The same reasoning holds for any of the imitations involving the face. How many newborns see themselves in a mirror? In view of the extremely small number who do, the experience does

Graph (facing page) summarizes the results of a study in which babies several days to six months of age were shown three objects—a painted stylized face, a scrambled face, and a control pattern devoid of facial features—paired in all possible combinations. The average time in seconds that the babies looked at each of the objects in a two-minute test (vertical scale) is plotted against the age of the babies in months (horizontal scale). Even the very youngest infants clearly showed a visual preference for human faces and facial features. The discovery that infants two to three weeks of age imitate adult facial gestures (above) suggests that babies somehow can compare what they see with sensory information from their own unseen faces.

not seem to be either typical or necessary for precocious imitation to appear. The evidence indicates both that newborn infants can see and that they can relate what they see to what they can only feel.

Other experiments have provided answers to another common question: "Can the baby hear?" The most technically uncomplicated one, done in 1961, involved an infant subject who was less than two minutes old by the end of the experiment because the experimenter had been fortunate enough to obtain access to a delivery room. The scientist, Michael Wertheimer of the University of Colorado, was also fortunate in that his subject was delivered without anesthesia (an important point since the subject needed to be as alert as circumstances permitted) and that the birth was free of complications. Just after the baby emerged she was presented with a series of sounds, randomly to her right or left. Initially, before boredom set in, the subject correctly turned her eyes in the direction of the sound source. This study shows that a newborn not only can hear and can localize the source of a sound but also has a built-in expectation that there will be something to see at the source of the sound. A variant of this experiment, carried out in 1966 by Gerald Turkewitz and his colleagues at the Albert Einstein College of Medicine, New York City, uncovered some "intelligence" in the responses of newborns. When the sound was made very loud and therefore unpleasant and biologically threatening, the baby would look not toward it but away, turning eyes, ears, and head from potential danger.

More complex auditory abilities have been shown in a conceptually simple but technically difficult study done by William S. Condon and Louis W. Sander of the Boston University Medical Center in the early 1970s. The study examined interactional synchrony, a characteristic of communication that most people are aware of to some extent. When two adults talk to one another, they do not simply speak words. Their bodies engage in a kind of dance, moving in step with one another. In particular, both persons move in synchrony with the units of speech, called phonemes, of whoever is talking. The movements are extremely slight and are very difficult to detect without the aid of a filmed record and its frame-by-frame analysis. With adults, movements in interactional synchrony begin and end on breaks between phonemes. Babies only hours old engage in interactional synchrony when adults talk to them, and their movements also begin and end on phoneme breaks—but seemingly in any language, something no adult could do. This result suggests an astonishing ability of newborns to analyze the flow of speech, regardless of language, into its component parts and to do it better than their elders. It also suggests a predisposition to behave in a social manner, to interact with other humans rather than simply to react to them.

What newborns can learn

These elementary abilities—elementary in the sense that they occur so early—have been used to demonstrate a higher order ability; namely, the ability to learn, to profit from regularities in the world. The studies described above showed that infants can localize sounds and have a special interest in voices. A very young baby tends to hear voices either in play or during feeding. While babies sleep mothers do not talk to them, and changing and

A simple six-word utterance is shown divided into 23 phonemes, or basic speech units. Babies only hours old have been found to synchronize their movements with the phoneme breaks in the words of adults who are talking to them, regardless of language.

bathing are usually too fraught for much speech. During feeding, the direction of the voice depends on whether the baby is breast-fed or bottle-fed. In breast-feeding the sound of the voice comes from the same side as the food; in bottle-feeding it tends to come from the opposite side. A recent study by Jesus Alegria and Eliane Noirot of the Free University of Brussels showed that this regularity of the world is detected by babies less than seven days of age. Hungry breast-fed babies of that age, laid on their backs, will turn toward a voice stimulus; hungry bottle-fed babies in the same situation will turn away from a voice stimulus. Clearly, the two groups of babies have learned the association between voices and food.

To answer a question of concern to almost all mothers—"How soon does my baby recognize me?"—more involved studies of learning have exploited the baby's interest in faces and voices. In one experiment done by Genevieve Carpenter at St. Mary's Hospital, London, in the mid-1970s, two-week-old babies were presented with their mother's face or a stranger's face, which they saw through a porthole over their cribs. The mother could be silent or talking to her baby; the stranger could be silent or talking to the baby. Each baby heard the voices only through a speaker arrangement, such that the mother's face might be presented with the stranger's voice and vice versa.

(Left) P. Kreye—The Image Bank; (right) Champlong-Arepi—The Image Bank

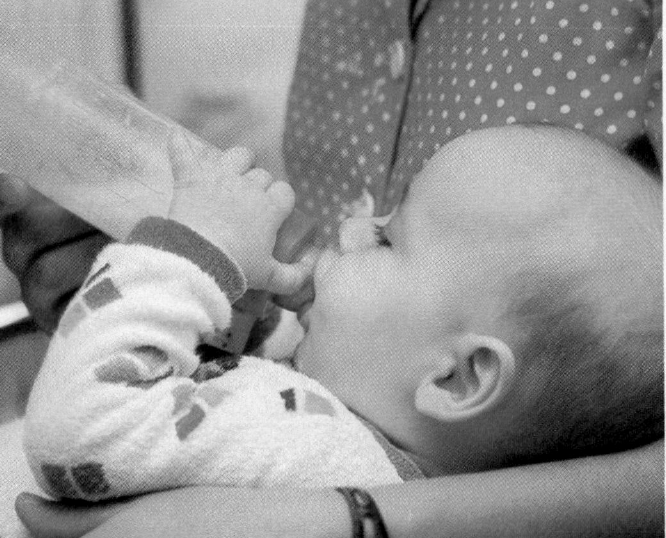

If a baby is breast-fed, the voice of the mother comes from the same side as the food (above); if the baby is bottle-fed, the voice tends to come from the opposite side (right). Recent experiments have shown that babies less than a week old learn this association and will turn appropriately either toward or away from a voice stimulus. (Facing page) When babies less than two weeks old were presented with the shadow projection of a rectangular object rotating in such a way that one end seemed about to strike (top sequence), they made defensive movements even though they had never before experienced such a visual event or the consequences that it suggested. By contrast, babies shown a control projection, an expanding and contracting rectangular shadow (bottom sequence), exhibited no defensive behavior.

It was quite apparent that these babies recognized their mothers, because they looked at their mothers much more frequently than they looked at the stranger. Furthermore, they seemed to know how their mothers should sound, because they showed what psychologists call gaze aversion whenever faces and voices were mismatched. The infants would turn their heads away in an attempt to avoid looking at the face with the incongruous voice.

Taken together, these studies suggest a new answer to the problem of the origins of sociability. Many complex theories have been offered to explain how the unsocial infant, born with no interest beyond his own comfort, can acquire the social motives that distinguish human beings. As ingenious as these theories have been, they now seem rather surplus to requirements. Instead, it appears that the newborn comes into the world with forms of behavior that are already social and that are not directed to personal comfort, unless personal comfort includes the satisfaction of social needs. The baby from birth is interested in people. Imitation is a social behavior. It is not directed to any purpose other than maintaining the "other" in view. Babies imitate, it seems, simply for the pleasure of interacting with another human. Interactional synchrony is even more obviously social. In adults it is a mark of sociability, and those few adults who do not show it are seen as abnormal. Newborn babies share this mark of humanity. If the foundations of sociability exist at birth then one no longer needs to ask how babies become sociable but rather how that sociability is shaped and directed.

Learning and perceptual skills in nonsocial situations

The above examples have focused on social responses to illustrate perceptual skills and learning ability. These skills and abilities have been studied in many other contexts as well. Together the studies show that newborns are competent organisms that are able to make a great deal of sense of the world around them. For example, working at Harvard University in the late 1960s, T. G. R. Bower, J. M. Broughton, and Moore discovered that if an infant is propped up in a vertical position and an object is brought toward her face,

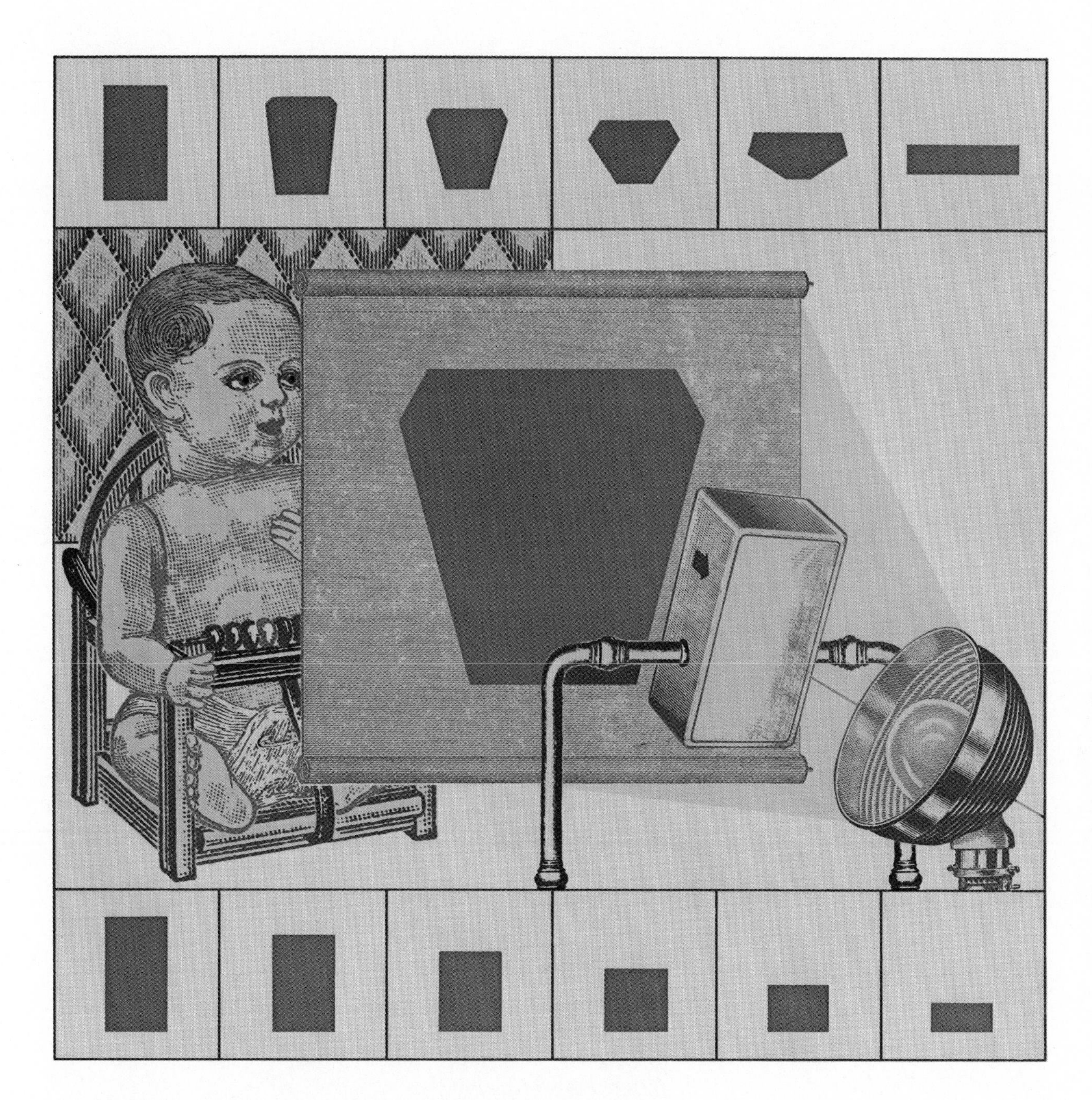

she will defend herself against the approaching object by pulling her head back and interposing her hands between face and object. This defensive movement has been observed in infants in the first week of life, although thus far it has not been detected on the first day of life.

A variant of this experiment shows that while defending themselves infants also can distinguish between an object approaching on a collision course with the face and one that will whiz harmlessly past their heads. Researchers William Ball and Edward Tronick, also of Harvard, showed that infants will defend themselves against the former and not against the latter. In this experiment the approaching objects were presented in the form of visual images on a screen. The arrangement eliminated the possibility that air movement might signal a physical object moving toward the baby's face. Consequently the defensive response in this case must have been elicited by visual cues alone.

This kind of defensive behavior could have been learned. A baby is quite likely to be struck in the face by her mother's breasts during feeding. While mother and infant are still unskilled, such an experience can be unpleasant for the baby and may even signal potential suffocation. Babies, in fact, protect themselves in this situation with a defensive response very much like that elicited by an approaching object. It is possible that babies may discover that a particular change in the appearance of an object—such as a sudden gain in size—signifies a potentially unpleasant event. It seems, however, that such an interpretation cannot be applied to a recent series of studies carried out in Edinburgh by Dunkeld. She found that the same defensive response could be elicited by a visual event to which an infant was most unlikely to have been exposed in the course of normal handling. This event took the appearance of a rectangular object rotating end over end such that its leading edge seemed about to strike the baby on the nose. Despite this "unnatural" situation babies in the second week of life defended themselves. As the object rotated on its axis, they moved their heads as far from it as they could, indicating that this particular change in the image signified an edge approaching their faces.

Motor abilities

Another question of great scientific interest is "Can newborns actually do anything?" As has been discussed, they can move head and eyes both to inspect and to defend against the environment. Newborns can also use their feet and legs if they are held. For example, if correctly supported they will march along a tabletop or similar surface. They can also use their hands and arms to surprisingly good effect, although they normally are given no opportunity to do so. A young baby spends much of his time on his back, and owing to the shape of his body, he will roll from side to side unless he uses his arms to maintain himself in one position. In research done in 1970 at Harvard, however, Bower, Broughton, and Moore showed clearly that if a newborn is supported in such a way that his trunk is steady and his hands and arms and head are free to move, one can observe quite surprisingly efficient arm and hand movements. Newborn babies can reach out and hit things and can occasionally even grasp them.

The newborn's reaching is not particularly accurate. And even if he is completely successful in catching hold of an object, he does not seem to know what to do with it. He is most likely to drop it after he gets it into his hand. Nevertheless, this motor skill is present and can occasionally be used. Because of the latency of the response (the time it takes to organize it), a newborn ordinarily does not get much chance to use his grasping behavior. Potentially graspable objects just do not hang around long enough for him to reach out and take them. This circumstance may be why in most cases the behavior fades out and disappears around three or four weeks of age, to reappear in a more controlled form around four or five months.

Babies do it differently

The newborn, then, can pick up information about the world and can act in the world in limited ways. Is there any evidence about the processes underlying these abilities? Scientists now know that newborns come into the world with a very high ability to learn. Indeed, one expert on infant learning, Lewis Lipsitt of Brown University, Providence, Rhode Island, has gone so far as to say that humans can learn better as newborns than at any later age. All of the basic types of learning are within the capacity of the newborn. The most spectacular demonstrations of this involve various forms of operant learning; that is, learning based on a system of rewards and reinforcement.

One of the most striking examples of a learning task that newborns can solve comes from a study done in 1966 by Lipsitt and E. R. Siqueland at Brown University. Babies were required to discriminate between two sounds and to respond to only one. When a bell sounded, the head was to be

Abilities of newborns, five-year-old children, and five nonhuman primates (clockwise, from top right: lemurs, gibbons, chimpanzees, gorillas, and orangutans) to deal with a learning task involving discrimination reversal are compared. Subjects were first taught that a correct response to only one of two cues produced a reward; then the values of the cues were reversed. Bars on graph indicate the percentage of correct responses the subjects made after undergoing ten trials with the cues reversed. Of the subjects tested, newborns scored highest.

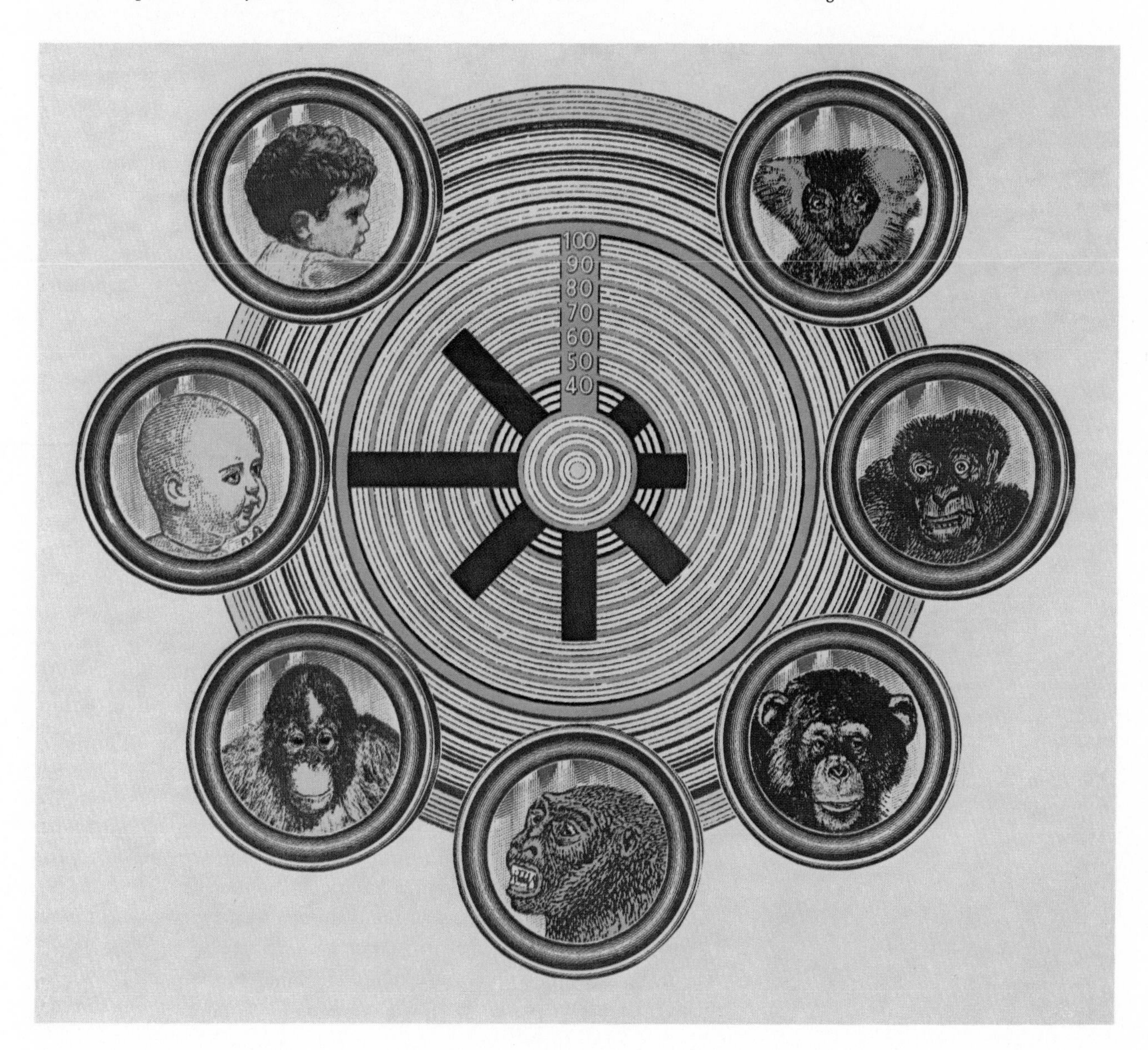

turned to the right to obtain reinforcement—the opportunity to suck a few drops of sugar solution. When a buzzer sounded, the head was not to be turned. The experimenters then reversed the contingencies, so that the bell meant no head turn and the buzzer a right turn. The babies were able to cope with this change (a change involving what is called discrimination reversal) very easily, more easily in fact than most other organisms do. This last ability can be put in an evolutionary context if one compares newborn scores with those of other species and older children.

Discoveries of this kind have changed the nature of the whole debate about infancy and, indeed, about human development. Historically the newborn was seen as sharing some abilities with the human adult. The focus of debate was how much was shared. The answers ranged from "a great deal" to "very little." The newborn was thus either a mini-adult or a partial mini-adult. The idea that a newborn could do something that an adult could not was not even considered. Research has discovered that newborns can do at least one thing that adults cannot; that is, segment the sounds of any language. There is some evidence that this ability may only be the tip of the iceberg. To understand the nature of that evidence one must consider a theoretical issue, that of how newborns can do as much as they do.

The question just introduced is one aspect of the century-old debate, often called the nature-nurture controversy, over the contributions of heredity and environment to human development. Nativists argued that human knowledge and human skill are built into the structure of the organism. Knowledge could be compared to the presence of lungs; it is something humans have because they are humans rather than fish. Skill could be compared to breathing; it happens inevitably, given the structure of human brains and bodies. Empiricists argued that human knowledge develops selectively as a result of specific encounters with certain types of environmental events. Human skills are formed as a result of successes or failures in coping with problems posed by the environment.

These two theories of human development, which could hardly have been further apart, had very different social philosophies associated with them. Nativists tended to be pessimistic about the perfectability of mankind. The unskilled and ignorant were held to be so simply because they are born unskilled and ignorant. Their lack of competence was seen as a characteristic comparable to eye color or skin color, something peculiar to native endowment and quite beyond the reach of human intervention. To the contrary, empiricists said that the unskilled and ignorant are so because their environment denied them the opportunity to develop skills and knowledge. If environments were sufficiently modified, ignorance would disappear and the whole community would achieve a high level of competence.

To this day these social philosophies heat the emotions to the extent that rational discussion becomes a near impossibility. Moreover, the debate has been given a new slant by recent discoveries in molecular biology. Scientists now know that the biological agents that program the development of an organism are genes made of DNA. They know about how much DNA there is in the human genome, the complement of genes that builds a human being. They can estimate how many genes there are and what each can contribute. From calculations based on these considerations, it would appear that there are not enough genes in total to construct the human visual system, much less the rest of the neural structures that underlie behavior.

The implication of this research is clear. If the genome cannot supply all of the information necessary to form the brain structures that mediate behavior, then learning—the taking of information from the environment—must do the rest. There are no other alternatives. Yet if learning is truly so important to neurological development, how then can newborns, who have had no time to learn, do so much? If there is not enough information in the whole genome to create a visual system, how is it that newborns can do so much with their visual systems, which can only have been shaped by a fragment of the whole genome?

From recent research has emerged the beginnings of an answer. At its core is the assertion that the newborn does not use the same information as the adult but rather information of a different order. An example will make this idea clearer.

Some of the investigations mentioned above showed that newborns can localize sounds. They can tell whether a sound is straight ahead, on the right, or on the left. The responsible mechanism depends on the detection of symmetry of stimulation between the two ears. A sound that is straight ahead is identical in both ears, and a sound on the right is louder in the right ear than in the left. A similar mechanism is involved in locating odors, another of the newborn's abilities, relying on symmetry of stimulation in the nostrils. A case can even be made for the dependence of visual localization on the detection of symmetry or asymmetry with respect to the nose.

The important point is that localization in these three kinds of senses, or sensory modalities, is mediated by common information; that is, information about symmetry or asymmetry of stimulation. It is possible that in infancy the three sources of information are handled by a common neurological structure rather than three separate structures. Thus the infant, presented with a sound on the right, might perceive that there is something on the right without even necessarily registering which sense is providing the information. Among evidence in support of this hypothesis is the fact that newborns invariably turn their eyes to sources of stimulation, whether it be a visual image, a sound, or an odor, and do so even in darkness. Indeed, even blind infants will turn their unseeing eyes to sources of stimulation, a behavior surely indicating a lack of sensitivity to the modality of information.

It obviously takes less genetic information to shape a neurological structure to respond to modality-free information than to form three or more modality-specific structures. Adults clearly have modality-specific structures, but they have had time to form them in interaction with the environment. Newborns could do what they do without these structures, but with a single "supramodal" structure. This suggestion, however, would seem to limit the perceptual world of the newborn. Some characteristics of the human world are modality-specific. The various colors, for example, are purely visual; red cannot be told from green by smell or sound. It may be significant that to date it has been difficult to demonstrate that the newborn is sensitive to such modality-specific variables.

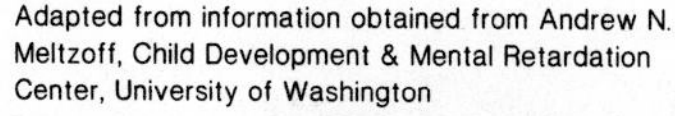
Adapted from information obtained from Andrew N. Meltzoff, Child Development & Mental Retardation Center, University of Washington

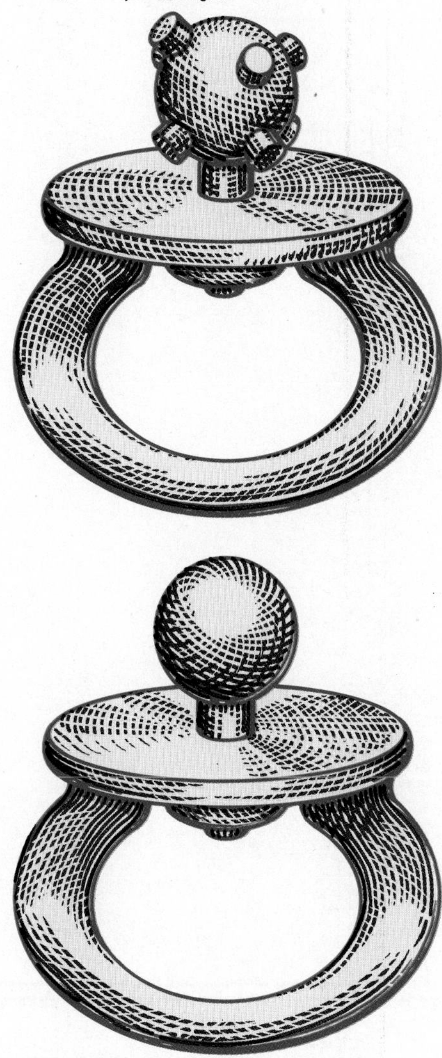

The belief that young infants handle incoming information from the various senses by means of a common neurological structure is supported by a recent study that tested their ability to match information received by touch and sight. Month-old babies were given, but not shown, either a studded or a smooth pacifier on which to suck for a short familiarization period. Immediately thereafter they were shown side-by-side models of the pacifier tips. The infants looked significantly longer at the model whose shape matched the pacifier that they had explored by mouth. These results suggest that infants have a surprising ability to store and use abstract, sense-independent information.

As an object approaches the eyes it produces a sensory change that can be described in terms of optical expansion: the size of its projection on the retina grows, and its image fills a progressively greater area of the visual field. This change can be expressed without reference to the sensory channel through which it is presented; for example, as a curve on a graph that plots distance to the eyes against retinal size. Once so freed of its modality, the change can be expressed through some other modality such as sound or touch.

Some benefits of research

This suggestion of a supramodal structure has interesting therapeutic possibilities. Consider the many infants born lacking a sensory modality; for example, sight or hearing. If newborns respond to supramodal information —modality-free information—then with sufficient technical ingenuity one should be able to supply the missing information to them through the remaining senses. Take speech comprehension in the deaf infant. As much as 80% of the information necessary to develop this ability can come through vision; in theory all of the rest can be transmitted by way of another sense; *e.g.*, through devices that produce a sensation on the skin.

As another case, consider visual distance information. When an object approaches the face of a sighted person, it produces a change in stimulation at the eyes. The change can be described in terms of optical expansion: the image of the object fills more and more of the visual field as it nears. However, if one considers the change not in visual terms but purely as a change over distance or time, one can give it a simple mathematical description that is modality-independent. Once generalized in this form the change can be re-presented to a person through any modality, not just vision. For

Photos, Ric Gemmel—NEW SCIENTIST

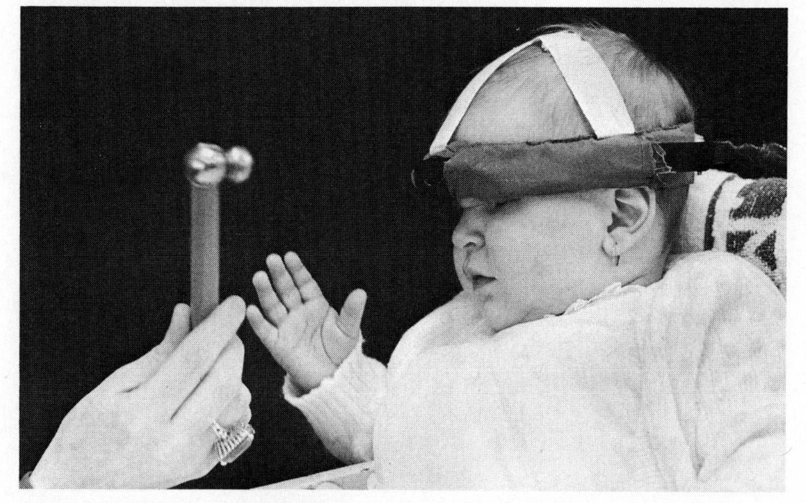

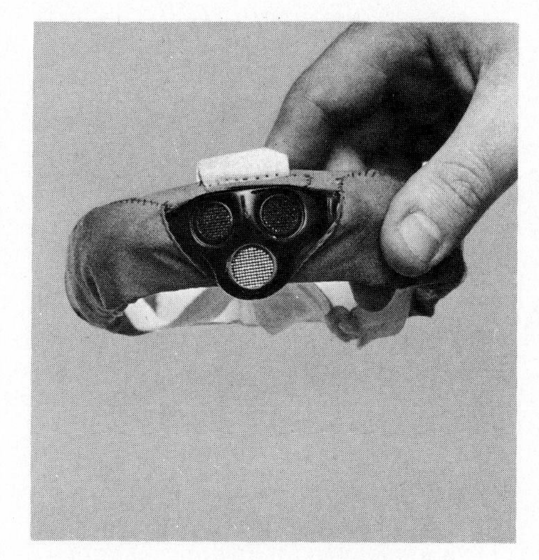

Blind baby uses modality-conversion device to locate objects held in front of her (left). The transmitter-receiver (above), which has been adapted to mount on the baby's forehead, emits ultrasonic waves and detects their reflections from nearby objects, much like a bat's echolocation system. A small external electronics package converts the reflections into audible sound conveying useful information. Pitch, loudness, and clarity of the audible signal relate to the distance, size, and texture of the detected object. Young infants seem to be able to make use of reflected sound as a source of information about objects much more readily than adults, who tend to perceive the sound as a property of the objects themselves.

example, decreasing distance to an object can be represented as increasing pressure on the skin or a rise in the pitch or volume of a tone. Indeed, devices for the blind have been constructed that present such visual information, plus directionality and a good deal more, in auditory form.

Recent work by Bower with blind infants in California and Edinburgh indicates that young infants can make use of this kind of information much more rapidly and extensively than any adult. Adults, in developing modality-specific structures, lose the ability to latch on to modality-free information. Most adults and even most older infants perceive sound as a property of objects rather than a source of information about objects. When given a modality-conversion device—for example, one that uses sound to provide information about objects—they misread its information because they interpret it within a modality-specific framework. A change in pitch no longer specifies an approaching object but a stationary object making peculiar noises. The success of infants with this kind of device bears on the nature-nurture debate. The device is a man-made artifact and thus an environmental modification. Its use can have no evolutionary history, unless evolution has operated to select the kind of modality-free structures that have been postulated. Yet it can redirect the way senses gather information and consequently the way in which neurological structures develop and the way learning takes place.

This article has concentrated on the newborn while leaving untouched the larger problem of change during infancy. Nevertheless, recent discoveries about the newborn are changing the ways in which behavioral scientists view the mechanisms that transform babies into competent adults. Until one has assimilated human beings at their "newest," one can hardly begin to approach these mechanisms of transformation.

FOR ADDITIONAL READING

T. G. R. Bower, *A Primer of Infant Development* (W. H. Freeman, 1977).

Rudolph Schaffer, *Mothering* (Harvard University Press, 1977).

Daniel Stern, *The First Relationship: Infant and Mother* (Harvard University Press, 1977).

AS THE BODY AGES

The aging process touches all levels of biological organization, from the entire organism to the individual molecule. Identifying its basic causes is proving a herculean task for biomedical specialists.

by Richard C. Adelman

Efforts to understand the causes and problems of aging are expanding at a dramatic rate, in large part owing to the rapidly increasing proportion of elderly people among the world's population. Because of the diverse nature of the subject, its investigators have found it necessary to tap a vast array of scientific subspecialties not traditionally in communication with one another, ranging from molecular genetics and organ and tissue interactions to zoology and demography. To date most of their research has dealt with two broad topics: the nature of genetic and environmental factors that influence biological aging and the search for underlying mechanisms. Major discoveries in these categories have been few and hard won, and answers to some extremely formidable puzzles are just barely in sight. Yet, only upon continued research rests any hope of gaining some control over the aging process in human beings.

Genetics and environment

Scientists who have looked at the mortality statistics of human populations have uncovered two fundamental principles: the environment in which people live can influence dramatically the distribution of individuals of different ages; and the apparent maximal life span, that is, the age of the oldest known members of a species, is determined by the genetic composition of the species. For example, consider the cases illustrated in figure 1. In British India between 1921 and 1930 the average person could expect to live about 20–25 years, whereas an extremely small proportion of the population survived as long as 90–100 years. By contrast, in New Zealand between 1934 and 1938 average life expectancy was 75 years, but still only an extremely small proportion of the population survived as long as 90–100 years. Statistical projections for the U.S. indicate that very soon as much as 15% of the population will be 65 years of age or older (figure 2), but only a handful of individuals will slightly exceed 100 years.

Cary Wolinsky—Stock, Boston

The state of medical care, sanitation, nutrition, and other environmental elements can dramatically affect the number of people in a population who live to old age. Yet these factors seem to have little influence on the limit of human longevity, which is genetically determined. Figure 1 compares survival curves for various human populations in three countries and for three time periods in the U.S. Figure 2 depicts the percentage of the U.S. population 65 years of age and older from 1900, with a forecast to the year 2030.

(Top) Adapted from A. Comfort, THE BIOLOGY OF SENESCENCE, 1956, Holt, Rinehart, and Winston, Inc., New York; (bottom) U.S. Census Bureau

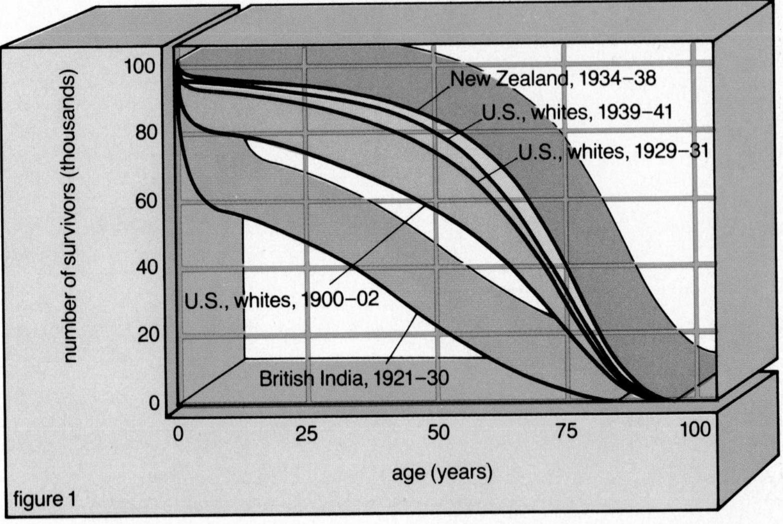

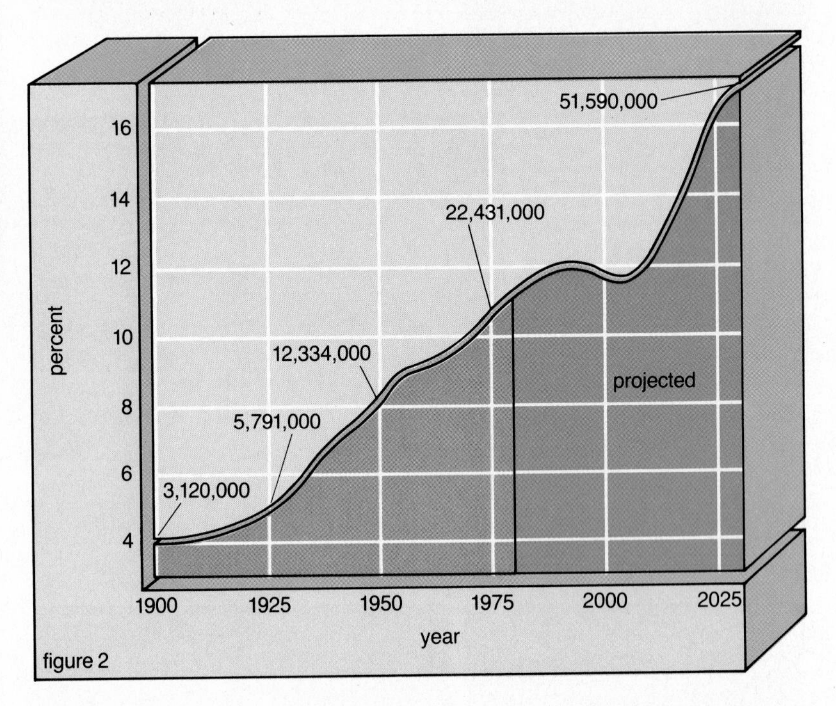

The message in those figures is clear. With such environmental influences as increased technology and improved medical care, sanitation, and nutrition, more and more people can be expected to survive in better health to greater ages. These same statistics, however, project life only until the clock strikes 12 midnight on one very predictable day, somewhat beyond a period of 100 years and not one instant longer. The implications in these figures are both striking and important. Although improvement in human survival during the last century has been substantial, the apparently finite biological limit to human longevity has remained essentially the same. Therefore, it is likely that the means which have allowed more people to approach their maximal life span do not contain the keys for extending that maximum.

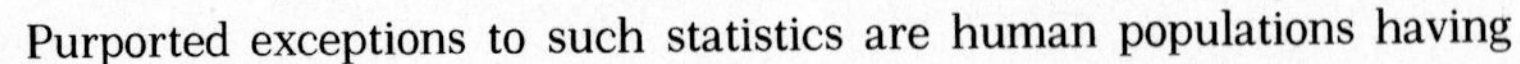

Purported exceptions to such statistics are human populations having

__RICHARD C. ADELMAN__ is Executive Director of the Temple University Institute on Aging and Professor of Biochemistry at the Fels Research Institute, Temple University School of Medicine, Philadelphia.

numerous members with unusually long or short life spans. Such people apparently show features typical of aging other than the rate at which they are expressed. For example, at a number of locations around the world are so-called geographic pockets of longevity. Among the more publicized of these is Soviet Georgia, in which large numbers of people are purported to live to ripe old ages of 120–150 years. Careful study of this situation has revealed both the negative influence of politics and a legitimate topic for research. The claims of superlongevity cannot be documented with certainty, given the questionable nature of the birth records involved. Historically, such claims apparently relate to the passing of identical names from one generation of males to the next in efforts to avoid conscription into the tsar's armies, as well as to the chauvinistic propaganda of Stalinism. Inexplicably, however, there certainly are extraordinarily large numbers of elderly Georgians, perhaps 60–80 years old, in remarkably good health who busily climb hills while tending their flocks in the virtual absence of sophisticated medical care, sanitation, technology, and counseling.

Perhaps even more puzzling are those unfortunate persons afflicted with symptoms of premature aging. For example, there are children who, although apparently normal at birth, develop very slowly and show many of the clinical symptoms and the physical appearance of extreme old age by age six. This illness, commonly called progeria, is extremely complex and generally fatal by the mid-teens; fortunately, cases are rare. Although many investigators suspect that progeria represents a precocious genetic expression of aging, they can offer neither satisfactory treatment nor real comprehension of its cause.

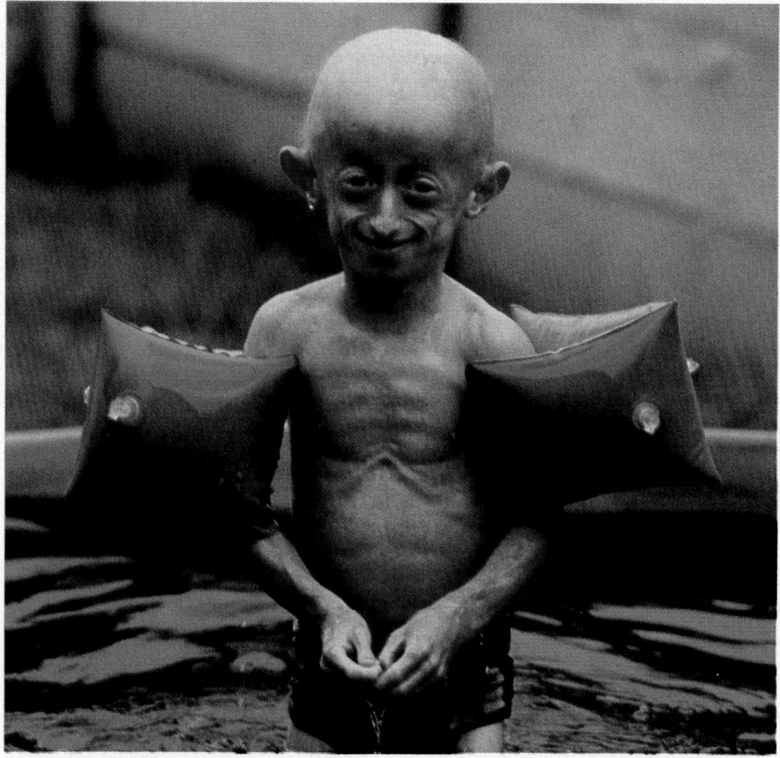

McCann—Gamma/Liaison

Investigators of the aging process are puzzled and intrigued by illnesses that produce many of the symptoms of extreme old age in children and younger adults. These afflictions may be evidence for a biological clock whose regulation of the genetic expression of aging is out of control.

Ample evidence exists that heredity contributes to longevity. Long-lived parents, for example, are more likely to produce long-lived children and grandchildren than parents who reach only the average life expectancy. Laboratory and domestic animals can be bred selectively to increase or decrease their life spans. However, appropriate perspective was provided by one investigator 30 years ago: "It may be well, as has been suggested, to seek advantages in longevity by being careful in the choice of one's grandparents, but the method is not very practicable. It is simpler and more effective to adapt the environment more closely to man."

Another aspect of genetic constitution that may influence longevity is the female advantage. Fish, spiders, water beetles, houseflies, and humans can be listed among many species for which the female sex is known to be the longer lived. Exceptions also exist, *e.g.*, pigeons, in which the male lives longer. Apart from genetics, however, the very fact of being either male or female may make quite a difference in the environment of a member of a particular species. Life for the male black widow spider, for instance, is decidedly more risky than it is for the female. It is not known to what extent such sex-influenced environmental differences contribute to the female advantage or its exceptions.

One conclusion that can be drawn from the results of investigations in both humans and other animals on environmental determinants of longevity is that moderation is preferable to excess or abstention with respect to such activities as physical exercise, alcohol consumption, and sexual activity. There are difficulties in interpreting such data, including the fact that the effects of these determinants appear to vary widely from species to species. In some species, for example, life span of the male is shortened by copulation. In others, regular mating improves general condition and longevity. There even is evidence that longevity in male rats is increased if one young female is introduced for purposes of grooming.

Pursuing basic mechanisms

Research has produced two fundamentally distinct schools of thought regarding the nature of biological aging. One concerns the general decline in physiological capabilities that accompanies the passage of time for all individuals of any particular population. The other addresses the vast differences in apparent maximal life spans—some prefer the phrase "rate of aging"—seen in various animal and plant species. Each research approach uses extremely diverse experimental model systems, draws from virtually the entire spectrum of biological disciplines, generates a plethora of descriptive literature, and has no firm grasp of the origin of the phenomenon.

The capabilities of the human body generally change during aging. Measuring them reproducibly, however, and then interpreting the results are probably more complicated in elderly subjects than in any other area of biomedical research. Many characteristics, such as filtration of blood through the kidney or transmission of nerve impulses, diminish progressively during aging. Others, such as pulse rate and the level of the sugar glucose circulating in the blood, are stable during aging except in response to specific challenges of the environment. For example, the recovery of pulse rate after

The task of measuring the changing capabilities of the human body during aging is complicated by the influence of such factors as general health and physical and social activity. Regular physical exercise, for example, can improve the ability of the aging body to use glucose. Again, elderly people who live isolated or sex-segregated lives may have lower levels of sex hormones than those who are more socially active.

a burst of exercise is sluggish in elderly individuals. So is the recovery of normal blood glucose levels following tests that measure the ability of the body to metabolize a given quantity of glucose.

Capabilities that change during aging are influenced by such factors as general health, diet, physical activity, behavior, and social convention. Some of these influences are difficult to define and allow for. For example, the declining ability to use glucose is accentuated by obesity and diabetes but is partially overcome by regular physical exercise. Similarly, decline in sexual prowess is much less apparent in certain healthy elderly individuals. Indeed,

Milton Feinberg—Stock, Boston

The desirability of age-associated change is in the eye of the beholder. Although mathematicians, physicists, and biologists are more likely to do their best work in early adulthood, politicians, statesmen, administrators, and religious leaders tend to excel in later years, despite declining physiological abilities. Many of the great conductors reached advanced ages and remained productive almost until death.

the reported finding of low levels of testosterone, the principal male sex hormone, in the blood of elderly males largely reflects measurements obtained in institutionalized, sex-segregated populations and may not be valid for elderly males living social, noninstitutionalized lives.

Investigators also confront other difficulties in monitoring age-associated changes in humans. Following changes within a single group of individuals as they age, called a longitudinal study, may result in the experimental subjects outliving the investigator. Monitoring changes within a group of persons of similar age, called a cross-sectional study, may give misleading results because of characteristics specific to the group members but not to the aging population as a whole; *e.g.*, dietary habit, susceptibility to disease, or institutionalization. Interpreting the value of age-associated change is yet another complicating factor. The fact that such changes lead to the grave does not make them undesirable in themselves; their desirability is in the eyes of the beholder. In one view, physiological decline in the elderly is an unmistakable sign of approaching ineffectiveness and uselessness. The following verse, however, exemplifies an alternative view:

> King David and King Solomon
> Led merry, merry lives
> With many, many concubines
> And many, many wives;
> But when old age crept o'er them
> With its many, many qualms,
> King Solomon wrote the Proverbs
> And King David wrote the Psalms.

Many of the incredible array of genetic and environmental modifiers of age-associated change in people can be avoided by using animal model systems. An important treatise on the suitability of animal models for research on aging recently was prepared by the Institute of Laboratory Animal Resources of the U.S. National Academy of Sciences. It addresses similarities and dissimilarities to human aging with specific detailed examples of advantages and pitfalls in the use of rats, mice, rabbits, cats, dogs, nonhuman primates, and other animals. For example, scientists can now raise colonies of mice and rats into old age while providing constant diet, living conditions free of disease-causing organisms, and other rigorously defined conditions of genetics and environment. Nevertheless, striking differences in vigor, mortality, or both seen among these colonies are the direct consequence of features that the investigator either deliberately or inadvertently included in the experimental design; for example, inbreeding versus outbreeding, unmated stock versus breeder stock, population density, caloric intake, or even selection of bedding material. Therefore, any attempt to draw conclusions, such as those discussed below, from these experiments demands comprehensive awareness of those genetic and environmental factors that are incorporated into the experiment through either design or accident.

Physiological changes

Investigators see changing physiological capabilities during aging at every level of biological organization. These include susceptibility to diseases that are major causes of death, functions of tissues and organs, and processes at the cellular and molecular levels. Statistics well document the rise in the death rate with age from cancer, heart and kidney diseases, upper respiratory infections, ulcers, hypertension, and a host of other afflictions. Controver-

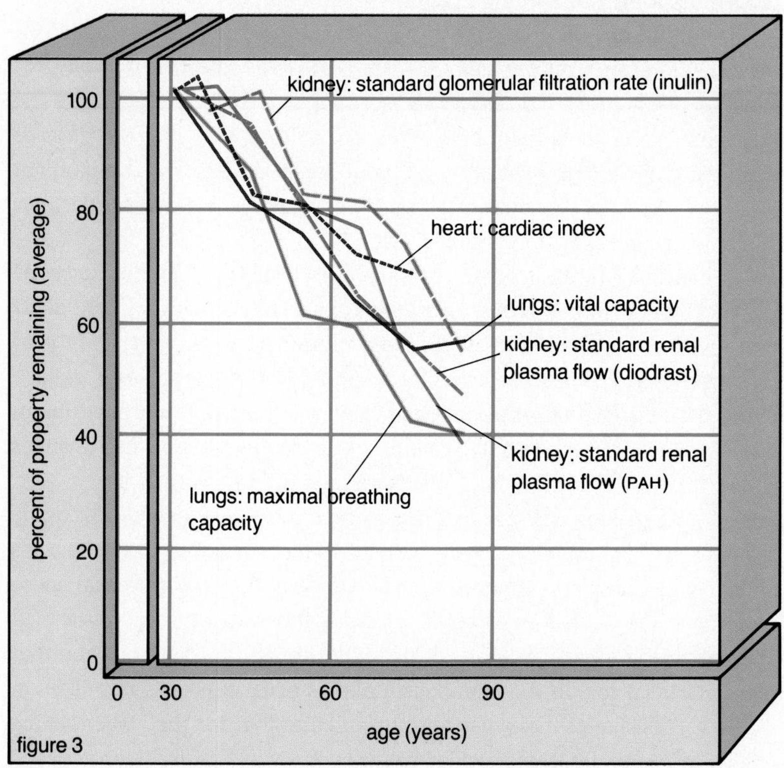

Adapted from Shock, BIOLOGY OF AGING, 1960, American Institute of Biological Sciences, Washington, D.C.

Figure 3 traces the decline in various human physiological mechanisms as a function of age. Their levels at age 30 are assigned values of 100%. *See page 96.*

Adapted from R. C. Adelman, "Age-Dependent Effects in Enzyme Induction—A Biochemical Expression of Aging," EXPERIMENTAL GERONTOLOGY, vol. 6, 1971

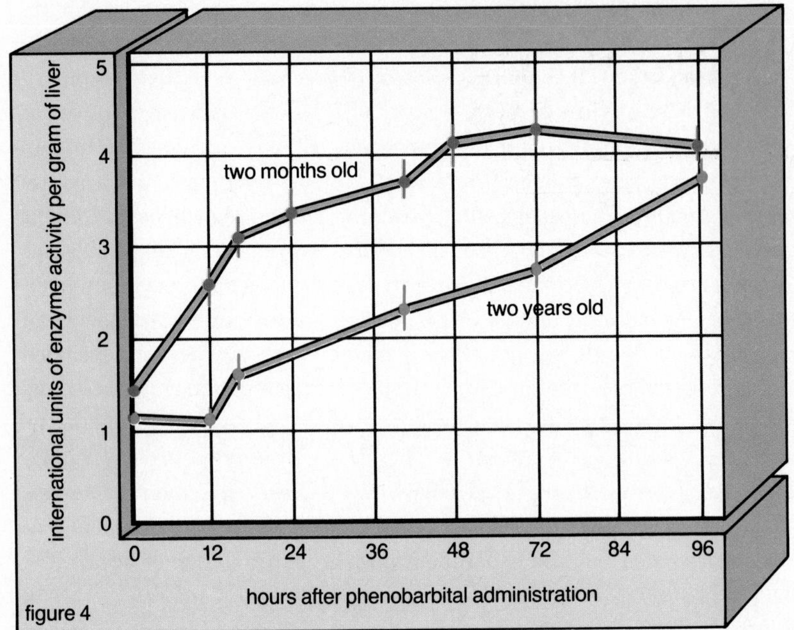

figure 4

One age-associated change on the molecular level appears as sluggishness in the cellular manufacture of new enzyme molecules. Figure 4 presents results of an experiment in which young and old rats received the drug phenobarbital; subsequently, the activity of a liver enzyme (NADPH: cytochrome c reductase) that helps metabolize this drug was traced for a 96-hour period.

sy exists, however, concerning whether aging is the cumulative effect of certain diseases or whether age-associated disease represents the penultimate expression of more subtle cellular and molecular changes that appear earlier in life. In any event, increased susceptibility to disease among the elderly is quite evident. For example, the onset of disease in the elderly probably is a major consequence of the age-associated decline in the body's predominant disease-fighting machinery, the immune system.

Virtually all physiological activities of tissues and organs exhibit sluggishness or inefficiency during aging. Figure 3 presents selected examples for function of the heart, kidneys, and lungs in aging people. In the case of the heart, for example, its function as a pump declines with age regardless of the method of measurement. Beyond about age 30 cardiac output falls about 0.7% per year, and the amount of work performed per heartbeat declines at a rate of approximately 1% per year.

One feature typical of aging at the cellular level is the loss of hormone receptor sites within the cells. These sites are the cell's "sensors" that allow it to respond to chemical messages from various parts of the body. Loss of these receptors has been detected for several different hormones in experiments with a variety of cell populations isolated from people and experimental animal models; it probably accounts for certain age-associated changes in the function of such cells as heart and fat cells.

The most striking changes that occur at the molecular level during aging relate to the regulation of enzyme activity. Aging can modify both the cell's ability to synthesize new enzyme molecules and the structure that those molecules will have. For example, figure 4 illustrates the age-associated sluggishness in phenobarbital-stimulated synthesis of a liver enzyme that plays a crucial role in the metabolism of this drug in people as well as in rodents. This sluggishness may account in large part for the longer lasting effects of the barbiturate in older subjects.

Better experimental approaches

One major problem that exists with these findings—and one quite aside from the issue concerning the difficulties of interpreting data—is that the descriptive nature of the information provides virtually no distinction between cause and effect. To know that older people generally possess wrinkled skin or wrinkled molecules sheds no light on the nature of underlying mechanisms. Therefore, the need to design an experimental approach that identifies sequences of responsible events has become a central issue of research on aging.

One example of such an experimental approach is based on the observation that all aging organisms show a progressively modified ability to adapt to changes in the environment. Some investigators feel that the causes of aging will be found by resolving the following kinds of problems: (1) In what cells of what tissue does there originate this decline in the adaptive capability of an aging organism? (2) What biochemical event is modified such that it forms the basis of this decline? (3) At what age in the life span of the organism does this critical biochemical modification appear for the very first time? (4) What is it that commits a cell to the emergence of such a biochemical modification at a particular point in life span? (5) How do these events relate to the potential vigor and maximal life span of the individual member of the population under study, of its species, and of other species?

One indicator of the adaptability of an organism to environmental challenge that can be easily measured is the fluctuation of enzyme activity in response to a broad spectrum of stimuli; that is, the phenomenon of enzyme adaptation. Thus, it has been reasoned, making measurements of the activity of these biological catalysts in animals of different ages, under suitable experimental conditions, may be a relatively straightforward way to assess the progress of aging.

Researchers have detected a decline in enzyme adaptation during aging in a variety of tissues from different species in response to diverse kinds of

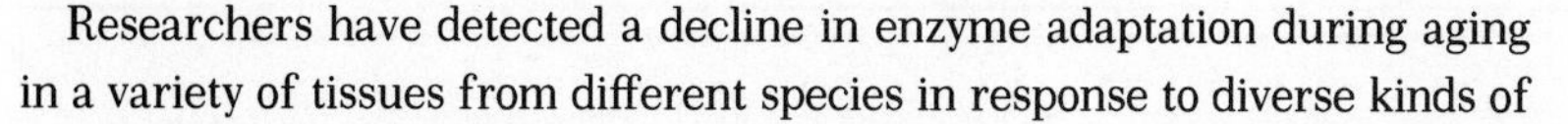

Michael Abramson—Black Star

"... the same kind of sluggishness that perhaps is more obvious in the impaired ability of an elderly pedestrian to avoid an oncoming car also is expressed at the level of molecular activity in aging animals." Figure 5 (below left) illustrates enzyme adaptation differences in young and old rats made to fast for 72 hours and then fed by injection into the stomach with glucose solution. The activity of the enzyme glucokinase, which catalyzes the first stage in the liver's use of glucose, was followed for 48 hours.

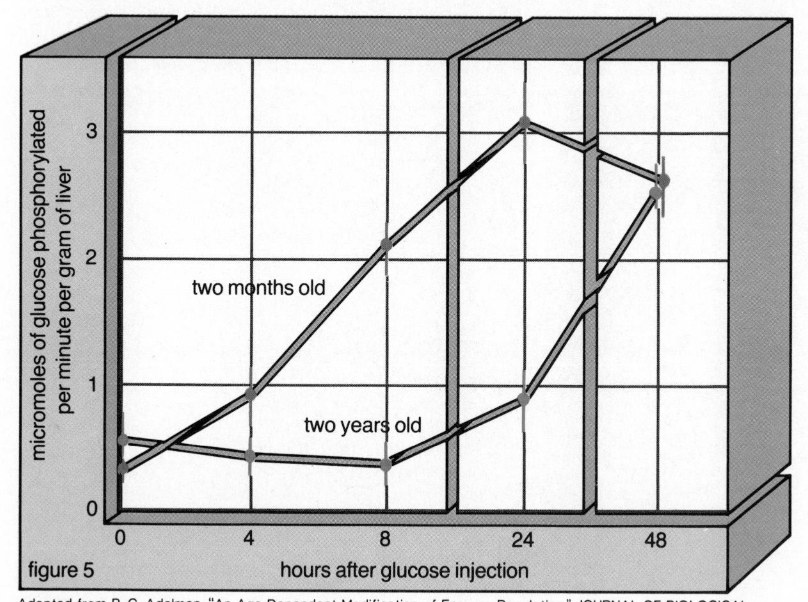

Adapted from R. C. Adelman, "An Age-Dependent Modification of Enzyme Regulation," JOURNAL OF BIOLOGICAL CHEMISTRY, vol. 245, 1970

stimuli. One specific example of an enzyme adaptation that is altered during aging is illustrated in figure 5. When a solution of glucose was injected into the stomachs of two-month-old male and female laboratory rats, the activity of the enzyme glucokinase, which catalyzes the first stage in glucose use by the liver, increased according to the indicated time course. When the identical experiment was repeated in 24-month-old rats, more time was required to adapt to the same degree. In other words, the same kind of sluggishness that perhaps is more obvious in the impaired ability of an elderly pedestrian to avoid an oncoming car also is expressed at the level of molecular activity in aging animals. Indeed, the time needed for the enzyme response to begin is directly proportional to chronological age, thus providing a convenient means of assessing the progress of aging in this animal model.

A biological clock

The progressive nature of biological aging also introduces a very new and fundamental concept. Although features of aging are expressed most blatantly in the oldest members of the population under study, at least certain

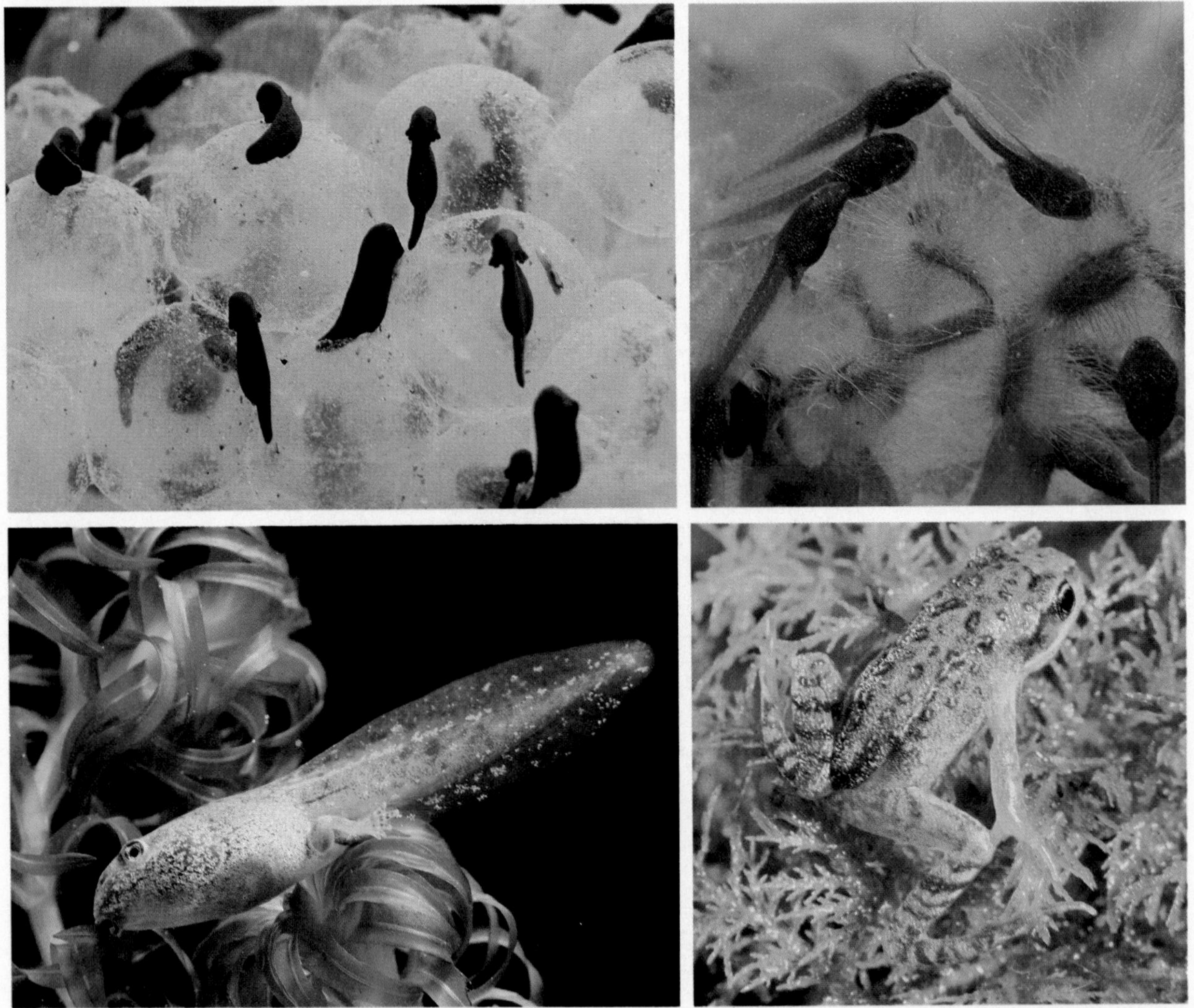

Photos, Jane Burton—Bruce Coleman Inc.

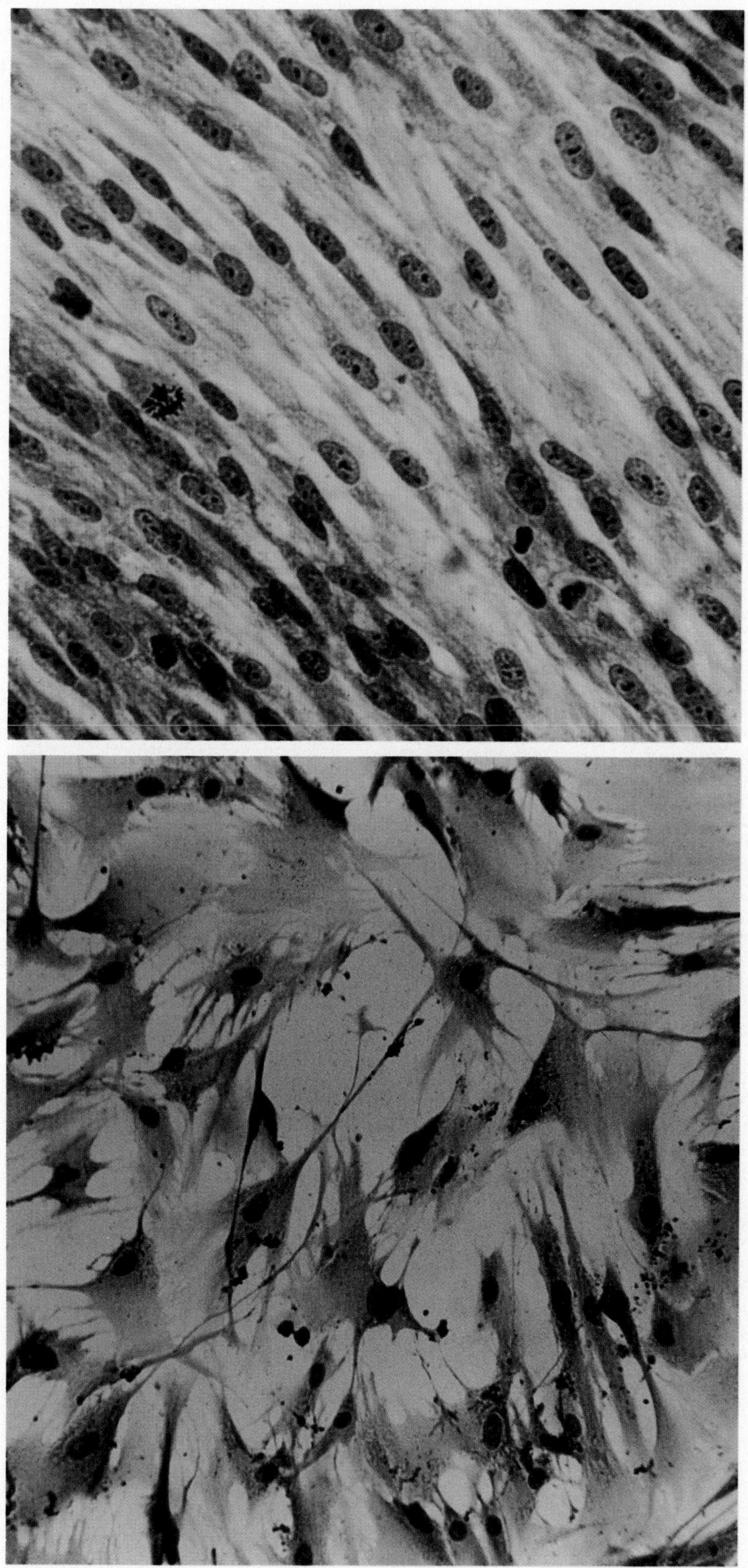

Photos, Leonard Hayflick, Children's Hospital Medical Center, Oakland, California

Disappearance of a tadpole's tail and gills as it matures (facing page) is a programmed sequence of cell degeneration and death governed by a biological clock. The suggestion that aging at the cellular level is controlled similarly comes from experiments that follow the ability of connective tissue cells called fibroblasts to multiply as they age. Spindle-shaped fibroblasts newly derived from a human fetus (left) proliferate actively in culture. As the number of cell divisions approaches 50, the cells gradually stop dividing, undergo various degenerative changes in structure and appearance (below left), and die.

features of aging have been discovered to emerge for the first time very early in life; these include sluggish adaptation for certain enzymes and sluggish cell division in certain body tissues. Therefore, the precise molecular basis of these age-associated changes ultimately must be explained in terms of the workings of a clock whose alarm punctually instructs one or more specific physiological capabilities to begin their relentless decline throughout the remainder of life.

Biological clocks are a normal part of developmental sequences in animals. One of their purposes is the deliberate limitation of life span for tissues and organs—*e.g.*, the tails and gills of tadpoles, the organs of larval insects, and human baby teeth—by means of the programmed degeneration and death of cells. The possibility that aging at the cellular level is an expression of such programmed cell death arises from experiments in the early 1960s involving certain connective tissue cells, called fibroblasts, that had been derived from human embryonic lung tissue and grown in culture. It was found that after a period of active multiplication these cells slowed and eventually ceased division, at the same time becoming filled with cellular debris. Ultimately the entire cell population degenerated and died.

This phenomenon now is generally regarded as an innate characteristic of all normal dividing cells grown in culture, as well as a possible intracellular basis for aging. Its relevance to aging in complete organisms is substan-

In figure 6 the life spans of four species of animals (top) are compared with the proliferative capacity of fibroblasts taken from the fetus or newborn of those species (bottom). The right portion of each bar in the bottom graph reflects the variance among fibroblasts of each species in the number of doublings that they may undergo before death. Such evidence suggests a direct relation between a species' life span and the ability of its cultured cells to survive.

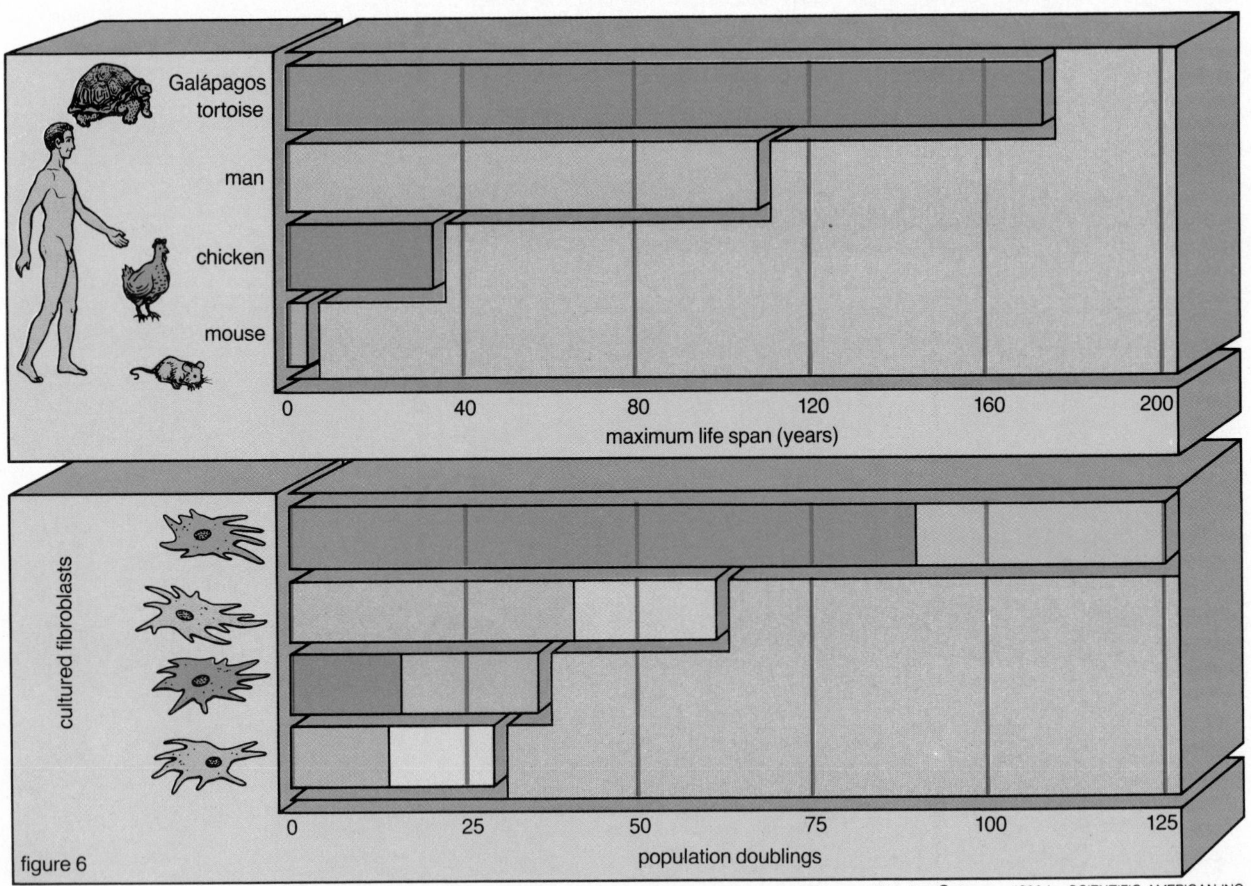

tiated by several lines of evidence. For example, the greater the age of a human donor, the more quickly does a culture of donor cells decline and die. Cultured cells derived from individuals afflicted with progeria-like illnesses decline and die more quickly than cultured cells from age-matched normal subjects. When cells are transplanted successively from old animals to young ones of the same inbred strain to avoid death of these cells when the host animal dies, the transplanted cells age and die as they do in culture. Although numerous changes in cell structure and function are known to precede the loss of proliferative capacity, the origin of aging in cell cultures is not yet understood.

Species-specific maximal life span

The horse and mule live 30 years
And nothing know of wines and beers.
The goat and sheep at 20 die
And never taste of Scotch or Rye.
The cow drinks water by the ton
And at 18 is mostly done.
The dog at 15 cashes in
Without the aid of rum and gin.
The cat in milk and water soaks
And then in 12 short years it croaks.
The modest, sober, bone-dry hen
Lays eggs for nogs, then dies at ten.
All animals are strictly dry:
They sinless live and swiftly die;
But sinful, ginful rum-soaked men
Survive for three score years and ten.
And some of them, a very few,
Stay pickled till they're 92.

Although more accurate and extensive information on the maximal life spans of animals has been tabulated, both the scientific literature and this tavern verse provoke the same question. Why should one species, such as the Galápagos tortoise, live 150 years, whereas humans live approximately 100 years and the laboratory rat or mouse only 3–5 years? The general conceptual approach to this issue seeks to explain why one species possesses a greater life span than another in terms of what it is that longer lived species do better. Work just after the turn of the century demonstrated that the higher the metabolic rate of a mammalian species, the shorter is its life span—although human beings live several times longer than predicted by this relationship. In the late 1950s it was shown that for numerous species of animals life span has a close functional relation to certain anatomic and physiological features of the species, such as brain and body weight. More recently it was reported that the ability of fibroblasts grown in tissue culture to survive is directly proportional to the life span of the species from which the cells were derived. (*See* figures 6 and 7.) Investigators also apparently discovered a precise relation between the life span of various mammals and the capacity of their cultured cells to repair genetic material—DNA molecules—that had been damaged by ultraviolet radiation. Other recent work showed that the greater the susceptibility of DNA in cultured cells to dam-

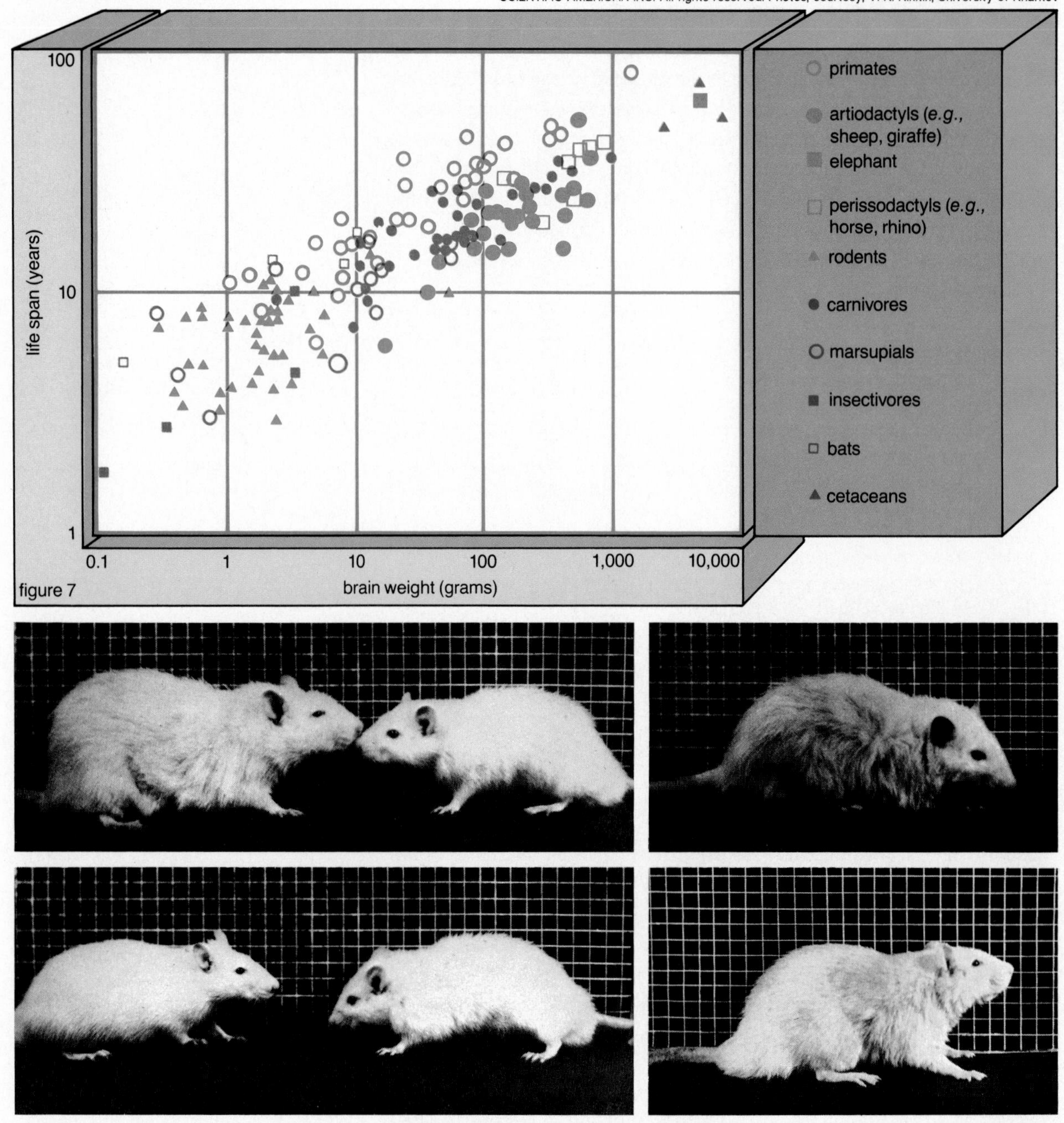

age caused by certain chemical cancer-causing agents (carcinogens), the shorter is the life span of the mammalian species from which the cells were derived.

As was described above for the nature of declining physiological capabilities in aging populations, available knowledge about species-specific maximal life spans also is extremely descriptive. The genetic basis of such differences is not remotely understood. Nevertheless, the evidence seems to suggest the evolution of longevity-assurance systems that confer on each individual of a species the life span and vigor necessary for development and reproduction.

The future

What will come from research on aging? Can one anticipate an extension of life span beyond that which already may be genetically endowed? Is there any assurance that a scientifically wrought increase in the length of life will be accompanied by mental alertness, physical strength, and other attributes that make a long life worth living? Preliminary evidence is encouraging. Consider the following two examples. (1) Rats and many other species that are placed experimentally on a selective restricted diet enjoy about twice the apparent maximal life span of identical animals that are provided with all of the food that they can consume freely. Although rats on the restricted diets generally experience the same kinds of age-associated diseases and declining physiological capabilities as the freely eating rats, features of aging appear in them for the first time at a much later age. (2) Mice that are treated with the steroid dehydroepiandrosterone (DHEA) seem to remain free of a distinct set of age-associated problems, such as certain types of cancer and obesity. The steroid-treated mice appear younger and live longer than untreated, age-matched controls.

Interpreting these results is difficult. For example, does dietary restriction extend life span, or does free eating shorten life? Are the effects of dietary restriction really nutritional in nature or, instead, a result of neural and hormonal changes related to the vastly different times of day at which food is eaten within the experimental design? Does treatment with DHEA extend apparent maximal life span or, instead, does it target only specific symptoms? Presently there are no answers to such questions.

Extrapolation to humans is tempting. For example, properly supervised nutritional therapy is indeed a generally acceptable approach for certain age-associated cancers and cardiovascular lesions. Similarly, circulating levels of DHEA are known to decrease in people during aging, as well as in individuals who are predisposed to breast cancer. Nevertheless, extending to humans the conclusions drawn from experimental animal studies can be accomplished only in carefully planned, ethically guided clinical trials.

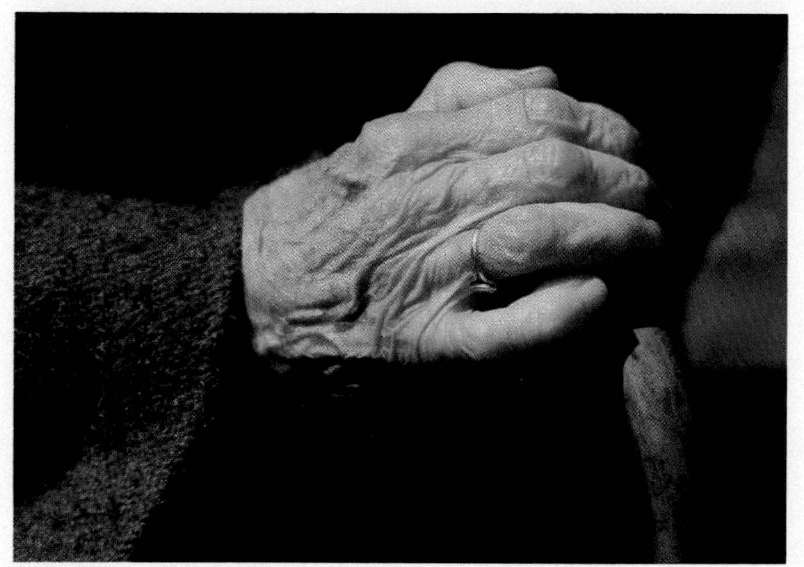

Figure 7 (facing page) plots brain weights of species in ten mammalian orders against their life spans. The loose grouping of data points shows a significant trend from lower left to upper right, indicating a positive correlation. Photos below figure compare the effects of restricted and unrestricted diets on rats. Upper and lower rats at center, raised on a calorie-restricted but nutritionally sound diet, are 24 months old, having lived about two-thirds of a normal life span. Both appear about the same age and size as the juvenile, three-month-old rat, fed a normal diet, at the lower left. By contrast, the 24-month-old animal at the upper left, fed normally, is larger and shows signs of old age. The rat at the lower right is 39 months old; it was first fed a restricted diet and later given additional food and allowed to grow to normal size. It appears as fit as the 24-month-old, normally fed rat above it.

SYMBIOSIS AND THE EVOLUTION OF THE CELL

Contrary to earlier belief, symbiotic relations are widespread in nature and figure centrally in evolution. Some vital components of plant and animal cells may well have had their start in microbial partnerships.

by Lynn Margulis

Symbiosis was defined by the botanist Heinrich Anton de Bary in the late 19th century as "the living together of differently named organisms." From the beginning it was recognized as a peaceful phenomenon and distinguished from pathogenesis and parasitism, in which living together occurs at the expense of one of the participants. In pathogenesis one of the organisms is severely damaged or destroyed, and in parasitism one partner is gradually debilitated.

Over the past century the definition of symbiosis has not changed very much; it is generally considered to be the association for a major portion of the life cycles of two or more partners that are members of different species. Scientists now know, however, that symbioses may transform into pathogenic or parasitic relationships and that parasitic relationships may become benign and even obligate; that is, necessary to both partners. This realization that associations change with time has led to new perspectives on symbiosis. The phenomenon is common rather than rare, and it is a product of an evolving interspecific relationship. It even seems clear on the basis of microscopic and biochemical studies that the nucleated, or eucaryotic, cells of plants and animals are themselves the products of symbioses between ancestral microbes.

A new perspective

Symbioses were thought to be limited to bizarre special cases. Often cited as examples were lichens, which are always composed of fungi in symbiotic partnership with photosynthetic cyanobacteria (blue-green algae) or green algae, and cow rumens, which are special stomachs harboring billions of cellulose-digesting microorganisms. It is now realized that symbioses are ubiquitous and central to the mainstream of cell evolution. The truth is that nature abhors a "pure culture"; that is, organisms living in the absence of members of other species. All organisms are associated with members of other species, some more and some less. It has been estimated, for example,

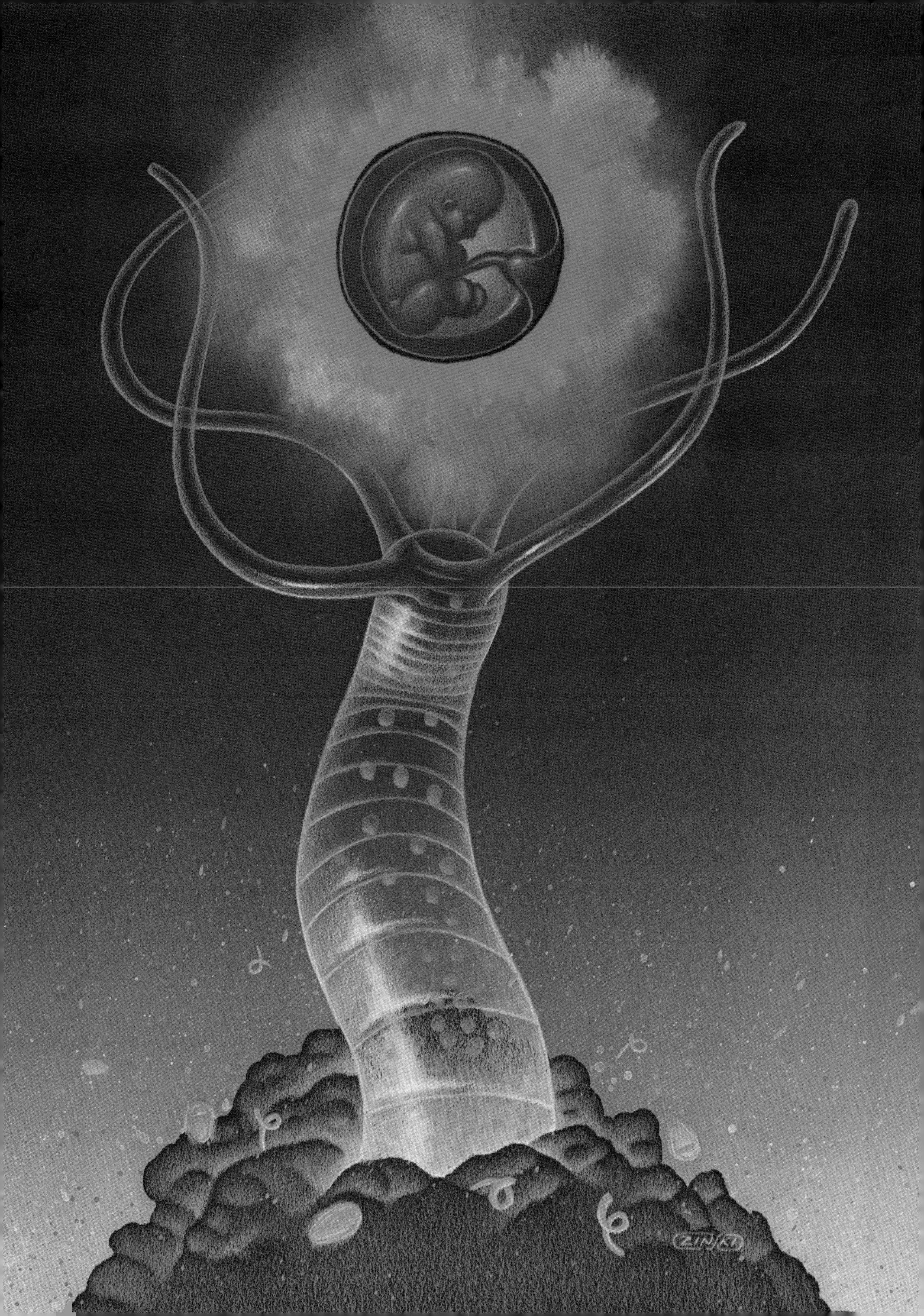

© L. West—National Audubon Society/Photo Researchers

Photos, Vernon Ahmadjian, Clark University

The lichen Cladonia cristatella, *commonly called British soldiers, is composed of a fungus in symbiotic association with a species of green algae. Microscopic examination shows the partnership to consist of algal cells woven into a matrix of fungal filaments, or hyphae (top right). Single algal cell (bottom right) is almost completely enveloped by hyphae.*

that even human beings are about 10% dry-weight symbionts. The vast majority of these organisms are housed in the small and large intestines and are bacteria. Symbionts on the skin, in the mouth, and in the genital tract add to the load. Because they make vitamins and in chemical ways enhance the digestive process, bacterial symbionts are indispensable to human life.

The nature of symbiotic associations change with times that are much shorter, in some cases, than the hundreds or thousands of years previously assumed. For example, Kwang Jeon at the University of Tennessee has studied a symbiosis between bacteria and amoebas that began as a destructive infection. At first no amoebas harbored these bacteria. Then unidentified bacteria, some 150,000 in each amoeba cell, were found to be growing within and killing their hosts. Not all of the hosts died, however, although the rate of growth for the survivors slowed tremendously, from a cell division about every two days to a division every two or three weeks. Over the next five years the amoebas recovered their normal growth rates. Instead of 150,000, the survivors harbored some 40,000–50,000 bacteria per amoeba. In ingenious experiments Jeon transferred nuclei from amoebas of the same strain that had never been infected with bacteria into amoebas harboring bacteria; likewise, he moved nuclei from amoebas with bacteria into amoebas that had never been in contact with the bacterial infection. As a result he was able to show that the bacteria, which once had been serious pathogens, now were required by the nuclei of the infected amoebas for their very survival.

LYNN MARGULIS is Professor of Biology at Boston University in Massachusetts.

Illustrations by John Zielinski

From EXPERIMENTAL CELL RESEARCH, Jeon and Lorch, vol. 48, pp. 236–240, 1967

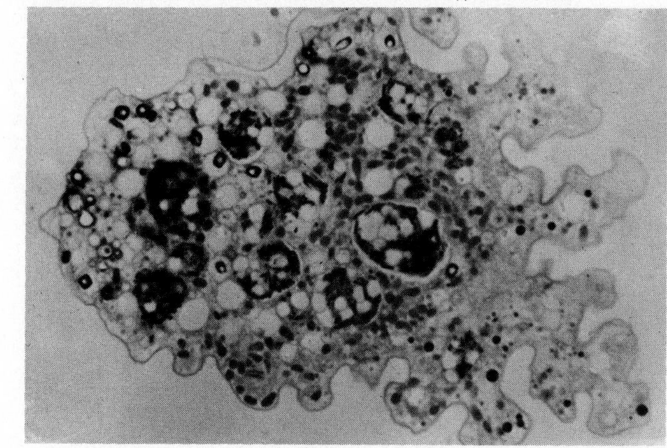

From JOURNAL OF PROTOZOOLOGY, Jeon and Hah, vol. 24, pp. 289–293, 1977

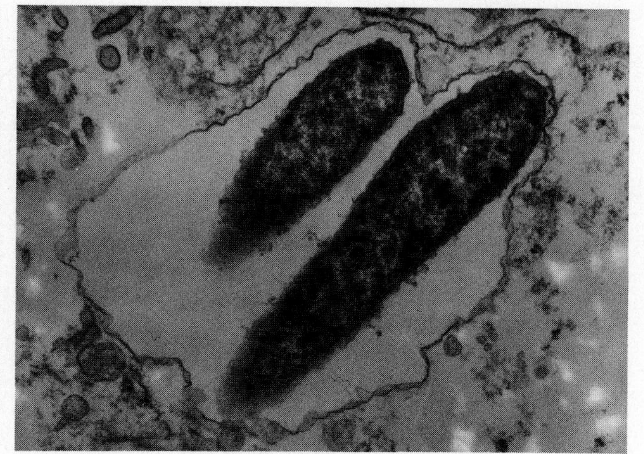

Section of amoeba (left) shows its infection with tiny bar-shaped bacteria clustered in vesicles of various sizes. At greater magnification can be seen two of the bacteria inside a small vesicle (above). In the course of five years, what began as a destructive bacterial invasion of these amoebas transformed into a symbiotic relationship.

Cases like this one, in which symbiotic relations are seen or inferred to change, have led to a far more dynamic view of the formation and evolution of associations. The terms pathogen, parasite, and symbiont are very relative. Mutualists are considered partners in a close, mutually beneficial relationship that may be necessary for their survival, whereas commensalists are partners that are associated in such activities as locomotion or feeding but not necessarily for survival. Examples of commensalists are the remoras that attach themselves to sharks and the tickbirds that both groom and live on the backs of cattle, rhinoceroses, and other large mammals, signaling their hosts when danger threatens. Yet parasites can become pathogens, and a parasite in one environment may be a mutualist in another. These terms refer to the relative selective advantage for each of the partners in the association. Selective advantage, however, can change with environment; hence the words do not get at the essential nature of the associations.

One way of identifying symbioses without having to define advantages is by counting up the numbers of different kinds of species and individuals of partners in any association. A terminology based on this concept has been developed by F. J. R. Taylor at the University of British Columbia. It is so recent that it has not yet come into general use. A single organism or a descendant of a single organism belonging to the same species is called a monad; for example, a pure culture of the colon bacterium *Escherichia coli*. An association throughout most of the life cycle of two individuals of different species, each with their own genetic composition, is called a dyad. An example would be *Bdellovibrio* and their hosts. These are bacteria that invade other bacterial species and grow inside at their hosts' expense. Another example would be consortia bacteria: associations between large bacteria that possess slender propulsive appendages called flagella and surface-attached bacteria that are able to photosynthesize. The result is a partnership that is both motile and photosynthetic. If three species of partners are involved the association is called a triad; if four, a tetrad; and so forth.

This terminology has many advantages. If the association changes—for example, from a pathological to a benign one—the terminology does not change because the gross nature and the number of participating species does not change. This terminology focuses not only on numbers of species

M. P. Kahl—Bruce Coleman Inc.

Rhinoceros and its symbiotic tickbirds are an example of a problem-solving commensalistic relationship. The birds rid their mammalian hosts of annoying parasites, while the hosts provide their partners food and a safe refuge.

but also on numbers of distinct genetic systems in the association, permitting the analysis of important questions. Who are the members? How many of each are there? How is the partnership maintained? What is the selective advantage of the association for each of the partners? How may the numbers of individuals in the partnership vary while still maintaining the association? Over what fraction of the life cycle of each member does the association persist? In many associations, such as that of cycad trees with the cyanobacteria of their special roots or of legumes like beans and peas with their nitrogen-fixing bacteria, the partnership is reestablished each generation. In others, such as many symbioses between insects and bacteria or the amoeba-bacteria association referred to above, the partners are permanently associated throughout the life cycles of both and into those of succeeding generations.

Symbiosis solves problems

Symbiotic associations offer their members workable solutions to many of the basic problems of health and survival. Often the factors that maintain symbioses are obvious. Tickbirds, for instance, feed on insects and other small animals while they groom their hosts. In exchange for grooming and warning cries, the hosts provide housing for the birds and even a place for the birds to copulate.

Many symbioses are maintained because one partner nourishes another. One good example is provided by the hydra, a small, tentacled, freshwater coelenterate related to jellyfish, corals, and sea anemones. The hydra is composed of two principal layers of cells, a protective and sensory outer layer called the epidermis and an inner, digestive layer, called the gastrodermis, that lines the animal's central body cavity, or coelenteron. One species, the green hydra (*Hydra viridis*, or *Chlorohydra viridissima*), maintains its characteristic color because of its association with *Chlorella*, a species of photosynthetic green algae that lives within the hydra's gastrodermal cells. Green hydras that harbor 20–30 *Chlorella* cells per gastrodermal cell do not starve if sunlight is plentiful. White hydras, which resemble them in every way except for the presence of symbiotic algae, starve within three days or so if no other food is provided. Green hydras have been kept alive without any food at all, except for the nourishment provided by symbiotic algae, for more than three months. If placed in the dark or experimentally "cured" of their algae these same unfed hydras die quickly.

Many marine fish have organs that emit light. These structures may be used as "flashlights" for seeking food in dark waters or for signaling members of the same species. In some cases the advantage of the light organs to the fish is not known. The source of the light has been identified in certain species: the fish have evolved chambers that harbor some 100 million bacteria per milliliter (0.034 ounces) of light organ fluid. When removed from the host and provided with ATP (adenosine triphosphate), the usual cash of energy currency in biology, the bacteria glow bright yellow-green. The bacteria are intrinsically luminous. The fish provide sanctuary and food for the bacteria, and the bacteria provide the fish with a steady, controllable source of light.

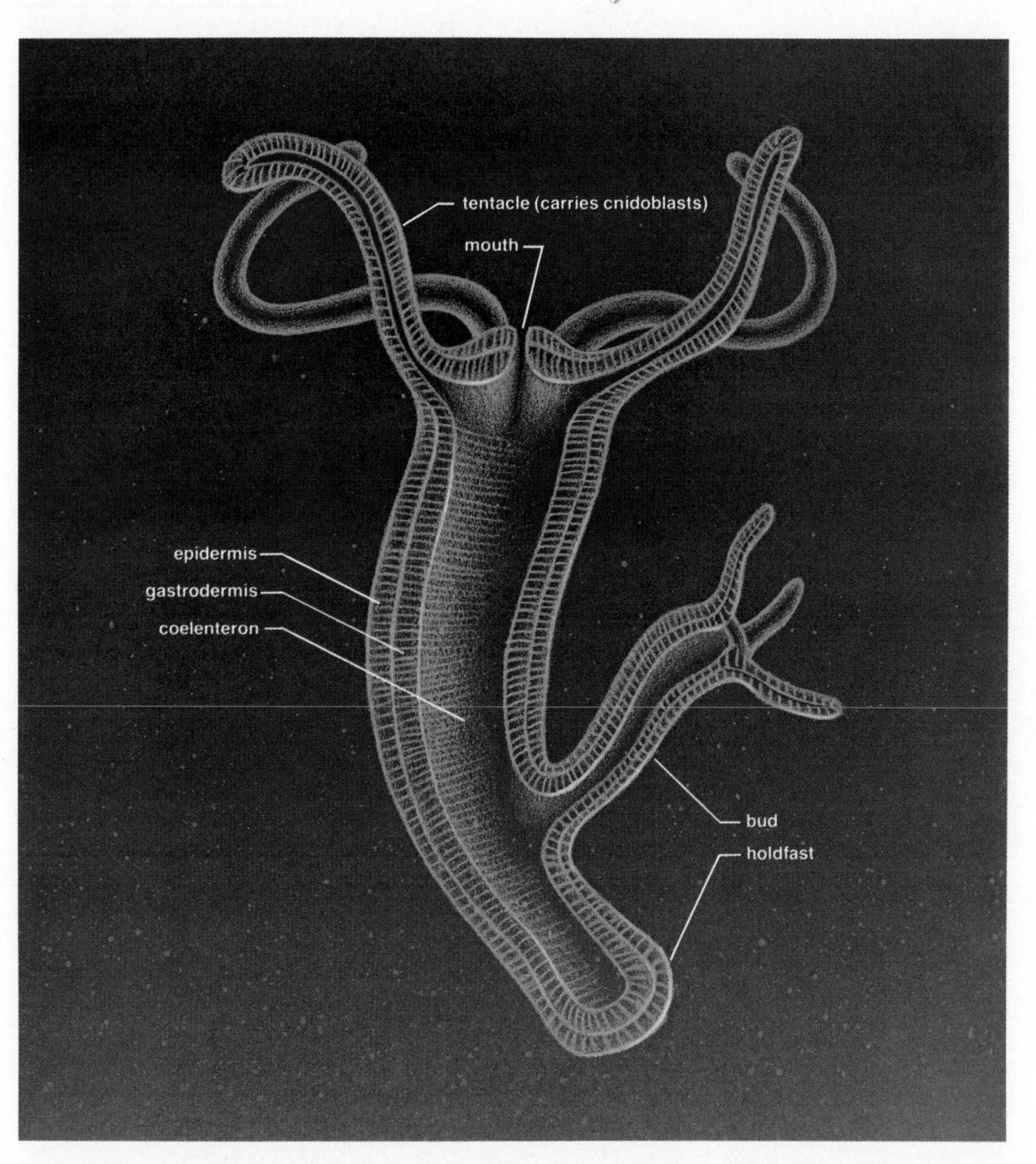

Principal parts of the hydra are identified in schematic cross section. In the green hydra, Hydra viridis, *symbiotic green algae live within the cells of the animal's gastrodermal layer.*

Other problems solved by symbioses include the breakdown of refractory foods. Most of the carbon in the aggregate of living organisms presently on Earth is in the form of lignin and cellulose, the long polymers that make up wood. Very few organisms can use these polymers as food sources; otherwise wood would not be the ubiquitous construction material that it is. Exceptions are the dry-wood and subterranean termites, which attack wood and digest the cellulose (although not the lignin, which can be attacked by only a few specialized fungi and perhaps some bacteria). More precisely, however, the cellulose of wood is only ingested by the termites. No species of termite alone can actually digest wood. Wood-eating termites harbor an

Ken Lucas—Steinhart Aquarium, San Francisco

Anomalops katoptron (right) and other flashlight fish possess below each eye a luminous organ that they use for seeing, communicating, luring prey, and deceiving predators. Electron micrograph of sectioned organ tissue (below) reveals numerous light-emitting bacteria housed in compartments surrounding a central vessel. The vessel carries nourishment to the symbiotic bacteria from the fish's bloodstream, while the bacteria serve as the light source of the organ.

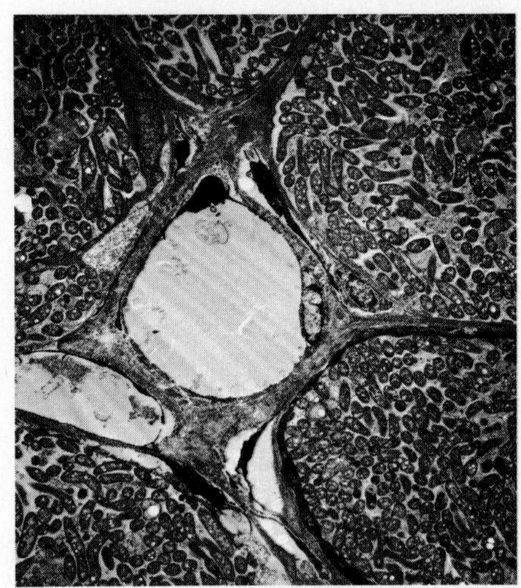

Steinhart Aquarium, San Francisco

(Facing page) Wood-eating termite (bottom), like all of its kind, relies on a large array of symbiotic protists and bacteria in its hindgut to digest cellulose. Protists from the gut of one termite species include (small photos, top and center) Trichomitopsis termopsidis, *a polymastigote with few undulipodia, and* Trychonympha campanula, *a hypermastigote well endowed with undulipodia. Internal view of* T. campanula *(small photo, bottom) shows red-stained wood fragments that the protist ingested from the gut of its host.*

enormous array of symbiotic bacteria and nonphotosynthetic protists in a portion of their intestinal tract called the paunch, a part of the hindgut. (Protists are eucaryotic microorganisms: protozoans, algae, slime molds, and others.) This special organ for wood digestion contains as many as a billion bacteria and a million protists per milliliter. If deprived of these hindgut microorganisms—for example, by antibiotic or heat treatment—the termites will die of starvation within two weeks. The termites may be repopulated with healthy symbionts, however, and if done in time they will be rescued and continue to live out their normal life spans. The first sign of malaise in a termite is depletion of its microbial community and a loss of its species diversity.

In some termites—for example, the common subterranean termite, *Reticulitermes flavipes*, which is well known in eastern North America—more than 15 different species of protists are known and more than 100 species of bacteria are suspected, although most have not been isolated and characterized. A serious problem in this work is the difficulty of growing the microbial symbionts outside the termite. Most are poisoned immediately by oxygen in the air, and for most their food requirements cannot even be guessed at. Some recent success has been reported, however. Michael Yamin of The Rockefeller University in New York showed that at least some protists from termite hindguts can grow with wood or purified cellulose fibers as their sole source of carbon and energy, under oxygen-free conditions.

Another problem solved by symbiosis is that of motility and large size. There are some protists, eucaryotic single-celled microorganisms such as polymastigotes (some of which are intestinal symbionts of termites), that have a set number of whiplike propulsive structures called undulipodia, or cilia, in a given relationship to the number of nuclei per cell. They are doomed to either a small number of undulipodia and nuclei or a large number of undulipodia and nuclei. Apparently selection pressure often acts against an increase in number of nuclei per cell and consequently against good motility. Members of this group frequently take on motile bacteria as external symbionts. Hundreds of these bacteria crowd the surfaces of polymastigotes and beat their bodies in synchrony. Several species of the larger polymastigotes have acquired rapid motility, conferred by huge populations of motile helical bacteria called spirochetes.

How microbial symbioses form

Microbial associations are everywhere in nature. In some cases the way in which the association arose can be inferred from the extant legacies that represent the sort of steps the partners went through as the relationship became more intimate. In all cases one must imagine that selection pressure favored the partnership over the individuals; that is, partners in association left more offspring than the unassociated partners, leading to greater frequency of the symbioses.

Evolution of the association of protists with spirochetes is a case in point.

Photos, Michael A. Yamin, Boston University Marine Program, Woods Hole

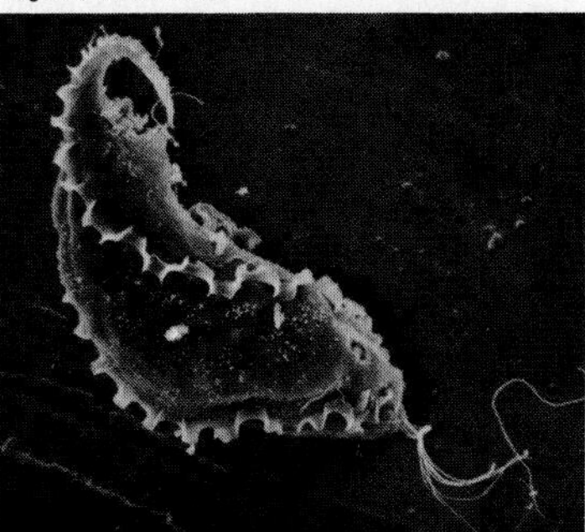

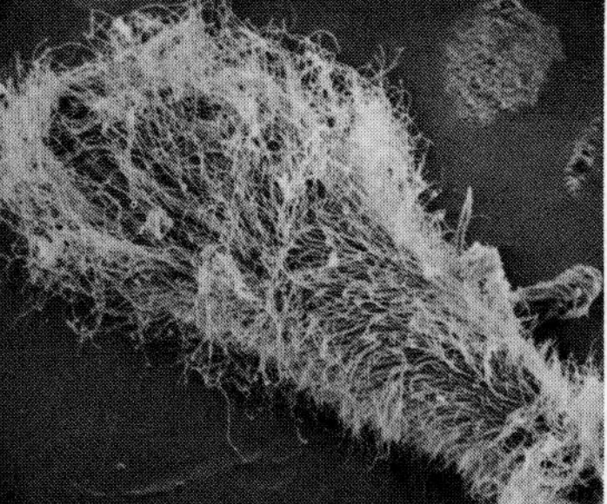

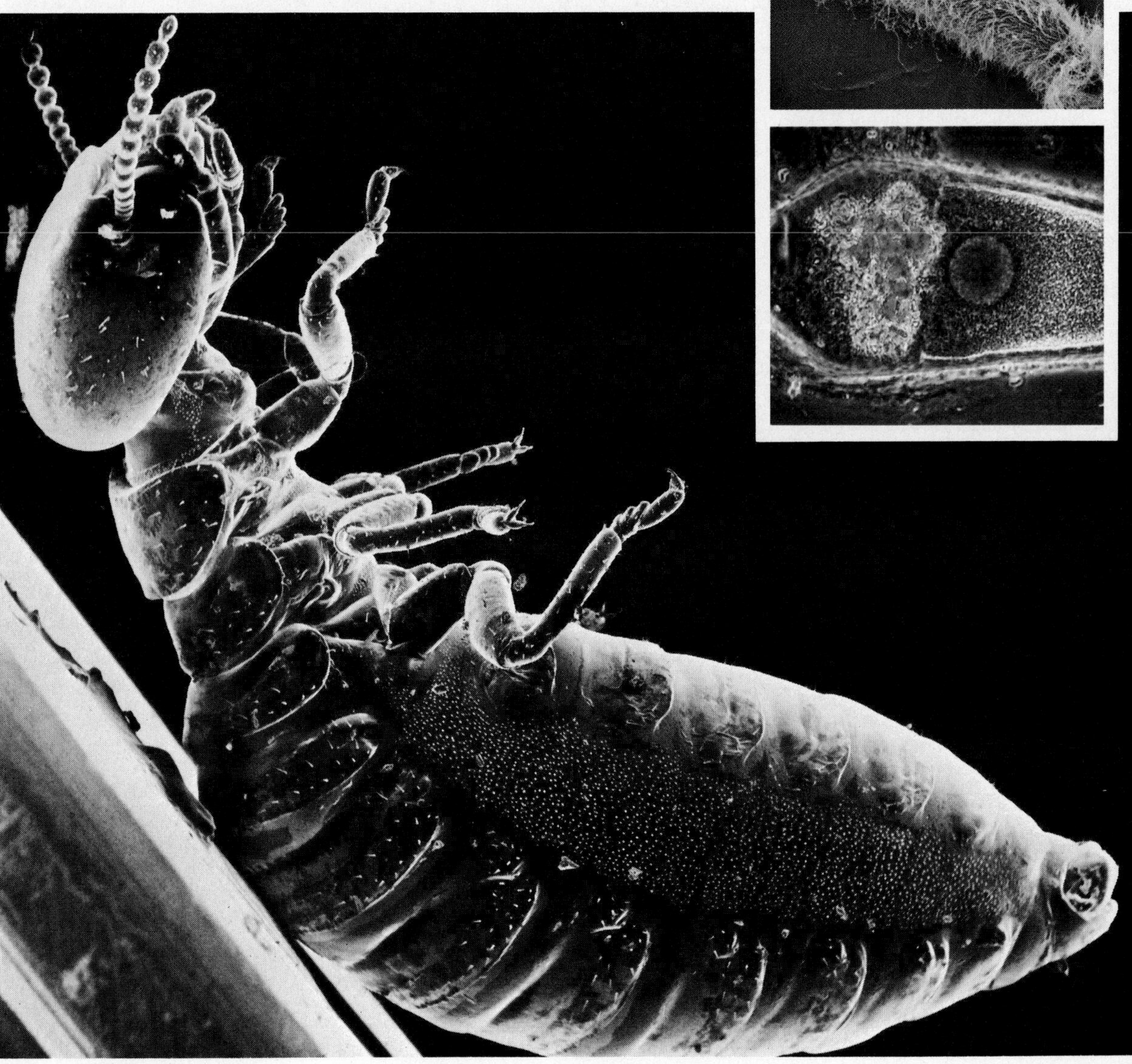

Lynn Margulis, Boston University

Photo (bottom, left) and drawing (adapted from), A. V. Grimstone, University of Cambridge; (bottom, right) David G. Chase, Cell Biology Research Laboratory, Sepulveda, California

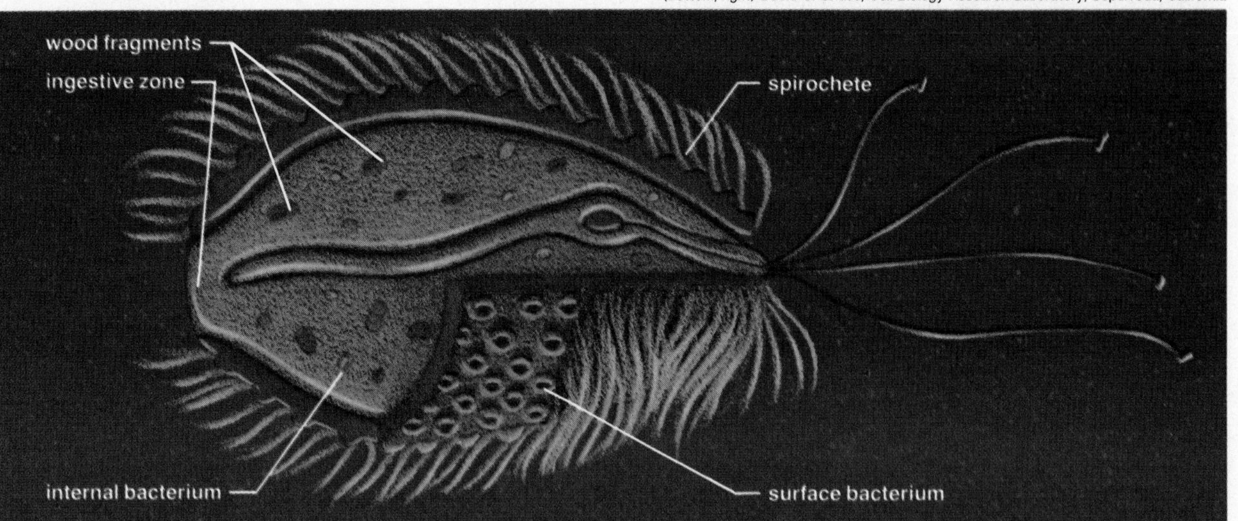

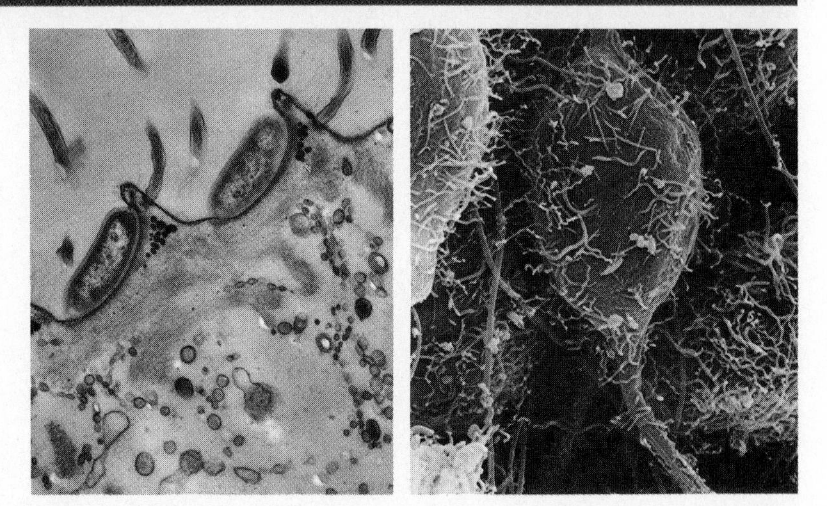

Myxotricha paradoxa (top), a wood-digesting symbiotic polymastigote in the gut of certain termites, hosts three symbionts of its own: surface-attached spirochetes, which propel the protist and which were first mistaken for undulipodia; other surface bacteria; and internal bacteria. In a magnified section of M. paradoxa near its surface membrane (bottom left) surface bacteria lie in pits in the membrane, and parts of spirochetes can be seen attached to narrow rises on the membrane. Microrhopalodina occidentis (bottom right), another termite-gut symbiont, also plays host to spirochetes and various other bacteria.

Illustration (facing page, top) depicts a possible sequence of events in the evolution of the association of protists with spirochetes: (a) completely independent organisms share the same environment; (b) some spirochetes orient toward protist to maintain closer contact; (c) spirochetes adhere to protist surface; (d) spirochetes penetrate protist surface. (Facing page, bottom) Hydra captures and ingests a tiny crustacean. The association of hydra and algae probably began when a nonsymbiotic hydra inadvertently ingested algae while devouring its primary prey.

Another case is the association between invertebrate animals and photosynthetic partners; for instance, hydras and green algae. In the first case motile spirochetes with requirements for complex nutrients simply were found in environments with protists. Those spirochetes that stayed in close contact with the protists grew more rapidly, presumably because they benefited nutritionally from materials—*e.g.*, membrane proteins and incompletely metabolized carbon compounds—that were exuded from the protists. Those spirochetes that developed fuzzy, adhesive materials on their surfaces improved their chances of retaining close relationships with their food sources.

With time, selection pressures drove the spirochetes to develop more and more elaborate mechanisms of attachment to their hosts. Some even evolved mechanisms to enter and feed within their host cells, avoiding digestion. In some populations the vigorous motility of the attached spirochetes was exploited by the hosts. Together the spirochetes and hosts could move more rapidly and farther in search of food than either could alone. In fact, some scientists believe that the undulipodia of protists originated in such a symbiotic partnership between spirochete bacteria and protist hosts.

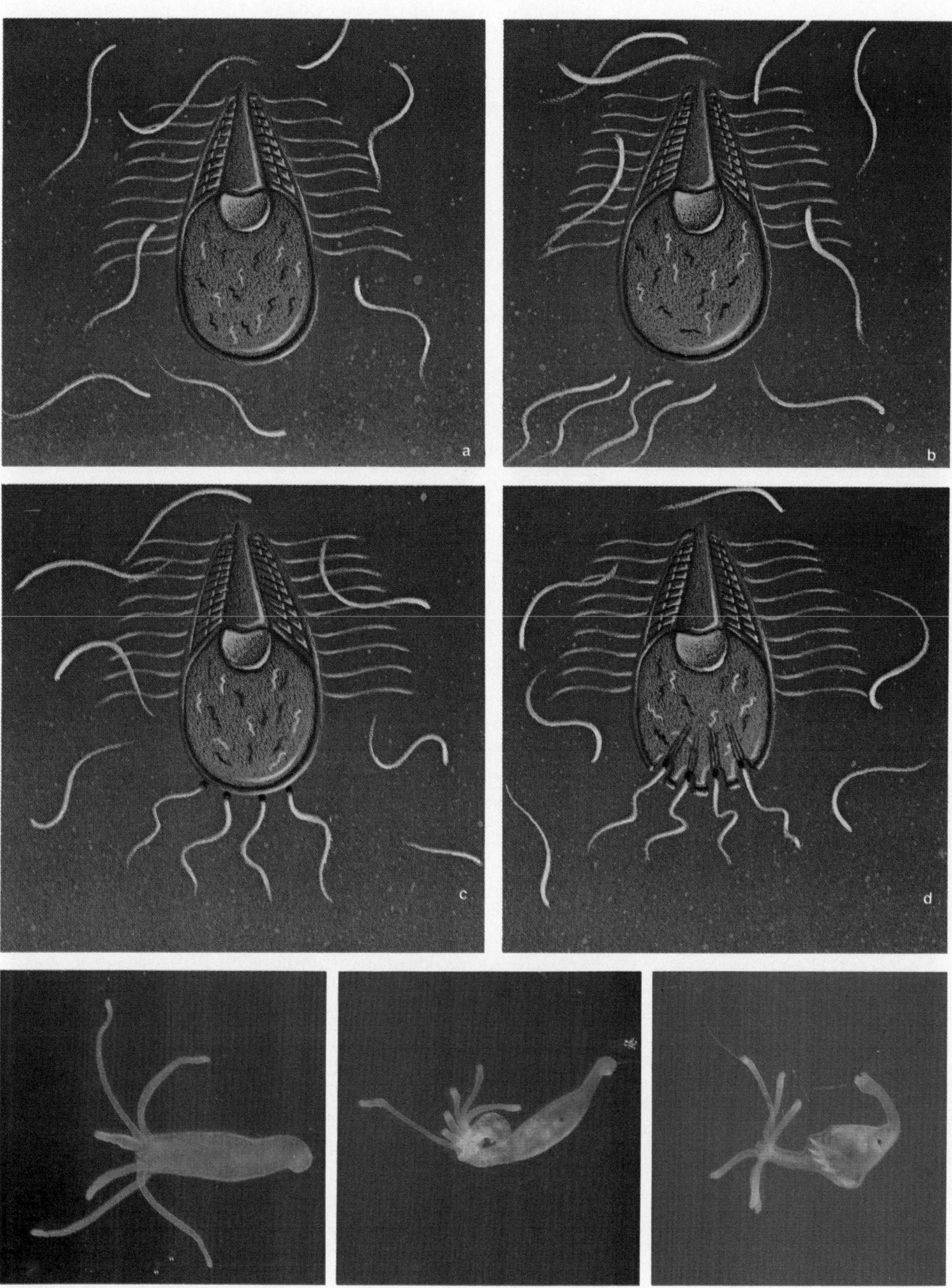

Photos, E. R. Degginger—ANIMALS ANIMALS

(Top, left) Walker—The Natural History Photographic Agency; (top, right and bottom) Lynn Margulis, Boston University

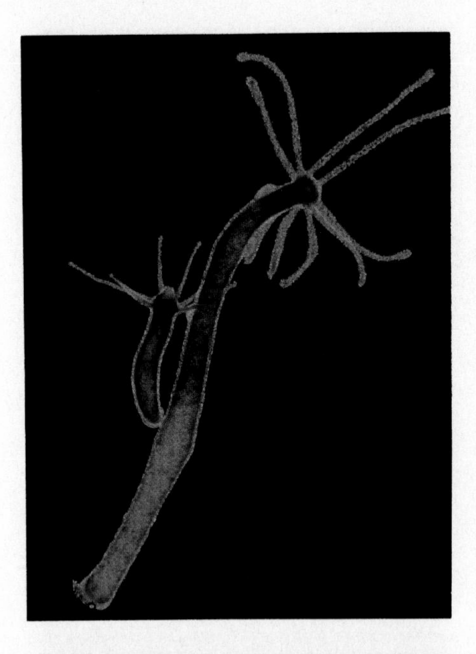

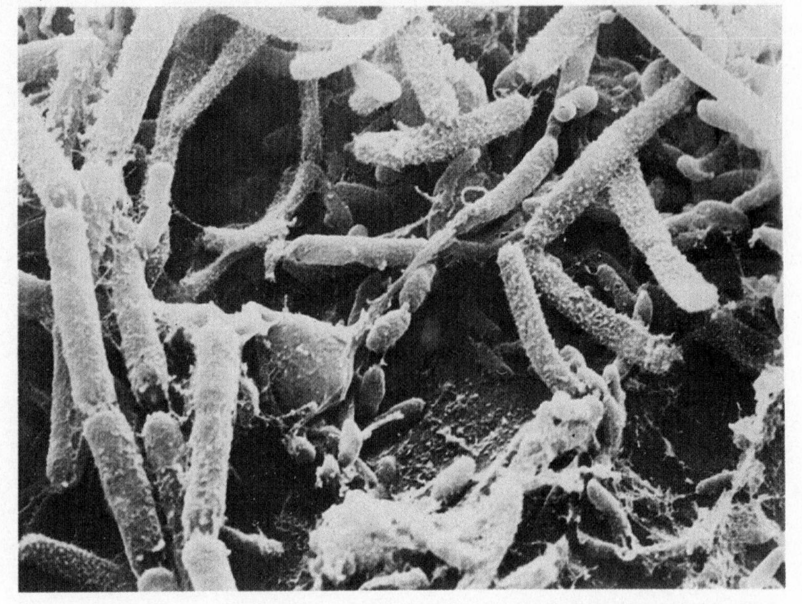

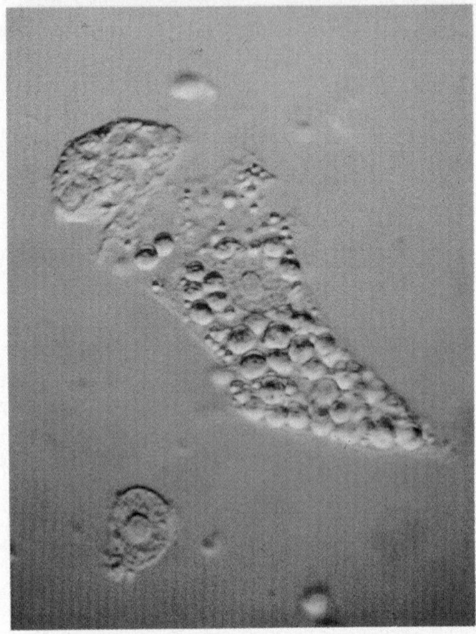

The green hydra (top left, reproducing asexually) harbors about 20–30 photosynthetic Chlorella *cells per hydra gastrodermal cell (above). In single gastrodermal cell (above), algae cluster near the end of the cell normally closest to the outside surface of the hydra, where sunlight is most intense. Most green hydras also host a species of rod-shaped, flagellated bacterium (top right) in their gastrodermis and ovarian tissue and on the surface of the egg.*

The association between hydras and green algae probably first began as a consequence of hydra feeding behavior. Nonsymbiotic and symbiotic hydras alike feed on water fleas, insect larvae, and other pond invertebrates. Hydras immobilize their prey with nematocysts, stinging apparatus borne in specialized cells called cnidoblasts on their tentacles. When prey are drawn to the mouth by the tentacles and stuffed into the coelenteron, other, smaller organisms such as bacteria and algae inevitably enter as well. In at least several strains of hydra, green algae probably entered this way and continued to function within the rather transparent animal, photosynthesizing and producing food compounds. These algae were taken across the membranes of the hydras' gastrodermal cells, as are particles of food, bacteria, and even indigestible materials (including styrofoam plastic spheres). Occasionally, however, these materials fail to be either digested or expelled. Such was the fate of some algal cells, and with time populations of algae established permanent associations with the hydras.

At least some strains of green hydra have fine-tuned their relationships with their symbiotic algae such that in either light or darkness large quantities of metabolic products flow in both directions: from hydra tissues to algae and from algae to hydra tissues. These substances include sugar products of photosynthesis; nucleic acid derivatives, or nucleotides, the components from which genetic material, DNA, is synthesized; and amino acids, the components of proteins. If placed in the dark, green hydras cannot survive for long on the metabolic products of their algal partners. Yet, as long as they are allowed to feed and are kept clean, these hydras retain their intercellular algae. Thus, in conditions of darkness, the algae must receive all of their food requirements for growth and cell division from their animal hosts. Indeed, the algae survive well in the dark inside the hydras. Although the absolute number of algae per gastrodermal cell declines (for example, in some strains, from about 20 to 2 or 3), it never falls to zero. As soon as the green hydras are returned to the light the number of algae per cell increases within hours to its normal value. Thus, although the detailed mechanisms are not known, the relative growth rates of the partners are closely controlled such that even under adverse conditions the symbiosis is maintained.

Only if the association is drastically disturbed—for example, by chemicals that poison photosynthesis in the presence of very strong light—do the hydras expel or digest their algae and thus permanently lose them. By such dramatic experimental conditions aposymbiotic (previously symbiotic) white hydras are formed. If these white hydras, however, are placed in the vicinity of normal green hydras, they will take up algae that are inevitably around, in, and on the green ones. They ingest the algae, recognize them by means of some biochemical system, avoid digesting them, and transport them to the periphery of their gastrodermal cells. In one or two days the symbiosis is reestablished.

Sex and the symbiotic hydra

Hydras and their algal partners serve as good examples for symbioses between photosynthetic animals because so much is known about them. They display another trait that is characteristic of obligate associations: they have developed methods for retaining the symbiotic relationship through the sexual cycle of the host animal.

Walker—The Natural History Photographic Agency

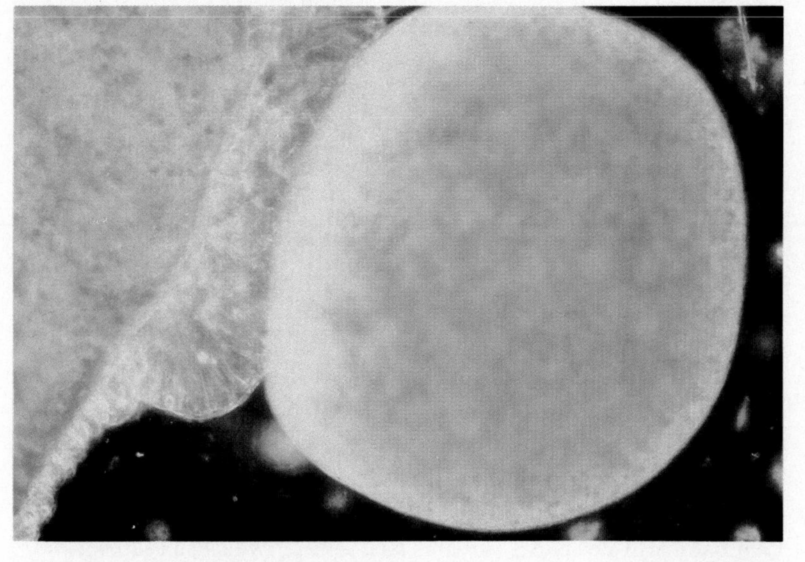

Oxford Scientific Films—ANIMALS ANIMALS

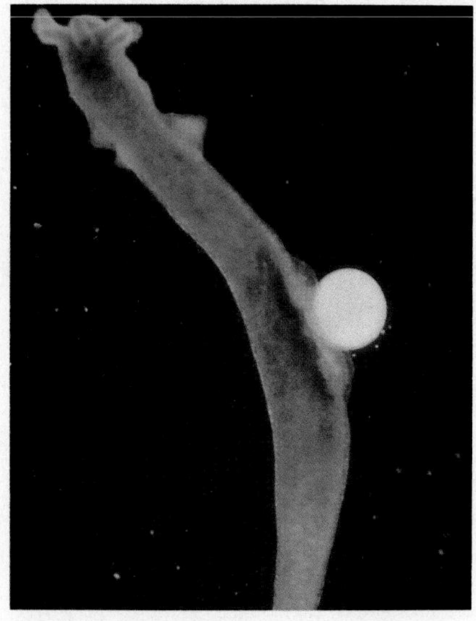

Green hydra in sexual phase (above) bears several spermaries just below the tentacles and a white, fully developed egg. Close view of extruded egg (left) reveals details of its attachment to the parent.

Most of the time hydras reproduce asexually by budding. Offshoot animals complete with tentacles and coelenteron develop on the sides of the parent. These gradually assume the size of the parent hydra, break away, and then quickly mature into adult hydras that form their own buds. When a hydra gastrodermal cell duplicates itself by cell division to supply the growing bud, the total number of symbiotic algae in the cell simply are segregated into the offspring cells such that each cell has about half of the algae immediately after division. As the gastrodermal cells grow prior to their next division the number of algae doubles to its normal figure. In this way algae are always associated in proper numbers in asexually budding hydras.

In the north temperate regions of western Europe and North America adult hydras develop sexual organs at least some time during the year (either late spring or late autumn or both). Usually a single hydra forms spermaries, which are bulbous male structures that release hundreds of sperm into the open water, and larger ovaries, one or two depending on strain, that produce a single huge spherical egg. Fertilization follows with the egg attached to the hydra. The fertilized egg forms a heavy dark coat, or theca; then it migrates down the body surface of the parent and eventually is released into the pond or stream. This egg resists the insults of hot summers and freezing winters, hatching when conditions permit.

In one strain of hydra that has been carefully studied, neither the enormous complex egg, which contains the cells that become the embryo and an abundance of lipid material for nourishment, nor the tiny sperm ever contain symbiotic algae. How then does the symbiosis persist through this sexual cycle? Speculation that algae actually enter the egg during development had been brought forward but was never documented. Recently this problem was solved by Glyne Thorington of Boston University. The green hydra has developed a remarkable way to insure itself that the algae persist through the sexual cycle and the overwintering. During the period when sexual organs are produced and sperm released, a week or two of intense activity, algae as well as sperm are spewed into the surrounding water. Special algal vesicles are released in such quantity and with such vigor that the hydras may look as if all their algae are gone; sexual hydras may be white to the naked eye. (Microscopic examination shows, nevertheless, that they always retain one or a few algae per gastrodermal cell.) The algal vesicles break, and the algae coat the surface of the unfertilized egg. The egg soon becomes fertilized, develops, and descends down the body to the base. A matrix of interwoven fibers binds at least some algae to the egg surface, and these algae persist in association with the developing egg.

Recent studies of the green hydra have uncovered the method by which its symbiotic algae are transferred to its offspring through the sexual cycle. During the week or two of the animal's sexual development (a), special algae-containing vesicles form from hydra gastrodermal tissue and spew from the mouth (b), often in such quantity that the hydra temporarily becomes white to the naked eye. The algal vesicles break, and some of the algae coat the unfertilized egg. After the egg is fertilized with sperm from a second hydra (c), it migrates down the body of the parent (d and e) and develops a dark protective coat of interwoven fibers, called the theca (f). The matrix of thecal fibers binds at least some of the algae to the egg surface (g). When the egg eventually hatches, green algae are present, ready to populate the young hydra.

The next season when young hydras hatch from the eggs, they ingest the associated algae but do not digest them. The algae are recognized and retained by gastrodermal cells, and within a few hours after ingestion they arrive at the well-lit periphery of the hydra cell, thus optimizing the quantity of sunlight they receive. The symbiotic algae grow and divide inside their hosts, and the life cycle persists. In short, the symbiotic algae of green hydras are maternally inherited but external to the egg. No algae have ever been seen to be paternally inherited or associated in any way with sperm.

In nature, in the ponds, streams, and lakes in which these hydras are found, this story is even more complicated. Most green hydras examined harbor yet another type of symbiont in large numbers: gram-negative, rod-shaped, flagellated bacteria, a new strain of the genus *Aeromonas punctata*. These bacteria dwell within the hydra but only in specific hydra tissues—the gastrodermis and ovarian tissue—and on the surface of the egg. Those in the gastrodermis apparently take advantage of the food leak from the algae to the hydra. They also take advantage of the "ride" to the next generation by their association with the ovarian tissue and the egg. When during sexual development algal vesicles are formed by the coelenteron membranes of the hydra, the bacteria enter the vesicles and become enveloped in vesicle mem-

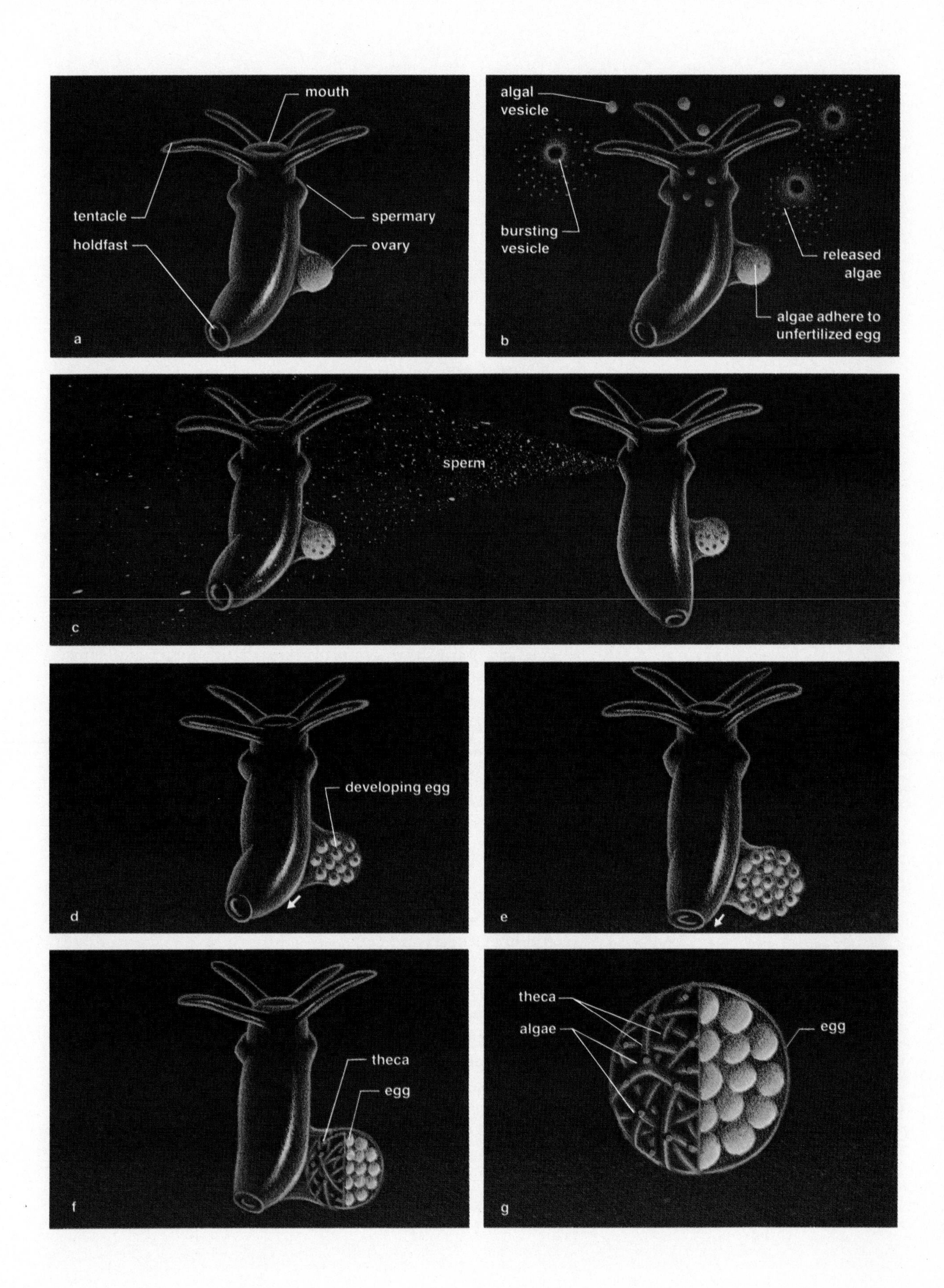
mouth
tentacle
spermary
holdfast
ovary
a
algal vesicle
bursting vesicle
released algae
algae adhere to unfertilized egg
b
sperm
c
developing egg
d
e
theca
egg
f
theca
algae
egg
g

brane with the algae. In fact, some vesicles contain only bacteria, and at their release the bacteria outnumber the algae at least ten to one.

These bacteria, unlike the many other kinds that are casually associated with the outer surface of the egg and the hydra in general, resist digestion. They are capable of growth and division inside the hydra gastrodermis. They are not obligate symbionts, however—not yet anyway—because they can grow easily on defined media outside the hydras. Hydras treated with antibiotics or forced experimentally to lose their algae also lose the symbiotic bacteria without much apparent consequence. Such aposymbiotic hydras can survive indefinitely if they are treated very gently: fed brine shrimp every day, cleaned six hours after feeding, and frequently resupplied with clean water. Because of the pattern of bacterial distribution and behavior, it is inferred that the bacteria-hydra symbiosis is a rather new one, with each partner mildly adapted to the association. It seems obvious that the bacteria penetrated and grew inside hydra gastrodermal cells only after the algae were already there as a source of food.

Hydra associations are good subjects for discussion because enough is known about them to provide insight into many general problems: how are symbioses begun, on what levels are the symbionts integrated, how do the associations persist through the development stages of the partners, what are the advantages to the partners relative to unassociated species, how is cell division of the symbionts regulated such that the ratio of partners in association stays relatively constant, and what substances are actually exchanged between the partners? Evolution of hydra symbiosis with both its bacterial and algal partners is particularly easy to reconstruct. Furthermore, because the detailed structure, developmental pattern, and biochemistry of both the hydras and the algae differ in the different green hydra strains (Carolina, Ohio, Frome, English, Jubilee, European, and others), it seems very likely that most, if not all, of these symbioses evolved independently. Low nutrient conditions abounded in the many places where both pondwater algae and hydras that required complex nutrients grew together, and such parallel conditions led to very similar but not identical symbioses. Yet all symbiotic hydras are called today by the same genus and species name: *Hydra viridis*. The differing strains are generally named after the location of collection of the animals.

Evolution of eucaryotic cells

The cells that divide by the process of chromosomal duplication known as mitosis and have membrane-bounded nuclei—the eucaryotic cells—may have themselves originated by microbial symbioses. Two sources of information, the present-day abundance of microbial symbioses coupled with the recognition that a plant or animal cell contains several, only remotely related semi-independent genetic systems, have led to the revival of ideas of the symbiotic origins of certain cell components called organelles.

One class of organelles, the mitochondria, are found in nearly all protist, fungal, plant, and animal cells. These organelles are responsible for the generation of chemical energy in the form of ATP at the expense of oxidation of molecules that have been obtained from food or generated through photo-

Keith R. Porter, University of Colorado, Boulder

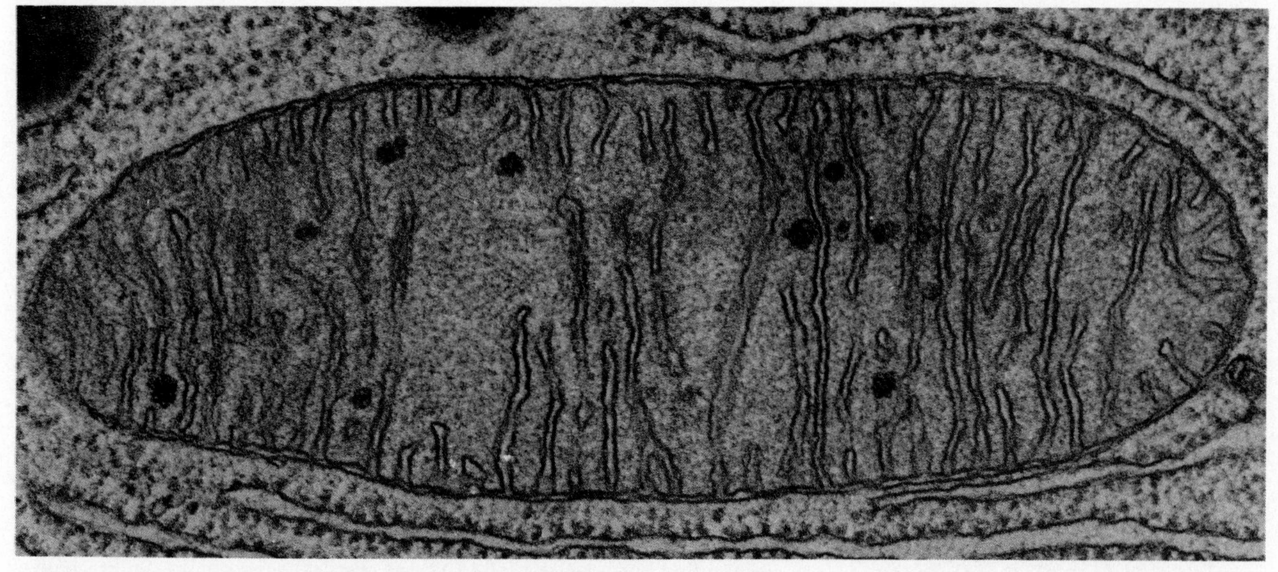

The mitochondrion, an organelle found in nearly all protist, fungal, plant, and animal cells, functions as the cell's "energy factory." It is in many ways structurally and biochemically similar to a bacterium and even contains its own genetic material.

synthesis. Mitochondria contain their own genes, usually circular DNA molecules somewhat like those of viruses or bacteria. Furthermore, mitochondria are the size of bacteria, they divide like bacteria, and they contain complex internal membranes composed of cytochrome proteins and certain lipids. Their physiological and biochemical features are much like certain free-living bacteria belonging to the genera *Paracoccus* and *Rhodopseudomonas*. Mitochondria, however, do not contain enough genetic material to produce all of their own components. They are dependent on the rest of the cell, the nucleocytoplasm (the nucleus and cytoplasm taken together) and its protein-synthesizing machinery, for many of their proteins. Yet, in the composition of their cytochromes and in other biochemical respects mitochondria resemble these free-living bacteria far more than they resemble the nucleocytoplasm in which they are found. Such observations have led to the hypothesis that mitochondria were acquired as symbionts and have persisted in intimate association with their nucleocytoplasmic hosts since the earliest days of the appearance of the ancestors to eucaryotic cells.

The concept that the photosynthetic plastids—*e.g.*, green chloroplasts, red rhodoplasts, and other colored, light-absorbing organelles of algae and plants—derive from once-independent photosynthetic monads has even more support. Not only do chloroplasts and other plastids have their own genetic machinery (DNA, messenger RNA's, transfer RNA's, and ribosomes), they are seen to divide very much like wall-less cyanobacteria. Plastids are bounded by double membranes; the outer membrane is synthesized according to instructions carried in the eucaryotic nucleus, whereas the inner membrane is synthesized from instructions contained in the genetic system of the plastid. The plastids of algae and plant cells are extremely similar in fine structure, macromolecular chemistry, and function to free-living oxygen-producing photosynthetic coccoid bacteria.

The rhodoplasts, which are the red-colored plastids of red seaweeds, and certain free-living coccoid cyanobacteria are especially alike. They both contain distinctive proteins called phycobiliproteins, bear their photosynthetic pigments on structures called phycobilisomes, and have many other features in common. It seems reasonable that, like the algae of hydras, eucaryotic, mitochondria-containing ancestors to red algal cells acquired symbionts that once had been cyanobacteria.

Recent discovery of a new sort of monad, a photosynthetic oxygen-eliminating microorganism, has also shed light on the possible symbiotic origin of organelles. These coccoid photosynthesizers, members of the genus *Prochloron,* are of great interest to evolution because they have the features that were predicted of the ancestors of the chloroplasts. These microorganisms are in the size range of chloroplasts, from 8 to 25 micrometers (millionths of a meter) depending on strain, and they contain two varieties of the chlorophyll molecule, *a* and *b*, in ratios comparable to those in chloroplasts. They are also grass-green in color and perform oxygen-releasing photosynthesis. Yet they are entirely bacterial in cell organization. Like bacteria and chloroplasts their genetic material is loose in the central portion of the cell, not bounded by membrane or organized into chromosomes. Today no *Prochloron* that has been studied can be grown in the laboratory. All *Prochloron* cells must be collected from the surfaces or the cloacal cavities of didemnid tunicates. These are chordate animals living in subtropical marine waters such as those off the Palau Islands of the western Pacific or the west coast of Baja California, Mexico. Today's *Prochloron* is always found in loose surface symbiotic associations with animal hosts. The discovery of these unique microorganisms, however, by R. A. Lewin at the Scripps Institution of Oceanography at the University of California at San Diego in La Jolla, is so new that future studies most likely will reveal more genera as well as ways of growing the cells under laboratory conditions for detailed comparisons with chloroplasts.

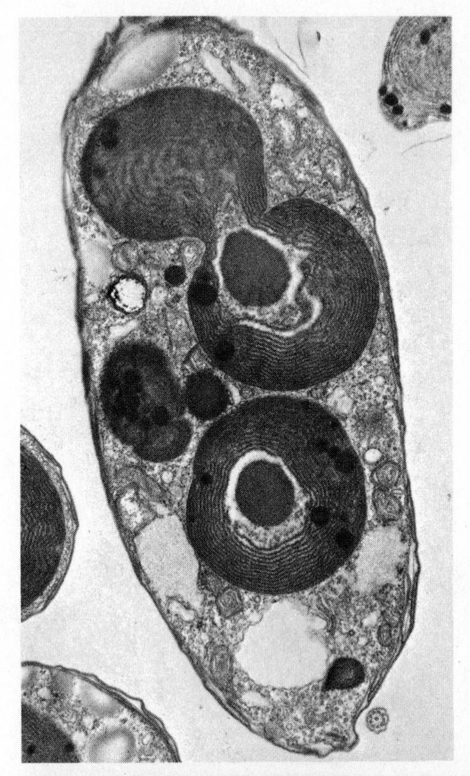

A multilevel phenomenon

Regardless of the details it seems evident that most "single organisms" are chimeras: they are composed of components which were once communities of organisms that interacted in complicated ways. The levels of integration of the partners vary. Some are loose and basically behavioral. Others, like

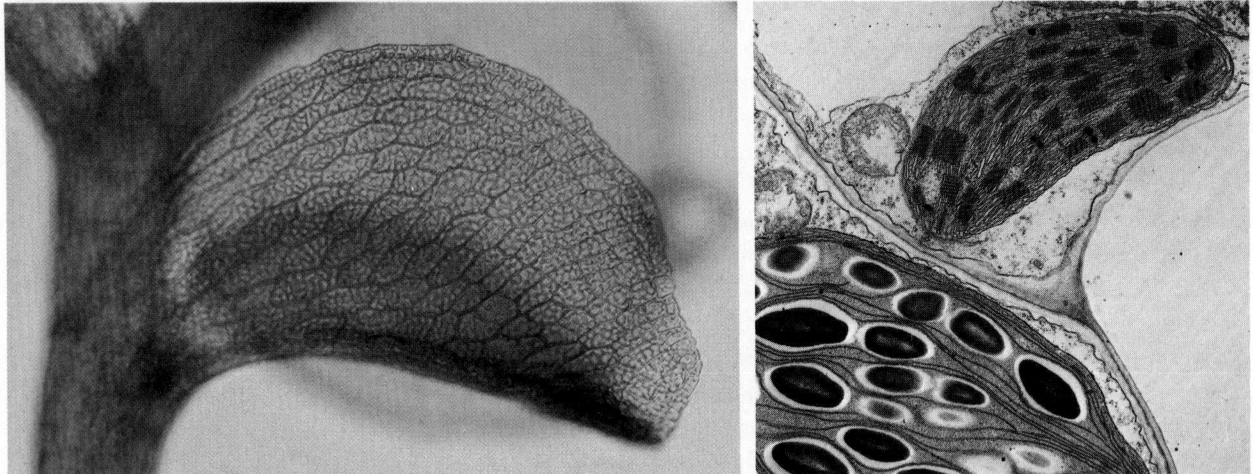

(Top) William T. Hall, National Cancer Institute, National Institutes of Health; (bottom, left) © Biology Media—Photo Researchers; (bottom, right) J. Whatley—Oxford Scientific Films

Ralph A. Lewin, Scripps Institution of Oceanography

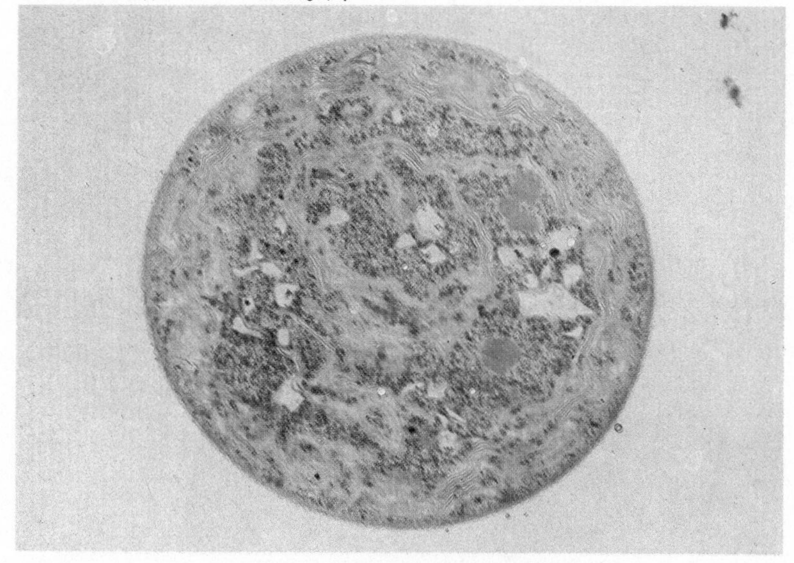

Prochloron, a recently discovered photosynthetic microorganism, possesses the features predicted of the ancestors of the chloroplast. It is found in loose symbiotic associations with animal hosts.

the green hydras and their algae as well as the spirochetes and their protist hosts, are metabolic in the sense that substances cast off by one of the partners are used as food and further metabolized by the second partner. Still other associations are more intimate: gene products (either proteins or ribonucleic acids of one partner) are incorporated and used in the metabolic activities of the second partner.

In some cases, indeed, it seems that the level of integration is still more intimate: the genes of one of the partners have actually been transferred to the genetic material of the other partner. For example, the synthesis of compounds called diterpene gibberellins, which are plant hormones, is strictly a metabolic virtuosity of vascular plants. These compounds are not produced by fungi at all. Yet a certain fungus that has long been associated as a parasite of plants, *Gibberella fujikuroi*, unlike all of its relatives produces these compounds. Because of the total absence of the metabolic pathway to these compounds in related fungi and because the fungus can synthesize these compounds even if removed entirely from the plant host, it is thought likely that the genes for the biosynthetic pathway to diterpene gibberellins were transferred from the plant to this fungal parasite during its coevolution with its host plants. When carefully examined with the methods of molecular biology, many, if not most, symbioses will probably reveal that the integration occurs on more than one level. As in the case of the green hydra, behavioral, metabolic, and developmental factors combine to optimize the chances that natural selection acts on the association rather than the partners as individuals.

Scientists now realize that the establishment and evolution of symbioses are not a peculiar phenomenon limited to such obscure cases as lichens and certain luminous marine organisms. Rather they play a central role in the emergence of new species with properties that neither partner itself had. As the likelihood grows that certain components of plant and animal cells originated as microbial symbionts, the study of symbiosis is moving from the fringes of biology to dead center.

(Facing page) Translucent cells of a moss plant (lower left) swarm with tiny green chloroplasts, the photosynthetic organelles of algae and higher plants. Under high magnification single chloroplast (top structure in photo at lower right) exhibits system of internal membranes and other machinery responsible for the production of ATP from light energy. Like mitochondria, chloroplasts have their own genes; their fine structure, biochemistry, and function closely resemble those of certain photosynthetic bacteria. The protist Cyanophora paradoxa *(upper left) contains organelle-like units (distinctive by their numerous concentric membranes) that appear to be modified photosynthetic cyanobacteria living as obligate symbionts in the protist. Similar partnerships may have been the ancestors of algae and higher plants.*

OF DINOSAURS AND ASTEROIDS

The discovery in Cretaceous geological strata of high concentrations of iridium suggests that the mass extinction of many plants and animals, including the dinosaurs, at the end of the Cretaceous Period resulted from a collision between the Earth and an asteroid.

by Stephen Jay Gould

Generations of students in geology courses have been forced to memorize the geological time scale and have objected, often vigorously, to this imposed rote learning. What good are all the names of eras and periods, students have asked, when they represent no more than an arbitrary subdivision of continuous time?

But the names are not arbitrary at all. The history of life is not a tale of smoothly increasing complexity and diversity of form. The record of life has been marked by episodes of mass extinction—short times during which significant percentages of all species have abruptly disappeared. The boundaries of the geological time scale record these episodes. The three major eras—Paleozoic, Mesozoic, and Cenozoic—are separated one from the other by the two greatest extinctions in the history of life. The monumental Permian extinction 225 million years ago ended the Paleozoic Era and initiated the Mesozoic Era. Up to 90% of all species of shallow-water marine invertebrates perished during this very short interval.

The extinction separating the Mesozoic and Cenozoic eras, some 65 million years ago, was not so extensive, but it has captured more popular attention for a specific reason: dinosaurs were among its victims. In this great Cretaceous extinction (the Cretaceous is the last period of the Mesozoic Era), probably half the species of shallow-water marine invertebrates disappeared. The extinction was selective in its effect—and this is one of the main phenomena that any adequate theory of its cause must explain. On land dinosaurs disappeared, while land plants, freshwater organisms, and small mammals were not greatly affected. In the sea the plankton (small, usually microscopic, floating organisms) virtually disappeared along with

many groups of marine invertebrates (the ammonites—relatives of the modern chambered nautilus—and the curious rudistid clams that built reefs and looked like corals, for example). Other groups of marine invertebrates were largely unaffected.

The coordination of all these extinctions across such a wide range of habitats and styles of life is the primary fact demanding explanation. There is no separate problem of "the extinction of the dinosaurs," and the various theories that attribute their demise to some special problem or inadequacy—becoming too big to hold up their own weight or small mammals eating their eggs, for example—must fail because they ignore the fact that the disappearance of dinosaurs was just one aspect of a coordinated event affecting the entire world of life.

Types of explanations of mass extinctions

Two traditions of argument have dominated discussions of mass extinctions. The first, usually favored by geologists, looks toward processes normally active on the Earth's surface. Some of these Earth-based mechanisms might be relatively catastrophic—worldwide mountain building, for example (a favorite of generations past, but now abandoned). But most are in the tradition of gradual and accumulative change, such as deterioration of climate or rise and fall of sea levels. At most, they call upon an increase either beyond the usual rate of change or the magnitude of its effect. The dinosaurs of Walt Disney's motion picture *Fantasia,* marching to their death across parched deserts of a drying Earth, reflect one common version of these gradualistic Earth-centered theories.

The second tradition calls upon some extraterrestrial influence. These theories have generally postulated true catastrophes. Explosions of nearby stars (supernovas) have often been invoked, as have a variety of impacting bodies from meteorite swarms to asteroids to comets. Most geologists have a long-standing, historically based preference for gradualistic explanations, but in candid moments most will admit that astronomical catastrophes are not inherently implausible. After all, stars do explode and large bodies must occasionally strike the Earth. During the 600-million-year history of complex multicellular life such catastrophes may have occurred a few times. Since the major extinctions of life can be counted on the fingers of one hand, perhaps these exceedingly rare astronomical catastrophes have contributed to their cause.

But for most geologists the problem with these catastrophic ideas has not been their implausibility in theory—all must admit that catastrophes can happen. Rather, the reason for strong prejudice against astronomical catastrophes has been the failure to devise, even in imagination, a way to test such theories. What evidence would a supernova or an impacting comet leave in the geological record? Heretofore, the answer has been "none, so far as we know." Mechanisms that leave no evidence, however plausible they are in theory, are of no use in science. For science, as British zoologist Peter Medawar says, is "the art of the soluble"—and theories that cannot be tested by evidence are useless, no matter how plausible. During the past few years all this has changed.

STEPHEN JAY GOULD *is Professor of Geology at Harvard University, Cambridge, Massachusetts.*

Illustrations by Ron Villani

The Alvarez hypothesis

Many, probably most, original ideas in science arise as unplanned and surprising by-products of studies done for other purposes. After all, one cannot set out explicitly to find the unexpected. The first scientifically respectable catastrophic theory of the Cretaceous extinction, proposed in 1979, arose in this way. Luis Alvarez, a celebrated physicist at the Lawrence Berkeley Laboratory of the University of California at Berkeley; his son Walter, a geologist on the faculty of the University of California at Berkeley; and analytical chemists Frank Asaro and Helen Michel were studying variations in the rate at which sedimentary rocks were deposited at the Cretaceous-Tertiary boundary. (The Cretaceous-Tertiary boundary marks the time of the great extinction. The Cretaceous was the last period of the Mesozoic Era; the Tertiary was the initial period of the Cenozoic Era.) One method for inferring rate of sedimentation involves calculating the abundance of the rare metallic element iridium.

The Earth's crust contains virtually no indigenous iridium. When the Earth formed, it probably contained iridium in normal cosmic abundances. But iridium is a heavy and highly unreactive element, and the Earth's original supply of it must have sunk into the central core when the Earth melted and differentiated some 4,000,000,000 years ago. Therefore, almost all the iridium now on the Earth's surface has arrived via extraterrestrial bodies; meteorites, for example, contain iridium in normal cosmic abundances.

Alvarez and his colleagues assumed initially that iridium reached the Earth in a relatively constant supply through the gentle rain of small meteorites and the cosmic dust that continually bombards the Earth. Thus, they reasoned, sediments with little iridium accumulated quickly and included very little cosmic debris. Sediments with higher concentrations of iridium formed slowly, and more cosmic iridium accumulated in them.

But Alvarez's data on iridium forced him to revise his perspective radically —indeed, to stand it upon its head. His research team initially studied sediments at the Cretaceous-Tertiary boundary in two European sites, Gubbio in Italy and Stevns Klint near Copenhagen in Denmark. Measuring variation in 28 elements in the Gubbio sediments, they found that 27 of them did not change greatly across the boundary. But iridium increased by a factor of 30 right at the boundary itself, and then dropped abruptly back to normal levels in the Tertiary sediments. The Danish results were even more surprising. There, iridium increased by a factor of 160, again right at the boundary. These enormous increases in iridium could not reasonably be attributed to a marked slowdown in sedimentation. Alvarez then reversed his assumption; perhaps, he reasoned, variation in iridium is not caused by differences in dilution of a constant supply by a varying amount of Earth-based sediment. Perhaps the rise of iridium at the boundary reflects a true and sudden increase in cosmic iridium—in other words, an abrupt influx caused by the impact of some extraterrestrial body.

Since all the evidence came from two European sites, critics could reasonably suspect that the iridium anomaly might represent some local event rather than a global catastrophe. But the original Alvarez report captured the imagination of many scientists and provoked a series of tests (still

An asteroid collides with the Earth, generating a huge cloud of dust and debris that rises into the atmosphere and darkens the Earth for several years to a level below that of full moonlight. Photosynthesis all but ceases, and the Earth's food chain collapses. Such an event has been hypothesized to explain the sudden demise of the dinosaurs and many other forms of life at the end of the Cretaceous Period 65 million years ago.

High abundance of iridium in the boundary layer of rock representing the end of the Cretaceous Period lends support to the theory that the Earth was struck at that time by an extraterrestrial object.

continuing) on sediments at the Cretaceous-Tertiary boundary in other widely scattered parts of the world. A marked increase in iridium was noted in all other samples examined, including two from northern Spain, one from New Zealand, and even one from a deep-sea core in the central North Pacific Ocean.

The Alvarez team then turned its attention to the most likely extraterrestrial source for the enhanced iridium. They rejected the granddaddy of catastrophic hypotheses, an exploding supernova zapping the Earth with lethal cosmic rays, because other elements that should have increased if a star blew up in the Earth's vicinity showed no variation in sediments across the Cretaceous-Tertiary boundary. They then turned their attention to actual impacts.

Most asteroids circle the Sun in a broad area located between the orbits of Mars and Jupiter. But a small number, the so-called Apollo objects, cross the Earth's orbit in their more erratic paths around the Sun. Astronomers estimate that about 700 Apollo objects are greater than one kilometer in diameter. The chance of any asteroid colliding with the Earth as it passes through the Earth's orbit is very small for each crossing (because the Earth's orbit is long and the Earth occupies only a tiny bit of it at any moment). But crossings occur continually, and over millions of years an impact is virtually inevitable. Astronomers calculate that an asteroid with a diameter of ten kilometers (six miles), or larger, should hit the Earth about every 100 million years or so, a figure roughly comparable to the average time between mass extinctions.

But why should a large asteroid engender a worldwide extinction extending across all habitats of life? It might hit a *Tyrannosaurus* or two, but why should it decimate the biosphere? The Alvarez team argued that a ten-kilometer object would excavate an enormous crater, one about 200 kilometers (120 miles) in diameter. No crater of the proper size and age has yet been found on the Earth's surface. There is an approximate $^2/_3$ probability that the asteroid fell in the ocean, and because its probable diameter of ten kilometers is twice the typical ocean depth a crater would be formed on the ocean floor and pulverized rock ejected. In that case it is unlikely that the crater will ever be found because much of the pre-Tertiary sea floor has been subjected to the process of subduction, the descent of one crustal plate under another. The dust thrown up into the atmosphere following such an asteroid collision, Alvarez contended, would be sufficient to darken the Earth for several years (until the dust settled) to a level well below the brightness of full moonlight. Photosynthesis would effectively cease, and the Earth's food chains, all ultimately dependent upon primary productivity by photosynthesizing organisms, would collapse.

The Alvarez group then tried to match the pattern of observed extinction with the presumed effects of such a darkening. The most profound extinction affected the photosynthesizing oceanic plankton, and this fact fit the scenario well. These tiny creatures, with their life-spans measured in weeks, could not wait out the great darkness until light returned. They perished completely, taking with them many of the larger animals dependent either directly upon them or upon smaller animals that ate plankton. But other

Abundance of iridium in acid-insoluble residues at Stevns Klint (SK) sea cliff in Denmark	
Sample*	Abundance of iridium (parts per billion)
+2.7 meters	<0.3
+1.2 meters	<0.3
+0.7 meters	0.36 ± 0.06
boundary	41.6 ± 1.8
−0.5 meters	0.73 ± 0.08
−2.2 meters	0.25 ± 0.08
−5.4 meters	0.30 ± 0.16

*Numerical values are the distances above (+) or below (−) the boundary layer.

patterns demand a more forced explanation and indicate potential trouble for the Alvarez argument. The Alvarez team argued that individual land plants died but their species survived because seeds and runners could have lain dormant for several years until the light returned. Large animals, including dinosaurs, were ultimately dependent upon land plants and perished with them. But small animals, including mammals, might have survived by eating seeds, insects, and plant debris. Freshwater creatures might have held on by eating the nutrients and remains of vegetation washed down from the upland debacle.

Doubts about asteroids and dust

The asteroidal hypothesis has been subjected to two general criticisms of markedly different type: doubts that an asteroid represents the best extraterrestrial scenario and doubts that an extraterrestrial impact is required at all. For the first, in order to throw up a cloud of dust sufficient to block photosynthesis the impacting asteroid must excavate a huge amount of Earth material and cast it skyward. The Alvarez team assumed that a ten-kilometer asteroid would dislodge 60 times its own weight in Earth-based dust, a value that matches observed levels of iridium in sediments at the Cretaceous-Tertiary boundary. But many experts on impact cratering believe that such an asteroid would send a far greater amount of terrestrial dust aloft. Figures of 1,000 to 10,000 times more Earth dust than the weight of the asteroid have been mentioned. If these higher figures are correct, then the Alvarez hypothesis is in trouble because asteroidal iridium would have been so diluted by terrestrial dust that the high levels of iridium actually measured in sediments at the boundary would not have accumulated.

Manicouagan Lake in Quebec, as seen from the Landsat 1 satellite at an altitude of 914 kilometers (568 miles), occupies a circular depression believed by many to have been formed by a collision with an extraterrestrial object. With a diameter of 66 kilometers (41 miles) the crater is about half the size of that hypothesized to have resulted from the Cretaceous asteroid impact.

Courtesy, NASA

Further trouble arises from the scenarios devised by Alvarez and his colleagues to explain the different responses of various animal and plant groups to the debacle. Their argument works well for the oceanic plankton, but it seems a bit forced for the large terrestrial plants that scarcely noticed the catastrophe. Land plants may have survived through the dormancy of their seeds, but Alvarez also wishes to bring mammals and other small animals through in part by eating those seeds and the surviving plants. You can't have it both ways.

These criticisms only cast a specific sequence of asteroid, dust cloud, and failure of photosynthesis into question; they do not diminish the significance of the iridium anomaly as indicator of an extraterrestrial catastrophe. Perhaps a different astronomical theory must be sought. Kenneth Hsü of the Geological Institute of the Swiss Federal Institute of Technology in Zurich, Switzerland, suggested that the impacting body was a comet rather than an asteroid. He wrote that "a cometary impact may kill by three different means." First, Hsü argued, a speeding comet may produce local temperatures so high that vast chemical explosions and even thermonuclear reactions might result. A sudden global increase in air temperature of 10–20° C (18–36° F) might occur and eliminate dinosaurs and other large animals by heat shock. Second, some comets contain highly poisonous cyanides that might have killed marine creatures if the comet fell at sea. Third, the comet's impact might have greatly increased the concentration of carbon dioxide in the ocean by two means: directly through input of carbon dioxide in the comet's head itself, and indirectly through later oxidation of the cyanides brought in by the comet. The more carbon dioxide in the ocean, the greater its power to dissolve carbonates. Since most shells and skeletons of marine organisms are made of carbonates, Hsü argued that this temporary change in oceanic chemistry might have made it difficult or impossible for organisms to secrete their hard parts.

The Hsü scenario has the advantage of not requiring a vast dust cloud that might dilute extraterrestrial iridium below the levels measured in sediments at the Cretaceous-Tertiary boundary. But it does not resolve all the problems of differential extinction that plague the Alvarez hypothesis as well: why did freshwater organisms survive; why did some marine invertebrates die, while other large groups scarcely noticed the debacle?

Doubts about extraterrestrial impact

A more fundamental criticism holds that the iridium might not represent an astronomical impact at all. Perhaps Alvarez's original assumption was right: variation in iridium reflects the rate of terrestrial sedimentation. Some geologists and oceanographers had once suggested that iridium might be high in boundary sediments because ocean currents winnowed the lighter clay particles away, leaving behind heavier meteoric particles derived from the ordinary non-catastrophic influx of small meteorites. But this explanation faded with the discovery of enhanced iridium throughout the world and in deep-sea cores as well as shallow-water sediments. Winnowing might suffice for local areas or even for shallow waters throughout the world if sea levels dropped rapidly, but it cannot explain high iridium in deep-sea sediments where the winnowed clays should accumulate.

Other researchers suggested that the high concentration of iridium may be an artificial consequence of the extinction itself and not a marker of its astronomical cause. If so many marine organisms with hard parts died (including virtually all the plankton, a major contributor to deep-sea sedimentation), then rates of sedimentation might drop abruptly because the large contribution usually made by organic skeletons would cease. High iridium might then represent the normal cosmic influx, undiluted by the usual rain of organic skeletons. But proponents of the impact hypothesis have pointed out that deep-sea sediments now forming slowly with virtually no input from organic skeletons contain levels of iridium far below those measured in the Cretaceous-Tertiary boundary sediments.

In conclusion, the idea that the iridium anomaly represents a catastrophic impact of an extraterrestrial body gained strength with the discovery that the anomaly exists worldwide and at all oceanic depths so far investigated. In addition, all attempts to explain the iridium excess by diminution of terrestrial sedimentation rather than true extraterrestrial enrichment have failed. Yet, while the basic notion of extraterrestrial impact looks better and better, investigators remain unsure of the actual scenario of extinction. Neither dust kicked up by an asteroid nor atmospheric heating and oceanic poisoning by a comet match all the paleontological facts of differential extinction.

Is extraterrestrial impact a complete explanation?

Investigation of the fossil record sheds some light on the Cretaceous extinction and the suddenness with which it occurred. Some aspects of the record are consistent with sudden termination. The extinction of the plankton, for example, occurs at a single geological stratum that professionals often call the "plankton line." But other aspects of the record—though the story is incomplete and subject to varying interpretation—indicate to most paleontologists that many groups exterminated in the extinction had been suffering a steady and marked reduction for millions of years. The decline of ammonites and other large marine invertebrates can be traced throughout the late Cretaceous. In fact, a steady decline of large invertebrates and a later sudden wipe-out of plankton is exactly the reverse of what the catastrophic theory predicts. For the plankton stand at the base of oceanic food chains, and their demise should be the trigger that generates the later and rapid extinction of larger forms ultimately dependent upon them.

As for the animals and plants on land, most paleontologists agree that dinosaurs suffered a long decline throughout the later Cretaceous and that the terminal asteroid (or whatever) could only have wiped out the last few lingering species (but other paleontologists disagree and view late Cretaceous dinosaur diversity as high). Paleontologists Leigh Van Valen and Robert Sloan believe that they can trace a steady dinosaur decline correlated with decreasing temperatures, the pushing of remaining dinosaurs into tropical areas, and a steady advance in domination of mammals at temperate latitudes. The same pattern of correlation with decreasing temperature is more clearly evident in the record of land plants. Species living in Alaska, northern Canada, and Siberia suffered heavy losses, but tropical plants were scarcely affected by the Cretaceous extinction.

In short, geological evidence for a general deterioration of climate during the late Cretaceous is good. The record indicates that sea levels were dropping steadily throughout this time, and falling seas generally mark decreasing temperatures and greater temperature differences between high latitudes and the tropics. (Indeed, falling sea level is the one well-established geological correlate of most other mass extinctions as well.)

This evidence for deteriorating climate and for a gradual decline of several groups of organisms at the same time does not negate the notion that an extraterrestrial impact strengthened and climaxed what might otherwise have been a relatively mild episode of extinction. But it does suggest that the Cretaceous extinction was a complex event with multiple causes. Its cause cannot have been so simple as an asteroid striking a healthy Earth, literally out of the blue, and wiping out large chunks of life suddenly. The biosphere was in decline for other reasons, and the asteroid (or whatever extraterrestrial body brought the iridium) enhanced a process that had already begun for other reasons.

But it would also be wrong to view the extraterrestrial impact as merely a coup de grace, a dramatic ending to what was happening anyway. For without the impact general climatic deterioration would probably have produced only a mild extinction, one that geologists might scarcely have noticed in the fossil record and surely not one profound enough to mark a major boundary of the time scale. If the impact had occurred during a time of favorable climates, its effect might have been quite modest. Perhaps the few great extinctions of the fossil record occur when catastrophes strike an Earth filled with groups of organisms already weakened, if not at the brink of extinction.

The Cretaceous extinction and human life

The tale of the Cretaceous extinction is usually told as a rip-roaring story from a distant and unaffecting past—high drama without a direct impact upon our lives (beyond our disappointment that we must meet dinosaurs in museums rather than in zoos). This comfortable perspective is almost surely incorrect. The Cretaceous extinction was probably the single most important event affecting our existence today. Without it we would probably never have evolved. The Cenozoic Era (the 65 million years from the Cretaceous extinction until the present) is conventionally called the "age of mammals." But few people know that most of the history of mammals is not the story of their Cenozoic success but the tale of their 100 million Mesozoic years, living as small and uncommon creatures in a world dominated by dinosaurs. As long as dinosaurs survived, mammals had little chance to evolve beyond their Mesozoic ratlike sizes and to radiate into the diversity of habitats that they presently occupy. Without this radiation large and differentiated forms like the higher primates could not have evolved.

Dinosaurs were in decline during the late Cretaceous. But how can we be sure that they would have perished without the great extraterrestrial impact? They might have rallied (as they had before, following similar declines) and might still dominate the Earth today. After all, they had lived successfully for more than 100 million years; why not grant them another 65 million or more? Thus, a great and unpredictable cosmic accident may have ended the reign of one group of animals and set the stage for the evolution of creatures that could find and understand the evidence that it left some 65 million years ago.

FOR ADDITIONAL READING

Luis W. Alvarez, Walter Alvarez, Frank Asaro, and Helen V. Michel, "Extraterrestrial Cause for the Cretaceous-Tertiary Extinction," *Science* (June 6, 1980, pp. 1095–1108).

Kenneth J. Hsü, "Terrestrial Catastrophe Caused by Cometary Impact at the End of Cretaceous," *Nature* (May 22, 1980, pp. 201–203).

Richard A. Kerr, "Asteroid Theory of Extinctions Strengthened," *Science* (Oct. 31, 1980, pp. 514–517).

D. A. Russell, "The Enigma of the Extinction of the Dinosaurs," *Annual Review of Earth and Planetary Sciences* (1979, pp. 163–182).

L. Van Valen and R. E. Sloan, "Ecology and the Extinction of the Dinosaurs," *Evolutionary Theory* (1977, pp. 37–64).

HEALTH OF THE ANCIENTS

Changing concepts of infectious disease and new analytical methods have launched a quiet renaissance in the study of health and nutrition, as well as of sickness, in ancient humans.

by Kubet Luchterhand

Every society that has left a historical record has shown interest in disease and healing—so much so that practitioners of the healing arts are at least in the running, if not the clear victors, in the competition for the oldest profession. But until quite recently very little in detail was known about the nature, prevalence, or importance of disease in ancient humans. During the past decade several developments have changed that picture and in particular have led to a better understanding of ancient diseases, prehistoric diet, and the relations between them. Although paleopathology, the study of disease in ancient skeletons and mummies, and paleoepidemiology, the study of the interactions between diseases and ancient populations, have existed for much longer than a decade, the study of health and nutrition in ancient peoples is just beginning.

Health and nutrition, in fact, are both relatively modern concepts, whereas disease and starvation have been around for a very long time. Health and nutrition are not precisely the absence of disease and starvation, however. Rather, they are ideas about the normal, natural, preferred, or best relations between humans and disease and humans and food. Famine and pestilence threatened human society in the Old Testament; adequate diet and relative freedom from the ravages of disease are hopes for the world's peoples of a more recent vintage. An important aspect of understanding modern human health and nutrition is knowledge of diseases and diets of antiquity.

Figurine of ill man (facing page), from Colima in Mexico, dates from the Classic Period of Meso-American culture, in the first millennium AD.

KUBET LUCHTERHAND *is Associate Professor of Biological Anthropology at Roosevelt University in Chicago and a Research Associate in Geology at the Field Museum of Natural History, Chicago.*

Historical beginnings

The study of disease in human remains came into existence as a scientific discipline during the second half of the 19th century. At the same time that pathologist Rudolf Virchow was advancing his cellular explanation of disease in Europe during the 1850s and 1860s, anatomists and pathologists began to apply their findings from autopsy of modern humans to the interpretation of both skeletons and mummified tissue from archaeological excavations. For example, Australian anatomist and anthropologist Grafton Elliot Smith and his colleagues, while working on human remains recovered

(Top) T. A. Reyman, Mount Carmel Mercy Hospital, Detroit; (bottom) Ny Carlsberg Glyptotek

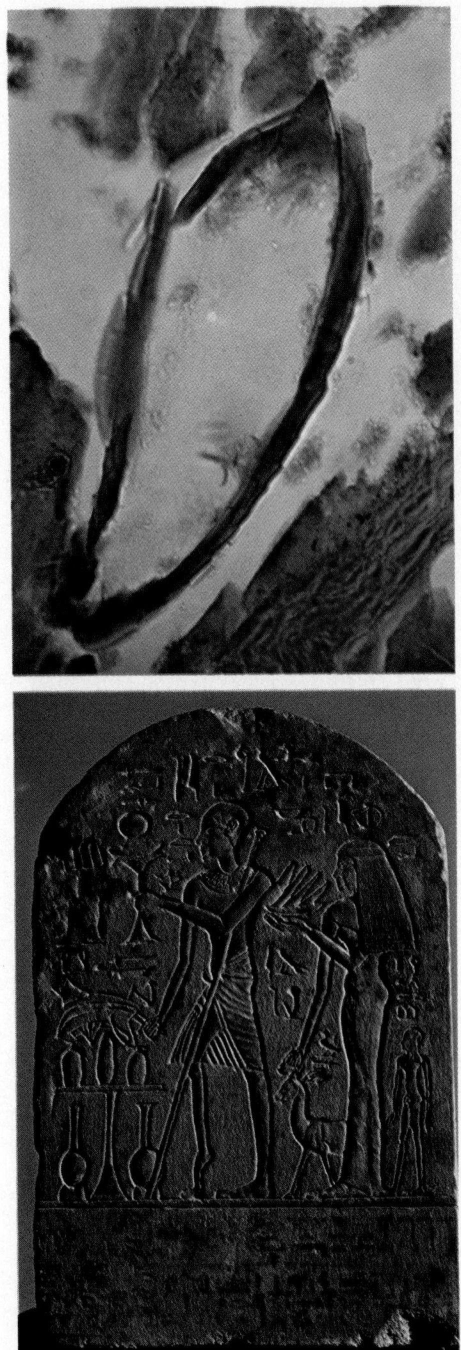

Instance of an ancient disease is revealed in photomicrograph of Schistosoma egg *(top) taken from an Egyptian mummy of the Royal Ontario Museum. Withered leg of deceased Egyptian on 18th-dynasty gravestone (above) strongly suggests childhood poliomyelitis.*

by the Archaeological Survey of Nubia in the late 19th century, pored over tens of thousands of specimens, describing lesions in bone and abnormalities in soft tissues and interpreting them in terms of the effects of modern diseases. Also, in 1876, U.S. physician Joseph Jones, a pioneer worker in forensic medicine and public health, described what he believed were signs of syphilis in human bone that he had excavated in the southeastern U.S.; these remains dated from pre-Columbian times, or before the mass arrival of Europeans in North America.

The idea that modern diseases could be discovered and studied in archaeological specimens captured the imagination of scientists and nonscientists alike, and between the 1860s and the 1930s many attempts were made to trace the antiquity of individual diseases and to increase the number of diseases that could be recognized in skeletal or mummified soft tissue. Since these efforts were often carried out by the most eminent medical specialists of the day, it is no surprise that advances in biological knowledge and laboratory techniques were almost as quickly applied to the investigation of ancient diseases as to the diagnosis and treatment of contemporary ones. Only two years after Wilhelm Röntgen's announcement of his discovery of X-rays in 1895, X-ray techniques were used to study prehistoric Egyptian bones in both England and Australia. During the first decade of the 20th century, when microscopic studies of human tissues and of disease-causing agents were on the verge of wide use in medical research, British physician Marc Armand Ruffer made microscopic examinations of diseased tissues from a series of Egyptian mummies of the 20th dynasty and demonstrated the presence of eggs of the same parasitic flatworm that causes schistosomiasis today.

Ruffer believed that he had coined the term paleopathology to refer to the study of disease in ancient human specimens, but Roy L. Moodie, a U.S. anatomist and another pioneer in paleopathological studies, discovered the word in a dictionary published in 1895. Moodie is an interesting figure in the history of the discipline in that he was one of the first practitioners to extend it to virtually the entire biological realm. His classic text, *Paleopathology: An Introduction to the Study of Ancient Evidences of Disease* (1923), ranges from his primary concern with vertebrates, including man, to include material on diseases seen in fossil plants and speculations on the nature of disease and the extinction of races. Moodie's catholicism of interest is not typical, however, and especially during the past 50 or 60 years, paleopathological studies have tended to be more narrowly focused.

From 1900 until about 1935, what might be called the classic tradition of human paleopathological studies was established. Museums built up, catalogued, and described large collections of human bones from archaeological excavations, and medical scientists established and refined the techniques by which diseases could be recognized and distinguished from one another. Perhaps the most important generalization that emerged from the work of this period concerned the limitations inherent in the nature of the materials under study and the diagnostic techniques then available. Only a relatively limited set of diseases leave traces in bones distinctive enough for diagnosis. Consequently, because the vast majority of specimens available for study

Photos, courtesy, R. Ted Steinbock, Harvard Medical School and Massachusetts General Hospital

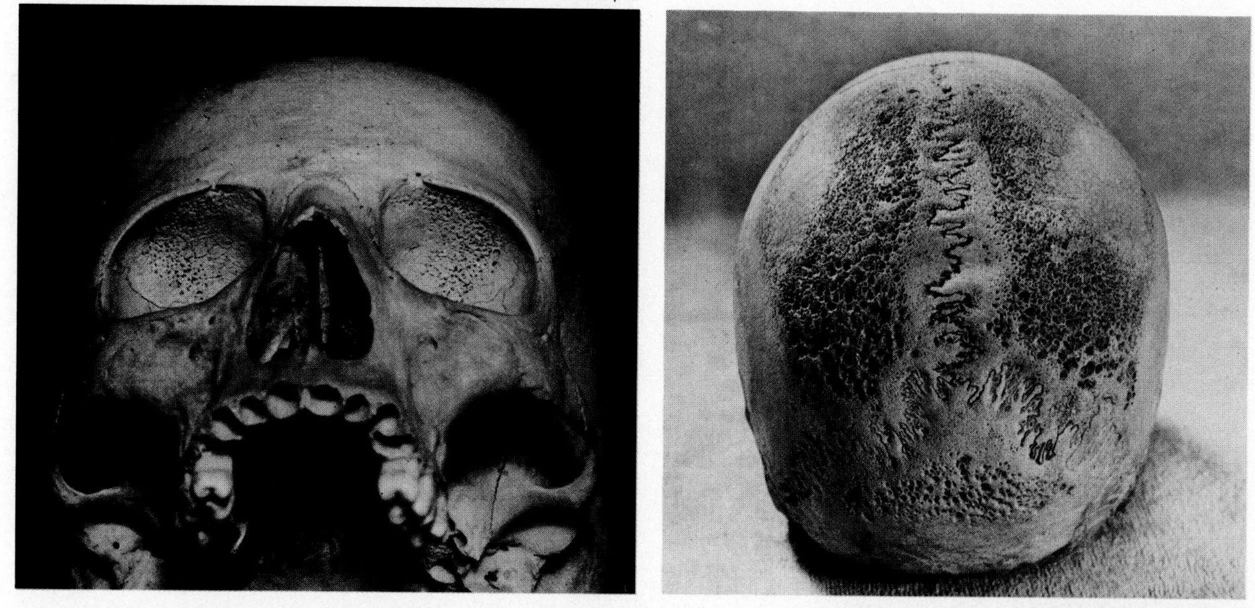

consist entirely of bones, only a relative handful of diseases could be studied in ancient populations—for example, degenerative joint disease; osteoarthritis (bony deposition in connective tissue); bone infections including spinal infections from the bacterium that causes tuberculosis (tuberculous spondylitis); loss of bone substance (osteoporosis) due to malnutrition, metabolic disorders, or other causes; and vitamin D deficiency disease (rickets). To make matters worse it was not always possible to distinguish such diseases from one another using only lesions in bone.

Connective tissue, which includes ligaments, tendons, cartilage, skin, and other tissue providing support and structure, forms the great bulk of soft tissue that is routinely preserved in mummies. Even the most sophisticated analytical techniques generally have failed to glean much from this material beyond such gross disorders as enlargement of individual organs, developmental asymmetries, arteriosclerosis (in some Egyptian mummies), and various sorts of lesions of the skin and other soft tissue, because most of the cell architecture of most tissues has been destroyed. Also, because mummies are relatively rare and tend to be a product of special circumstances—natural desiccation or cultural embalming practices—they do not provide a good sample of entire ancient populations. So, even when diagnosis of unusual diseases is possible, it is unsound to generalize beyond the local population in which the condition is found. Recent discoveries of embalmed ancient Egyptians, dehydrated (and thus preserved) American Indian "mummies," and "freeze-dried" remains from the Scythian burial tombs in the Eurasian steppes may give dramatic glimpses of the external appearance of large specimens of ancient tissue, but they do not contribute much to an overall understanding of the way in which disease affected the lives of ancient humans.

These limitations were all clearly understood by the mid-1930s, and human paleopathological studies settled into a somewhat routine cataloging of the occurrence of diseases that could be diagnosed in skeletal populations.

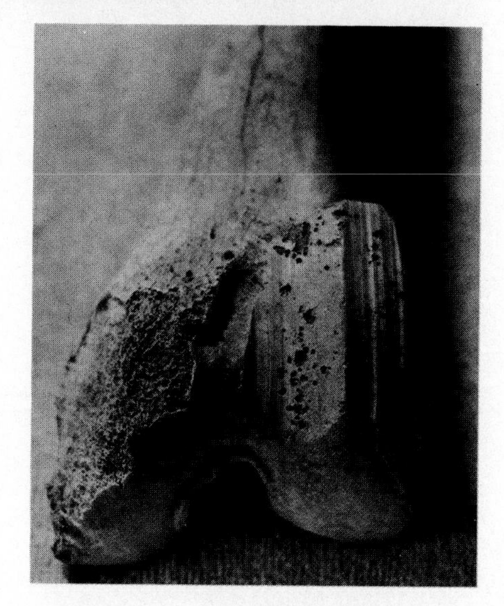

Although far separated geographically and culturally, adult skulls from predynastic Egypt and pre-Columbian Peru (top, left and right) share features of spongy bone development. Distinguishing among the causes of such abnormalities is difficult; possibilities include several kinds of anemia and nutritional disorders, including iron deficiency. End of femur from an ancient Peruvian (above) bears striking evidence of osteoarthritis. Grooved, polished appearance was caused by bone-on-bone friction.

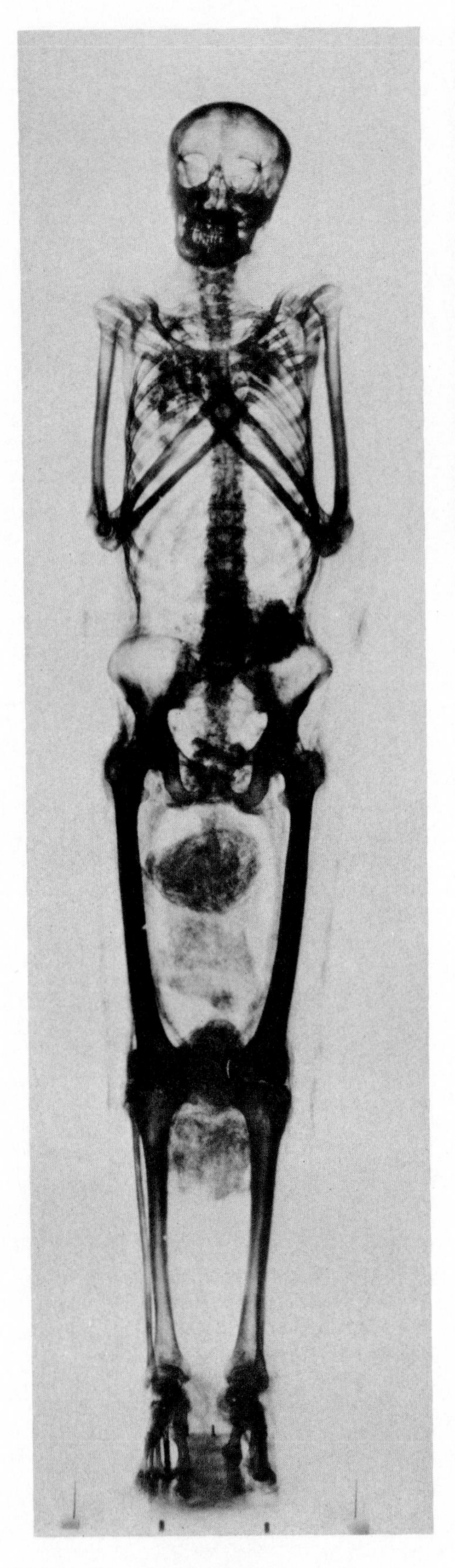

Well-preserved mummy of Harwa (above), an Egyptian who lived about 3,500 years ago, shows marked narrowing of hip and knee joints under X-ray examination (left), a symptom of a genetic disease called alkaptonuria. Presence of an abnormal dark pigment in the mummy's hip bones confirms the diagnosis. Serendipitously embalmed head of a 2,000-year-old Dane (below), complete with facial hair stubble, resulted from natural tanning action in a peat bog. Although these cases are dramatic, they add little to an overall understanding of health and disease in ancient communities.

(Left and top) Courtesy, Field Museum of Natural History, Chicago; (bottom) Silkeborg Museum

New information came primarily from study of new populations rather than from discovery of additional diseases in populations already known.

While tabulating disease frequencies in different populations, however, physical anthropologists could study the geographic distribution of diseases, and they began to dwell both on the ecology of the diseases themselves and on cultural influences upon their distribution and prevalence. This led to study of the age structure of ancient populations and of the relative frequency of diseases in the two sexes and in different age classes. Thus, the focus of the discipline gradually changed from one exclusively on diagnosis of disease in individual remains to one of considering overall patterns of health and disease of entire communities. In a phrase, paleopathological techniques and data began to be used to study paleoepidemiology.

New concepts for old diseases

Since the end of World War II the widespread use of antibiotics and vaccines, the enormous advances in biochemistry and microbiology, and the growth of information about the nature of the immune system have all contributed to modern ideas of what an infectious disease actually is. Where it had once been possible to think in terms of a simple invasion or infestation of a host by a particular microorganism, it is now clear that the relation between a microorganism and its host population is quite complex.

Populations of microorganisms relate to and evolve in terms of populations of hosts, and the interaction between a particular infection and a host's immune system has important consequences for the rest of the host population, over and above the outcome of the particular infection. The virulence of a strain of disease-producing organisms, or pathogens, is a property of the interaction between that particular strain and a given host population. In the microorganism population, virulence is subject to an intense natural selection that tends to produce progressively less virulent strains as pathogen and host populations live and evolve together. From the point of view of the microorganism, the most successful adaptation is one that maximizes its ability to spread and reproduce while interfering least with the host population's general state of health. The host population is the microorganism's source of livelihood. Given enough time to work out a balanced relationship, microorganisms and their host populations can be expected to coexist, with the microorganisms present at some level that balances their immediate tendency to infect new hosts against their long-term dependence upon the unimpaired survival of the host population. (*See* Feature Article: SYMBIOSIS AND THE EVOLUTION OF THE CELL.)

The perception of disease as an ongoing and dynamic interaction between populations of pathogens and hosts is underlined by the results of antibiotic therapy against some infectious bacteria. Rather than producing complete extermination, the use of antibiotics sometimes simply selects for an antibiotic-resistant bacterial strain. Furthermore, the fact that some bacterial strains can transfer genes for this resistance to other strains suggests the existence of balancing mechanisms between disease organisms as well as between disease organisms and hosts that act to maintain their patterns of coexistence even in the face of changed environmental circumstances.

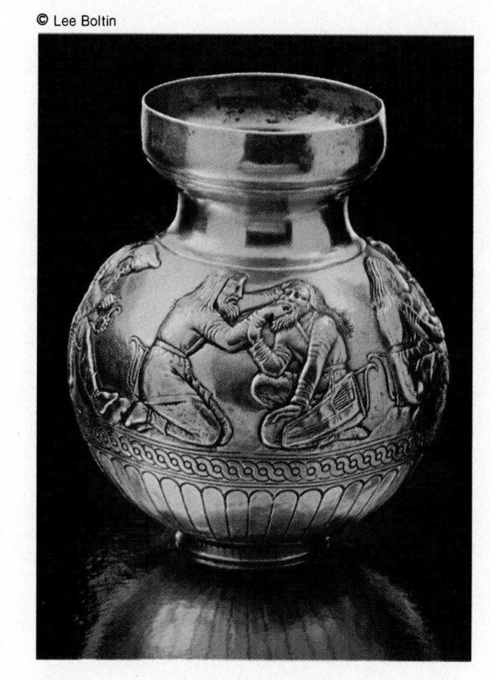

Gold bottle from Scythian burial of the 4th century BC (above) carries a mythological scene depicting human pain and the human impulse to relieve the sufferer. Green fluorescent glow appears in section of bone taken from a Nubian cemetery of the 4th to 6th centuries AD (below), an indication that its owner had been exposed to tetracycline, perhaps as a result of eating grain contaminated with Streptomyces *microorganisms. Natural exposure of ancient populations to antibiotics could well have affected the evolving relation between these communities and their diseases.*

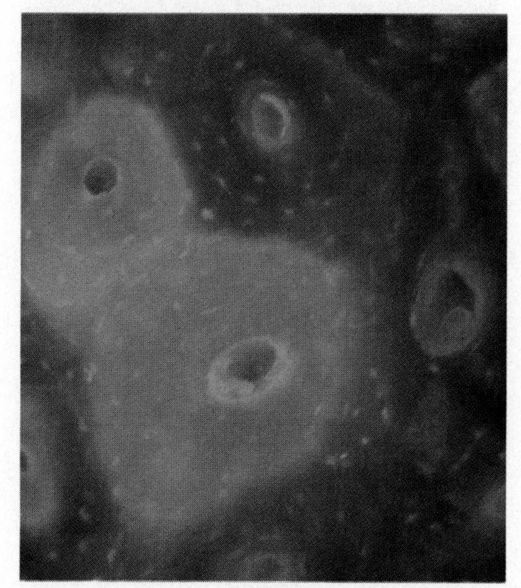

Everett J. Bassett and Margaret S. Keith

© Dan McCoy—Black Star

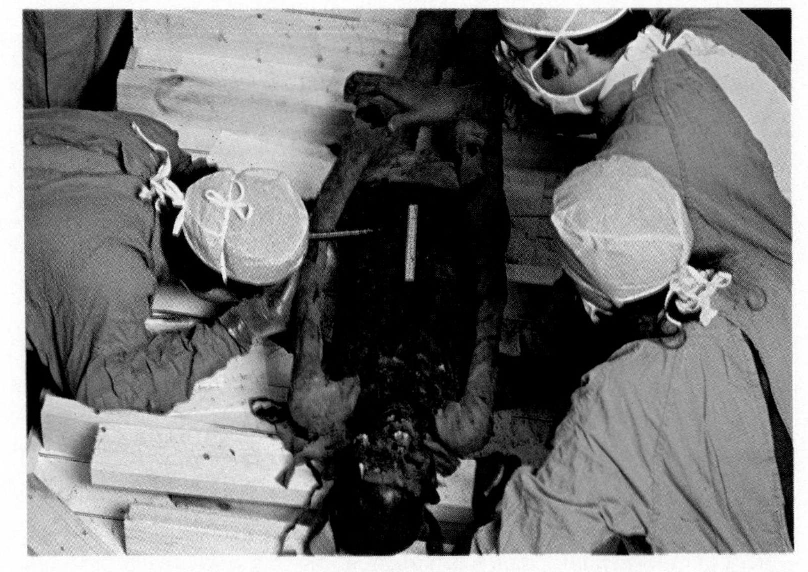

Medical investigator uses sigmoidoscope to see inside the abdomen of an Egyptian mummy autopsied at Wesleyan University in Connecticut in 1979. The examination showed the body to be that of a young adult in good general health, with bones and teeth indicating excellent nutrition. An inscription on the bandages identified the person as a teacher of grammar or literature. Masks worn by the investigating team are to prevent inhalation of infectious mold from the bandages.

Vertebra of Caribou Eskimo (below) bears lesions that were shown by means of differential diagnosis to have been caused by tuberculosis. Differential diagnosis chart (facing page) organizes patterns of expression for more than a dozen diseases which affect the skeleton in a way that identifies each disease or group of closely related diseases with a unique set of diagnostic characteristics. Tracing a pattern of disease found in a given skeletal population through the network of variables yields a conclusion representing the most likely cause of the abnormalities.

Jane E. Buikstra, Northwestern University

The relevance of all this for the study of paleopathology and paleoepidemiology is that both the specific response of individual hosts to a disease and the overall interaction between populations of host and pathogen are now known to be important in interpreting the symptoms of a disease in ancient humans. Knowledge of the "typical" response of host to pathogen is as important in diagnosing a case as the specific set of symptoms seen in that case.

It is commonplace in modern diagnostic practice to examine many different aspects of the body's response to a disease in order to narrow the range of possible causes of a given set of symptoms. In so doing, a physician will often run tests of both general physiological states and very specific biochemical or physiological factors. The physician will also check for involvement of various organs and tissues and determine the locations of all symptoms that might be involved in a particular case. For example, presence of local inflammation and swelling with no fever leads to a different conclusion than a general swelling of lymph nodes accompanied by a low-grade fever. Again, localized aches and pains in a limb suggest a different cause than general body aches.

Although paleopathologists cannot run tests of physiological responses in their "patients," they can use clinical observations and autopsy materials from present-day humans to build up an overall profile of the patterns of expression of different diseases both in individuals and in segments of populations. For example, some degenerative bone diseases tend to afflict particular parts of the body while others are more generally distributed. Again, some diseases are common only in certain age or sex groupings. Such profiles can then be compared both with the pattern of disease seen in individual skeletons or mummies and with the way that symptoms are distributed among the various members of the population. They often make it possible to distinguish among diseases in prehistoric populations that would be very difficult to separate on the basis of gross appearance or tissue analysis alone.

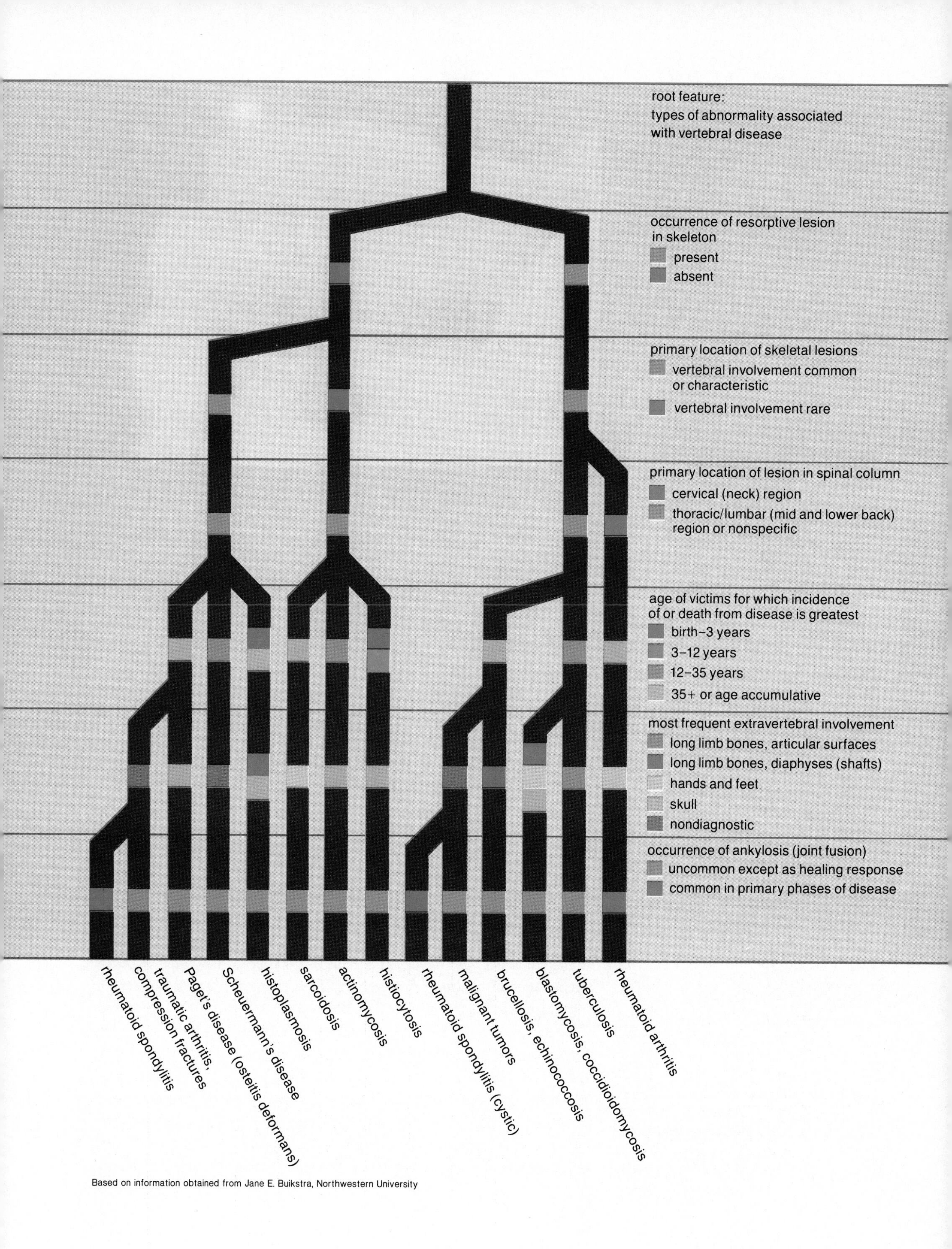

Based on information obtained from Jane E. Buikstra, Northwestern University

Prehistoric human bone is sawed into samples for tests intended to determine nutritional intake. Nutritional studies, coupled with information derived from investigations of environmental stress and from direct estimates of the effects of diseases besetting a community, may eventually allow researchers to develop a comprehensive picture of the general state of health of that community.

In the mid-1970s anthropologist Jane Buikstra of Northwestern University, Evanston, Illinois, made one of the first attempts to apply this principle of "differential diagnosis" in a formal and rigorous way to paleopathology in a study of resorptive lesions—*i.e.*, defects owing to reabsorption of material—in the bones of Caribou Eskimo skeletons. Buikstra used data on the patterns of expression of more than a dozen diseases that affect the human skeleton to conclude that the pattern seen in her population of Caribou Eskimo burials most closely conforms to that produced by tuberculosis. This study allowed far greater confidence in the conclusion than would have been possible without recourse to differential diagnosis. It also graphically demonstrated that tuberculosis was apparently a rather common affliction of Caribou Eskimos.

(Facing page) X-rays of the left and right femurs (right, top and bottom) of an Indian from the Archaic Period of Kentucky (8000–1000 BC) reveal series of closely corresponding Harris lines. Lines farthest from the ends of the bones formed first, when the bones were short. X-ray (left) made during investigation of an unusual case of arrested growth in the right leg of a contemporary child two years after a trauma offers a dramatic example of Harris line formation. Pair of arrows in the right-hand image—of the left knee—point to Harris lines laid down in the left femur and tibia following fracture of the child's right femur. Their distances from the epiphyseal growth surfaces near the bone ends reflect a normal two-year growth period since the fracture. In the left-hand image—of the right knee—almost no growth is seen in the femur and tibia. The Harris line that formed in the right tibia, noted with an arrow, has barely moved away from the growth surface.

Buikstra's study thus provides a striking example of the fact that disease can be and should be considered a normal aspect of ancient human communities. Such studies may eventually go much further toward clarifying the history of tuberculosis in humans than would ever have been possible using only older techniques. Moreover, the ability to assess the nature and prevalence of tuberculosis—or any other disease—in a given community also makes it possible to consider more general aspects of that community's state of health. If the effect of even some of the diseases besetting a community can be estimated directly and if such factors as nutrition and environmental stress can be studied, a relatively comprehensive picture of the general state of health of members of that community can be developed.

The importance of considering disease to be a normal part of human existence was emphasized from another point of view by a study of the paleopathology of Neanderthal skeletons found in France that are more than 50,000 years old. In 1957 William Strauss of Johns Hopkins University, Baltimore, Maryland, and A. J. E. Cave of St. Bartholomew's Hospital Medical School, London, showed that the old "Alley Oop" caricature of Neanderthals as stoop-shouldered, hunchbacked, incompletely bipedal, and somewhat apelike was due in part to earlier workers' failure to realize that the skeletons upon which this reconstruction was based showed evidence of extreme osteoarthritis and joint degeneration. In this dramatic and unusual case failure to appreciate the diseases present in a population led to a fundamental misunderstanding of the nature of an entire stage in human evolution. Taking account of the abnormalities present, Strauss and Cave demonstrated that the skeletons of Neanderthals, excluding their skulls, were quite similar to those of modern humans.

Skeletons and stress

Studies of modern human skeletons have revealed certain features that have been interpreted as general indicators of stress. Stress in this case refers to any factor or combination of factors that disturbs an individual's normal physiological equilibrium. Because it has not yet been possible to link most of these features conclusively with specific causes, they are termed nonspecific indicators. Physical anthropologists are now studying the occurrence of Harris lines, patterns of dental defects, and various aspects of demography as nonspecific indicators of stress in ancient humans.

Harris lines are transverse patterns of increased bone density near the epiphyseal surfaces (regions of growth) near the ends of long bones, and they are usually taken to indicate past interruptions of normal growth. Thus, any factor that can interrupt growth, such as seasonal famine, bouts of infectious disease, periods of psychological stress, or periodic dietary deficiencies, might produce Harris lines.

Harris lines have been interpreted in different ways in different studies, although they seem most often to correlate with periodic dietary stress in ancient populations. In the study of Caribou Eskimos mentioned above, Buikstra observed many Harris lines and interpreted them as probably having been due to seasonal famine. On the other hand, Calvin Wells, a British physician and paleopathological specialist, suggested that the higher frequency of Harris lines in female skeletons than in male skeletons from Anglo-Saxon burials may indicate preferential treatment of males during

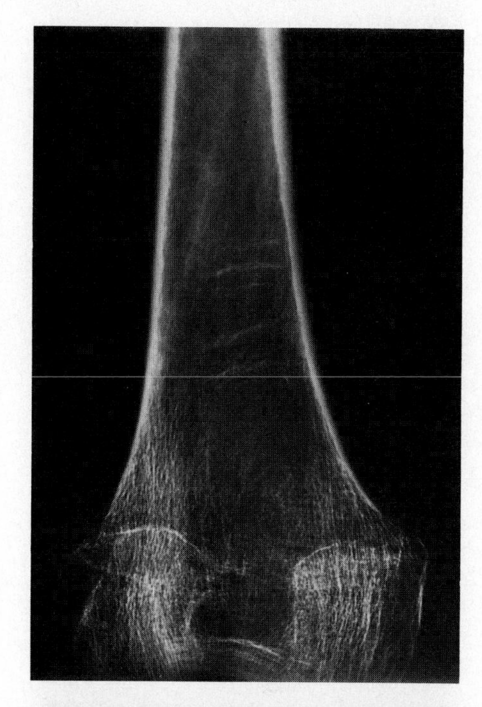

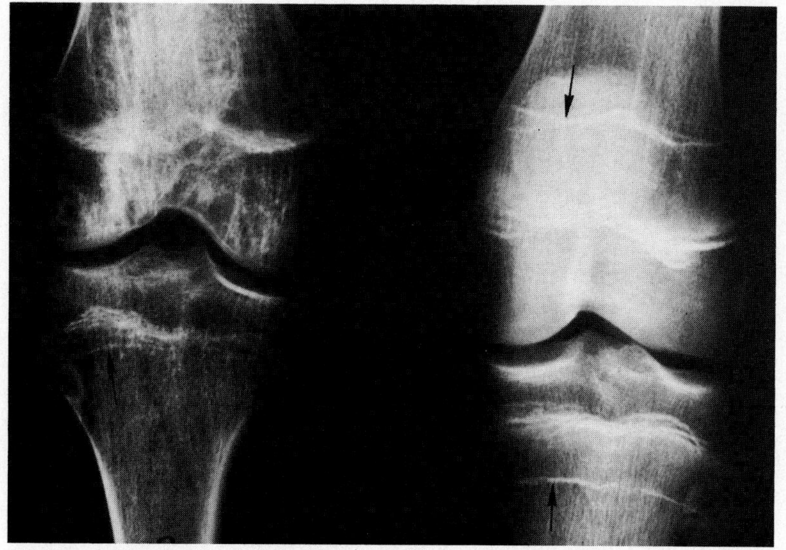

A. K. Poznanski, "Journal de L'Association Canadienne des Radiologistes," vol. 29, March 1978

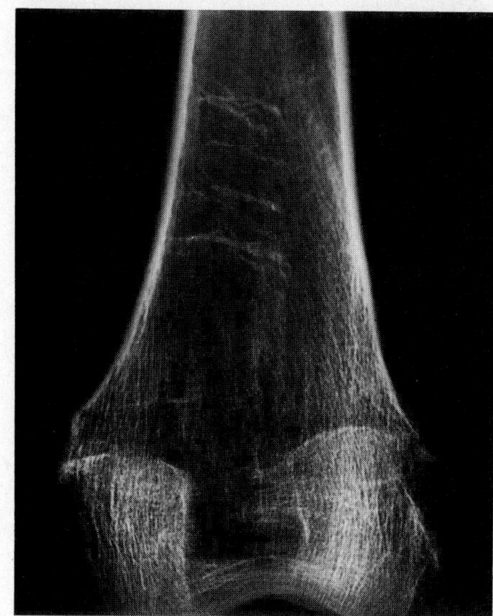

Photos, courtesy, R. Ted Steinbock, Harvard Medical School and Massachusetts General Hospital

Photograph of Indian skeletal population of the Mississippian culture (AD 800–1500) records its original lie during excavation of the Dickson Mound in Illinois.

infancy and early childhood. There is some evidence that Harris lines form most often during early stages of life when growth rates are at their maximum and that lines formed later in life are more likely to be resorbed than those formed earlier, although Harris lines can form throughout life.

Dental defects ranging from plainly visible pits in enamel to microscopic abnormalities in the patterning of enamel crystals have also been studied in modern humans as nonspecific indicators of stress. Large defects that occur together with abnormal crystalline structures and growth lines in dental enamel probably indicate disruptions of metabolism during the time that the tooth crowns were forming, and some types of defects can be associated with specific nutritional problems or diseases. Interpretation of dental defects in ancient humans is still at an early stage, but simple counts of the frequencies of defects of different types can be used as a gross measure of stress. Patterns of associations between different types of defects in individual skeletons and the distribution of dental defects by age and sex within a community may eventually allow diagnosis of specific diseases.

Overall growth rates and the average gross size of skeletons provide another nonspecific indicator of stress. Studies on living populations have shown a clear relationship between overall health, skeletal growth, and adult size as well as between nutrition, skeletal growth, and adult size. Since dental development responds less readily to stress than does bone development, stages of dental development can be used to determine developmental age and thus to calibrate the degree of bone and skeletal development. Consequently it is often possible to obtain data on average body size for different age groups and for the two sexes in ancient communities.

Most studies of skeletal growth and development have been simple comparisons between groups, but some recent work has tried to use average growth patterns as a means of comparing the degrees of dietary or disease stress in populations that succeeded one another in time in the same geographic region. For example, the average rate of growth in the skeletons of Indians living in the central Illinois River region decreased after the cultivation of maize (corn) became important in that region about AD 1000. This suggests that the switch to maize in that time and place was accompanied by increased dietary stress, increased incidence of disease, or both.

Life tables, which simply group numbers of deaths in a population by age, are also used to investigate stress in modern human communities. There are important problems that must be overcome when compiling a life table from a death assemblage. One is the tendency to underrepresent younger groups owing to the greater durability of more mature skeletons. Another is the tendency to overlook such cultural practices as the killing of female babies or the differential treatment of the bodies of some age groups that would affect their representation in the death sample. A third is the possibility of introducing sampling errors due simply to chance. Errors may also come about owing to population growth or decline; for example, in an expanding population young age classes will be relatively overrepresented in the population and hence also in the death assembly. Nevertheless, life tables can be important, useful tools in assessing the health of a population if they are used carefully. For instance, an unusually high number of deaths at weaning age might indicate increased nutritional or disease stress in a community, since infants lose both an important source of nutrition and some antibody protection when they are no longer receiving mother's milk. An odd pattern of deaths may indicate an epidemic, mass accident, or other catastrophe. A significant lengthening or shortening of average life span in a region over a long time span probably indicates a change in the average level of stress experienced by the population. When used with other nonspecific indicators of stress and with all the data on specific diseases and causes of death that can be gathered for the community, the life table can provide a framework for assessing the overall health status of its members.

Increased nutritional or disease stress in an ancient community may manifest itself as a rise in the number of deaths of children of weaning age. Stirrup-spouted water bottle is from the Mochica culture in Peru, about 250 BC to AD 750.

Isotopes and trace elements

Depending on their environment and species, plants fix, or incorporate, carbon from atmospheric carbon dioxide during photosynthesis via differing metabolic pathways. In particular, maize and millet fix carbon by means of a series of biochemical reactions called the C_4 pathway, while most of the terrestrial plants of the eastern part of the U.S. fix carbon through the C_3 pathway. One of the important differences between the two systems is that they fix different ratios of the natural stable isotopes of carbon, carbon-12 and carbon-13, present in carbon dioxide. Thus, these ratios differ in the tissues of C_3 and C_4 plants, and animals that feed upon those plants (or upon animals who have fed upon them) retain those carbon isotope ratios in their own tissues. By studying such ratios in a protein—collagen—in prehistoric bone, it is possible to assess the proportion of the animals' diet that was derived from each of the two types of plants.

Courtesy, R. Ted Steinbock, Harvard Medical School and Massachusetts General Hospital

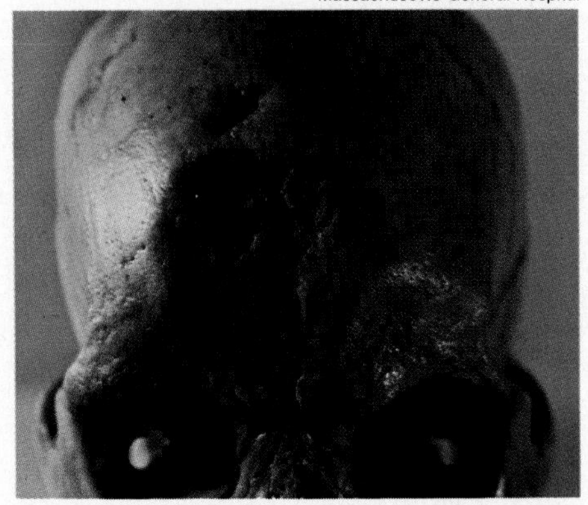

Skull with frontal lesions caused by syphilis is of an adult male Indian from Arizona and dates from about the 16th century AD. Paleopathological work has documented past distributions of syphilis in pre- and post-Columbian times.

In 1978 South African investigators Nikolaas van der Merwe and J. C. Vogel showed that the carbon isotope ratios in collagen from human skeletons in the eastern woodlands of the U.S. change markedly, beginning at about AD 1000. They interpret the change as having been due to the introduction of maize as a significant dietary item in the region at that time. As was mentioned above, this dietary change is accompanied by a rise in several indicators of stress in those populations.

The carbon isotope ratios in marine food chains is also different from that in C_3 plants, and so it may eventually be possible to use them to document the relative importance of marine resources in prehistoric diet. This technique is in an early stage of development, however, and many problems must be worked out before it can yield widespread reliable results.

Trace elements, those that make up less than 0.01% of the body, can also provide evidence of ancient diet. For example, levels of nonradioactive isotopes of strontium tend to be higher in the bodies of plant-eating animals (herbivores) than in those of meat eaters (carnivores). Strontium levels in skeletal remains can thus be used to gauge the consumption of animal protein. But there are complications here, too. Some types of freshwater shellfish contain high levels of strontium and could distort strontium readings if these shellfish were a significant item in a given diet. And, because it is not yet clear exactly why strontium levels are lower in carnivores than in herbivores, other complicating factors may well be discovered in the future. Other trace elements are being investigated as measures of different dietary constituents and also as indicators of specific diseases. Zinc deficiency, for instance, may well be implicated as a factor in various kinds of sudden, severe bone infections. These investigations eventually may yield specific information on aspects of disease and diet that can not now be studied in ancient humans.

Analyses of carbon isotope ratios and trace elements have already been used to attempt to document differences in diet between populations of different ages in the same region, between populations of similar antiquity in different regions, and between different age and sex groupings within populations. Preliminary results suggest that dietary differences were probably greater than anyone previously thought. As the techniques are further developed, more and more data on specific aspects of paleonutrition should be forthcoming.

Photomicrograph of diseased lung tissue reveals deposits of fine sand characteristic of sand pneumoconiosis, a hazard shared by both ancient and modern humans living in regions prone to frequent sandstorms. Although the disease has been detected in Egyptian mummies, its prevalence in ancient populations is not known.

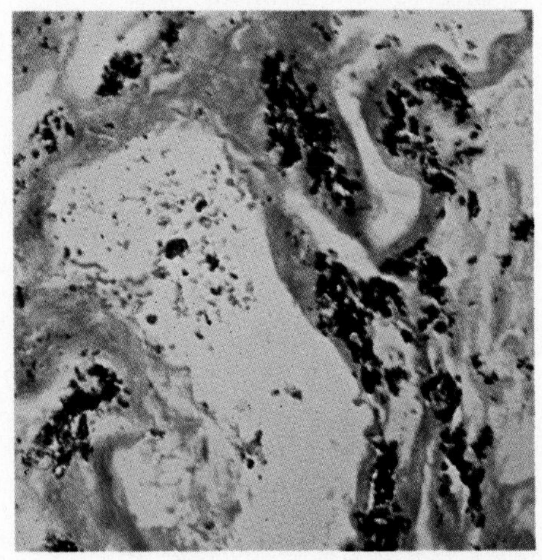

Courtesy, P. J. Turner, East Birmingham Hospital

The future

The work of the past few years in paleopathology and paleoepidemiology has begun an entirely new stage in the study of ancient disease, health, and stress. Where once scientists worked to diagnose particular diseases in individual specimens and to generalize statistically to communities, they are now able to assess the general state of the health of ancient communities directly, using nonspecific indicators of stress. This information then can help to interpret the patterns of disease that are present and to consider the relative importance of nutrition and disease in the health of the population. Analyses of trace elements and carbon isotope ratios, once they are more widely applied to the study of ancient diets, will further this effort.

Paleopathological work in the past century not only has enlightened humans about their own history but also has had some practical benefits for the modern health sciences. In documenting past distributions of such diseases as tuberculosis, as discussed above, and of syphilis in pre-Columbian times, it has provided modern epidemiology with information obtainable in no other way. As its quality improves and as it draws on techniques from an ever-widening range of disciplines, it should have even greater relevance in the future. Paleopathological results can supplement historical and clinical information concerning the evolutionary history of various diseases as they have run their courses in human communities. For example, it eventually may be possible to learn a good deal about the importance of malaria in ancient populations from the study of bone lesions related to hereditary anemias that give protection from malaria.

It is very important, however, to bear in mind some of the limitations on the kind of information that can be obtained reliably from paleopathological studies. Even when such studies yield very specific results and conclusions, it may be both difficult and misleading to generalize from them. From the study of mummies, for instance, it is known that atherosclerosis and sand pneumoconiosis (a lung disease caused by inhalation of fine sand) occurred in ancient Egyptians. But without information about the frequency and distribution of those afflictions in the overall population and about other health and nutritional factors, it is risky to say much beyond noting the presence of the diseases in each mummy in which they occur.

Perhaps the most promising and exciting contributions paleopathologists can expect to make to the modern health sciences are insights into the "normal" relationships between ancient human populations and their diseases. Their work can be invaluable in clarifying the balance between human diseases and populations that existed in the past—before modern medicine, the environmental changes wrought by the industrial revolution, and the removal of geographic barriers to disease dispersal brought about by modern transportation made it virtually impossible to find either a disease or a human population sufficiently isolated and undisrupted to study. Clearly such information would be of great value to epidemologists attempting to understand the effects of environmental pollution, medical intervention, and human mobility patterns on the current state of human health and disease.

Meso-American ceramic figurine from Tlatilco in Mexico, dating from 1700 to 1300 BC, personifies the dualities of death and life and of disease and health that have concerned human beings since long before the attentions of modern medicine.

FOR ADDITIONAL READING

Jane E. Buikstra and Della C. Cook, "Paleopathology: An American Account," *Annual Review of Anthropology 1980,* pp. 433–470 (Annual Reviews, Inc., 1980).

Aiden Cockburn and Eve Cockburn (eds.), *Mummies, Disease, and Ancient Cultures* (Cambridge University Press, 1980).

Saul Jarcho, *Human Paleopathology* (Yale University Press, 1966).

R. Ted Steinbock, *Paleopathological Diagnosis and Interpretation* (Charles C. Thomas, 1976).

N. J. van der Merwe and J. C. Vogel, "^{13}C content of human collagen as a measure of prehistoric diet in woodland North America," *Nature* (Dec. 21/28, 1978, pp. 815–816).

Greetings from
ISLAND
for

SCIENCE
Described as "a small body of land surrounded
by rich blue opportunities," a private venture
into alternate sources of food, energy,
medicine, and recreation is offering
impoverished Caribbean nations a measure of
self-sufficiency.
by Neil P. Ruzic

Photos, © Nicholas DeVore III—Bruce Coleman Inc.

Although blessed with a bounteous ocean, dependable winds, and a sunny sky, the people of the Caribbean depend heavily on imported fuel, food, and fresh water. Energy scarcities in the region contribute to high unemployment and feelings of hopelessness.

The Caribbean Sea is a painting in tranquillity. The ocean flashes in the short end of the visible spectrum: violet, deep blue, and turquoise streaked with the white of wind on water. Pastels reveal shallow banks. Islands of all sizes and shapes poke their ragged brown heads out of the sea.

That tranquillity is surface-deep. The Caribbean today is rife with revolution, drug rings, and piracy. Yachts are seized, their owners killed, the boats used for smuggling. The alcoholism rate is among the highest in the world. The people of the islands are restless.

An overall thermodynamic explanation, if such can be applied to human activity, is that the entropy of the region is increasing too fast. Entropy is a measure of the amount of energy no longer capable of conversion into work. The Caribbean derives all of its electrical and locomotive energy from burning imported petroleum. Most of the food and much of the fresh water also must be imported. As the dissipated energy, or entropy, increases, available energy decreases. It becomes ever harder to make a living. Anxiety and attitudes of hopelessness lead to alcoholism, crime, and revolution.

The problem certainly is not unique to the Caribbean. The entropy of the whole world is increasing, together with its by-products: pollution, crime, scarcities, inflation, and political instability. While perhaps not the complete answer to these evils, finding enough energy for the foreseeable future will attack the root cause of the problem. The worldwide answer, to the extent that technology can contribute to social problems, is to substitute unlimited energy sources for limited sources. Rich, populous nations can develop fusion energy as their ultimate goal, meanwhile conserving and exploring for more fossil fuel. But this is not a solution for poor countries.

Surprisingly, with all its problems, the Caribbean is in an especially favorable position to make the transition to unlimited energy that all nations ultimately must make if they are to survive. First, the population pressures of such countries as India and Mexico are absent. Race relations are among the best in the world. Second, traditions of democracy and free enterprise

NEIL P. RUZIC is President of Island for Science Inc., Beverly Shores, Indiana, and founder of Industrial Research *magazine. He is the author of eight books on the practical application of science.*

Illustrations by John Youssi

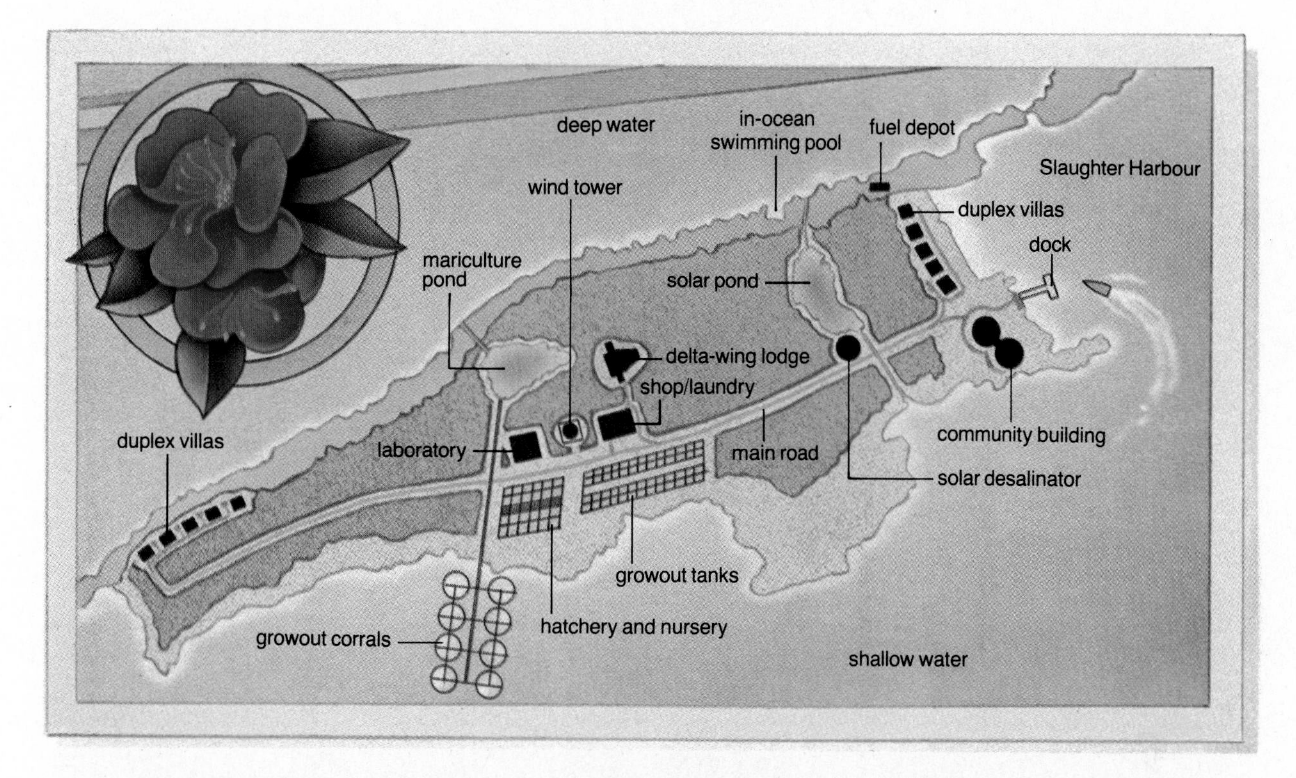

are well established. Communist countries like Cuba are exceptions rather than the rule. Third and most important, the natural resources of the region, while largely unrecognized by the governments in the Caribbean, are among the most abundant in the world. Although the region is devoid of such limited energy sources as oil, coal, and uranium, it is rich in unlimited energy sources: plentiful sunlight, trade winds, and productive warm seas.

It was for these reasons that a site near the Caribbean, in The Bahamas, was chosen as the home of a unique enterprise. This project, named the Island for Science (IFS), has been under construction since May 1979. As it grows it will become a technological base for the region where practical ideas for energy and self-sufficiency can be worked out, demonstrated, and later applied to improve the human condition. Its working philosophy is that a single major technological breakthrough can mean more to humanity than all the laws and treaties between nations.

The Bahamas was picked as the IFS base for several reasons. It is physically similar—with islands in a warm sea—and geographically adjacent to the Caribbean. Yet it lacks most of the Caribbean's problems. The government is a stable democracy with close ties to the U.S., Canada, and Western Europe. The country has a high literacy rate. Its population of only 232,000 will help it avoid the future instabilities and political strife of more crowded nations. And, like IFS's founders, the people speak English.

The geography

The Island for Science is called Little Stirrup Cay on navigation charts. It is 198 kilometers east of the south Florida coast, roughly between Nassau and Freeport (one kilometer equals 0.62 miles). Only about a kilometer and a half in length, the island is unique. It has deep water on one side and

Map (above) depicts general features of Little Stirrup Cay and locates IFS's proposed and completed construction. Local legend holds that Blackbeard (below) buried treasure on the island, and the name of its harbor, Slaughter Harbour, seems to betray some dark pirate conflict.

The Bettmann Archive

Photos, Neil P. Ruzic

shallow on the other. A high central ridge reaches 18 meters (one meter equals 3.28 feet). The island contains two small lakes, freshwater wells, plentiful vegetation, a natural harbor, and some of the best fishing in the world. In addition, it is near the island of Great Harbour Cay, which has both a work force that commutes to the island daily and a commercial airport.

It takes about three hours to walk around Little Stirrup Cay. The interior is a lush, subtropical jungle filled with thatch palms, gum elemi, buttonwood, Australian pines, berry trees, and wild tamarind. The animal life consists primarily of birds and crabs. Gulls, terns, pelicans, shearwaters, and sand pipers patrol the shallow-side ocean. Wood ducks inhabit one of the lakes. Mockingbirds, mourning doves, and white owls live in the jungle. Ghost crabs dig holes on the beach, and large land crabs, whose legs make excellent eating and fortunately regenerate, stride along the road at night.

Little Stirrup Cay divides deep from shallow water, as does the entire 45-kilometer-long Berry Island chain of which this island is the northernmost. The deep side is eight meters deep right up to the rocky coast and about 800 meters deep a kilometer and a half out to sea. The shallow side, consisting of numerous coves of white-sand beaches rimmed with coconut groves, is waist deep. The twice-daily tide is one meter.

In the interior, near the east end, there are thousands of meters of "slave walls" or fences winding endlessly into the jungle. They were built by slaves before England freed them in the 1830s. The islanders of years past, as those of today, ignored the potential of their main asset, the sea, as a place for cultivating crops. Instead, they made their slaves build unmortared walls of native stone. These became corrals for pigs, goats, and sheep in unsuccessful attempts to transfer mainland agriculture to these islands.

Off the deep side of the island the sea has dug a dungeon into the side of the coast, lighted underwater only by blue-scattered rays from the Sun. Giant groupers live in this submarine cave, guarded by a living curtain of thousands of sardine-like fish. Barracuda, parrot fish, rays, and nurse sharks are also about. Fanning out from the cave lies a vast field of coral heads, exposing giant "brains" and sprouting "staghorns."

Construction at the east end of the island (facing page, top) includes a dumbbell-shaped community building, concrete riprap dock, and a row of five duplex villas. Slaughter Harbour lies in the foreground, and one of the island's two lakes is visible behind the villas. Community building (bottom, right) houses a kitchen, library, and two dining rooms: one enclosed with glass doors for nighttime use and a second, open area for buffets. Interior of enclosed dining room (bottom, left) features a central fireplace for cooking and a copper hood, which will be lowered into place during later construction.

Construction and funding

Virtually all of these physical assets contribute to IFS's pursuit of alternate sources of food, energy, medicine, and recreation. The seaside cave is the site for an underwater habitat to aid in screening marine organisms for pharmaceuticals. The depths far off the northern coast allow the production of energy and fresh water using equipment that exploits the temperature differences between surface and deep ocean water. At least one of the island's lakes will become a maturation pond for cultured seaweed, fish, shrimps, and other marine crops. The central ridge of the island is suitable for several types of structures that focus the wind to generate electricity. Even the harbor and scenic beauty have unique uses: the east end of the island is being developed into a daytime cruise-ship port of call where guests can view wind and solar demonstration projects, eat the produce of the shrimp farm, sun and swim in the shallow ocean, and contribute both funds and publicity to the enterprise. Beginning in late 1981 the island will be visited twice weekly by a large cruise ship leaving from Miami.

Under construction on the multipurpose island are roads, several complexes containing five villas each for staff and tourists, a large dock with slips for boats, and a fuel depot. In addition, there is an office and laboratory building; a community building containing dining room, library, and kitchen; a shop and laundry building; a multipurpose tower for wind generators, observation, and water pressure; and a seafarm hatchery, nursery, and growout facility.

Neil P. Ruzic

Gravid female shrimp carries thousands of eggs on her body. Whereas in nature perhaps a dozen offspring at most will reach adulthood, on a controlled seafarm 60–95% of the spawn will mature.

Funding, which by early 1981 stood at $2.4 million, has come equally from Charles C. Worthington and Neil P. Ruzic. Each is a 50% shareholder in Island for Science Inc., a for-profit U.S. corporation registered to do business in The Bahamas. Worthington is founder of Worthington Biochemicals Corp., now a subsidiary of Millipore Corp. Ruzic is founder of *Industrial Research, Oceanology,* and other scientific magazines; his publishing company is now a division of Dun & Bradstreet Corp.

Seawater farming

The water surrounding IFS is far away from industry, and the ocean is unpolluted. These conditions contrast markedly with those of the Gulf of Mexico, one of the world's largest natural shrimping grounds, where oil spills and industrial pollutants are adversely affecting the ability of shrimps and fish to spawn. Because of such contamination, most commercial shrimp and fish farms are placed in South America or Asia, away from industry. Most of these seafarms, however, treat their three-dimensional ponds or tanks like a field and grow a single crop, a practice that wastes the majority of the volume. When a mariculture system is intended for high profit, its planners inevitably are led to polyculture, the simultaneous farming of two or more kinds of organisms. After all, the same capital investment can be amortized over additional cash crops, and if chosen correctly, these crop organisms can contribute to the well-being of each other.

Borrowing from centuries-old Chinese aquaculture, in which bottom, mid-level, and top dwellers are grown in a polyculture finfish system, the IFS plan is one of intensive, controlled culture in large, eight-meter tanks on the island with fresh seawater pumped in and out. In winter a large roof of translucent fiberglass will cover the tanks, hundreds of them eventually, to provide solar heating. In summer an opaque white material will cover the fiberglass for shade. The roof will double as a rain catchment for the island's freshwater needs. One or both lakes on the island will be used as additional sites for seaweed, fish, and other mariculture combinations. Corrals in the open shallow water will be built later. Costs and growth rates of these three growout methods will be compared and expansion planned accordingly.

Prototype breeding trays and hatchery, nursery, and growout tanks soon will be built on the island. Some tanks will contain only one crop such as shrimps or seaweed. Other tanks will be used for polyculture: conchs on the bottom for their meat and shells, large shrimps (prawns) at all levels, herbivorous finfish called tilapia or vertically supported trays of oysters in the middle zone, and seaweed attached to monolines stretched across the tops of the tanks. Red seaweeds are farmed for their valuable gel extracts of carrageenan and agar. These gels are added as thickeners and stabilizers and for other purposes to thousands of food products.

The IFS polyculture concept is an augmentation of nature, whose plants and animals live in symbiosis. Red seaweed adds photosynthesized oxygen to the water and removes metabolic wastes from other organisms. In open-sea culture the seaweed also provides shade, hiding places, and protection from birds for the shrimps. Tilapia produce some of the nutrients required by the seaweed. Oysters, which are filter feeders, use shrimp feed that has dissolved in the water and otherwise would be wasted. Conchs contribute to the system by cleaning solid, potentially sludge-forming wastes that sink to the bottom.

The food web so designed is typical of ecological systems in the real world. Food webs in nature usually are organized into a hierarchy of three or four trophic levels. The first level consists of primary producers—fertilized phytoplankton, seaweed, and commercial shrimp feed in the IFS plan. The second contains herbivores—conchs, oysters, and tilapia. And the third level comprises carnivores—in IFS's case, shrimps. Shorter food chains in nature cannot fill available niches, whereas longer food chains, according to theories of evolution, may result in population fluctuations so severe that species at the highest trophic level cannot persist.

To be economical the IFS system does not parallel nature exactly. Experiments have been conducted to determine which species of these various crops will emulate nature in a productive manner; the prototype described will determine which of several growout methods is optimum.

Shrimps for all seasons

Peneid shrimps, which can be grown from juveniles in only three months, were chosen over other, expensive crustaceans such as crabs or lobsters. The three-month period not only permits growing three to four crops a year but also allows fitting different species of shrimps to seasonal differences in temperature.

For instance, following a strategy developed by shrimp systems engineer Larry Elam of Panama City, Florida, IFS plans to grow *Peneus vannamei* and *P. monodon* in the half year that includes summer, since these species grow well at 35°–24° C (95°–75° F), and *P. stylierostris,* whose growth range spans 32°–20° C (90°–68° F), in the winter. In colder climates *P. aztecus,* a brown shrimp, could be grown in the winter since it prefers temperatures of 26°–16° C (79°–61° F) and will survive, but not grow, at temperatures as low as 5° C (41° F).

P. vannamei, a large, aggressive, extremely hardy white shrimp, will be grown in the same tanks with *P. stylierostris,* which is almost as hardy, when the temperature permits good growth, because competition between these two fighters enhances food intake and rapid growth. Both species come from South America, primarily on the Pacific side. *P. monodon,* the largest saltwater shrimp in the world and indigenous to the Indian Ocean, will be grown separately for comparative determinations of such qualities as feed-to-growth ratio over identical time periods and disease resistance in controlled growout conditions.

Species of *Tilapia* were chosen as the finfish candidate over more truly herbivorous saltwater varieties such as milkfish and mullet because of better taste and higher market prices. Tilapia, indigenous to the Near East, supposedly are the biblical fish that Jesus multiplied with the loaves of bread. Research conducted for IFS has revealed that male hybrid tilapia—various crosses between two species—grow the fastest. It also was found that both tilapia and shrimps grow faster with seaweed than without it. Moreover, it was discovered that one could prevent overbreeding in produc-

(Top) Roger Archibald—ANIMALS, ANIMALS; (center) © Tom McHugh—Photo Researchers; (bottom) Douglas Faulkner

Crassostrea oysters, red seaweed, and the queen conch (top to bottom) are included in one trilevel polyculture system planned for IFS.

tion pens—a major problem that produces unusable runts instead of full-grown fish—simply by growing them in seawater. Tilapia do not breed in saltwater. Researchers working for IFS learned how to grow them just as well in seawater as in fresh water and consequently avoided the overbreeding problems encountered elsewhere.

The species of red seaweed chosen for the IFS seafarm are rich in carrageenan and agar. These include several species of *Eucheuma, Hypnea,* and *Gracilaria.* They grow very fast, about 5% a day, in the warm, nutrient-rich waters of a polyculture system. Much of the expertise contributed to IFS's plans for seaweed has come from James DeBoer, a phycologist (algae specialist) at the University of Texas Marine Science Institute.

Several species of the genus *Crassostrea* will provide the oyster crop alternative to tilapia in some of the island's polyculture tanks. The pollution-free waters of The Bahamas eliminate a major problem of oyster culture in the U.S. Pollution is especially harmful in oysters because the animals concentrate pollutants in their flesh and so become dangerous for human consumption.

The native queen conch, *Strombus gigas,* completes the seafarm population. These beautifully shelled mollusks take three years or longer to grow to marketable size, but they are hardy and easily withstand being moved about while the faster growing crops are harvested. Especially perfect shells and occasional pink conch pearls can be sold in addition to conch meat, which is eaten in the Caribbean islands and Florida as "cracked conch" or conch chowder.

Focusing the wind

The natural resources of the region can be made to yield not only food but energy, and one such major resource is the wind. Just as optical lenses or mirrors focus diffuse solar radiation, diffuse wind energy also can be collected and focused. If the wind could be concentrated sufficiently to produce a small, controlled tornado, the output of conventional wind generators would increase by whole orders of magnitude.

Two wind-focusing systems are planned for testing and use on the Island for Science. One, called a toroidal accelerator rotor platform (TARP) by its inventor, Alfred Weisbrich of Windsor, Connecticut, is designed to form the top level of high-rise buildings, silos, water towers, or other tall structures. On the island three vertically stacked TARP units will be positioned atop a combination water and observation tower.

The hollow toroid of a TARP is shaped somewhat like a tireless automobile wheel. The hole in the center is positioned vertically for access from below. Two wind-turbine rotors spin in the outer channel of each platform formed by the flanges of each "wheel." The two rotors are linked and together move freely on a track around the platform as the direction of the wind changes. In operation the ambient wind is entrained and accelerated around the channel in which the rotors are placed.

Forcing the wind to flow through the channel increases its velocity more than 50%. Even a small increase in velocity results in a large increase in power because the kinetic energy in the wind increases as the cube of its

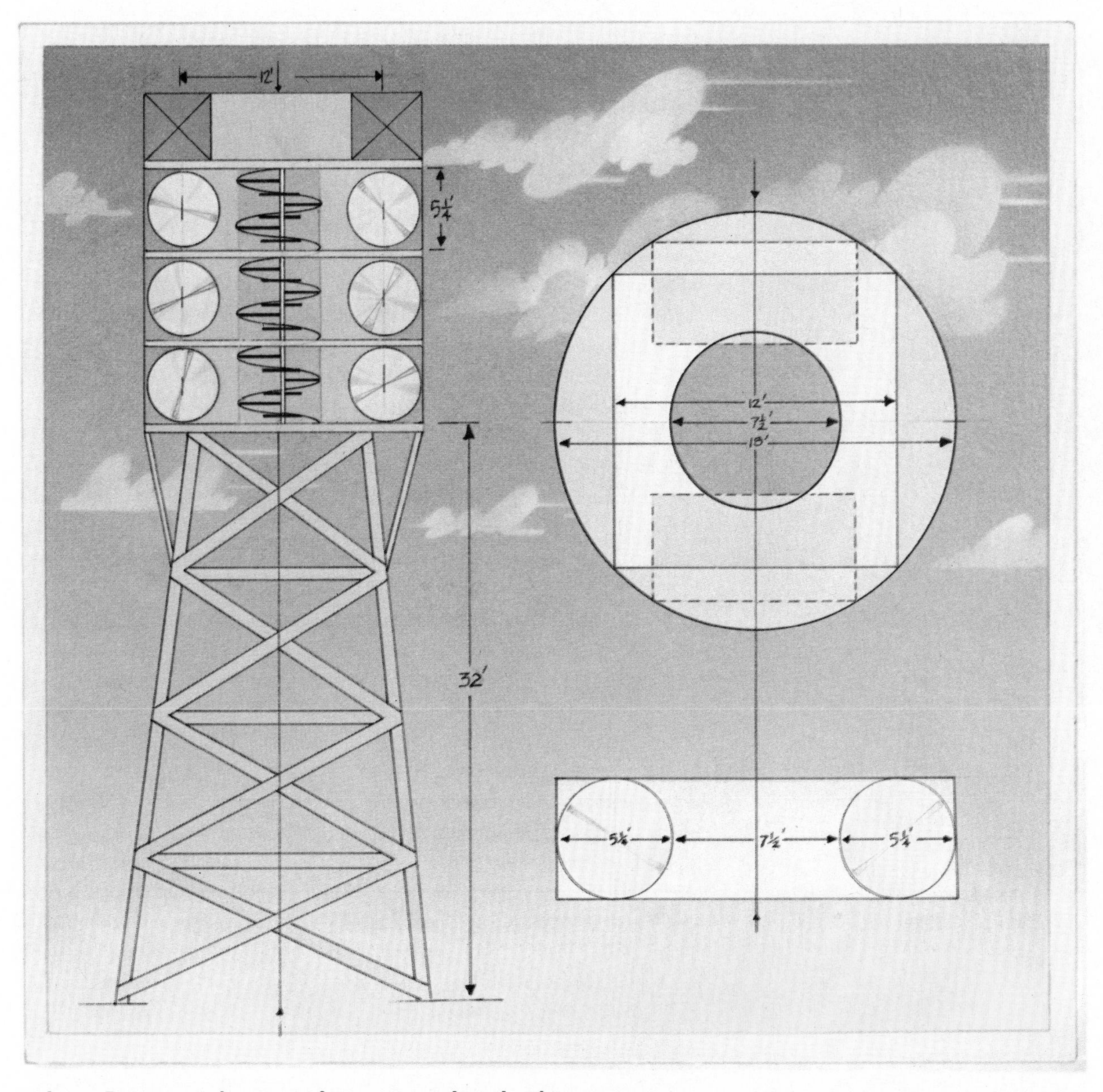

Design and approximate dimensions of combination water tower, observation platform, and TARP *wind-focusing system planned for* IFS *is sketched above. One foot equals 0.305 meters.*

velocity. By contrast, the power of a conventional wind turbine is proportional only to the square of the diameter of the rotor, and so it is clear that speeding up the wind pays greater dividends than increasing rotor size. Power augmentation factors of more than 4.5 have been demonstrated for a TARP system in wind-tunnel tests at Rensselaer Polytechnic Institute, Troy, New York.

Using small rotors means that propellers can be made simpler and stronger. For instance, there is no need to feather the propellers (turn the blade edges into the wind) in gales or hurricanes, so fewer moving parts are necessary. Feathering is not required because the linked-rotor unit automatically revolves in its tracks around the channel to place the propellers at right angles to the wind when a brake is applied to one of them above a set wind velocity.

Neil P. Ruzic

Section of main road across Little Stirrup Cay passes the site of IFS's combination water and observation tower, which will incorporate three vertically stacked TARP units. The tower's observation platform will rise more than 15 meters (50 feet) from the top of a 12-meter hill, for a view of the entire northern Berry Islands.

The tower installation planned for the island will incorporate three TARP layers with six 1.6-meter propellers. A free wind of 15 miles per hour (24 kilometers per hour) will produce 47 kilowatts; a 17-mile-per-hour (27-kilometer-per-hour) wind will produce 69 kilowatts.

Another way to focus the wind is to employ a delta wing, typically in the shape of an isosceles triangle. A lodge for the island has been designed with such a delta roof, the apex of which is 30°. The roof is inclined 22° to the ground, the point aimed up and into the prevailing east wind. When the wind blows over the delta roof, it creates vortices along the rear outer edges in opposite directions. It is in these miniature tornadolike regions that wind rotors are placed. The design not only increases the velocity of the wind, but it also channels the wind downward along the roof, thus eliminating the requirement for a tower.

Pasquale M. Sforza of the Polytechnic Institute of New York in Brooklyn has pioneered the delta wing, or vortex augmentor concept (VAC), for wind power. As a professor of aeronautical engineering he has taught that winged aircraft derive their lift from the lowered pressure created by vortices on the upper wing surfaces. Why not, he thought, use these vortices on a stationary wing for a wind generator?

The author has built several miniature buildings to fit under Sforza's wings. These were tested in the institute's wind tunnel before a full-size VAC building for the island was designed. Baffles to spill the wind in a hurricane and other design improvements were developed in this way. The wind-tunnel tests also showed that the apex of the delta wing could be cut off to form a long trapezoid without loss of efficiency.

The augmentation factor of the VAC is about the same as that of the TARP. In the Caribbean, where true trade winds blow from the east most of the time, a fixed VAC system such as a roof becomes feasible. Both systems serve to concentrate the energy of the wind into a small region, making much smaller turbines usable. Reductions in the size and weight of rotating parts mean lower construction costs and greater reliability. In addition, because the blades rotate faster than conventional wind generators, the rotational speed of the turbine is closer to the operational speeds of available electrical generators, lessening the requirement for step-up gearboxes.

Solar desalination

Like wind energy, solar power currently is wasted in the Caribbean. It is ironic that all of the energy in the entire region comes from burning petroleum, while the tradewinds blow and the Sun shines brightly almost every day. Most Caribbean nations also have chronic water shortages, again in the midst of a plentiful resource: the sea that surrounds them.

The pursuit of solar desalination thus becomes not merely another economic opportunity for technological innovators but a vital necessity. And, while The Bahamas and the Caribbean are of first interest to the Island for Science, effective solar desalination techniques are important on every continent. By the year 2000 the U.S. itself will require an estimated 110 billion liters (29 billion gallons) of desalinated water a day in addition to its existing freshwater supplies. On the Island for Science for only 3.5 days out of the

Delta-roof lodge designed for the island features a wind augmentation system that exploits basic principles of aerodynamic lift. The roof is inclined 22° to the ground, with its truncated apex pointing up and into the prevailing wind. Wind blowing across the roof creates vortices along its rear outer edges. In these tornadolike wind flows are placed small rotors for electric power generation.

year is visibility less than one kilometer. For an average of 130 days of the year clouds cover less than 30% of the sky, and for only 85 days do they cover more than 80%, normally when it is raining.

One IFS approach borrows from two known and fairly simple concepts: direct solar heating and vacuum distillation. In the first, solar radiation passes through a clear plastic cover and then through another clear film that floats on about five centimeters (two inches) of seawater. Sunlight is absorbed by a dark-colored layer of material beneath the seawater, and the resulting heat warms the water. The clear film on the water's surface prevents evaporation and thus eliminates a source of heat loss. Both smaller solar tents and larger solar ponds may be constructed along these principles. On the island one of the one-hectare (2.5-acre) lakes may be covered with a polyethylene sheet for a large-scale source of hot seawater.

Water heated in this manner to about 65° C (150° F) can be boiled without further heating in vacuum stills, which take advantage of the fact that liquids boil at increasingly low temperatures as the pressure on them is reduced. Hot water from plastic solar tents or solar ponds can substitute for more expensively heated water in conventional, electrically operated vacuum stills. An IFS cost analysis shows the hybrid system to deliver distilled water at less than 83 cents per thousand liters ($3.25 per thousand gallons) plus cost of electricity (which admittedly can be high in remote areas).

A more economical vacuum still has been invented by Philip Youngner, a physicist at St. Cloud (Minnesota) State University. Youngner, who plans

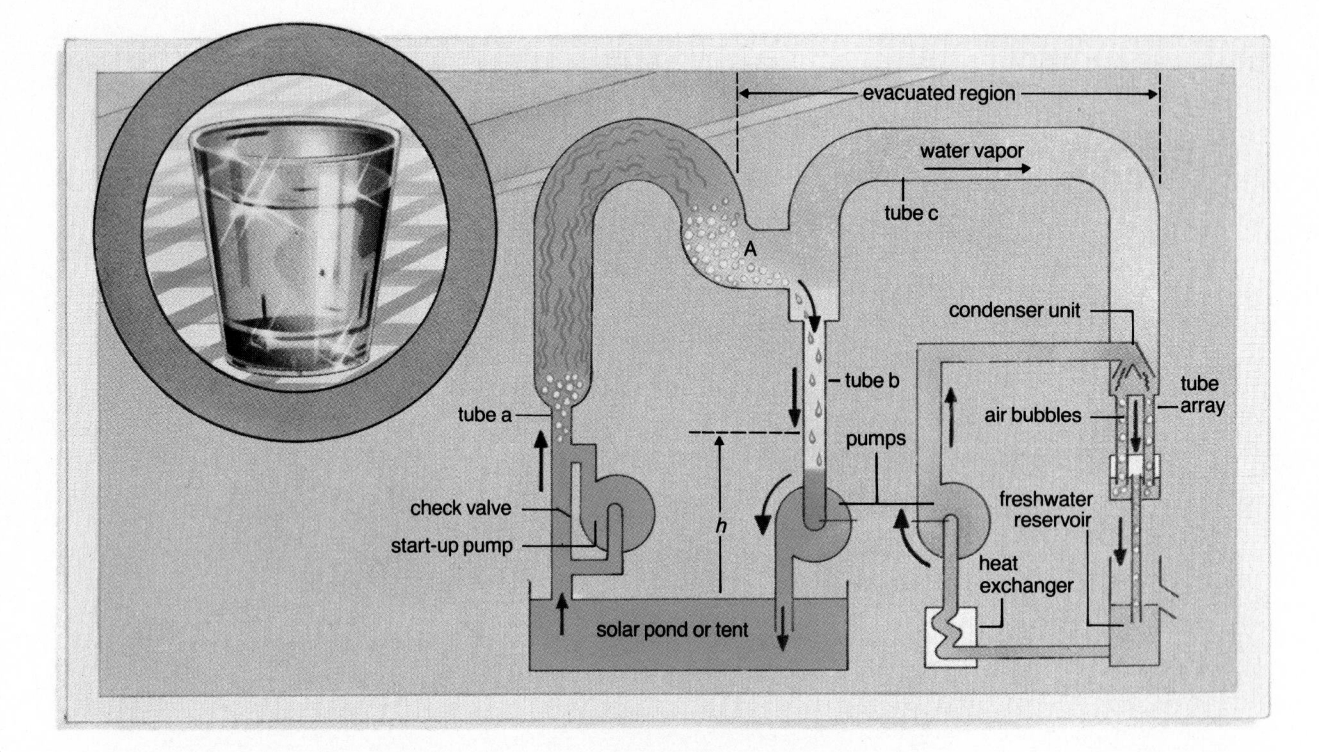

Vacuum still planned as a demonstration project for IFS (above) comprises two connected circulating loops. On the right a pump draws fresh water from a reservoir through a heat exchanger, where it is chilled indirectly with cold seawater, and into a condenser. The water then falls through an array of small tubes, trapping air bubbles and driving them down faster than they tend to rise. The vacuum thus produced in tube c causes the water in tubes a and b, on the left, to rise to height h. The start-up pump forces solar-heated water up tube a, over the gooseneck, and into the evacuated region at A, where some of the water evaporates. The temperature differential between the condenser and A drives a flow of vapor toward the condenser through tube c. The vapor condenses on the cold water flow in the condenser, adding fresh water to the reservoir.

(Facing page) Multistage OTEC plant envisioned for the island produces both electricity and fresh water. *See text on pp. 160–161.*

to build a demonstration model on the Island for Science, creates a vacuum in a vertical piping system by the falling motion of entrapped air bubbles in water. A conventional source of energy is required for the primary pump in the system, but only in start-up until a siphon effect is established. Afterward, little energy is required other than the warmth of the solar pond.

Youngner has built a small-scale pilot model that produces 75 liters (20 gallons) of distilled water per day with only a small expenditure of energy. His is the least expensive method IFS has seen to produce distilled water. Using standard amortization rates and following construction of a demonstration unit with a capacity of a thousand liters or more per day, IFS estimates that distilled water may be produced for less than 52 cents per thousand liters ($2 per thousand gallons) including electricity.

Ocean thermal and wave energy

Another alternative energy concept, called ocean thermal energy conversion (OTEC), can produce both electricity and desalinated water. The major prerequisite is a nearby deep subtropical ocean where the bottom water is at least 15° C (27° F) colder than the Sun-warmed surface water. The temperature difference between the deep-side bottom water $1^1/_2$ kilometers from IFS, where the depth is 800 meters, and the shallow-side surface water, where the prototype seafarm is to be located, ranges from 17° C (31° F) in the winter to a munificent 29° C (52° F) in the summer.

The OTEC plant envisioned as an IFS demonstration project is similar to a steam plant; the major difference is that the energy to vaporize the water comes from the water itself rather than from burning fuel. Again, as with the desalination system described above, the method used to boil warm seawater—typically at 28° C (82° F)—is low-pressure vaporization. To make boiling practical at this temperature, the considerable amount of air that is normally dissolved in seawater must be removed; at IFS this will be accomplished in a multistage, energy-efficient process that is still undergoing refinement by developers J. Hilbert Anderson and his son, James Anderson, of Sea Solar Power Inc., York, Pennsylvania. Once deaerated the warm water flows into a vacuum chamber where a fraction of the water evaporates to low-pressure steam. The steam is passed through a turbine, which extracts energy from it in the form of rotary motion. Afterward, the now somewhat cooler vapor flows across a condenser surface chilled by cold bottom seawater. Fresh water that drips from the condenser is stored.

Not only can the Island for Science use the electrical and freshwater output of an OTEC plant, but the seafarm can benefit from two other byproducts. The deep water pumped up for the OTEC plant is rich in nutrients that can replace fertilizer in the seafarm; the cold water also promotes seaweed growth during hot summers.

Wave power is another project for IFS investigation. A floating, hinged raft moored to the seabottom in the water 8–12 meters deep right off the northern coast could be used to power a pump when the seas are rough. Such a wave pump could run a reverse-osmosis desalination unit or a refrigeration compressor by direct hydraulic energy. A prototype system planned for IFS will have a raft three meters in diameter and should produce as much as 5,700 liters (1,500 gallons) of fresh water per day or cool three tons of fish.

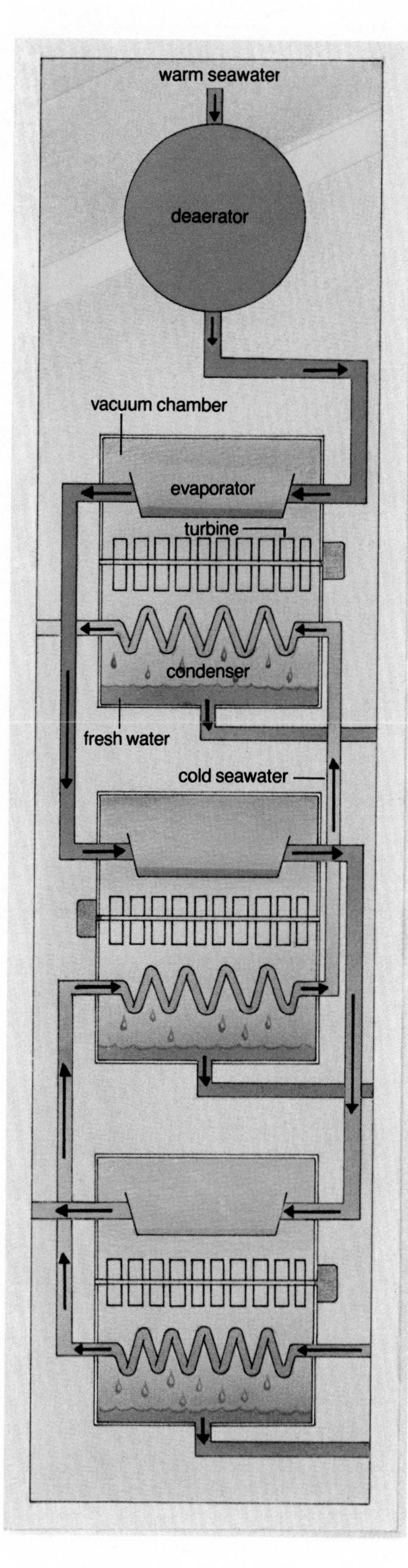

Pharmaceuticals from the sea

Another virtually limitless resource of the ocean is its prolific variety of life. Some 80% of the Earth's animal life, representing more than a half-million species, live in the water. Thousands of marine organisms are known to contain toxins, enzymes, steroids, amino acids, and other useful substances. Yet less than 1% have been examined for pharmacologic activity.

Although marine pharmacology is still in its infancy, recent work has uncovered a variety of anticancer substances in sea invertebrates; insecticides; antibiotic activity in sponges and coelenterates; a substance from a tunicate that possesses antitumor and immunosuppressive activity; many toxins from sea cucumbers, sponges, sea slugs, and cyanobacteria (blue-green algae); and biologically potent substances called prostaglandins from the common sea whip, *Plexaura homomalla*, which grows all over The Bahamas. Extracts from red seaweed of many of the species to be cultured at IFS for commercial extraction of their gels are known to possess antibacterial and antifungal properties. University of California researchers have discovered that a polysaccharide in the seaweed blocks the site in human cell membranes at which viruses normally enter the cell. The drug is effective in stopping viral multiplication in types I and II herpes infections.

Marine species are so extremely plentiful and incredibly diverse in their chemistry, morphology, function, adaptability to environment, and evolutionary progress that they represent a rich natural reservoir for new drugs—in fact, the richest on this planet. Yet testing hundreds of thousands of marine organisms for possible pharmacologic activity poses a formidable

(Top) Jane Burton—Bruce Coleman Inc.; (center) Bob and Clara Calhoun—Bruce Coleman Inc.; (bottom) Zig Leszczynski—ANIMALS ANIMALS

Tim Rock—ANIMALS ANIMALS

challenge. One inexpensive random screening, or "heterotargeting," assay system planned for the Island for Science was worked out by Ivor Cornman while at the Marine Biological Laboratory, Woods Hole, Massachusetts. Cornman's method uses nine living organisms as a battery of screens or "biofilters." These include *Artemia,* or brine shrimp; the sea urchin *Diadema* and its eggs; the small mosquito fish *Gambusia;* gram-positive and gram-negative microorganisms and a mold; and two varieties of plant seeds.

A sample of a marine plant or animal to be assayed is ground up and extracted with water or an organic solvent. A dilution of the extract is then placed with each of these living biofilters and various effects noted. The sea urchin, for instance, is responsive to substances affecting the human nervous system. Effects on the cell division of urchin eggs by a marine extract may reveal a member of one of two large groups of anticancer drugs. Effects on the fertilization mechanism of these eggs may signify a potential human contraceptive or an antiviral agent. If brine shrimp are killed or their normal activity disrupted, the extract most likely contains a potent insecticide. The microorganisms reveal antibiotic activity. The seeds yield information pertinent to economical poisons.

Using the method described, Cornman has been able to screen thousands of substances for further study in laboratories of sponsoring drug companies. His results to date are impressive. Fully 95% of the organisms shown by his biofilters to be active subsequently have been shown active in mammals in one way or another.

A variety of challenges

Other potential enterprises abound in IFS's subtropical environment. One is raising rhesus monkeys in the island's jungle. Rhesus monkeys are in increasing demand for medical research even as the supply from India diminishes. Monkeys raised in the free-ranging state, as opposed to indoor cages, are healthier both physiologically and psychologically.

Biological insect control also can be investigated more easily on a small, isolated island than on mainlands where ecological factors are too numerous. In many cases chemical insecticides have not effectively controlled insects owing to the occurrence of resistance in the intended targets. One modern strategy uses insect hormones to prevent larvae from maturing into reproductively active adults or, in the case of mosquitos, to cause them to die in their pupal stages. Sex pheromones, chemicals emitted by insects to help them identify mates, can be synthesized to confuse insects and prevent mating. If insect "languages" can be learned, radio or chemical signals may be able to disorient such social insects as termites, wasps, or ants, inducing them to destroy their own colonies. Experiments along these lines will be readily isolated from control groups by the island itself.

Among other opportunities on the Island for Science are developing solar air-conditioning, raising pearl oysters, producing barnacle cement as a commercial adhesive, investigating saltwater food crops both on land and in the ocean itself, conducting archaeological treasure-diving expeditions, training the dolphins that abound in the region, educating students, capturing and culturing tropical fish, and establishing a collection center for marine biosamples. The last would include moray eels, rays and skates, stonefish, and other exotic creatures for schools and research laboratories.

The challenges are illimitable. An island for science is a small body of land surrounded by rich blue opportunities. The idea of "your own island" is hardly new. It seems almost ingrained in the minds of retirees. The problem such people face in realizing that dream, either alone or with a few others, is what to do after they have made their island paradise as self-sufficient as possible—a job, incidentally, more suited to a mechanic than to the typical retired executive who can afford it. Thousands of retired executives-turned-yachtsmen end up docked in the world's tropical harbors seeking some of the excitement and challenge of the society they left behind. An island for science, as opposed to an island for oneself, provides the kind of stimulus most humans require, a blending of worthwhile goals, personal freedom, practicality, and romance:

> Follow the white sand beach
> Around a tropical island,
> But reach for the infinite sea,
> Else end, where you started.

The living ocean is an enormous reservoir of biologically active substances and potential new drugs. Although marine pharmacology is still in its infancy, work in the past two decades has uncovered useful toxins from shell-less gastropods commonly called sea hares (facing page, top left), toxins and antileukemic agents from sponges (top right), and heart stimulants from the sea anemone Anthopleura xanthogrammica *(center left) and from the skin venom of toads, including the marine toad* Bufo marinus *(bottom left). An inexpensive assay system for screening marine organisms for biological activity has been planned for IFS. Another IFS project involves establishing a collection center for moray eels (below), rays and skates, mollusks, starfish, lobsters, and other marine animals valuable in research and education.*

A NEW WORLD OF GLASSY SEMICONDUCTORS

After weathering a decade of controversy, amorphous semiconductors have begun yielding their basic secrets. Concurrently they have found niches in such diverse fields as solar energy, computer memory, and photography.

by Arthur H. Seidman

The word semiconductor may summon to mind such well-known devices as the transistor and the integrated circuit (IC), a tiny chip of material containing thousands of transistors. Both devices have been used in countless products ranging from electronic equipment aboard the Voyager space probes to the hand-held calculator that is having prodigious effects on human life-styles. Pivotal to the operation of transistors and IC's is the element silicon (Si), specifically silicon in crystalline form. Recently, however, considerable progress has been made in using silicon in its noncrystalline, or amorphous, state for solar cells. Because crystal growing and other processes allied to crystalline silicon technology are eliminated, amorphous silicon solar cells promise to be less costly, making them especially attractive as a possible alternate energy source.

In crystalline silicon the atoms of the element are arranged in a well-defined geometric pattern, or lattice, that is repeated throughout the crystal. This structure is distinguished by a long-range periodicity of atoms in their crystal sites, a characteristic important to the rapid advances made in semiconductor devices in the past few decades. Quantum theory, initiated in 1900 by the theoretical physicist Max Planck and developed during the 1920s by Erwin Schrödinger and Werner Heisenberg, is the key in dealing with semiconductors and such other solid-state devices as the laser. Application of quantum theory to crystalline silicon, with its periodic structure, permits an order of mathematical simplification that makes it relatively easy to understand the electronic behavior of silicon-crystal semiconductors. On the other hand, amorphous semiconductors do not exhibit the periodic structure of crystalline silicon. Whereas atoms in a silicon crystal are oriented in an array of rows and columns, the atoms in amorphous silicon occupy random positions. At best the latter substance may possess some short-range order that does not persist for more than a few atoms in any direction within a sample of the material.

Test bed holds 729 amorphous silicon solar cells arrayed for performance checks at RCA Laboratories in Princeton, New Jersey.

In addition to silicon, solid-state physicists have shown interest in another kind of amorphous material called chalcogenide glass. This is glass contain-

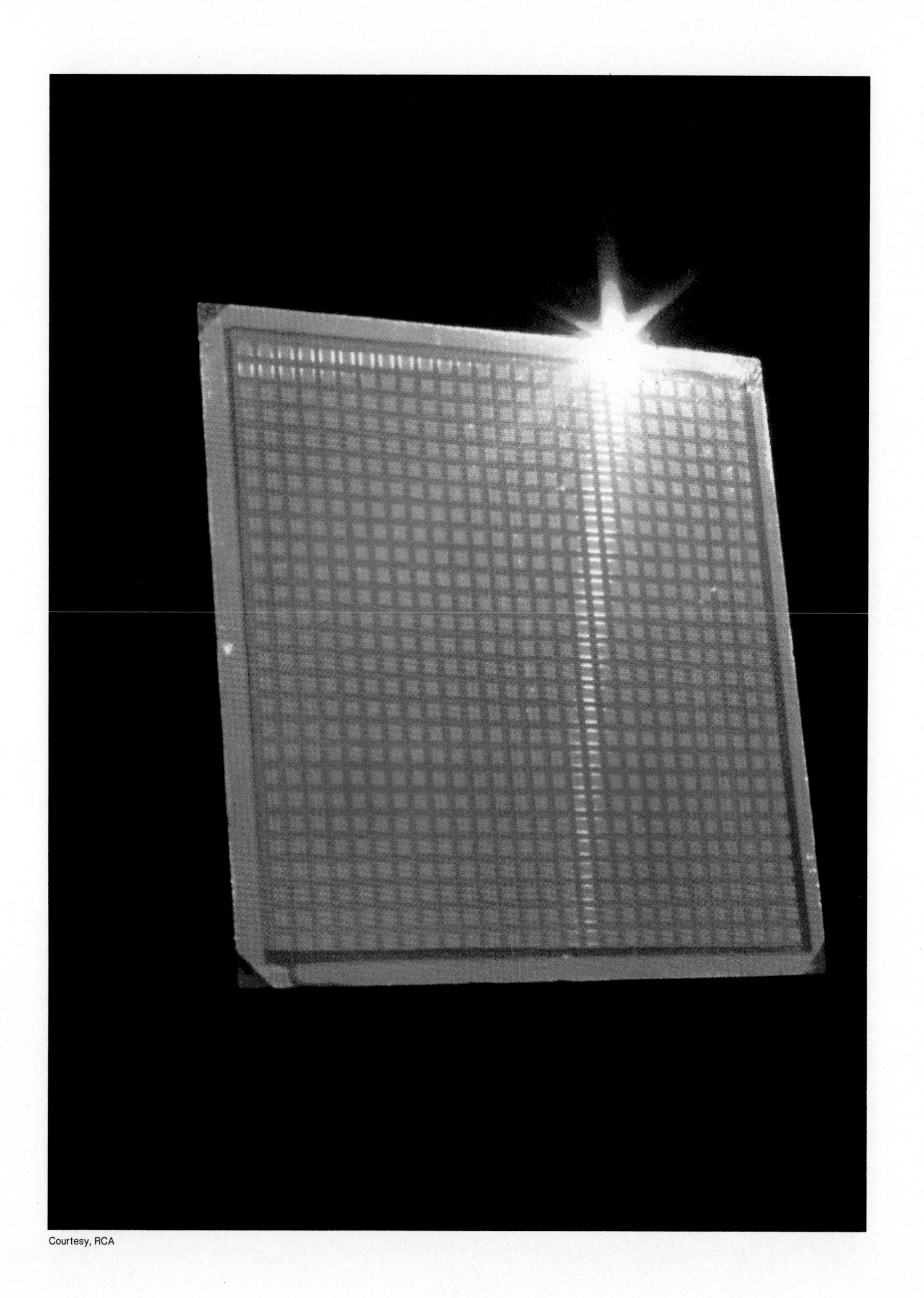

Courtesy, RCA

ing a large percentage of one or more of the chalcogen elements—*e.g.*, sulfur, selenium, or tellurium, which reside in column VI of the periodic table of chemical elements—linked with such other elements as arsenic, germanium, and silicon. In 1958 Stanford R. Ovshinsky, a controversial self-educated physicist and founder of Energy Conversion Devices Inc. (ECD), in Troy, Michigan, began working with amorphous glass. The results of his efforts, eventually published in the erudite *Physical Review Letters* in 1968, created a furor among workers in the field. Many thought that the glass switch described in his paper would challenge the supremacy of the conventional switch made of crystalline silicon. Others questioned the merits of Ovshinsky's claims. Soon after its publication in the *Letters*, Ovshinsky's work made the front pages of the *New York Times*, the *Wall Street Journal*, and other major U.S. newspapers.

In some quarters expectations ran high that a new revolution in electronics was in the making. But it never materialized. For one reason, it was difficult to understand the exact nature of the switching effect in glass described by Ovshinsky in this paper. Quantum theory is not easily applied to the disordered structure of amorphous materials. Another reason for the declining interest was the host of significant improvements made in the processes used for manufacturing transistors and IC's based on crystalline silicon. These improvements led to greater yields, increased reliability, and lower costs for these devices.

In the 1970s workers in the U.S. and other countries began making progress in understanding the behavior of amorphous semiconductors. Their efforts reached a milestone in 1977 when Nevill F. Mott of the University of Cambridge and Philip W. Anderson of Bell Laboratories in the U.S. shared the Nobel Prize for Physics for their work on amorphous materials. Although experimental work in the field had taken place in the early 1970s, once a clearer fundamental understanding of amorphous materials was established, workers were encouraged to apply this new knowledge to practical applications. Owing to the energy shortage, one important application has been the development of inexpensive solar cells. Others include computer memories and photography. After a decade-long hiatus, scientists may be on the brink of a new technology having many interesting ramifications.

Laying the groundwork

To understand the differences between crystalline (also called single-crystal) silicon and amorphous silicon (a-Si), one must first consider the atomic structure of crystalline silicon. Located in the fourth column of the periodic table, silicon is an example of a tetravalent element. A tetravalent element possesses four valence electrons, ones that can be shared with other atoms to form chemical bonds. Other examples of tetravalent elements include carbon in the form of diamond and germanium.

In absolutely pure, or intrinsic, silicon covalent bonds are formed by the sharing of its valence electrons among neighboring atoms (*see* figure 1a). Because of these covalent bonds silicon at temperatures near absolute zero (−273° C, or −460° F) acts as an insulator. There are simply no charged carriers available for the conduction of electric current. At temperatures

ARTHUR H. SEIDMAN *is Professor of Electrical Engineering at Pratt Institute, Brooklyn, New York.*

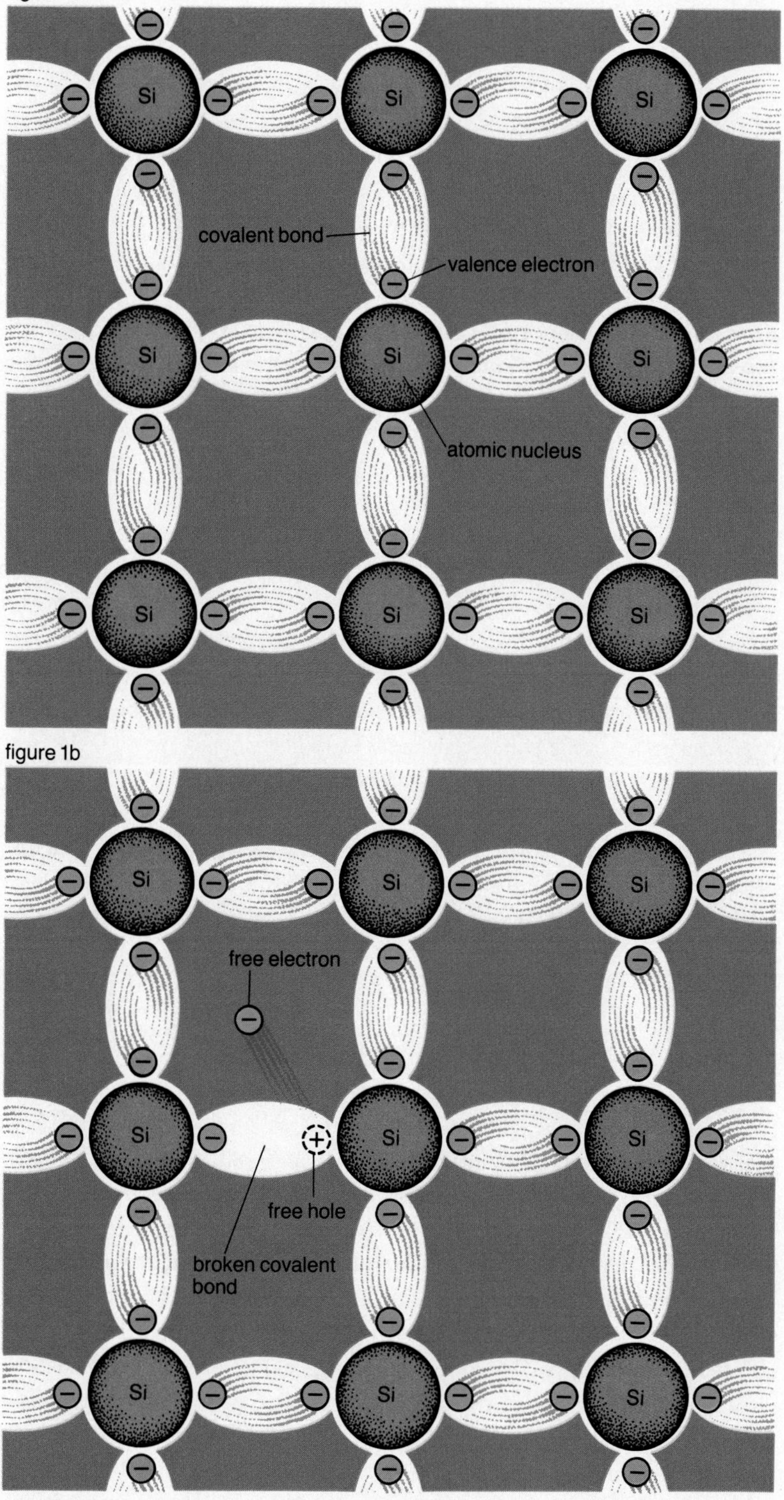

Two-dimensional schematics depict covalent bonding in crystalline silicon. At temperatures near absolute zero (a) virtually all valence electrons are shared among neighboring atoms, and no current carriers are available. At warmer temperatures (b) two types of carriers appear: an electron and the hole it leaves behind, which together are called an electron-hole pair.

figure 2

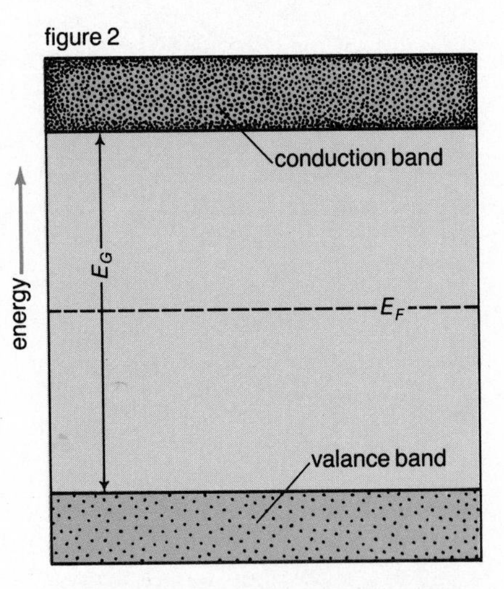

Energy band diagram for intrinsic silicon places the Fermi level, E_F, midway in the energy gap, E_G, between the valence and conduction bands.

One technique for doping crystalline silicon makes use of a diffusion oven, in which silicon wafers are heated close to their melting point in an atmosphere containing atoms of the impurity to be introduced. The impurity condenses on the silicon surface and tends to diffuse into the crystal, changing its surface layer into a doped region.

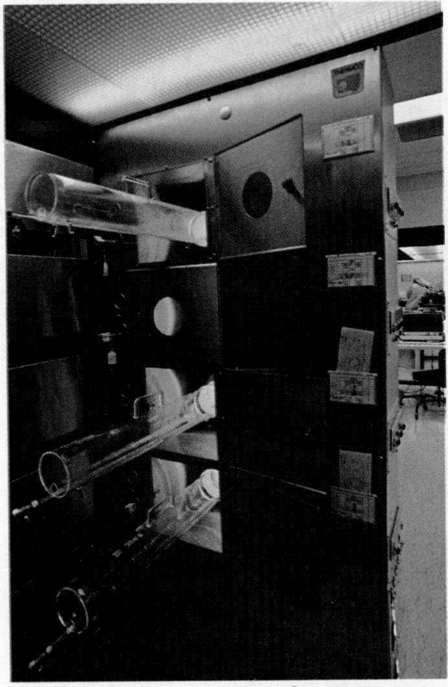

somewhat above absolute zero some of the covalent bonds are broken. When this occurs, two types of current carriers appear: an electron and a hole, which together are called an electron-hole pair (figure 1b). The hole, previously occupied by an electron, is essentially a vacancy. Nevertheless, quantum theory allows the hole to be considered an actual entity, a positive charge equal in magnitude to the negative charge of an electron.

To understand this behavior more thoroughly, it will help to introduce the important concept of the energy band diagram. Such a diagram for intrinsic silicon is illustrated in figure 2. According to quantum theory, valence electrons united in covalent bonds can possess only specific quantities of energy in a certain energy range; another way of expressing this idea is that they must occupy specific energy levels contained in the valence band of energies. For an electron to reach one of the allowed energy, or quantum, states in the conduction band of energies, where it becomes available as a carrier of current, it must acquire an energy equal to or greater than that of the energy gap, E_G. For crystalline silicon the value of E_G is 1.1 electron volts (eV), in which one electron volt equals 1.6×10^{-19} joules of energy.

At room temperature some electrons in the valence band of energies acquire enough energy (owing to the ambient thermal energy in the room) to leap over the energy gap and occupy one of the available quantum states in the conduction band of energies. When this occurs, electron-hole pairs are created. If a voltage is impressed across the silicon, both carriers—electron and hole—would contribute to the resulting current flow.

According to the tenets of quantum theory no electron can find itself in the energy gap, aptly called the band of forbidden energies. The dashed line labeled E_F, called the Fermi level, is a probability function. For intrinsic silicon it is located in the center of the energy gap. It signifies that the probability of finding an electron or a hole is 50%; that is, for each electron in the conduction band there exists a hole in the valence band of energies.

Doping

Intrinsic silicon cannot be fashioned into useful devices like the transistor. What is required is the addition of impurities, called doping, such as phosphorus (P) or boron (B) to intrinsic silicon. Phosphorus resides in column V of the periodic table and therefore contains five valence electrons; it is called a pentavalent element. When a minuscule amount, on the order of one phosphorus atom for every 100 million silicon atoms is introduced, intrinsic silicon is transformed to n-type silicon, which is rich in free electrons (figure 3a). Because of the extremely small doping ratio, when an atom of phosphorus assumes its position in the silicon crystal, it will be surrounded by four silicon atoms. One of the five valence electrons of phosphorus consequently is not united in a covalent bond and is free to carry current.

If boron, found in column III of the periodic table, is added to intrinsic silicon, free holes are created (figure 3b). Boron has three valence electrons and is referred to as a trivalent element. When it takes its place in the crystal, it forms only three covalent bonds; the missing bond is a hole. Silicon doped with boron is rich in holes and is called p-type.

In doped silicon the Fermi level shifts from its midposition in the energy

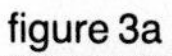

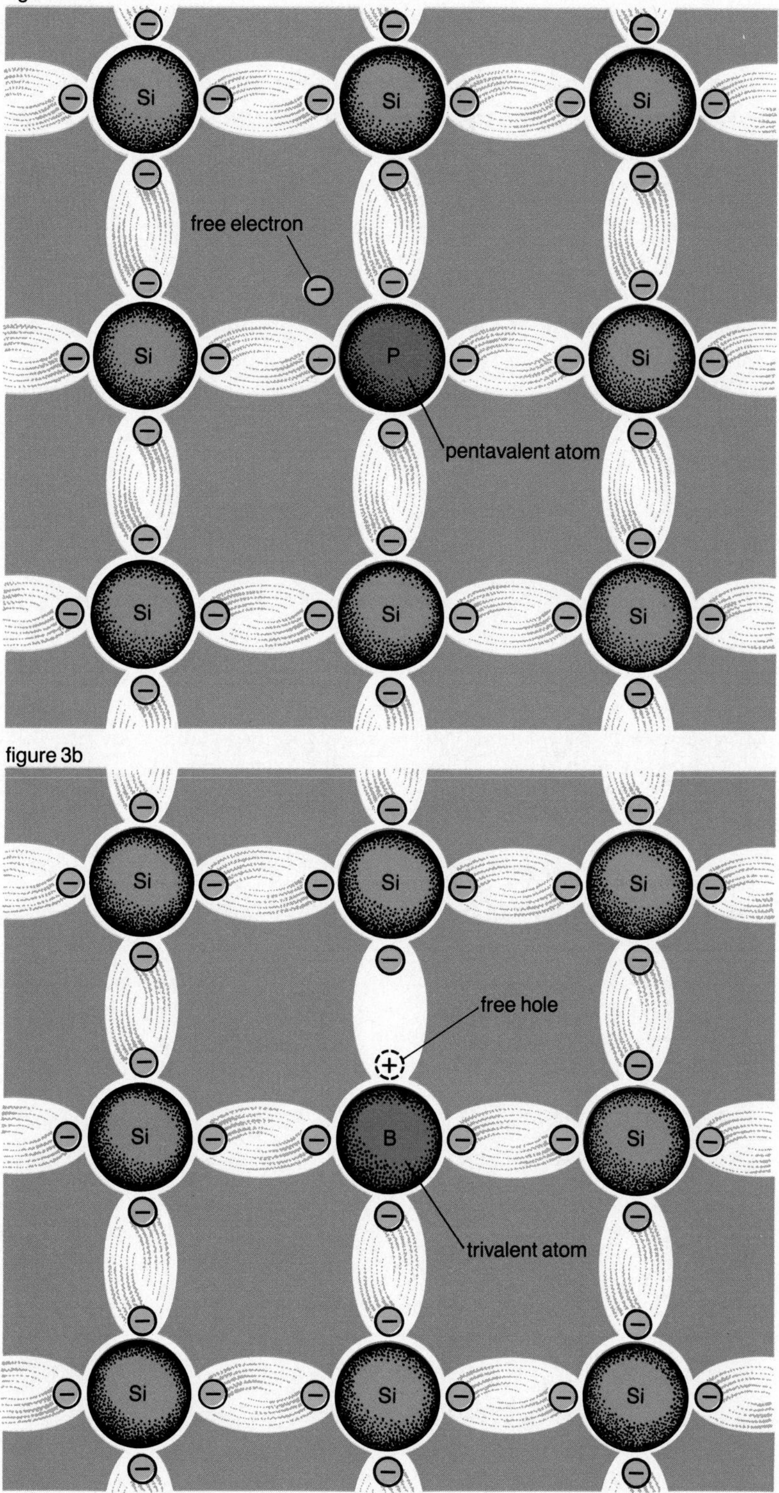

Schematics similar to those in figure 1 illustrate the effect of adding impurities to intrinsic silicon. When a pentavalent atom replaces an atom of tetravalent silicon in the crystal (a), one of its electrons is free to be a carrier of current. Substituting a trivalent atom for a silicon atom in the crystal (b) results in a current-carrying free hole.

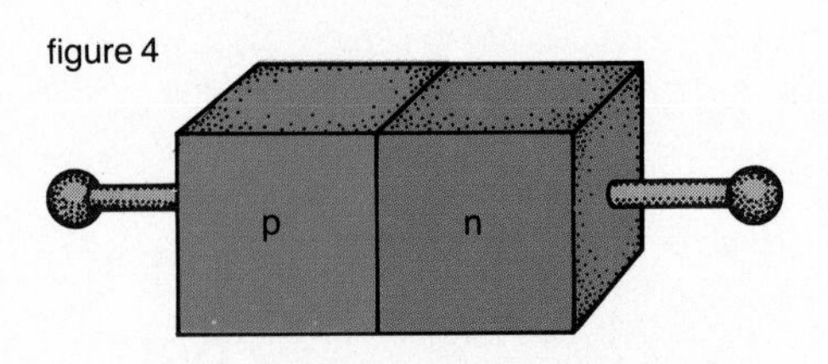

Doping adjacent regions of crystalline silicon with p- and n-type impurities results in a p-n junction, the basic structure of solar cells and other semiconductor devices.

gap for the intrinsic variety (figure 2). In n-type silicon the Fermi level is located just below the conduction band. This indicates a high probability of finding many free electrons and a low probability of finding free holes. In p-type silicon the Fermi level descends to just above the valence band. In this case, the probability of finding free holes is much greater than the probability of finding free electrons.

Now that it is possible to dope intrinsic silicon with n- and p-type impurities, a p-n junction (figure 4) may be formed. This structure is created by doping adjacent regions of a silicon crystal in a way that the transition from p-type silicon to n-type silicon occurs over a very short distance. The p-n junction is fundamental to the functioning of transistors and integrated circuits as well as the solar cell.

Amorphous silicon

Before the mid-1970s, scientists working with amorphous silicon were thwarted in their efforts to dope the material to form a p-n junction. In 1975, however, Walter Spear and Peter Le Comber of the University of Dundee in Scotland succeeded. The key to their success was the use of hydrogen (H) in forming the basic amorphous silicon material. To understand how this was possible, it is necessary to consider the concept of density of states. The density of states specifies the number of allowable quantum states electrons can occupy per unit volume of silicon for a specified value of energy; it is denoted by the symbol $N(e)$ in figure 5.

Figure 5a shows the density of states diagram for crystalline silicon, in

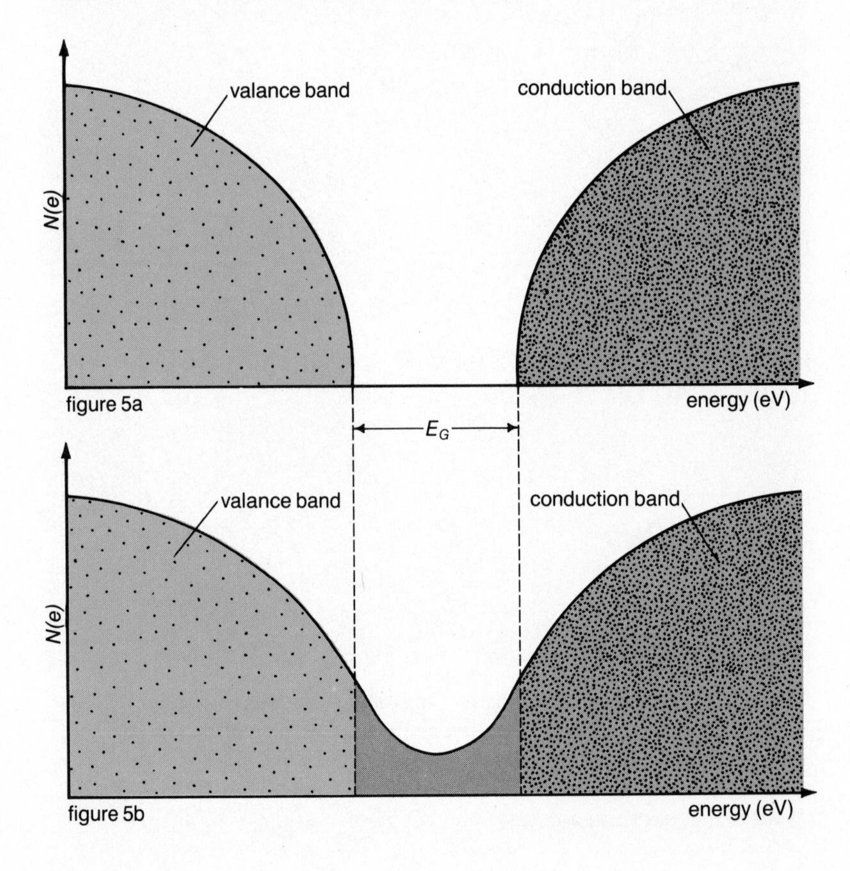

Density of state diagrams are shown for (a) crystalline silicon and (b) amorphous silicon. The energy gap, or band of forbidden energies, is denoted by E_G. N(e) represents the number of allowable quantum states that electrons can occupy for a specified energy.

Fritz Goro

Gross differences in crystalline and amorphous semiconductor material are evident in side-by-side comparison of polycrystalline germanium (left) and a chalcogenide glass made of an alloy of germanium, selenium, and tellurium (right).

which can be seen the well-defined edges of the valence and conduction bands. For amorphous silicon in figure 5b, however, the bands lack the sharp edges and instead tail off into the energy gap region. Many of the allowable quantum states, therefore, are now located in the band of forbidden energies, which means that correspondingly few current carriers are available.

According to a recently developed theory, the protrusion of these bands in the energy gap region stems from dangling bonds. These are bonds that, because of the poorly structured arrangement of atoms in the amorphous material, arise from the incomplete sharing of valence electrons among the silicon atoms. The method used by Spear and Le Comber for correcting this is to coerce the dangling bonds to form covalent bonds with atoms that have a single valence electron, like hydrogen. In this situation hydrogen ties up the dangling bond. Amorphous silicon containing hydrogen is designated a-Si:H. Other candidates for tying up bonds are fluorine and chlorine.

The manner in which hydrogen is introduced to form a-Si:H is basically simple. Silane gas (SiH_4) enters a quartz tube in which a substrate, an inert support for the amorphous silicon, is held at a temperature between 200° and 300° C (390° and 570° F). Controlled amounts of gases that contain boron (as diborane) or phosphorus (as phosphine) can be added for doping before radio-frequency current is applied to coils surrounding the tube. Upon excitation of the coils, the resulting electromagnetic field produces an electrical discharge through the gas, causing the molecules of silane to break down. Once freed, atoms of silicon and hydrogen (and dopant) then deposit on the substrate to form a stable layer of a-Si:H.

By contrast, many steps are required to produce crystalline silicon structures. These include growing single-crystal ingots from polycrystalline silicon. In the commonly used Czochralski method, a seed crystal is brought into contact with the molten silicon and slowly withdrawn. Afterward, the ingot must be cut into very thin wafers, polished, and doped. All these steps add to the cost of the finished device.

In addition to lower processing costs, amorphous silicon is superior in

Courtesy, General Signal

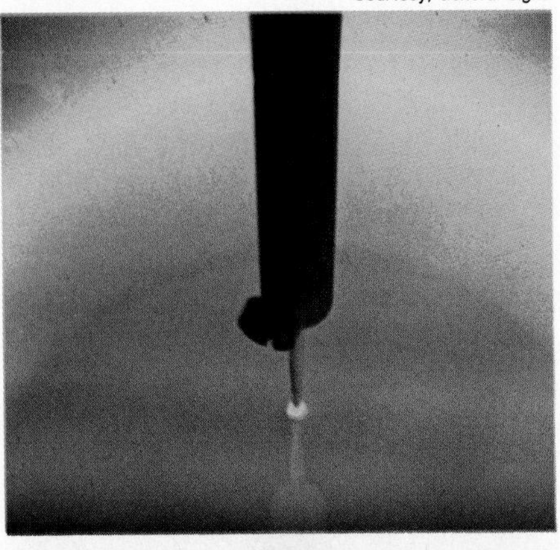

Manufacture of a crystalline silicon solar cell (above right) commonly begins with the pulling of a single-crystal ingot from a crucible of molten silicon kept just above its melting temperature, 1,410° C. A small seed crystal of silicon held in a clamp is touched to the surface of the melt and then slowly withdrawn (above, top to bottom). The molten silicon crystallizes around the seed to form a circular ingot that may reach a meter in length.

© Dan McCoy—Rainbow

some respects to crystalline silicon. For one, amorphous silicon has an energy gap of 1.6 eV, compared with 1.1 eV for crystalline silicon. The value of 1.6 eV is close to the energy of the most intense photons (packets of energy) emitted from the Sun; this is a significant advantage in using amorphous silicon for solar cells. Also, owing to the inherent disorder of an amorphous material, a-Si appears to be more resistant to radiation, such as cosmic rays, than conventional crystalline silicon.

Amorphous silicon solar cells

A solar cell is essentially a p-n junction device fabricated to allow its junction to be exposed to light. If the energy of a photon striking the junction is at least equal to the energy gap, a valence electron absorbs it and is carried across the energy gap; consequently an electron-hole pair is created. As a result, an open-circuit voltage (the p-n junction is unconnected to a circuit) appears across the junction. This phenomenon, called the photovoltaic effect, converts light energy into an electrical voltage. If the p-n junction is now connected to a circuit, current will flow.

In the practical realization of a-Si:H solar cells three basic structures are used: the Schottky barrier, the p-i-n, and the metal-insulator-semiconductor (MIS). Cross-sectional views of these structures are provided in figure 6. In the Schottky-barrier cell (figure 6a) a thin layer, less than 0.1 micrometer thick (one micrometer, μm, is one-millionth of a meter), of heavily doped n-type a-Si:H (designated by n^+ a-Si:H) is deposited on a steel substrate. This is followed by deposition of undoped a-Si:H (0.3 to 1 μm thick) on top of the first layer. A 50-angstrom film (one angstrom, Å, is a ten-billionth of a meter) of platinum (Pt), called the Schottky barrier, is evaporated onto the undoped layer. For an electrical connection a small pad of platinum may

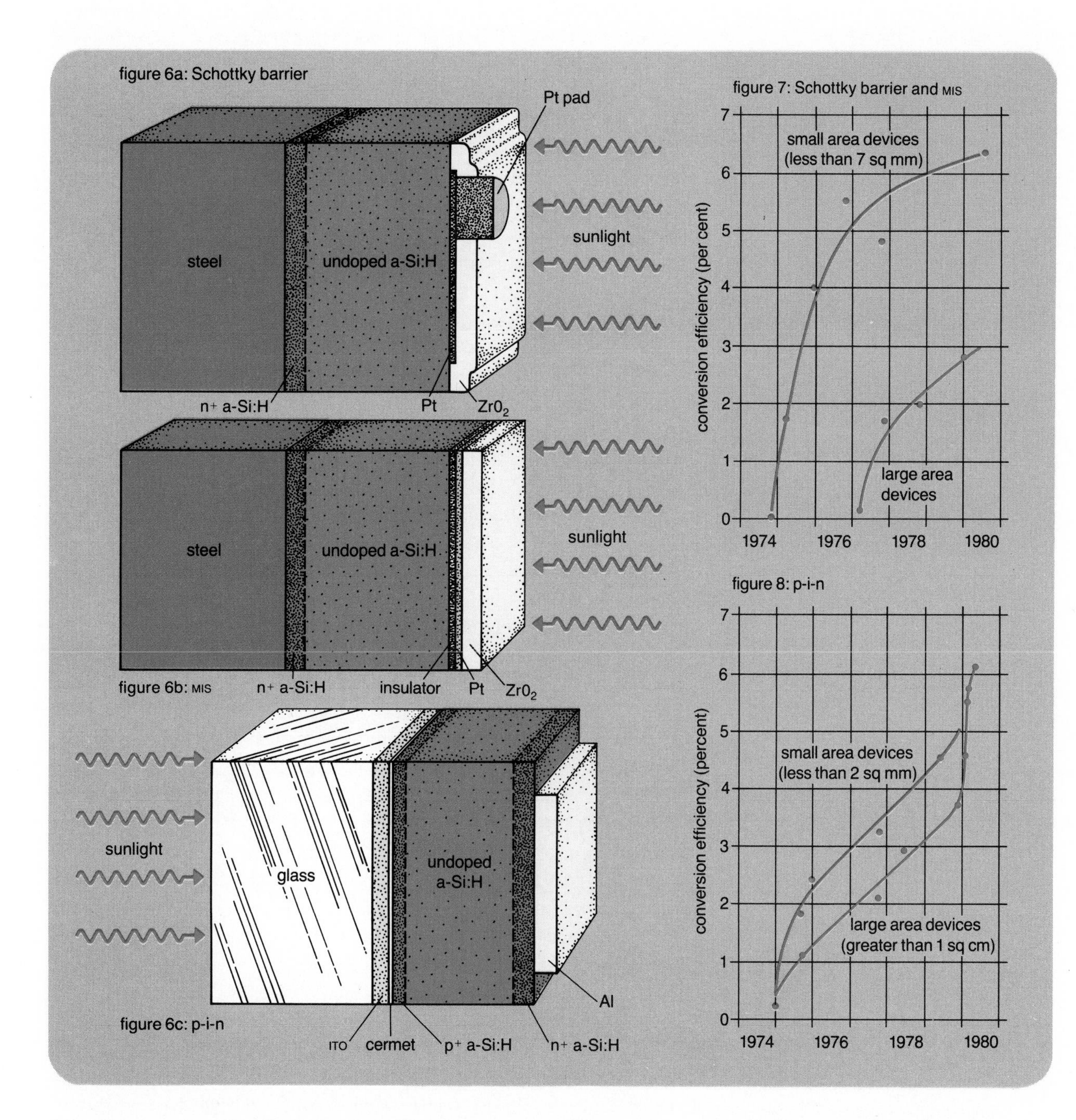

be added near the edge of the platinum film. Finally, an antireflection coating of zirconium oxide (ZrO_2) is deposited on top of the platinum.

If during this procedure a thin insulating layer, on the order of 25 Å, is formed on top of the undoped a-Si:H layer and then followed by a deposition of platinum, the result is an MIS solar cell (figure 6b). An advantage of the MIS cell over the Schottky-barrier type is its greater open-circuit voltage.

In one type of p-i-n cell (figure 6c), a thin layer of a mixture of platinum and silicon dioxide, called a cermet, is deposited on glass that has been coated with indium-tin oxide (ITO). This is followed by a thin p^+ (heavily doped

Basic structures of three types of amorphous silicon solar cells are diagrammed in figure 6. Figures 7 and 8 plot performance history of these cells between 1974 and 1980.

figure 9

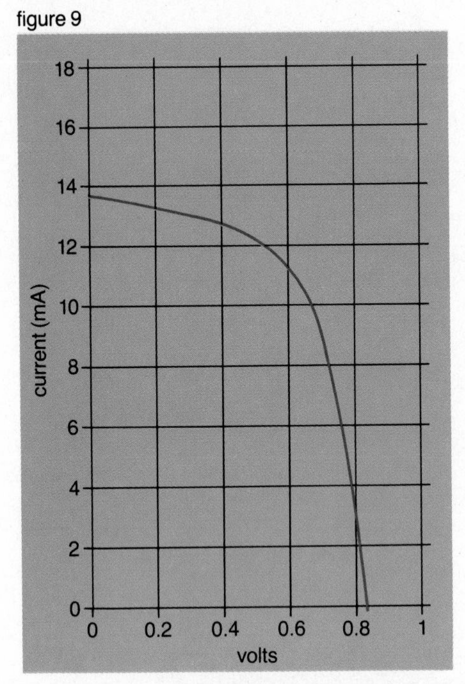

Curve depicts current-voltage characteristics of an amorphous silicon p-i-n solar cell. One mA (milliampere) equals a thousandth of an ampere.

p-type) a-Si:H layer and deposition of undoped a-Si:H. An n^+ layer of a-Si:H is then placed over the undoped layer. Finally a back electrode, such as aluminum (Al), is deposited on the n^+ layer. The cermet ensures good contact of the p^+ layer to the glass. The p-i-n cell exhibits a number of operating characteristics that are superior to those of the Schottky-barrier type. One is that the p-i-n cell is not affected by humidity or handling. In addition, because the p-i-n structure is deposited on glass, which is an insulator, many small p-i-n regions can be configured on a single sheet of glass and then easily connected in series to yield higher output voltages.

Performance characteristics of amorphous cells

In the evaluation of solar cells, whether they be amorphous or crystalline, a key criterion is their conversion efficiency. Conversion efficiency tells how much incident sunlight, when the Sun is directly overhead the cell, is converted into electrical energy. Calculations indicate that the maximum theoretical conversion efficiency of a-Si:H cells is between 15 and 20%. By 1980 conversion efficiencies in the vicinity of 6% had been achieved. Pundits in the field believe that a conversion efficiency of at least 10% is necessary to make the a-Si:H cell economically viable. For comparison, in 1980 crystalline silicon solar cells on the commercial market had a conversion efficiency approaching 15%.

Progress achieved in improving the conversion efficiency of a-Si:H solar cells is illustrated in figures 7 and 8. Figure 7 shows the rapid increase of conversion efficiency for small-area Schottky-barrier and MIS cells from 1974 to about 1977 and then a declining rate of increase from 1977 to 1980. The maximum efficiency attained in 1980 was about 6.3% for these devices. Efficiency of large-area devices reached only 3% in 1980.

The performance history of p-i-n cells is traced in figure 8. Significant in this case is the rapid increase in the conversion efficiency of large-area devices during 1980, with efficiency peaking at 6%. This notable improvement, achieved at RCA's Energy Systems Research Laboratory, Princeton, New Jersey, is attributed to optimization of such factors as doping levels and layer thickness in the fabrication of cells. David Carlson of RCA, a pioneer in the development of a-Si:H solar cells, also claims that it is somewhat easier to manufacture large-area p-i-n cells than Schottky or MIS types.

Although a-Si:H solar cells are currently less efficient than crystalline silicon cells in sunlight, they perform as well as or better than crystalline cells under fluorescent light. This improved efficiency stems from the close match of the spectral response of the a-Si:H cell with the spectrum of fluorescent light. Exploiting this phenomenon, Sanyo Electric Co. of Japan made plans in 1980 to market electronic watches and calculators containing an a-Si:H cell for their power source.

Variations in current and voltage occurring for a p-i-n cell connected to a circuit is illustrated in figure 9. One notes that as current increases, output voltage of the cell tends to fall. At zero current, voltage (open-circuit voltage, V_{OC}) is at its maximum. For the cell represented in figure 12, $V_{OC} = 0.83$ volt, approximately one half the voltage of a flashlight cell. To obtain greater voltages the cells are connected in series. In such configurations the total voltage is the output voltage of one cell multiplied by the number of cells in the array.

By early 1981 a number of companies were establishing facilities for the manufacture of amorphous solar cell arrays. Arco Solar Inc., a subsidiary of Atlantic Richfield Co., signed a $25-million contract with Ovshinsky's ECD to be supplied with amorphous solar cells. These are the Schottky-barrier type and contain fluorine (F) in addition to hydrogen; they are designated a-Si:H:F. In Japan Sanyo designated $50 million for a plant to produce a-Si:H cells for a range of consumer products including radios and tape recorders. Others entering the field include Solar Power (owned by Exxon Corp.) and Solarex. U.S. experts in the Department of Energy are sanguine that a 10% conversion efficiency for amorphous solar cells can be reached before 1985.

Glassy semiconductors

When Ovshinsky announced in the late 1960s that he had developed a device that could transform the electronics industry, that device was made with chalcogenide glass. As mentioned above, this is an amorphous material that contains one or more chalcogen elements in addition to such elements as arsenic, germanium, and silicon. The interesting property of chalcogenide glass is that it can function as a switch, called an ovonic switch (after Ovshinsky).

A striking difference between a-Si and chalcogenide glass is the much

Courtesy, RCA

Amorphous silicon p-i-n solar cell developed by RCA contains 17 cells connected in series. With an open-circuit voltage of 0.83 volts per cell, the array yields a total 14.11 volts. Its active area is 63 square centimeters, a bit more than the area of an ordinary playing card.

greater disorder of atoms in the glass. Atoms in the latter material can exist in various locations, and it takes very little energy to urge them to move to new sites. When a voltage below a certain minimum value is impressed across a thin film of chalcogenide glass, no current flows; the material is said to be in its high-resistance state. But when this minimum, or threshold, voltage is reached, current flows, and the device is said to be in its low-resistance state. In one type of switch the device remains in its low- or high-resistance state even after the impressed voltage is removed. This behavior, as is explained below, is the basis for using the ovonic switch in computer memories.

In the late 1960s there was considerable, at times acrimonious, debate regarding the exact nature of the switching mechanism in chalcogenide glass. One group of physicists contended that switching in glass was purely a thermal phenomenon; current flow heated the glass, which resulted in permanent structural changes in the material. This pronouncement was ominous because such a device would be unreliable and therefore unsuitable for use in products. Ovshinsky and others maintained that switching was electronic and that no permanent structural changes occurred. Although by 1981 there was still some disagreement, further work in this field appeared to have vindicated Ovshinsky.

The exact operating mechanism in the ovonic switch is still debatable. One key concept about which there seems some accord is that of the "lone pair" of electrons. When chalcogen elements like selenium form covalent bonds with their neighboring atoms, two of their valence electrons, called a lone pair, are not shared in the bonding. These electrons become free agents, and in response to a suitable impressed voltage will flow in a circuit connected to the switch.

Whereas the properties of a-Si, like those of crystalline silicon, may be altered by doping, for a time it did not seem possible to change the properties of chalcogenide glass to any appreciable extent. In the late 1970s, however, Ovshinsky and his colleagues succeeded by the addition of chemical modifiers such as tungsten, nickel, or boron to increase the conductivity of chalcogenide glass some one million times. This ability of modification allows both electrical and optical properties to be tailored for specific applications of chalcogenide devices.

Glass memories for computers

One can perceive the computer as containing upwards of hundreds of thousands of ON-OFF switches. Because a switch has two states, either ON or OFF, it is called a bistable device. Symbolically and mathematically the two states can be represented by a 0 when the switch is OFF or a 1 when ON. Since there are only 0s and 1s to be concerned with, data and programming instructions are represented by strings of 0s and 1s. Arithmetic is performed on binary (base 2) numbers. This is indeed all to the good because there exists a host of devices, such as a punched card or a transistor, that exhibit bistable states which are reproducible and reliable.

A major section found in computers is the memory. The memory stores data, partial results, and the instructions that tell a computer how to proceed

Fritz Goro

Chalcogenide-glass memory, manufactured by Burroughs Corp. in the U.S., has a storage capacity of 1,024 bits of information. The memory elements form the wide horizontal pattern located about one-third of the way down from the top of the chip. The entire device measures 5.6 millimeters along an edge.

in solving problems. In the late 1970s Sharp Co. in Japan and Burroughs Corp. in the U.S. signed contracts with ECD to develop an electrically alterable read-only memory (EAROM) made of chalcogenide glass. The ROM portion of the acronym indicates that data can only be read out of the EAROM; no information can be written into the memory when data is processed. (Another type of memory used in computers is the read/write, or random-access, memory, called a RAM.) The read-only memory is used to store information that is not going to change for a given problem, such as the computer's programming instructions. The EA portion of the acronym signifies that, when the memory device is removed from the computer, the contents in the ROM can be electrically altered and another program stored in the memory.

One Burroughs unit is a 1 K memory, in which 1 K equals 1,024 bits (0s or 1s) of information. Thus, the memory can store 1,024 zeros or ones. It is a fast memory: the time it takes to read out a 0 or a 1, called the access time, is on the order of five nanoseconds (one nanosecond is one-billionth

Construction of a basic glass memory switch is diagrammed in cross section.

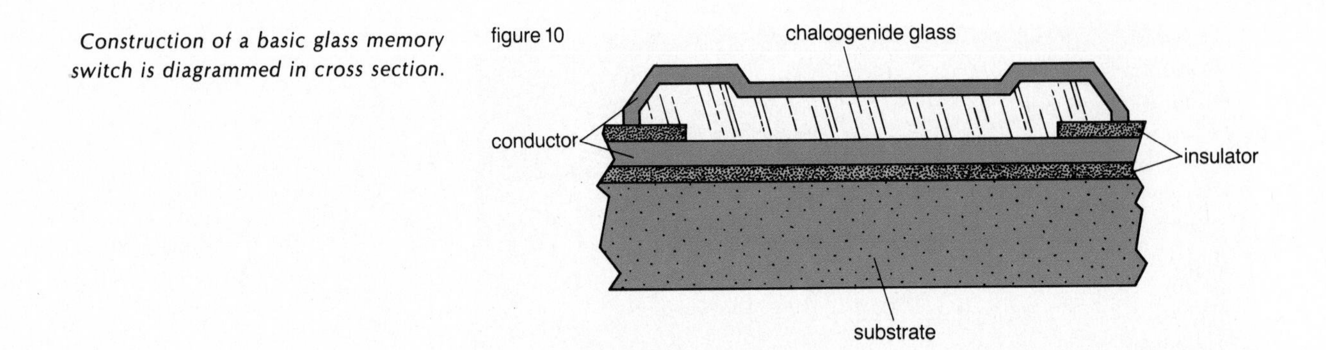

Electron micrographs show the surface of glass-switch memory material in its ON and OFF states (below and bottom). Appearance of a localized crystalline region is believed to be responsible for the persistence of low resistance in the material in its ON state.

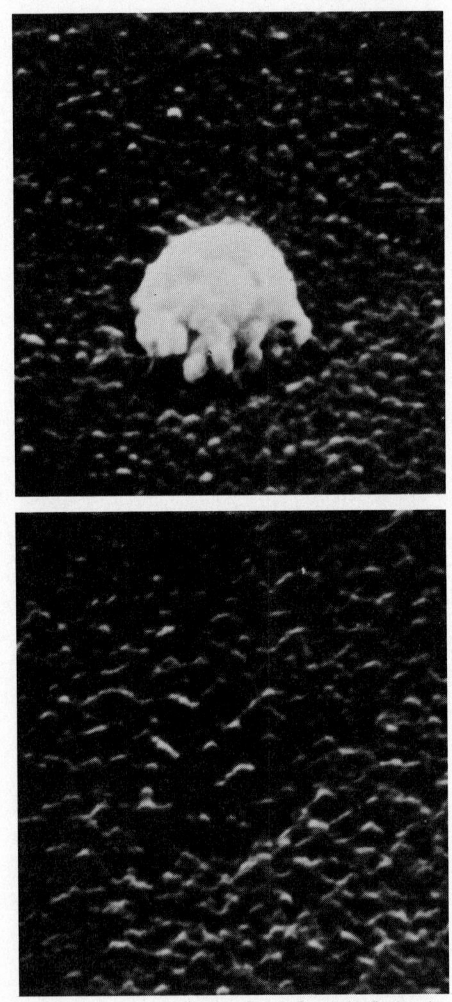
Courtesy, Energy Conversion Devices, Inc.

of a second). The writing in of data, however, requires a few milliseconds (one millisecond is one-thousandth of a second).

A cross-sectional view of a basic glass memory switch is shown in figure 10 and its electrical characteristics in figure 11. In figure 11 current is plotted along the vertical axis and voltage along the horizontal axis. Upon application across the unit of a voltage that exceeds the threshold voltage, V_T, the switch is turned ON and is in a highly conductive (low-resistance) state. If the applied voltage is allowed to remain for a few milliseconds and then removed, the switch will stay in its ON state. To switch to the OFF state a suitable pulse of current is applied, following which the material returns to a low-conductive (high-resistance) state.

The memory aspect of the memory switch may be explained in terms of rapid heating and cooling of the material following its initial switching action. It appears that after the switch attains its ON state, heat generated by the flow of current causes a portion of the amorphous glass to crystallize. This results in an ordered structure, leading to a reduction in resistance that persists even after current flow ceases. The current pulse applied to turn off the switch melts the crystalline structure, which cools rapidly, and the device reverts back to its amorphous, high-resistance state.

The memory switch contains germanium and antimony in addition to the chalcogens tellurium and sulfur. By contrast, the nonmemory version of the switch contains tellurium, germanium, silicon, and arsenic. The arsenic in the latter version seems to link the tellurium atoms and results in a strong structure that inhibits crystallization of the glass. Its electrical characteristics are shown in figure 12. A voltage whose value exceeds the threshold voltage, V_T, turns it on. As soon as the voltage drops below the holding voltage, V_H, the switch turns off. This action is that of a memoryless switch.

Photographic and other applications

The most common commercial application of chalcogenide glass has been in a device that is taken for granted: the office copier. In the xerographic process, selenium glass on a metal substrate (figure 13a) is positively charged to about 1,000 volts. For charge equilibrium an equal negative charge is produced on the substrate, and an electric field between the positive and negative charges is thereby established. When a document to be copied is illuminated by a light source, light striking the letters and other dark images on the document is absorbed. Light striking the white areas of the document, however, is reflected to the selenium glass. This pattern of reflected light creates a corresponding pattern of electron-hole pairs near the glass surface, and the electric field existing across the selenium-metal structure separates the electron and hole. The electron neutralizes a positive charge on the surface of the selenium, and the hole travels to the substrate to neutralize a negative charge (figure 13b).

Negatively charged black particles of the toner (effectively a powdered ink) used in the process are attracted to the positively charged areas on the selenium. The particles are then transferred to a sheet of paper, and a permanent image emerges upon the application of heat. By employing three layers of different chalcogenide glasses and three toners of different hues, Xerox Corp. has produced a machine that can copy images in color.

More recent use of chalcogenide glass has been made in the development of a new kind of microfiche material. A conventional microfiche consists of a sheet of silver-based photographic film that contains a number of optically reduced photo images. The chalcogenide material differs in that it is made of a specially tailored glass deposited on a thin Mylar sheet. Whereas glass chosen for memories exhibits a huge change in conductivity when switching states, glass for photographic applications is modified to have good absorption and other optical properties over a specified region of wavelengths for a given application. Upon being illuminated with a fluorescent light source, the glass undergoes selective, localized changes in optical transmittance. Variations in optical transmittance provide images made up of a continuous scale of gray tones.

Because the imaging process of the glass microfiche film is reversible and does not require chemical development that would permanently destroy its sensitivity to light, this new material offers a number of innovative features that should make the recording of information easier than in the past. These include selective exposure and development of individual frames at times that may be separated in years, the ability to edit and erase recorded images

Electrical characteristics of memory and nonmemory versions of a chalcogenide glass switch are compared.

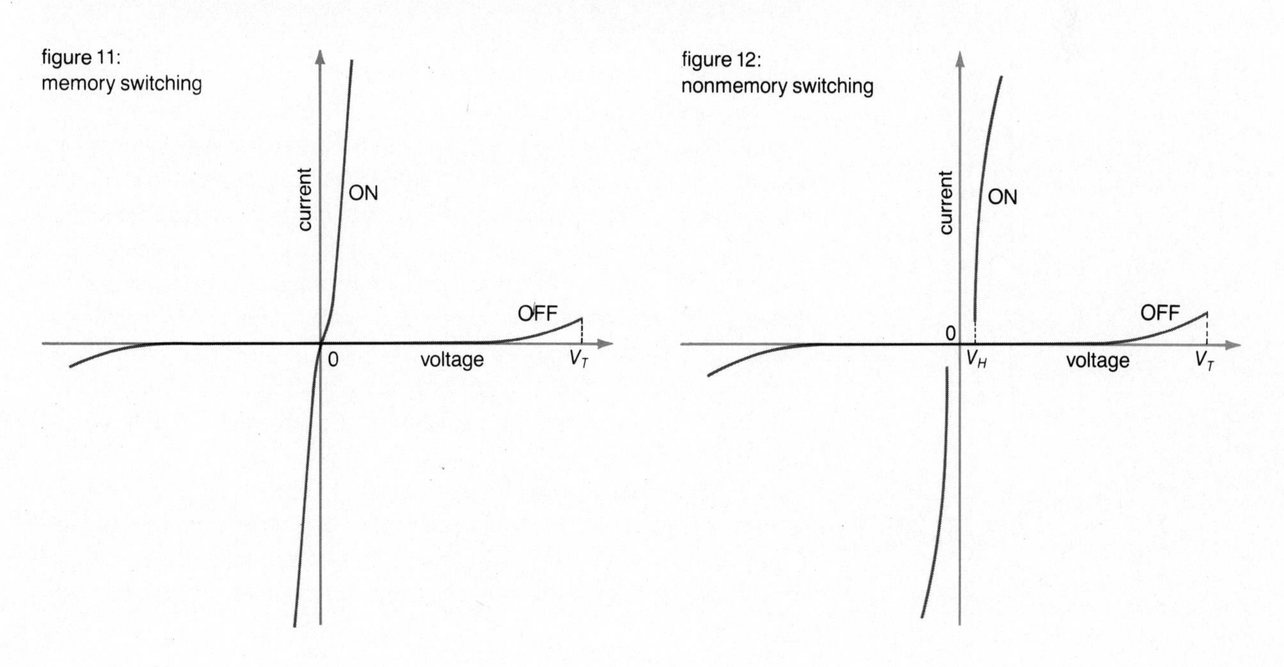

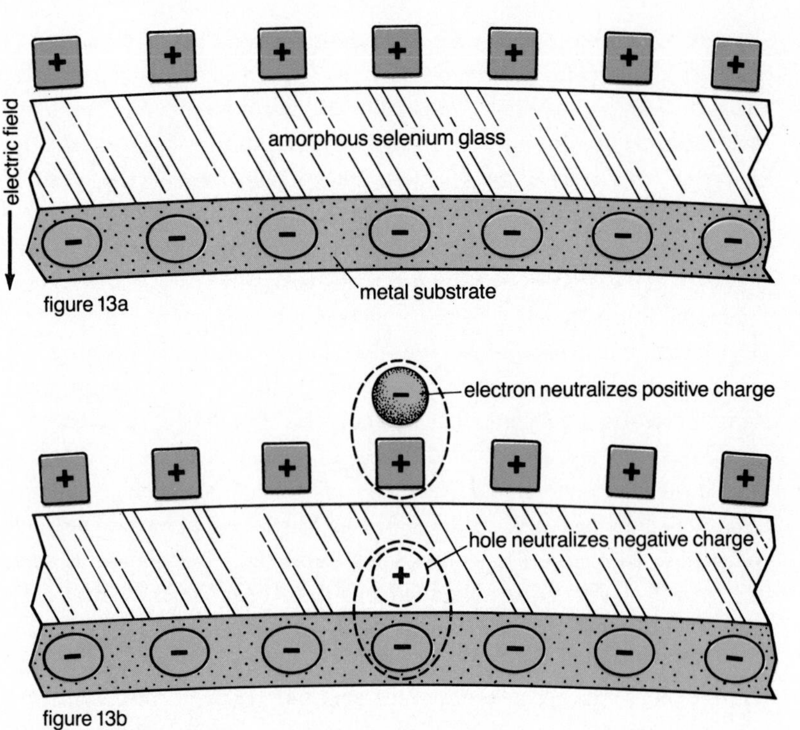

Two stages in the basic xerographic process are diagrammed schematically. At the beginning of the copying cycle, amorphous selenium glass is given a positive surface charge (a). For charge equilibrium an equal negative charge is produced on the substrate, and an electric field is established between the layers of opposite charges. Patterns of light striking the selenium create a corresponding pattern of electron-hole pairs in the material, and the electric field separates the electron and hole (b). The electron neutralizes a positive charge on the selenium surface, and the hole neutralizes a negative charge in the substrate. See text on pp. 178–179.

after exposure, and elimination of the need for processing chemicals. A desktop unit that allows microfilm recording and reading has been offered by ECD under the trademark name MicrOvonic file. In July 1980 ECD and A. B. Dick Co. formed a joint venture to manufacture and market the equipment.

Owing to the rising cost of silver, strong motivation exists among manufacturers of photographic film to develop nonsilver films. Progress has recently been made in the field of graphic arts film. Such a film, based on chalcogenide glass and invented by ECD, was demonstrated by Agfa-Gevaert AG in April 1980. The film, called Rapi-Lux, is for contact and duplication applications in the printing industry. Fuji Photo Film Co., Ltd., also plans to manufacture the film in Japan.

Although a nonsilver film called diazo has been used for years in microfilm and office copy equipment, it has had only limited application in the graphic arts. A reason for its restricted use is that diazo film is positive working; that is, it yields a positive image from a positive original. What is needed in the graphic arts industry is a film that is both positive and negative working (one that yields a negative of the image being copied). Again, ECD has produced such a film, called Ovonic Graphic Arts Film, using chalcogenide glass. Other developments based on chalcogenide glass include a nonsilver film that is developed by heating and nonsilver film for amateur and professional photographers.

David Adler of the Massachusetts Institute of Technology and others have performed a "marriage" in the laboratory of amorphous glass and crystalline silicon that has resulted in a transistor having interesting properties. When switched by a suitable pulse, this new transistor assumes and remains in its high-gain state after the pulse is removed. The action accomplished with this new device would normally require a number of conventional transistors. In addition, the device may serve as the basis for a logic system with three, instead of two, logic states. Such a system may lead to a highly efficient method for storing and processing data in a computer.

Photo by Virginia Hanchett; courtesy, David Adler, Massachusetts Institute of Technology

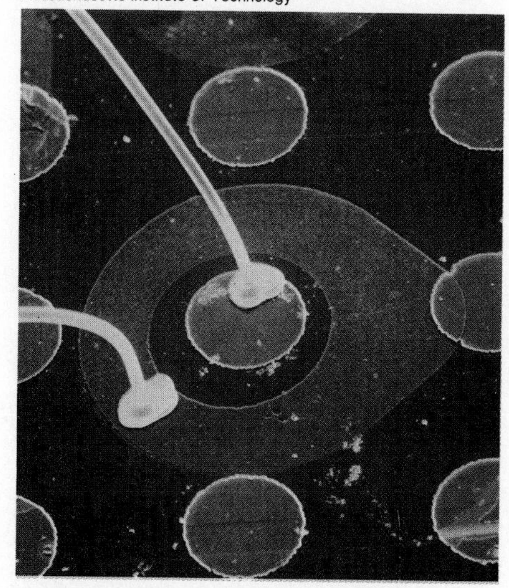

Novel glass-hybrid transistor employs conventional crystalline silicon for two of its elements, the base and collector, and a chalcogenide-glass switch for its emitter. The ability of the glass emitter to switch from high to low resistance—and thus the transistor from a low- to high-gain state—offers some exciting possibilities for future solid-state devices.

Future outlook

During the 1970s substantial progress was made in understanding the behavior and application of amorphous semiconductors. Whereas these efforts are having an effect on such disparate technologies as solar cells, computer memories, and silverless films, such is not the case for the electronics industry as a whole. The industry is based on crystalline silicon for the manufacture of large-scale integrated (LSI) circuits, such as the microcomputer, and other semiconductor devices. It has perfected its processing techniques to the extent that the devices it produces provide more operating functions per dollar than ever before. In fact, this industry appears to be bucking the inflation spiral; many components and finished products such as the microcomputer and the hand-held calculator were cheaper in 1980 than a decade earlier. In light of these achievements, except for special-purpose applications it appears unlikely that amorphous semiconductors will have an appreciable effect on the electronics industry.

One field of great promise for amorphous semiconductors is the realization of an economical and efficient solar cell. Experts in the field claim that to be economically viable, a conversion efficiency of 10% is necessary; currently, only a 6% plus efficiency has been achieved. Theoretical studies point out that efficiencies as high as 20% may be possible. With the world's limited resources of nonrenewable fossil fuel, this application is worth watching.

A second promising application for amorphous material is in photography. The availability of a nonsilver film that could be developed by the application of heat would be a boon for photographers. Whether such a film, both inexpensive and capable of yielding good picture quality, will become a reality remains a question for the future.

FOR ADDITIONAL READING

D. Adler, "Amorphous-Semiconductor Devices," *Scientific American* (May 1977, pp. 36–48).

D. E. Carlson, "Amorphous Silicon Solar Cells," RCA reprint RE-24-5-5 (1979).

D. E. Carlson, "Recent Developments in Amorphous Silicon Solar Cells," forthcoming in the periodical *Solar Energy Materials*.

L. E. Murr, *Solid-State Electronics* (Marcel Dekker, 1978).

S. R. Ovshinsky, "Amorphous Materials as Optical Information Media," *Journal of Applied Photographic Engineering* (winter 1977, pp. 35–59).

A. L. Robinson, "Amorphous Semiconductors: A New Direction for Semiconductors," *Science* (Aug. 26, 1977, pp. 851–853).

A. L. Robinson, "Chalcogenide Glasses: A Decade of Dissension and Progress," *Science* (Sept. 9, 1977, pp. 1068–1070).

NO PLACE TO HIDE

Wiretaps, bugs, spy beams, voice analyzers, night viewing devices, and other tools for personal surveillance form a two-faced technology that is at once an invaluable aid to law enforcers and a threat to privacy.

by Francis Hamit

With 1984 fast approaching, it seems a good time to ask how much of a seer George Orwell was in his famous novel of that name. A dominant theme in the work was an all-pervading surveillance by the government upon the lives and actions of private citizens. The good news appears to be that, in the Western world at least, such extensive surveillance has not happened. The bad news is that the technology to carry out such a task exists. Moreover, in light of continual disclosures about the activities of the FBI, the Central Intelligence Agency, and other U.S. government agencies, the feeling one may have of a subtle but prevalent disregard within society for one's privacy cannot be totally dismissed as paranoia.

In the past, wiretapping, bugging, clandestine mail opening, and other forms of surveillance were commonplace and often politically motivated. Today, in the U.S. such laws as the Omnibus Crime Control and Safe Streets Act of 1968 and the Privacy Act of 1974 place heavy restrictions on the legitimate use of surveillance technology and make illicit use of bugging and wiretapping a felony. Nevertheless, because these laws do allow for licit surveillance and because no law alone can deter someone whose needs are sufficiently desperate, it seems safe to conclude that not everyone now enjoys greater protection from violations of privacy. On the other hand, it is also reasonable to presume that the average person is not a surveillance target. Simple economics of time, money, and material resources dictate that a given person will not come under even the most rudimentary form of surveillance unless he or she is suspected to be involved in espionage or criminal activity or holds a high-profile position in politics, government, industry, or the media.

Indeed, members of the crime syndicate in the U.S. are under constant and unremitting surveillance by the FBI, the Drug Enforcement Agency, and

Establishing a wiretap requires that the eavesdropper have access somewhere within the telephone system to the wires that carry the signals of the phone to be monitored (right and facing page). One principal kind of tap, the induction tap (below), does not need direct electrical contact with the system. It uses a coil of fine wire to detect the fluctuating magnetic field produced by current surging through the phone line.

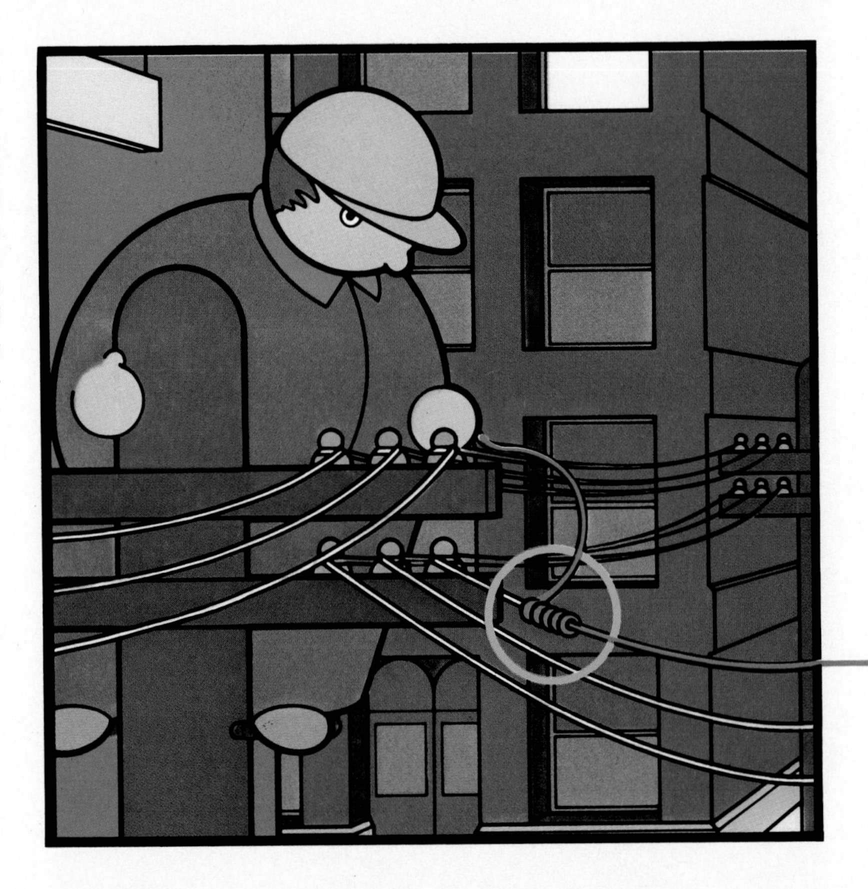

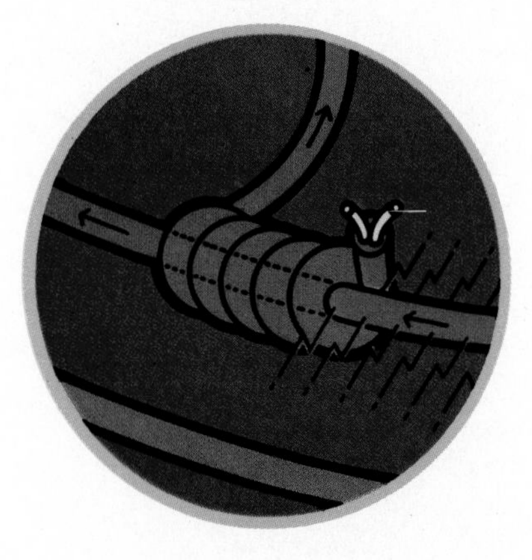

local district attorneys and federal prosecutors, but each act of surveillance is done only after a federal warrant is issued. In Chicago, for example, all of these investigations are maintained in a special secret "civil suppressed" docket before the chief judge of the U.S. District Court. The only agencies exempted from this system are the telephone companies, which are permitted under the Omnibus Crime Control and Safe Streets Act to listen to conversations of customers. The rationale for this blanket permission is to assure good customer service and to prevent fraud. There apparently have been abuses of this power, however, notably in Texas, where in 1974 the security staff of Southwestern Bell Telephone Co. was accused of involvement in illegal wiretapping operations in collusion with the Houston Police Department.

Most instances of surveillance involving a high degree of technology fall under one or more of the following categories: intrusion into one's conversations, both on the telephone and within the home and office; intrusion into other communications, including the transmission of computer data; the use of data banks to collect files of information on citizens without their knowledge; the analysis of voice characteristics as a form of lie detection; and the use of visual surveillance devices to monitor movement and activities. There is also the disturbing suggestion that technologies are being developed that may intrude upon the very processes of human thought. All of these activities have one thing in common: they can be carried out without the knowledge or consent of the persons against whom they are directed.

FRANCIS HAMIT is a free-lance writer and consultant on communications and the media.

Illustrations by John Craig

Wiretaps

One of the two most widely known forms of surveillance is wiretapping, which refers broadly to the interception of telephone conversations and other telecommunications. The other, bugging, is electronic eavesdropping within someone's home, office, vehicle, or other private place by means of unseen listening devices, commonly called bugs.

Wiretaps take advantage of the basic operating principles of the telephone system, which consists essentially of a pair of energy-converting devices—a microphone and a speaker—at each end of the conversation plus some means to connect them, such as wire or a radio link, and an external source of power. The microphone in the telephone mouthpiece converts the energy of sound waves into corresponding variations in electric current; the speaker in the earpiece performs the opposite function of converting the transmitted signals back to sound. This principle applies in the same manner whether the signal is transmitted entirely by wire or whether it is further converted for part of its journey into some form of electromagnetic radiation, such as radio waves, microwaves, or even laser light.

To establish a wiretap the perpetrator needs access somewhere within the telephone system to the wires that carry the signals of the phone to be monitored. A direct tap—one that makes direct electrical contact with the telephone circuit—can be made within the body of the telephone itself or by splicing into the wires that connect the instrument to the telephone network. This tap can also be made at the main switching terminal or at any

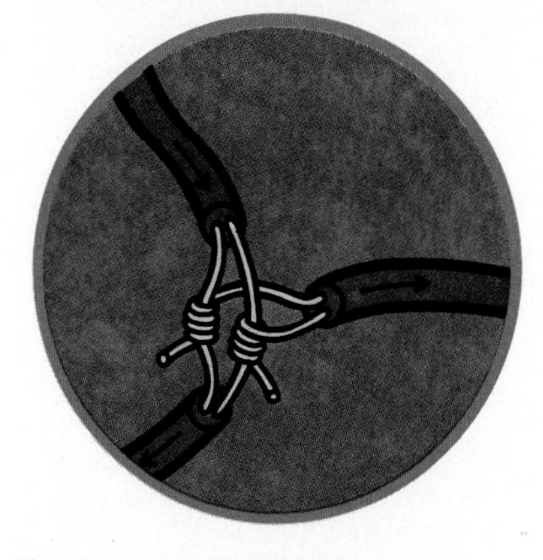

The direct tap must be spliced or directly connected in some other fashion to the telephone line (above). It drains current from the line and thus can be detected rather easily unless the drain is counterbalanced with input from additional circuitry. Both direct and indirect taps require other equipment, such as a recorder or earphones, for the eavesdropper to listen to the conversation (top).

One kind of long-range microphone makes use of a parabolic reflector to direct distant sounds to a microphone located at its focal point (below). Among bugs that rely on a microminiaturized microphone-transmitter is one version that substitutes for the microphone element in a telephone handset (bottom). Some broadcast in the commercial FM band and can be received on an ordinary pocket radio (bottom right and facing page, left).

intermediate terminal. Once out of the immediate vicinity of the telephone instrument, however, the wiretapper must have "cable and pair" information indicating which two wires of the myriads in the network are connected to that telephone. A major drawback of the direct tap is that it can be detected unless the current it drains from the system is replaced by additional circuitry.

A second kind of tap, the induction tap, does not connect directly to the system but picks up the weak magnetic field generated by the current in the wire. Its sensor, basically a coil of fine wire, need only be placed near or around the line to be tapped. Sound-modulated variations in the line current produce corresponding magnetic fluctuations outside the wire, which the sensor converts back into current. One commonly available example of such a device is the telephone pickup coil often supplied to consumers as a tape recorder accessory.

In addition to extracting the signal from a telephone wire, a tap must enable the eavesdropper to listen to or store the conversation. Consequently the output of the tap is connected to another telephone, an earphone, or a recorder hidden nearby or to a miniature radio transmitter linking it with the eavesdropper or a recorder at a remote location.

The problems of telephone security are further complicated by a device called a pen register. Usually connected at a remote terminal such as the central office of the telephone company, this device reads the dialing signals of the telephone and records the number called, as well as the time and

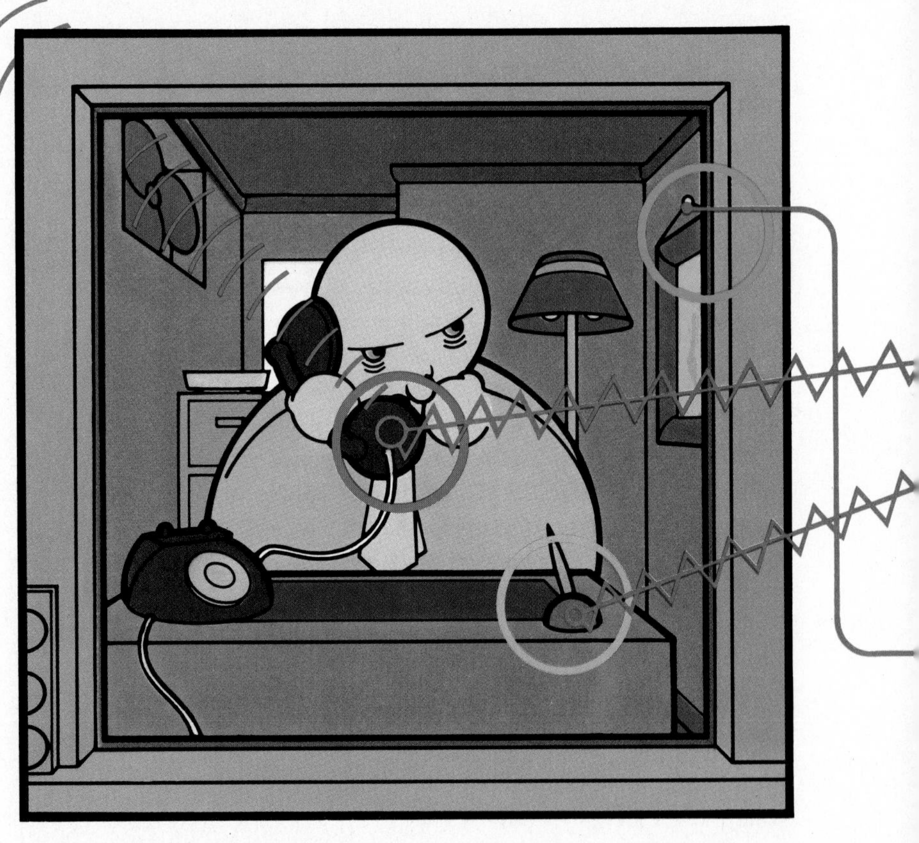

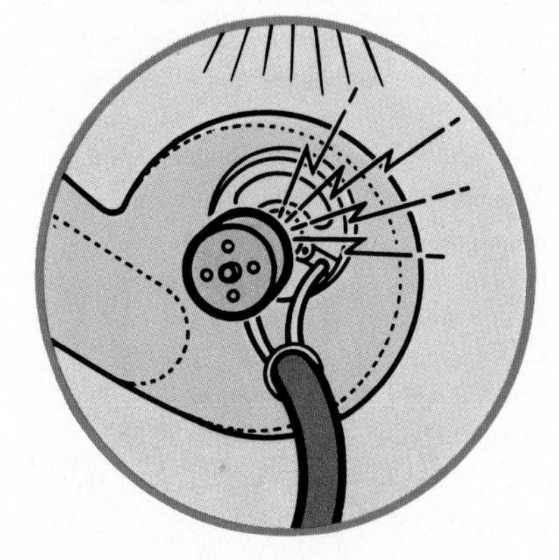

length of the conversation. It permits the user to know whose telephone number was called and for how long, and it often contains provisions for listening to or recording the actual conversation. Some law enforcement agencies have found that this device, combined with even the most nebulous telephone conversations, enhances their ability to discern what, if anything, is being perpetrated, to develop this intelligence, and to use it effectively as evidence or as the basis for gathering evidence, making arrests, and gaining convictions. More controversial has been the use of pen registers or other call-recording devices by some businesses to learn if employees are using company phones for personal calls.

The spike microphone (below), designed to be driven through a wall or baseboard from an adjoining room, is a comparatively crude device, yet it sometimes succeeds where more sophisticated devices cannot be used. (Bottom) Miniaturized transmitting bugs may take up no more space than an ordinary marble. Their place of concealment is often a small item of the victim's own belongings.

Bugs

Probably the type of bug most closely associated by the general public with cloak-and-dagger espionage is the combination microphone-transmitter. It is this microminiaturized device, often measuring less than two centimeters (three-fourths of an inch) in its longest dimension, that spy stories commonly portray lurking in a lamp or ashtray, behind a curtain, or at the bottom of a wastebasket, or secreted in the pen or cufflink of an unsuspecting party. Whereas a range of variations exist, the less sophisticated varieties can be tuned to broadcast on an unused frequency of the standard commercial FM band, and hence conversations in a bugged room can be picked up by an eavesdropper equipped with an ordinary—and unincriminating—pocket radio. One version of this device closely resembles the microphone element in

a telephone mouthpiece; an eavesdropper can make the substitution during a few seconds of privacy.

The distance over which miniaturized bugs can transmit is generally quite short and limited by the power source, which is usually a tiny battery. Consequently their effective range may be no more than 100 meters (330 feet). In most urban environments, however, a sphere of that radius can hold a hundred hiding places. One type of microphone-transmitter that avoids the use of a battery altogether employs a piezoelectric crystal as its microphone element. Certain natural and synthetic crystals exhibit the piezoelectric effect: they generate an electric current when subjected to mechanical stress, even the tiny stresses caused by sound waves striking them. The piezoelectric bug uses this current to power a minuscule silicon-chip transmitter and hence needs no power other than sound. Although its range is extremely short, such a device can be made small enough to be concealed in the swizzle stick of a martini.

Another class of bugs includes the "harmonica microphone" and similar devices that use the telephone system as a carrier. One such device, once installed in the body of the phone, allows the telephone mouthpiece to be used as a hidden microphone even if the handset is in place. To use the device the eavesdropper dials the number of the bugged telephone from a second phone, which can be virtually anywhere in the world. As the connection is made the caller sends a special tone over the line, a sound that may be generated by a simple appliance like a harmonica or an ultrasonic whistle. This tone activates the bugging device, which suppresses the first ring at the called telephone and then closes the switch that is normally closed only when the handset is lifted from the receiver. Every word spoken from that moment on in the immediate vicinity of the bugged telephone is now available to the eavesdropper.

Bugging can be accomplished not only with state-of-the-art electronic equipment but also with more primitive items such as conventional microphones concealed behind walls during construction—as the Soviets apparently did on at least one occasion at the U.S. embassy in Moscow. Another simple device commonly used in hotels and apartment buildings is the "spike" microphone, which is driven like a large nail through the walls or baseboards of adjoining rooms. These devices often succeed when access by other means proves impossible. Conventional microphones may also be concealed upon someone's person or in a briefcase or other large item that might be legitimately carried into and out of a room where confidential conversations are held. Again such bugs are combined with a recorder or transmitter.

Conversations held out-of-doors are not immune to interception. The parabolic microphone, a hand-held device employing a concave dish and a directional microphone at its focal point, can be used to pick up speech from as far as several hundred meters. Another highly directional microphone, the tube or "shotgun" microphone, resembles and is aimed somewhat like a rifle. Such microphones are widely employed in the broadcasting industry for radio and television coverage of news and sports and are legally available to the general public.

Spy beams

The cutting edge of eavesdropping technology belongs to a device that few admit even exists. The "spy beam" of science fiction is fiction no longer. It has been feasible since the invention of the laser, and for this reason the U.S. government has made extensive use of heavy curtains and blinds on the windows of the offices of its military, intelligence, and law enforcement agencies during the last decade. Called the laser microphone, this piece of equipment relies on the fact that any windowpane will act as a diaphragm, vibrating in response to the sound waves of conversations within a room. If an intense, highly directional source of light, such as an infrared laser, is aimed at the window, the vibrations of the window glass will produce comparable variations in the frequency of the reflected light. These frequency variations arise from the Doppler effect, in which the observed frequency of a wave changes because of relative motion between the object emitting or reflecting the wave and the observer. The returning reflections carry information much like the modulated carrier frequency in FM radio transmission. They can be received with an infrared-sensitive detector and translated back into sound using suitable electronic circuitry. Even before the invention of the laser, this same principle was used with reflector-focused beams of ultrasound and microwaves, but with less success and accuracy because of difficulties in preventing unacceptable diffusion of the radiation. The laser overcomes this problem because its light waves are naturally collimated into a tight beam.

The laser "spy beam" microphone treats the windowpane of a room as a diaphragm that is vibrating in response to spoken conversation within the room. The returning reflections of a laser beam aimed at the window carry the imprint of these vibrations in the form of changes to the basic laser frequency. With proper equipment these frequency modulations can be converted into intelligible speech sounds.

Even the laser microphone, however, has some serious flaws. It must rely upon a thin, rigid piece of material such as glass or sheet metal for a diaphragm. Interposing a soft, muffling material of sufficient thickness between this diaphragm and the source of the sound acts as an effective countermeasure. In addition, this method can only be used along clear line of sight, which means that both range and opportunity for its use are extremely limited. Unlike other eavesdropping technology, the laser microphone is prohibitively expensive and unavailable through commercial channels. (On the other hand, a talented and knowledgeable hobbyist could build one from common components.) Nevertheless, one great advantage to the eavesdropper is its virtual undetectability in use, because a beam of infrared laser light is invisible and literally pencil thin.

Voice stress analysis

Whereas electronic eavesdropping presents a threat to the privacy of one's words, another product of high technology threatens the privacy of the intent behind those words—the psychological stress evaluator (PSE). The PSE, also known as the voice stress analyzer, was conceived by two U.S. Army intelligence officers during the Vietnam war. It was designed as a simple and covert alternative to the polygraph, but unlike the polygraph it does not require a physical link between the subject and the machine. In fact, it can analyze voices on tape, telephone, radio, or television and without the subject's knowledge or consent. The PSE works by sensing and comparing minuscule vibrations called microtremors within the human voice. Controlled by muscles in the throat, these vibrations are altered by stress, and telling a lie or attempting to deceive will almost always produce stress. (One major exception is found in the case of sociopaths, persons suffering from a personality disorder that allows them to believe that the truth is what they say it is.)

The psychological stress evaluator monitors certain vibrations in the human voice to detect stress, which in turn may be a sign of untruthfulness. Although marketers of the PSE concede that it is not a lie detector, some of their advertisements claim that "it can help you tell the difference between truth and falsehood" and suggest that it can be used "to gain insights about a person's stress levels when talking about certain topics." Its capability for use without the knowledge and consent of the subject makes it a potential threat to privacy.

According to many critics of the PSE, many other phenomena in addition to instances of untruthfulness will also produce the same type of stress readings. For this reason, companies that market the PSE as a revolutionary type of lie detector are at pains to ensure that the same precautions—*i.e.*, a controlled environment, knowledge of its use by the subject, multiple readings of response to the same question—are taken that apply to a polygraph test. Nevertheless, many of these firms still exaggerate the capabilities of their product. Even after 60 years of use the polygraph itself is not a generally accepted device for lie detection and is outlawed in several U.S. states for any purpose except as proof of innocence. Studies of the PSE by several federal agencies indicate that its use for detecting the truthfulness of statements that were recorded in less than ideal circumstances is chancy at best. To use it without the informed consent of a subject as proof of truthfulness is not justified by the known capability of the technology.

Currently PSE's are not outlawed in the U.S. Available to the ordinary citizen are several models incorporating a variety of electronic sophistication. These range from suitcase-housed units priced in the thousands of dollars to hand-held versions selling for several hundred dollars to a microminiaturized edition built into a wristwatch case. The real threat to the privacy of the average person comes not from the use of the PSE but from the very act of intruding on his or her conversations. Employing a PSE to "analyze" such conversations simply compounds the problem. It is for reasons of privacy that current U.S. laws on wiretapping and bugging without a warrant are so strict. This effectively limits the legal power to eavesdrop to law enforcement and intelligence organizations with official constitutional sanction, as provided by law.

If the PSE were to be used in combination with sanctioned surveillance, it would most likely be done by an intelligence organization. Intelligence data analysis consists of weighing not only facts but trends, rumors, and feelings as well. Voice stress analysis, used as an indicator, might be applicable in some cases. Whereas such indicators would never be admissible evidence in a court of law, the intelligence services are not courts of law but rather players of a game in which a single clue often demands action.

Visual surveillance

Hand in hand with wiretapping and bugging operations is visual surveillance, which broadly includes such activities as invading the privacy of one's mail, credit records, checking account statements or, more narrowly, actually watching a person or a location. To assist surveillance efforts in the last category, technology has contributed several old workhorses—miniature cameras, telephoto lenses, and telescopes, as well as infrared-sensitive film, which when used with invisible infrared spotlights or flashguns allows still and motion pictures to be taken in what seems to be total darkness.

Among more recently developed night-viewing aids are image-intensifying devices, which in their present form are another technological offshoot of the Vietnam war. Originally, high-speed available-light photography was used for night surveillance. This technique relied greatly upon extremely sensitive film combined with special developers and techniques that pushed the film's ability to record an image to the limit. During World War II and the Korean War the U.S. military made use of infrared and ultraviolet sniperscopes for night observation and warfare. Sniperscopes employ an electronic device called an image-converter tube, which transforms invisible radiation from an object into visible form, much the way a fluoroscopic screen creates a visible image from invisible X-rays. Unfortunately, bulky spotlights are needed to emit the radiation required by the scopes. Such light can be detected easily with another sniperscope.

In the 1960s the surveillance technology of image conversion was extended to image intensification to produce a night-viewing device that requires no artificial illumination and that until recently was a classified defense secret. This item, often called a starlight scope, permits the viewer to see enlarged images at night with almost daylight clarity. The working element of the starlight scope, termed a photomultiplier tube, receives reflected moonlight or starlight and converts it into a pattern of electrons that are accelerated and then focused on a fluorescent screen for viewing or recording. Early starlight scopes were usually built with three photomultiplier tubes arranged in series for stepped light amplification, but newer one-tube models of improved sensitivity have replaced them. Some starlight scopes

Computerized image enhancement can increase the clarity of dark, low-contrast, or blurred photographs and visual displays. It is a useful aid to low-illumination photography, thermal imagery, and image intensification techniques.

are no larger than a small hand-held telescope and require only flashlight batteries for power.

An even more recent application of imaging technology is found in thermal imagers, which may eventually supersede starlight scopes at least in the military. Instead of receiving reflected radiation, thermal imagers detect the infrared radiation that is emitted by all objects at a temperature above absolute zero. In general the warmer the object, the brighter will be its image in the imager, and for this reason such objects as human beings, powered vehicles, and heated buildings make good targets. At the heart of the device is an array of solid-state detectors made of an infrared-sensitive semiconductor. Changes in the intensity of the radiation striking the detectors varies their conductivity. These variations across the surface of the array are converted to visible images on a small television display.

Surveillance equipment making use of image conversion, image intensification, and thermal imagery are all commercially available although prices are high. This technology is also integrated into closed-circuit television cameras used in security applications. It can likewise be used in conjunction with portable videorecorders. Because fiber optics can keep light images coherent while channeling them around curves, this technology also raises the possibility of using a fiber optic cable as the front element of a starlight scope. The cable could be introduced into a room by way of an entry too restricted for a human observer or a conventional imaging device, for example, a heating vent, a minute hole in a wall, or an overhead light fixture.

Unlike electronic devices the fiber optic cable does not generate an electromagnetic field that can signal its presence. It also allows the starlight scope to work more remotely, again lowering the chance of discovery.

Complementing visual surveillance equipment is a comparatively new processing technique called computerized image enhancement. Probably best known for its scientific application in clarifying pictures transmitted from Earth-orbiting satellites and interplanetary space probes, this process can dramatically increase the visual information content of a dark, murky, or otherwise indistinct image, be it a photograph or a television or imager display. An optical scanner first examines the image, feeding data to the computer about regions containing minute contrast differences. The computer then exaggerates these differences such that, for example, two bordering shades of gray that are barely distinguishable to the eye are assigned new shade values nearer black and white. The image is finally reconstituted and displayed or reproduced as a photograph. The computer enhancement process can also add color to further emphasize contrast. In addition, it can be used to sharpen images that are either out of focus or blurred because of relative motion between the imaging device and the subject.

Data processing technology

If asked what technology most threatened one's privacy, many people would probably answer "computers." Prominent among reasons might be the feeling of Orwellian depersonalization often associated with computerized services. But less well considered is the fact that the computer can function as a useful aid to the physical act of electronic eavesdropping. In addition, as a communications device it presents another target for eavesdropping. Still a different threat comes from the computer's obvious abilities in the assembly, comparison, and ordering of data for record-keeping, which allows those wishing to gather information on others to operate with greater efficiency—although not necessarily with greater accuracy.

Computers can enhance the clarity of recorded conversations by means of a technology commonly employed in the music business when making studio recordings. The object is to eliminate extraneous noise that would otherwise interfere with the clarity of the audio signals. This is done using "out-of-phase noise cancellation," a computerized process in which unwanted parts of the signal are detected and then electronically inverted. The inverted portion consists of a pattern of polarities and voltages made opposite that of the original. When the two signals are mixed the original and inverted patterns cancel, removing the noise.

A more sophisticated route to the same end makes use of electronic speech recognition. In this method a computer takes as its base a general profile of what speech sounds like: its wave patterns, tones, phonemes, and the general pattern of speech sounds. By comparing this information with the characteristics of a sound recording, it is able to discriminate between sounds that match its speech profile and those that do not. Only matching sounds are amplified, resulting in clear speech recording without background noise or other interference. Theoretically, even whispered conversation could be made clear and loud by this method.

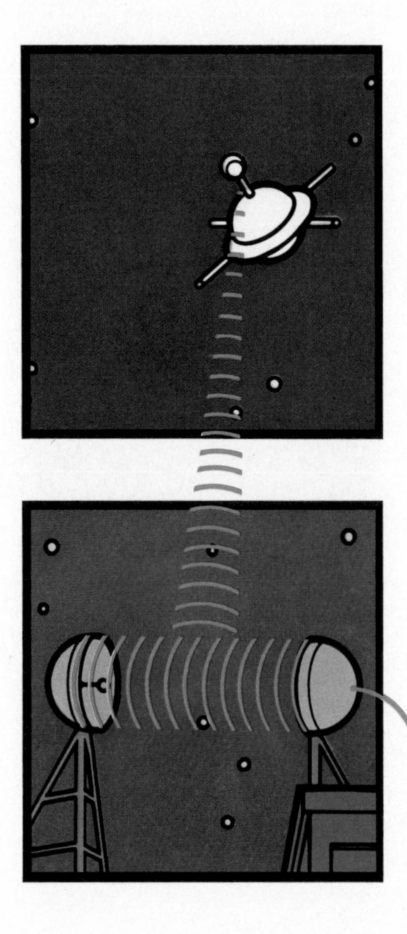

A related technique could also be used to select individual conversations out of a crowd, a pack of phone cables, or a mass-data microwave transmission. For example, the characteristic patterns of certain words (such as "gun," "bomb," or "political") might be programmed to act as a trigger that starts the analysis of a conversation on which the computer subsequently will focus for as long as the conversation lasts. A computer could separate and enhance several such conversations simultaneously.

Although there is no evidence that this technique is being used, it is technologically feasible given the current state of the art. It is noteworthy that the Soviet intelligence service, the KGB, is allegedly using orbiting satellites to "scoop" long-distance telephone communications carried by microwave in the U.S. and elsewhere. Only computer technology would allow an intelligence service to sift such wholesale lots of material for useful information.

As computers verge upon the status of standard office equipment, many businesses are learning to their chagrin that the computer's limitations are those of its human masters and that its complexity carries with it a high degree of vulnerability both to eavesdropping and to outright tampering. Indeed, the computer age has brought forth the computer criminal. Anyone whose business makes use of a computer needs to be aware of the very real problems implicit in computer security. And as the computer invades the home in the roles of record-keeper, information center, and electronic postal service, such problems will spread to the household members as well.

Conversations and computer data carried by phone lines may be transmitted by microwaves or satellite relay for part of their journey. In this mode they can be intercepted by a specially equipped orbiting satellite.

Some of these problems exist because many large computers used by government agencies and private businesses as well as by university research facilities are part of "distributed data systems" and are "on phone." This means that the main body of data can be read from a remote terminal attached to a common telephone line. The U.S. Social Security Administration, for example, operates a vast computer network of about 4,000 remote terminals and a central data bank that maintains files on almost all American citizens. The terminals, distributed in offices across the country, allow one not only to gain access to individual records but also to change them and to create new ones.

Naturally, some effort is made to protect any data base. In addition to physical security for the terminals, special access codes or "keywords" are used to permit entry into the system, and the data may require encryption for transmission. (This is essentially the same process as "scrambling" a spoken conversation.) Nevertheless, what can be done with computer technology usually can be undone with the same technology. Given the time and a readily available terminal (any of the current crop of personal computers would serve), a talented person can eventually invade a data bank, cause it to display a file at the remote terminal, and then erase any record of the entry. The invader could also maliciously rearrange the data in the file or erase the file. The term file should be understood to include the basic program that determines how the file is put together as well as the compilations of data within it.

A related weakness of computer systems is one that applies to telephones and to communications systems in general. Computers generate electronic signals that can be wiretapped or that may leak between physically unconnected telephone lines by means of overlapping magnetic fields, an induction phenomenon known as crosstalk. Moreover, transmissions of data through the telephone system at any great distance are likely to be carried by microwaves or satellite relay and are therefore as vulnerable to interception as voice conversations.

To date, most computer crime has been perpetrated for monetary gain by altering programs as well as data. Most notable was a 1973 scandal involving the Equity Funding Corporation of America, in which corporate insiders created thousands of bogus insurance policies that were then sold to other insurance companies. In another example, a computer genius barely out of high school swindled Pacific Telephone and Telegraph Company out of at least a quarter of a million dollars in the early 1970s. He subsequently was convicted, served 40 days in jail, and then hired himself out to various corporations as a computer security consultant.

Such threats diminish, however, when compared to the issue of the type and quantity of information available in computer data banks. Most data bank material on private citizens consists of insurance and credit history matters, law enforcement files, and income tax and military service records. To the credit of U.S. lawmakers much of this information is protected from disclosure by virtue of its exempt status in the 1966 Freedom of Information Act, which otherwise requires federal agencies to make available to any citizen, on request, a wide range of unclassified documents and records.

The possibility of psychic surveillance has been investigated both in the U.S. and the Soviet Union. What conclusions were reached and what regard the intelligence communities of these countries have for the potential of extrasensory perception is not known.

Ironically the existence of this legislation has created an unfortunate opportunity for privacy violations of a different sort. Because the federal government collects data on private companies, many firms have begun to use freedom-of-information requests as a source of intelligence on the confidential business of their competitors. Although the act specifically forbids disclosure of trade secrets, the tremendous processing capabilities of the computer allow sophisticated intelligence analysis that can reconstruct valuable information from seemingly innocuous data contained in government files. Combined with the intuition and the ability to weigh strategies that good business leaders possess, computer analysis can be used to lay bare the most confidential plans and communications of a competitor.

Of more concern to ordinary citizens is the information about themselves that is possessed by private firms. Most credit-reporting agencies rely upon the reports of subscribers and of part-time "investigators" who may receive so little remuneration per report that the quality of investigation is very low. Everything said about an individual, whether true or not, will go into the data base where it will gain the strength of truth simply because it comes from the "unimpeachable" computer. A lie told by a malicious neighbor will stay on the record without even the cautionary "alleged" qualifier used by journalists and law enforcement officials. Very little effort is made to check the truth of the statements, and the subject of the file is never asked because the investigation is "confidential." Such sloppy workmanship perpetuated by major insurance companies and credit companies is normally a minor irritation, but some people have suffered substantial damage from these practices. Nor has the U.S. government been immune to compiling files of unsubstantiated material. During the anti-Communist hysteria of the 1950s people were branded as subversives solely because their car was parked on the same block at the time of a Communist Party meeting.

The future

Because of the highly secret, intensely competitive nature of surveillance, one may well wonder if the technology surveyed above—the devices and methods that for various reasons have become common knowledge—represent the mere tip of the iceberg. What other fantastic techniques, presently described only in science fiction and perhaps hinted at in news stories, are being used to invade the privacy of human beings? Might microwaves, for instance, be involved in a new form of surveillance known only to the highest echelons of the intelligence community? In 1972 it became widely publicized that a low-power microwave beam, apparently too weak to be of use as a spy-beam microphone, had been irregularly bombarding the U.S. embassy in Moscow since the mid-1950s. Although U.S. State Department officials had known about the radiation, its embassy personnel had not. Shortly after the revelation, diplomats working in the building began to fear that the beam was some kind of "radioneurological" or "psychotronic" weapon intended to increase fatigue, confusion, and illness, and consequently countermeasures were taken, including the installation of aluminum screens on the embassy windows. Two government studies, however, apparently undermined that theory by failing to find any evidence of a health hazard. Today, two intriguing questions remain. Why did the State Department keep the existence of the beam concealed even from its diplomats? And, if one accepts the conclusions of the health studies, what was the true purpose of the beam?

In 1977 Robert Toth, a correspondent for the Los Angeles Times, was arrested in Moscow by the KGB after he had received a study on extrasensory perception from a Soviet scientist. Although his detention and interrogation on charges of espionage were decried by Western officials and the media as harassment and a frame-up, could the Soviets have had more serious reasons? It is well known that Soviet interest in parapsychology is high, in marked contrast to the low exposure such research receives in the openly published literature in the West. There are also suspected instances of practical application, such as the appearance of a well-known Soviet hypnotist and ESP researcher at an international chess match allegedly to "psych out" an exiled Soviet player. In leaving no stone unturned U.S. intelligence agencies have conducted their own investigations of the possibilities of psychic surveillance and reportedly have sought evidence for remote-viewing abilities and other exotic powers from U.S. academic researchers. Unfortunately all that is available to the ordinary citizen on this topic are rumors, anecdotes, and more questions.

1984? Perhaps one need not be concerned about it after all. If surveillance technology is being developed that can invade the privacy of human thought, then 1984 is long past.

THE PROMISE OF SYNTHETIC FUELS

Stimulated by the great increases in the cost of oil and natural gas, scientists and engineers are working to achieve the large-scale production of synthetic fuels.

by Charles A. Stokes

The world's industrialized nations are facing one of the greatest technological challenges ever to confront them—the need to create rapidly and effectively a capability to produce enough synthetic fuels to counter in a meaningful way the oil dominance of the Organization of Petroleum Exporting Countries (OPEC). The raw material of choice for these first synthetic fuel plants is coal for several reasons: the vast world coal reserves, including lignites and peat; the existence throughout the world of a highly developed coal mining industry; the susceptibility of coal to conversion to various types of ready-to-use, clean fuels; and the existence of enough already proved technology to launch the industry.

Faced with the need to move quickly because of the ever-rising petroleum prices, the industrial countries have little choice but to use the existing technology. This technology has been kept alive and improved over the many years since its origin in Europe and the U.S. Synthetic fuels technology first emerged from laboratory to practice in the years between the world wars. But extensive research, ongoing since the mid-1950s in the U.S. and more recently begun again in Europe and Japan, has been greatly accelerated since OPEC began to restrict the supply and raise the price of oil in 1973. This research promises considerable improvement in cost and efficiency of synthetic fuel manufacture when the early commercial plants, now in the planning and design stages, are revamped and expanded.

Sources of synthetic fuels

Coal is by no means the only raw material available for synthetic fuels. There are vast deposits of heavy hydrocarbons, ranging from very heavy crude oils to even heavier bitumen-coated sands known as tar sands. Intensive efforts are under way to use these materials; the most important projects are in Venezuela on heavy crudes and in Canada on tar sands. In addition to those sources, there are the even larger hydrocarbon deposits in shale rock, also well-distributed throughout the world. The U.S. oil shale resource is larger than the known reserves of oil in the Middle East and is the largest yet to be found in any country. The processing of heavy crude oils, tar sands, and oil shale yields what is usually called a synthetic crude oil that can be refined in conventional refineries. Conversion of tar sands into synthetic crude is commercial now.

Table I. Basic Reactions in Fuel Synthesis from Coal

1. $C + 2H_2 = CH_4$ + heat
2. $CH_y + nH_2 = CH_{y+2n}$ + heat
3. $C + O_2 = CO_2$ + heat
4. $C + \frac{1}{2}O_2 = CO$ + heat
5. $C + H_2O = CO + H_2$ (absorbs heat)
6. $CO + H_2O = H_2 + CO_2$ (nearly neutral thermally)
7. $CO + 3H_2 = CH_4 + H_2O$ + heat
8. $2CO + 2H_2 = CH_4 + CO_2$ + heat
9. $nCO + 2nH_2 = C_nH_{2n} + nH_2O$ + heat
10. $2nCO + nH_2 = C_nH_{2n} + nCO_2$ +heat
11. $CH_4 + 2H_2O = CO_2 + 4H_2$ (absorbs heat)
12. $CH_4 + H_2O = CO + 3H_2$ (absorbs heat)
13. $CO + 2H_2 = CH_3OH$ + heat
14. $CO_2 + 3H_2 = CH_3OH + H_2O$ + heat

C, CH_y	= coal
H_2	= hydrogen
CH_4	= methane
CH_3OH	= methanol
CH_{y+2n}, C_nH_{2n}	= hydrocarbons
CO	= carbon monoxide
CO_2	= carbon dioxide

CHARLES A. STOKES *is Chairman of the Stokes Consulting Group, Naples, Florida.*

Even renewable resource materials such as starchy grains and natural sugars are once again being used to make ethyl alcohol or ethanol (C_2H_5OH), an excellent fuel for internal-combustion engines. Efforts are being made to revive the use of cellulosic materials for the manufacture of ethanol. Wood was a commercial source of alcohol in Europe prior to and for some time after World War II.

Depending upon how one defines synthetic fuels, crudes derived from heavy natural hydrocarbons and oil shale and the ethyl alcohol derived from renewable resources can be considered as such materials. However, the new fuels that are being derived from coal are, in every sense of the word, synthetic.

Direct liquefaction

In oversimplified terms the preparation of synthetic fuels from coals of all kinds can be represented as involving the two apparently straightforward chemical reactions listed as 1 and 2 in Table I. In actual practice these reactions are difficult to achieve at high enough yields and efficiencies to be commercially feasible. High temperatures and pressures are required. There are by-products and co-products to contend with due to thermal decomposition and synthesis reactions. Catalysts can greatly facilitate both reactions, but maintaining a practical level of catalyst activity at a reasonable cost is difficult. Material handling problems are also severe. While the listing of reactions in Table I is useful to convey a general idea of what is happening in fuel synthesis from coal, these reactions represent end results and do not, by any means, describe the hundreds of highly complex intermediate steps that are involved.

An alternative approach is to first convert coal to suitable intermediate gases (mainly by reactions 3, 4, and 5 in Table I) and then convert that gas, after purification, composition adjustment by reaction 6, and removal of carbon dioxide, to gaseous or liquid synthetic fuels. The intermediate gas also provides a way to make hydrogen (reaction 6) for some of the other reactions such as 1, 2, 7, 8, 9, 10, 13, and 14. By combining some of these reaction steps, one can arrive at an idealized result: $2C + 2H_2O = CH_4$ (methane) + CO_2 (heat effects in close balance).

The existence of all of these reactions has been known for 50 years or more. Most of them have been used in one process or another during that time, but it is only in the last few years that massive efforts have begun to exploit all of these reactions in order to make synthetic fuels. These fuels range from methane to heavy fuel oil and include such important products as gasoline, diesel fuel, jet fuel, and methanol as well as by-products which serve as raw materials for the chemical industry.

Coal has been converted into coke, tars, oils, by-product chemicals, and fuel gases since the late 1700s—usually by thermal decomposition in the absence of air in retorts and coke ovens, but also by reaction of coal and coal-derived coke with air and steam (producer gas and water gas). However, these relatively crude and nonselective conversion steps and their products are not considered to belong in the category of synthetic fuels. The synthetic fuel industry is generally considered to have started a little prior to World

W. B. Finch—Stock, Boston

Coal is a major source of synthetic fuel. At the left are hand-dug peat bricks in Scotland, while below a dragline shovel strip mines for lignite in North Dakota.

© Allen Green—Photo Researchers

War I with the work of the German chemist Friedrich Bergius. He patented a process for making liquid fuels from powdered coal suspended in a carrier liquid by reaction with hydrogen under very high pressure in the presence of a catalyst. This process offered many difficulties when trying to make such fuels on a large scale, but it was, nevertheless, used successfully to produce important amounts of military and industrial fuels in Germany during World War II.

The Bergius approach was augmented in the period between the wars by another German coal-liquefaction process (Pott-Broche), which also succeeded in supplying significant quantities of heavy fuels and electrode coke during World War II. In this process coal was extracted at 415–430° C and pressures of 100–150 atmospheres with a solvent recovered from coal tar distillation (100° C = 212° F). The solvent was distilled off and returned to the process, leaving a small amount of liquid fuel and a larger amount of solid fuel of low ash and sulfur content that could be melted and handled as a liquid. It was found that the addition of hydrogen assisted the process by bringing more coal into solution. The liquid, after removal of ash and undissolved coal by filtration, was separately hydrogenated (combined with hydrogen) catalytically, as in the Bergius process, in order to produce upgraded liquid fuels.

These two processes laid the foundation for what is today called the direct liquefaction approach to making liquid fuels from coal. Since World War II extensive research in the U.S. has led to two processes for making liquid fuels from coal that combine the basic findings of Bergius and Pott-Broche. These two new processes are described as solvent refining of coal and are designated SRC-I and SRC-II.

SRC-I closely resembles Pott-Broche in that it uses a recycled solvent under temperatures of about 440° C and typically at 115 atmospheres to break down the coal; but, like Bergius, it adds hydrogen in the dissolution step. This allows time for considerable uptake of hydrogen, catalyzed to

(Top and bottom, right) © Hans J. Burkard/PLATA; (bottom, left) © Doug Wilson—Black Star

some extent by the iron compounds in the coal ash. The hydrogen treatment eliminates most of the sulfur as H_2S (hydrogen sulfide) and some nitrogen as NH_3 (ammonia), thus improving the resulting fuel quality. In the early 1980s the process was to be commercialized in a $1.5 billion plant, with a capacity equivalent to 20,000 barrels per day of crude oil, to be built in Newman, Kentucky, on the Green River. The builder of the plant, International Coal Refining, planned a fivefold expansion by 1990 to achieve a size comparable to a modern petroleum refinery. Funding for the initial plant was to be shared among several commercial partners and the U.S. government, with the government investing most of the money.

The SRC-II process stems from the post-World War II efforts of a modest-sized coal company, later acquired by Gulf Oil Corp., to improve on the Pott-Broche process. The U.S. Department of Energy and predecessor agencies sponsored much of the work. This process is generally similar to SRC-I in the liquefaction-hydrogenation step, but it uses vacuum distillation for product separation rather than filtration or solvent precipitation. (Vacuum distillation is carried on at a reduced pressure so that the substance being distilled boils at a comparatively low temperature and consequently suffers less loss from decomposition.) As in SRC-I, the undissolved residue is used to produce hydrogen and plant fuel, this time by means of a partial combustion process adapted from petroleum refining. The fuels produced are somewhat lighter and more versatile as to end use than are those from SRC-I, but, as with the former, power plant fuel is the major use contemplated.

In comparison with low-sulfur fuel derived from petroleum, the notable differences are a favorably lower viscosity for the coal-derived fuel but also an unfavorably high nitrogen content and a lower heating value on a weight and volume basis. Nevertheless, the SRC-II fuel oil is quite satisfactory for power generation. A 6,000-ton-per-day (coal feed) module of a commercial plant making about 20,000 barrels per day of liquid fuel is scheduled to be constructed in West Virginia so that it will go into production in late 1984 or early 1985. The project is a milestone in international cooperation, involving funding from the Japanese and West German governments along with U.S. and foreign companies and the U.S. government.

As with SRC-I there are plans to expand the SRC-II plant fivefold on the same site. The products expected from the expanded plant are shown in Table II. The thermal efficiency of 72%, while relatively high for synthetic fuels, is well below the 90% level expected for a modern high-gasoline-output refinery. This reflects the increased difficulty in converting coal, with its very low hydrogen content, to synthetic fuels compared to refining crude petroleum already rich in hydrogen. Ammonia is a unique by-product; petroleum refining also generates sulfur and tar acids, the latter to a much lower degree.

Not content with SRC-I and II, which are significant improvements over the Pott-Broche process, chemists and engineers are in the late stages of developing two advanced processes, known as H-coal and Exxon Donor Solvent. H-coal, essentially a vastly improved Bergius-type process, uses an effective catalyst to hydrogenate a bed of ground coal suspended in a liquid, thus providing better contact and heat transfer as well as allowing the use

Combination mine and refinery in Alberta, Canada (opposite page, top), recovers oil-laden tar sands (opposite, bottom left) at depths of up to 60 meters (200 feet) and from them produces 45,000 barrels of light crude oil per day. Steam is pumped through the pipes of such a plant (opposite, bottom right) to clean out accumulations of sludge. Below, a technician checks an underground retort in Colorado used for extracting oil from shale.

Courtesy, Occidental Petroleum Corp.

Table II. Products from a Typical Commercial SRC-II Plant
(Using 33,500 tons per day of West Virginia coal)

Major Products	Quantity per day
Methane	50 million cubic feet
Ethane/propane	3,000 tons
Butane	300 tons
Naphtha	17,000 barrels
Fuel Oils, light and heavy	56,000 barrels
By-products	
Sulfur	1,200 tons
Ammonia	180 tons
Tar acids	240 barrels

One process for obtaining liquid fuels from coal is described as solvent refining of coal (SRC). Two variants, SRC-I and SRC-II, have been devised. The products per day of an SRC-II plant are shown in the table above.

of lower pressures. The process will be flexible as to operating mode and the resulting products will range from predominantly heavy to predominantly light fuels. (Methane and LPG will be by-products in either case.) A commercial plant with an equivalent capacity of 50,000 barrels per day of crude oil is in the preliminary engineering stage. To be located in Kentucky, it is owned by Ashland Synthetic Fuels, Inc., and Airco Energy Co., Inc.

The Exxon Donor Solvent process separates the liquefaction and hydrogenation functions of the SRC-type processes into two steps. This allows the use of a more active catalyst in the hydrogenation step. The fully hydrogenated solvent literally donates active hydrogen to the coal in the liquefaction step. The products are separated from the residue by distillation as in the SRC-II process, but the residue is further processed by a special technique adapted from the petroleum industry known as Flexicoking®. This converts the residue into additional liquid products and fuel gas from which hydrogen for the process can be made. This use of Flexicoking demonstrates a striking feature of the advancing technology of coal conversion to liquid and gaseous fuels, that is, the adoption of technology already used by the petroleum industry, particularly that used in the processing of residual oil. The H-coal process was, in fact, adapted in its entirety from an earlier process called H-oil that upgrades petroleum residuals.

Coal gasification

Two other coal-conversion technologies have also come to the forefront. The first is the conversion of coal to various types of fuel gas: methane (substitute natural gas or SNG); low-BTU (British thermal units) gas (LBG), with roughly 15% the heating value of methane; and medium-BTU gas (MBG), having about 30% of the heating value of methane. The second is the conversion of the MBG into liquid fuels by reactions 9 and 10 in Table I—the Fischer-Tropsch synthesis—and by reactions 13 and 14. These two technologies are termed coal gasification and indirect liquefaction.

In coal gasification the Germans were again in the forefront originally, but there has been a strong U.S. input since the 1950s and by 1980 the work in the U.S. was far greater than anywhere else. Three basic gasification processes resulted from pioneering work in Germany, greatly improved versions of which are used today. They are the Lurgi (fixed-bed), Koppers-Totzek (entrained-bed), and Winkler (fluid-bed) gasifiers.

In the Lurgi process coal is fed in a direction opposite to the flow of hot gases rising in a vertical, cylindrical pressure vessel. At the bottom of the vessel a combination of steam and air or steam and oxygen is injected; either of these reacts with the coal as indicated by reactions 3, 4, and 5 in Table I, yielding intermediate gases and heat. As these gases rise through the bed at pressures of 20 to 30 atmospheres, other reactions, mainly 1 and 6, occur. In the upper, cooler part of the bed, the coal is distilled to produce tars, oils, ammonia, carbon dioxide, and methane and other light hydrocarbons.

The desired result of the process is a large quantity of methane and of hydrogen and carbon monoxide. The greater the proportion of hydrogen in relation to that of carbon monoxide, the better is the gas for subsequent synthesis of methane or of liquid fuels. The by-products distilled from the

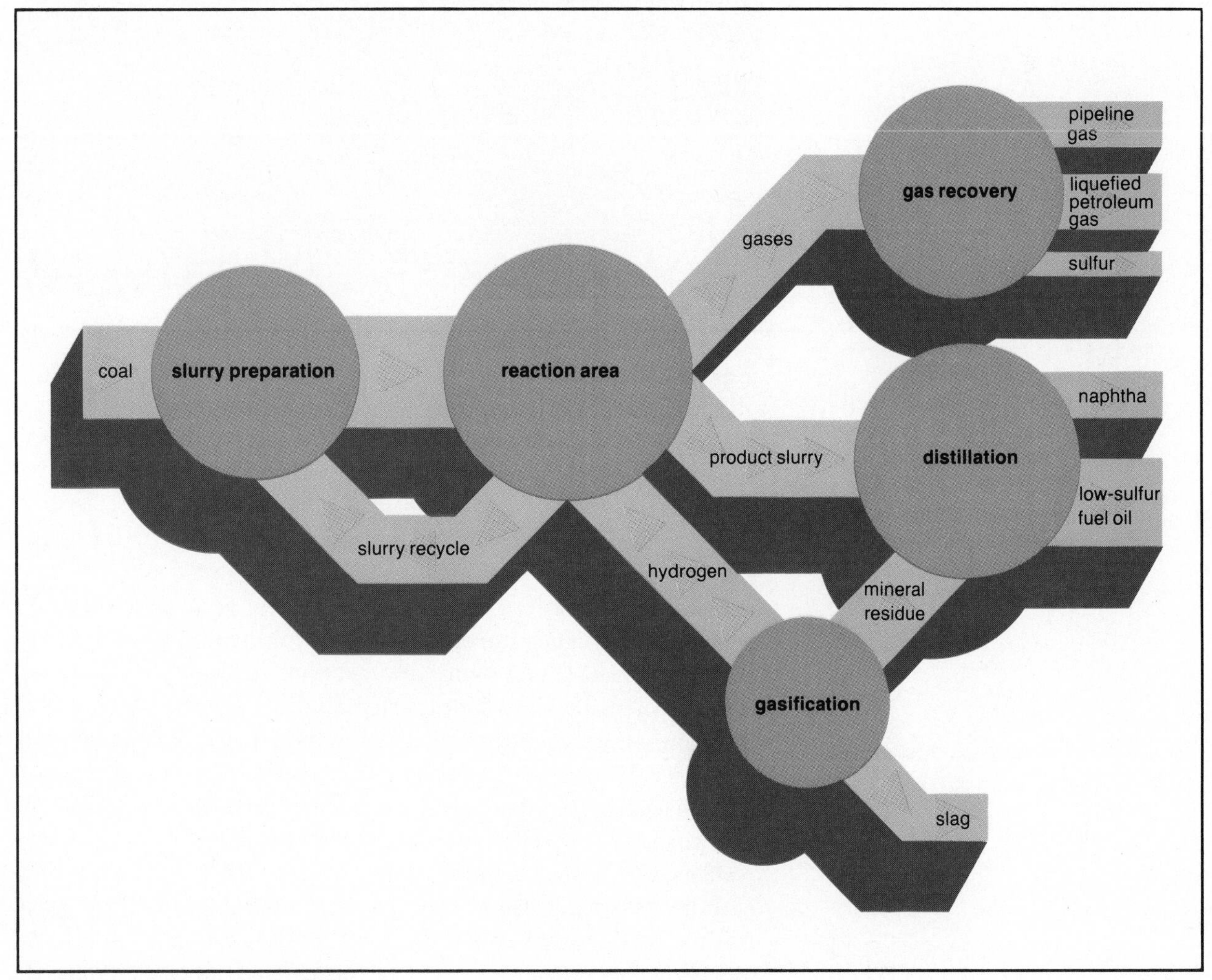

Adapted from information obtained from Gulf Oil Corp.

(Top) Courtesy, Fluor Corp.; (bottom) H. E. Nuttall, University of New Mexico

The Sasol complex in South Africa, the first large modern synthetic fuels operation in the world, uses the Lurgi and Fischer-Tropsch processes to produce fuels from coal by indirect liquefaction. At the top are the 36 Lurgi gasifiers of the Sasol Two Gasification Unit. Below are the raw material silos, cooling towers, and grass-stabilized ash mound at Sasol One.

coal are recovered separately and either burned as fuel, gasified separately, upgraded for various chemical uses, or used in a combination of these.

During World War II the process was used in Germany as a source of coal tar and oils for upgrading to military fuels. The gases were further converted to synthetic fuels by reactions 6, 9, and 10. Considerable by-product fuel gas remained which was largely methane. This could be used as such or converted via reaction 12 of Table I into more synthesis gas. The conversion to liquid fuels was primarily by the Fischer-Tropsch process.

The Lurgi process can be operated to produce primarily methane by catalytically converting carbon monoxide and hydrogen (reaction 7). Primary by-products are recovered and used as noted above. A large amount of CO_2 from reaction 6 is rejected from the process. This is being considered as a source of gas for increasing the recovery of crude oil from old wells.

In South Africa in the 1950s a large Lurgi and Fischer-Tropsch indirect liquefaction complex named Sasol was put into operation successfully. Further expansions are underway, making Sasol the first truly large modern synthetic fuels operation in the world. When the latest expansion is complete, the three Sasol plants will supply nearly 50% of South Africa's liquid fuel needs by the mid-1980s, more than 130,000 barrels of oil per day, together with a substantial amount of fuel gas and by-product chemicals.

In the South African complex, which was engineered and built by several U.S. contractors, a U.S.-developed Fischer-Tropsch synthesis process called Synthol is being used for all except a part of the original plant. This is because Synthol yields more of the desired hydrocarbon fuels and has better operating characteristics. Interestingly enough, the process was developed after World War II when there was an excess of natural gas and shortage of crude oil in the U.S.

As a starting point for the production of methane or liquid hydrocarbons, the Lurgi process has important drawbacks. These are being overcome by development work in Great Britain, in the U.S., and in West Germany. The improvements center on: (1) use of higher pressures and higher bottom-of-the-bed temperatures, leading to greater throughput and higher methane content; (2) removal of ash as a molten slag; (3) less by-product production; (4) lower steam usage; (5) recycling of by-products to make more methane and synthesis gas; (6) use of a wider range of coals, particularly the U.S. caking coals, which are coals that become plastic when heated; (7) higher output per dollar of investment; and (8) improved thermal efficiency.

Under present plans a test module of a commercially sized plant will be on-stream in the U.S. by about 1985, the end product being methane. It will be a cooperative project between the U.S. government and a consortium led by Conoco Coal Development Co. When this unit has met its expected performance, it will provide a basis for rapid expansion in the use of the improved technology. Many expect the improved Lurgi process to become a workhorse in coal gasification during the late 1980s and the 1990s.

The existing Lurgi technology is being embodied in a large new commercial plant in North Dakota that was designed to make methane to be fed into an existing natural-gas pipeline system. One of the first modern fully commercial synthetic fuel plants in the U.S., it will use 4.7 million tons per year of North Dakota lignite, a relatively low-grade coal that cannot be shipped economically any great distance from the mine. The cost of the gas in 1983 has been estimated at $9 per million BTU, as compared with a world crude oil price of roughly $6 per million BTU delivered to the U.S. in late 1980. The small particles of coal, not usable in the Lurgi gasifier, are to be conveyed to a power station. This sort of integration with power generation is expected to be a typical characteristic of the synthetic fuel industry in order to achieve maximum utilization of the coal. The gasifier by-products except for sulfur, which will be sold, will be used as fuel for the North Dakota plant.

If the objective of the coal gasification is to produce clean gaseous fuel for direct use as medium-BTU gas, or for further synthesis or for hydrogen production, two other existing processes may be used. They are the Winkler fluid-bed and the Koppers-Totzek entrained-bed gasifiers. A third, the Texaco entrained-bed gasifier, is under intensive continuing development for use with coal. The advantages of these processes, all of which gasify normally with steam and oxygen but which can also operate with steam and air, are the almost complete absence of by-products other than H_2S, COS (carbonyl sulfide), and NH_3; these arise from sulfur and nitrogen that are chemically combined with the coals used.

In the Winkler process small fragments of coal measuring about 6.35

millimeters are gasified at one to four atmospheres of pressure in a fluidized bed with steam and oxygen or air. (A fluidized bed is produced when a gas is forced through a bed of particulate matter at sufficient velocity to suspend the particles yet not blow them out of the container. This bed behaves much like a fluid; hence the name.) As the coal is used, more is fed so as to maintain an equilibrium state at the 1,000 (± 150)° C level, depending upon the position in the bed. Gases leaving the bed are CO_2, CO, H_2, and H_2O (water vapor) with small amounts of H_2S, COS, NH_3, CH_4, and N_2. These hot gases pass through boilers in order to recover high-pressure steam and then are cooled, washed, compressed, changed chemically by reaction 6 in Table I, and finally highly purified for the synthesis of fuels or chemicals.

The "second generation" version of this process is under development in West Germany. It will operate at pressures of ten atmospheres and at temperatures 150° C to 250° C higher than the original. This yields a higher content of methane in the gases and up to 5% better carbon-conversion efficiency.

In the Koppers-Totzek process ground coal is gasified at atmospheric pressure and a temperature high enough to remove the coal ash as molten slag. A number of commercial plants throughout the world use this method, and in 1981 a dozen or so new plants were being considered. An improved Koppers-Totzek entrained-bed gasifier was being developed in Europe by Shell and Koppers AG. This process feeds coal dry under pressures as high as 30 atmospheres, which will save appreciable capital and boost overall thermal efficiency somewhat.

Several other fluid-bed gasifiers are under development, most notable of which are the Institute of Gas Technology's U-Gas process and the Westinghouse gasifier. They operate at pressures up to 20 atmospheres, use caking coals, and remove ash by a new agglomerating technique intended to make such removal somewhat easier. Other fluid-bed gasification processes under development feature pressure operation at 20 atmospheres or higher, use of air instead of oxygen, and partial removal of sulfur in the bed by the addition of alkaline materials so that the fuel gas can be burned hot.

If the fluid-bed gasifier principle is extended to achieve complete combustion with heat recovery in the form of steam, one then has a fluid-bed boiler that can burn all kinds of coal, petroleum coke, wood, and municipal solid waste. Again, sulfur, if contained in the feedstock fuel, can be removed in the bed by the addition of a reagent. Such boilers promise to be among the most widely used direct-coal-burning devices.

A considerable part of the overall cost of making fuel and synthesis gas is the required capital and operating expense of the facility for producing oxygen. Scientists are seeking a way around this by simultaneously catalyzing the reaction of carbon with steam and carbon with hydrogen as in reactions 1 and 5. Reaction 6 is going on at the same time. The product gases are separated to remove CO_2, which is rejected, and to recover and recycle CO and H_2 with methane, the overall product desired. The net reaction then becomes $2C + 2H_2O = CH_4 + CO_2$ (neutral heat effect). This is known as the Exxon catalytic gasification process. If the pilot plant work on a 10-inch by 80-foot (25-centimeter by 24-meter) reactor, due for completion

Courtesy, General Electric Research and Development Center

First-of-a-kind research facility is the integrated gasification/combined cycle (IGCC) plant built by General Electric Co. at Schenectady, New York. Coal is gasified and then cleaned of pollutants and burned to produce electricity. IGCC facilities are cleaner and have the potential to be 20–25% more efficient than coal-burning power plants with stack gas scrubbers.

in 1981, is successful, this process could become the basic coal gasification method of the future. It can use any type of coal, even highly caking types; makes no by-product tars and oils; requires no outside source of heat; operates under pressure (about 35 atmospheres); makes methane directly; and requires no oxygen plant. It could be used simultaneously as a source of hydrogen for direct or indirect liquefaction with methane as a by-product. At present it is difficult to visualize any likely improvement in coal gasification that could go beyond the results promised by the Exxon process.

Indirect liquefaction

The manufacture of CO and H_2 as has been described above is the key to a third general approach to the production of synthetic fuels, indirect liquefaction. This most versatile of the fuel synthesis methods involves three steps: preparation of a purified mixture of about two parts of H_2 and one of CO; catalytic combination of these two gases at pressures from one or two to as much as 100 atmospheres to produce intermediate products; and refining or conversion of the first-stage products into finished fuels. The best-known example is the Fischer-Tropsch synthesis used at the South African Sasol complex described above. The main reactions are represent-

(Top) Bob Kolbrener—Uniphoto; (bottom) Battelle Pacific Northwest Laboratories

Forests such as that in Utah (top) are the sources of the wood chips used in the biomass conversion reactor of the U.S. Department of Energy (bottom). The reactor can gasify about 50 pounds of wood chips per hour. Biomass is one of the major alternatives to coal in the production of synthetic fuels.

ed by equations 9 and 10 in Table I. The product is a highly complex mixture of hydrocarbons together with alcohols, ketones, and other oxygenated compounds. Considerable refining is required to finally produce gasoline and jet and diesel fuels with by-product methane and liquefied petroleum gas (LPG). Little heavy fuel is produced.

The range of fuels produced by this process is an advantage, but the overall plant is costly to build and operates at low thermal efficiency. In recent years a breakthrough made by Mobil Corporation and the U.S. Department of Energy promises a simpler way to achieve gasoline from coal by indirect liquefaction. In this method coal is converted to a mixture of H_2, CO, and CO_2 by a direct route that avoids by-products. The gases are then compressed and catalytically converted at about 100 atmospheres to methanol as indicated in reactions 13 and 14 in Table I. This methanol, containing some water, is revaporized and converted again catalytically, at relatively low pressure, to high-octane gasoline that requires essentially no refining. The yield of gasoline is about 85% of the theoretical top limit, and this is increased to 90% or more when by-product gases are combined by using a petroleum-refinery process known as alkylation. The only residual products are methane and LPG, which are readily marketable. Several large plants, one for 50,000 barrels of gasoline per day, are in the preliminary design stage in the U.S. There is great worldwide interest in the process, which can also be used to make gasoline from cheap natural gas now being flared for lack of market.

Methanol itself is attracting great interest as a fuel. In an engine with a compression ratio in the range of 14 to 18, it is a more efficient and cleaner fuel than gasoline. Ethanol, made by fermentation of such renewable resources as grain and sugarcane, behaves similarly. While these alcohols can be blended with gasoline, their use as a sole motor fuel is more promising and is being evaluated by several countries. Methanol is also expected to become an important power-generation fuel for use in fuel cells.

Future prospects

Looking back on the efforts of scientists and engineers to create a viable synthetic fuel industry—an effort stretching over a span of 65–70 years—one can make two general conclusions: all of the processes involved in the current and emerging manufacture of synthetic fuels from coal can be generalized into a few common operations; and these operations, in turn, will eventually evolve into an ultimate scheme for processing coal to synthetic fuels, which will be achieved by a steady stream of improvements rather than by dazzling breakthroughs to entirely new ground.

With regard to the first observation, these common operations in the case of liquid fuels are: coal preparation; production of carbon monoxide and hydrogen via gasification with oxygen and/or steam; liquefaction, direct or indirect, to produce raw hydrocarbons or such oxygenated compounds as methanol; conversion and refining of the raw liquid hydrocarbons or oxygenated compounds; conversion and refining of the raw liquid hydrocarbons (usually by some form of processing with hydrogen as a reactant) to make finished fuels and petrochemical raw materials which meet the many required end use needs; generation of steam and power to run the self-contained plant; and effluent processing to meet environmental restrictions with recovery of sulfur in salable form. If the objective is only methane, liquefaction is replaced by direct methane production from coal and methanation of carbon monoxide; fuel refining is eliminated because there are no net liquid products. All other common operations remain the same. The production of medium-BTU gas for industrial and power-plant fuel is inherent in the basic gasification step; the production of low-BTU gas for the same purposes is simply a matter of using air in the gasification step instead of oxygen.

The ultimate processing scheme combining these common operations might be called a "coalfinery" by analogy to an oil refinery. Each type of refinery takes a crude natural product and converts it to clean fuels without undesirable discharges to the environment. In the "coalfinery" the step most significant from an economic point of view is apt to be the basic gasification procedure that makes from coal smaller molecules out of which the desired fuels can be synthesized. This step may be carried out by reaction of coal and steam directly with the aid of a recycled catalyst, thus avoiding an oxygen plant that requires large amounts of capital and energy.

There is no question of whether such a "coalfinery" can be built. It is only a matter of developing, through the operation of pilot and small commercial plants, enough good engineering design information to construct plants of maximum efficiency, flexibility, and reliability at an acceptable range of product sale price. Indeed, the question of synthetic fuels is now mainly a matter of price. Because of the large inventory of coal in the U.S.—100 years or more—it appears certain that the nation can afford to move not only with assurance of controlling costs at an acceptable level but also with great strategic gains.

And so, at last, the world is moving on conversion of coal and other raw energy sources, nonrenewable or renewable, to clean and convenient fuels. This accomplishment is vital to buy the time for developing such ultimate solutions to energy supply as the breeder atomic reactor, nuclear fusion, and the use of such renewable resources as solar energy. However, even with fusion-produced electricity, synthetic fuels will be needed for the foreseeable future in transportation, for peak-load energy supply, and as raw materials for the chemical industry. The emerging synthetic fuels industry in the U.S., based largely on coal, shale, and tar sands, could absorb on the order of $1 trillion of capital over the period from 1980 to 2010.

The ability of the U.S. to carry out this program is believed to be far more influenced by the overall political management of the undertaking than by technical limitations. The early German accomplishments, based on technology relatively crude by modern standards, together with the improvements that have been made by U.S. technologists along with contributions from Europe and Japan, show what can be done when strong nations have the will to do it. If there could be any real doubt of this, one only has to look at a nation that has done it already—South Africa—now approaching self-sufficiency in liquid and gaseous fuels by means of producing synthetic fuels based on coal.

(Top) Courtesy, Department of Energy/Archer Daniels Midland; (bottom) © Dennis Brack—Black Star

Grain is delivered to a plant in Illinois for conversion into alcohol (top). A bumper sticker (bottom) recognizes the increasing use of crops for the production of fuels.

THE NOT SO GENTLE RAIN

Hundreds of lakes in North America and Europe are now "dead"—devoid of fish and other marine life—because of acid rain, precipitation that has been altered chemically by pollutants in the atmosphere.

by Gene E. Likens

Interest in the chemistry of rain and snow is at an all-time high. Chemicals in precipitation may be an important source of plant nutrients, or they may serve as pollutants and toxicants for forests, agricultural areas, wetlands, and lakes, depending on the chemical and the amount involved. The magnitudes and the historical trends of these inputs are poorly known for large areas of the Earth, even for the most populated regions. As a result large-scale programs to monitor the chemistry of precipitation have recently been initiated in the United States, Canada, and Europe.

Sources of ions in precipitation

A diverse assemblage of dissolved and particulate substances are found in rain and snow, and they originate from a variety of sources, including atmospheric gases, terrestrial dust, oceanic spray, volcanic emissions, and industrial emissions. The water itself comes from evaporation and transpiration, as a part of the Earth's hydrologic cycle, and initially it is essentially pure. Once water vapor reaches the atmosphere, however, it condenses on solid particles and eventually reaches equilibrium with atmospheric gases. One of these gases is carbon dioxide (CO_2), which forms carbonic acid (H_2CO_3) when it dissolves in water. Carbonic acid is a weak acid (*i.e.*, one that dissociates only slightly in distilled water), so that at normal atmospheric concentrations and pressures enough of it is dissociated to lower the pH of rain and snow to about 5.6. The term pH is defined as the negative logarithm of the hydrogen ion concentration. The pH scale ranges from 0 to 14, a value of 7 characterizing a solution with 1×10^{-7} g of hydrogen ions (H^+) per liter. There is an equal concentration of hydroxyl (OH^-) ions at pH 7, and so the solution is considered to be neutral. A pH value below 7 indicates increasing acidity, and a value above 7 indicates greater alkalinity. Because the pH scale is logarithmic, unit changes may be misleading relative to the actual changes in amount of hydrogen ion or hydroxyl ion involved. For example, a solution at pH 4 contains 100 micrograms of hydrogen ions per liter, whereas at pH 3 it contains 1,000.

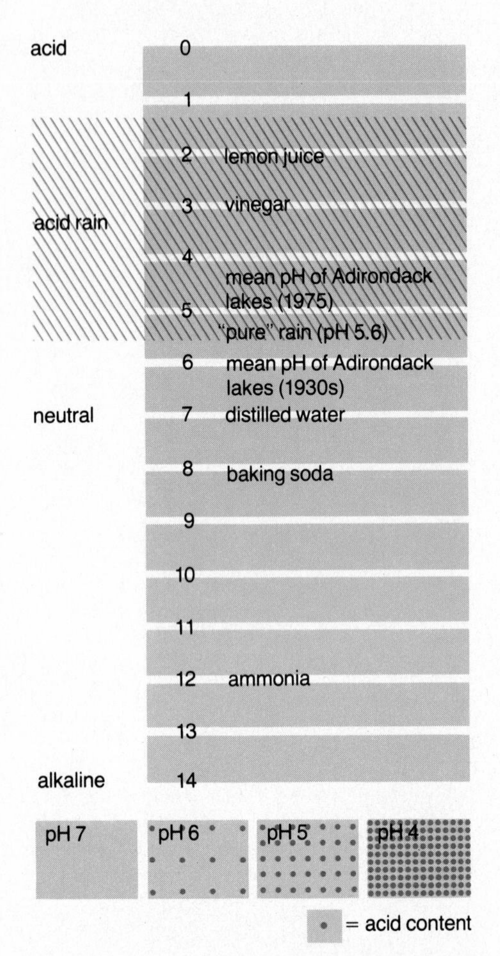

The pH (potential hydrogen) scale (above) measures the concentration of hydrogen ions. When a substance contains more of these positive ions than negative ions, it is acidic and has a pH of less than 7. The combustion of fossil fuels (opposite page) forms sulfuric and nitric acids in the atmosphere.

***GENE E. LIKENS** is Professor of Ecology at Cornell University, Ithaca, New York.*

(Overleaf) Illustration by John Zielinski

Other substances in the atmosphere may change the pH of these dilute solutions. For example, small amounts of dust and debris may be swept from the ground into the atmosphere by the wind, resulting in occasional brown rains and snow. When dissolved in water, soil particles are usually slightly alkaline in reaction and release basic cations (positively charged ions) such as calcium (Ca^{++}), magnesium (Mg^{++}), potassium (K^+), and sodium (Na^+), with bicarbonate (HCO_3^-) usually the corresponding anion (negatively charged ion). Ammonia gas (NH_3) originates largely from the decay of organic matter, and the ammonium ion (NH_4^+) in rain and snow tends to increase the pH. Dissolution of organic debris can contribute small amounts of dissolved organic compounds and such important plant nutrients as phosphorus, nitrogen, and potassium to rain and snow. In coastal areas sea spray may dominate the chemistry of precipitation, and the important ions are those most abundant in seawater, namely Na^+, and Mg^{++}, and chloride (Cl^-). Carbonate and bicarbonate are minor components in seawater.

In areas with high winds and sparse vegetation, such as deserts, precipitation may contain relatively large amounts of base cations. Precipitation in regions with calcareous soils often contains calcium and bicarbonate, presumably as a result of the incorporation of dust from the soil, and the pH of rain and snow in such areas is usually well above 6. Recent evidence suggests that dust from storms in arid regions of northern China and Mongolia may be transported great distances over the Pacific Ocean and even to the North American Arctic.

Gases such as sulfur dioxide (SO_2) and hydrogen sulfide (H_2S) that come from volcanoes and from other natural sources also can alter the chemistry of precipitation. Hydrogen sulfide and sulfur dioxide are transformed by oxidation and hydrolysis reactions in the atmosphere to sulfuric acid. Thus, the chemistry of natural precipitation depends on the relative amounts of these substances in the atmosphere.

In urban, industrial, and agricultural areas human activities often contribute large amounts of particulate and gaseous materials to the atmosphere and thus pollute the concentrations of various substances in precipitation. Some manufacturing operations (such as the production of cement products), smelting and refining, and the combustion of fossil fuels by electrical utilities are good examples. The emission of huge quantities of sulfur and nitrogen oxides from the combustion of fossil fuels has resulted in the formation of sulfuric and nitric acids in the atmosphere and thereby polluted precipitation over large areas of North America and Europe.

Precipitation of atmospheric substances to the surface of the Earth may occur through "wet deposition" of materials dissolved in rain, dew, snow, sleet, and hail or directly through "dry deposition" by gravity (for example, dust particles settling back to the ground surface). It also takes place by means of impingement of small particles on surfaces, and by gaseous adsorption on surfaces or uptake by organisms. Dry deposition of chemical compounds is difficult to quantify in natural ecosystems but cannot be ignored. Detailed ecosystem studies in New Hampshire and Birkenes, Norway, suggest that at least one-third of the total annual input of sulfur occurs by means of dry deposition.

Frequently the concentration of substances in rain and snow will be highest at the beginning of a storm. During some storms, however, the concentration will peak at maximum values during the middle or even at the end. These depositional patterns depend upon the rate of atmospheric transport of the chemical to the area, or the rate of its formation in the atmosphere, or both. The deposition of substances that originated or were formed in the clouds is called "rain-out," while the deposition of gases and particles that are scavenged below the cloud by falling raindrops or snowflakes is called "washout." Larger dust particles that might be rapidly "washed" from the atmosphere at the beginning of a rainstorm would show decreasing concentration with time, whereas, if polluted air masses (aerosols and gases) are swept into an area during a convective storm, their concentrations in precipitation could rise as the storm continues.

The process that forms acid rain often begins with industrial emissions into the atmosphere of such pollutants as oxides of sulfur and nitrogen. These gases mix with the water vapor in clouds to form sulfuric and nitric acids. When rain or snow falls from such clouds, it is highly acid. Such precipitation is particularly harmful to fish and other marine life in susceptible regions with thin soil and a lack of such buffering agents as limestone, and also has been found to damage some vegetation.

Concentrations also can vary significantly by season. For example, frozen soil and snow cover during winter greatly decrease the input of soil dust and ammonia to the atmosphere. Agricultural activity and seasonal patterns in the use of gasoline, natural gas, and home heating oil affect emissions of, for example, calcium, sulfur, and nitrogen to the atmosphere and correlate with seasonal changes in precipitation chemistry.

A vast variety of chemicals are found in precipitation, including PCB's (polychlorinated biphenyls), formaldehyde, lead, mercury, and pesticides. It has been calculated that some 70% of the total amount of PCB's added to Lake Michigan each year comes in through the atmosphere. In some areas of the northeastern U.S. the concentrations of lead and other metals in rain and snow exceed those recommended by the U.S. Public Health Service for drinking water. These chemicals are largely emitted to the atmosphere from human activities.

Acid rain

Acid rain is a popular term applied to rain, snow, sleet, and hail that has a pH value of less than 5.5–5.6. A pH value of about 5.6 is the lowest that can be caused by carbonic acid when carbon dioxide at normal atmospheric concentrations and pressures dissolves in pure water. Recent careful studies by French scientists of ice cores from Greenland and Antarctica show that the pH of snow falling on those remote areas prior to the Industrial Revolution was about 5.5. Major volcanic eruptions can be seen in this record, and when they occur they may lower the pH slightly, even in those remote locations.

It is now clear that human activities have drastically altered the natural chemistry of precipitation on a continental, if not global, scale. The enormous emissions of sulfur dioxide and nitrogen oxides from fossil fuel combustion and base-metal smelting operations have created a new source for ions in precipitation. Some of the sulfur dioxide and nitrogen oxides are deposited rapidly as gases or very small particles. But if the smokestack is tall enough and meteorological conditions are suitable, these materials may reside in the atmosphere from one to five days and be transported from hundreds to thousands of miles. During this residence in the atmosphere both SO_x and NO_x may be converted into sulfuric and nitric acid, respective-

© Freeman Patterson—Miller Services

Snow in the northeastern U.S. is acid with pH values of about 4.2 to 4.3. Most of the measurable acidity is contributed by sulfuric and nitric acids, which are formed in the atmosphere by industrial emissions.

ly, and these strong acids are added to the natural chemical composition of precipitation.

Small amounts of these strong acids are neutralized by the basic substances in the atmosphere, but in the highly urbanized and industrialized regions of North America and Europe the emissions of these acidifying compounds greatly exceed the very small buffering capacity in the atmosphere and the result is to produce precipitation that is decidedly acidic. Strong acids dissociate completely in dilute aquatic solutions and lower the pH to values well below 5.6.

On an annual basis rain and snow over large regions of the world are currently 5 to 30 times more acid than the lowest value expected (pH 5.6) for unpolluted atmospheres. Individual storms may be several hundred to several thousand times more acid than expected. Much of the eastern U.S. and western Europe is subjected to acid rain. In general, summer rain in the northeastern United States typically has a pH of about 3.7 or 3.8, while the current winter pH value is about 4.2 to 4.3. At a number of locations in the U.S. and Europe, pH values between 2 and 3 have been recorded, and concern about the environmental effects of such acid precipitation has led to major research efforts in Sweden, Norway, the U.S., and Canada.

There is considerable controversy about the transport of sulfur in the atmosphere, much of it relating to the crossing by air pollutants of political boundaries. In the northern temperate zone air masses tend to move from west to east. Thus, sulfur and nitrogen oxides that are produced in one area (or country) may be deposited as acid rain in another. Because of the large

Tom Myers

Rain falls on young corn near Sacramento, California, in one of the areas of the western U.S. where the precipitation has a high acid content. The pH values of rain in some parts of North America and Europe are less than 3. Experiments are under way to determine the effect of acid rain on such crops.

amounts of money that are involved in both generating and controlling emissions, political considerations recently have become important. In an attempt to reduce the impact of air pollution downwind of the emitters, legal issues have been raised by Norway and Sweden with the more urbanized and industrial countries of Europe, between the U.S. and Canada, and between various states in the U.S.

A variety of acids could potentially contribute to acid precipitation. There are strong acids (those that dissociate completely in aqueous solutions) such as sulfuric, nitric, and hydrochloric. The concentration of these acids can be measured directly with a pH meter. There also are weak acids (those that generate protons in a solution if conditions are suitable). Carbonic acid, a variety of organic acids, clay particles, ammonium, and ionic forms of aluminum, iron, and manganese are examples.

In the northeastern U.S. strong acids dominate the acidity of rain and snow falling on the landscape. For example, a storm on Oct. 23, 1975, in Ithaca, New York, had a pH of 4.01. At that level of acidity neither carbonic acid nor ammonium, aluminum, iron, or manganese contributed to the free acidity. A variety of organic acids, such as citric, isocitric, fulvic, and malic, were found in the sample but in very low concentrations. These organic acids contributed only about 2% of the total amount of dissociated hydrogen ions in that particular rainstorm. In contrast, 57% of the measurable acidity was contributed by sulfuric acid and 39% by nitric acid. On the average about 60–70% of the measurable acidity in rain and snow in the eastern United States is from sulfuric, and about 30–40% is from nitric acid.

Inco smelter near Sudbury, Ontario, is dominated by a 380-meter "super stack" that has been estimated to produce 900,000 of the 5.5 million tons of sulfur dioxide that Canada emits into the atmosphere each year. The resulting acid rain is killing fish in many Ontario lakes. A combination of dry deposition and acid rain from the Falconbridge smelter near Sudbury has killed much of the vegetation in the immediately adjacent area (bottom).

Historic trends

The historical record of acid precipitation in the United States is difficult to reconstruct. Prior to 1930 precipitation was not acid at various locations in Tennessee, New York, and Virginia. The presence of bicarbonate in those samples shows that they must have had a pH appreciably greater than 5.6. However, by the mid-1950s much of the eastern United States was subjected to acid precipitation. In contrast the pH of rain and snow in the western U.S. was higher than 5.6. Thus, sometime between 1930 and 1955 there was a large change in the chemistry of precipitation in the eastern U.S.

Today it appears that most of the eastern U.S. is being subjected to acid precipitation. Since 1955 there has been an intensification and a spread of the acidification, particularly in the Southeast. Data on increased atmospheric haziness in the Southeast and on increased sulfur concentrations in precipitation in Florida support this change in pH values. Even though the density of sampling points upon which these distributions are based is not great, the number of points that do not fit the pattern is very small.

There are only six sites in the eastern United States with data on precipitation chemistry during the period 1955 to 1980. Three of these areas—Washington, D.C.; Nantucket Island, Massachusetts; and Tampa, Florida—show upward, although highly variable, trends in acidity. The other three sites show no statistical trend. Additional, high-quality data are badly needed on a long-term basis to evaluate trends in the acidity of precipitation.

In Europe an extension and intensification of the problem during the last 20 years is apparent, and the record of data from Europe is of longer duration and probably of much better quality than that for North America. The emission of sulfur dioxide in Europe has doubled since 1940 to a value in 1980 in excess of 50 million metric tons per year. In contrast, the emission of sulfur dioxide in the U.S. increased much less with emissions of about 27 million metric tons per year in 1980. Yet, precipitation in the U.S. is generally more acidic than in Europe. One explanation may be the large increase in the emissions of nitrogen oxides in North America, from about six million tons in 1940 to a value that by 1980 was only slightly less than the sulfur dioxide emissions.

The increase in the emission of nitrogen oxides appears to have had a large effect on the nitrogen chemistry of rain and snow. Fortunately, in this case the record goes back to about 1915 for upstate New York. The nitrate concentration in precipitation was relatively low there until about 1945, after which it increased severalfold. A similar pattern of increase was observed during the last 10 to 15 years at several other locations in the northeastern U.S. The proportion of nitric acid in precipitation also increased during this same period with a corresponding decrease in the contribution of sulfuric acid.

Ecological effects

The most obvious effects of acid rain on the environment have been on aquatic ecosystems, particularly on those in regions with thin soils and granitic rocks, which provide little buffering to acidic inputs. In Scandinavia thousands of lakes no longer contain viable populations of fish. This finding

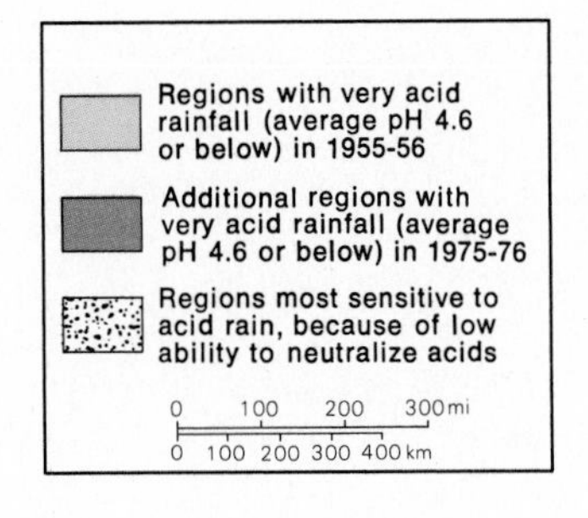

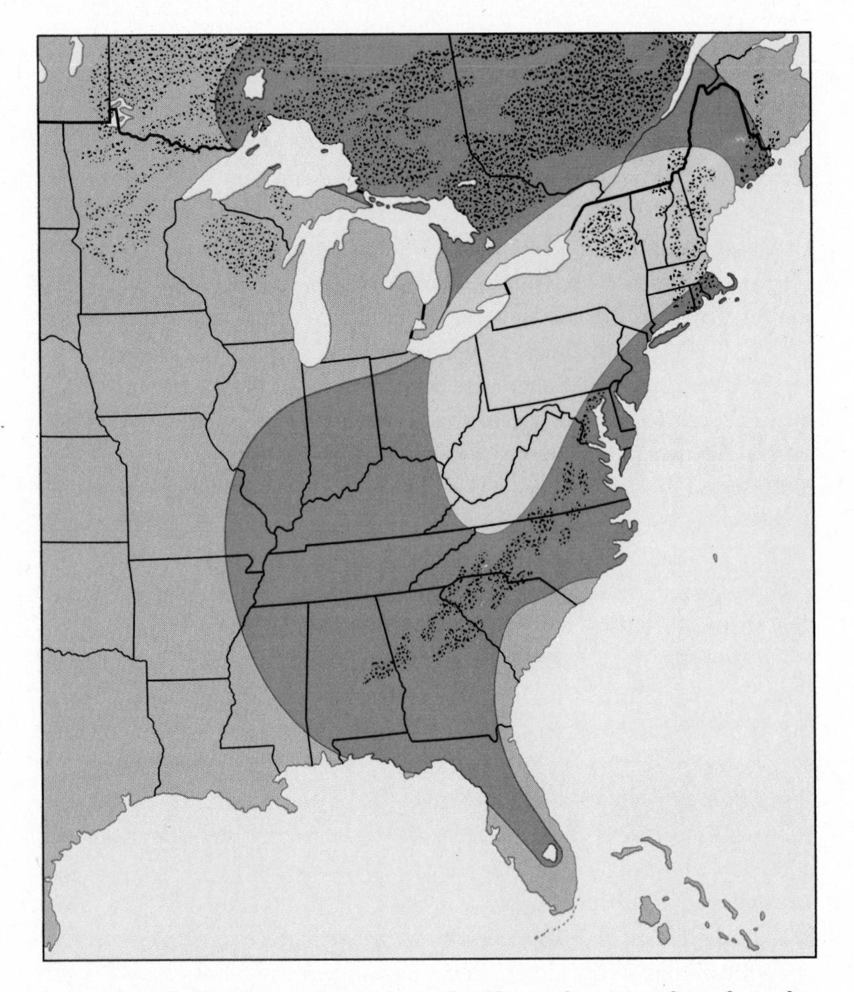

is correlated with increased emissions of sulfur to the atmosphere from the burning of coal and oil, with the increased acidity of rain and snow, and with the subsequent acidification of lakes. In one example from an area in southern Norway with 1,679 lakes, those with a pH of less than 5.0 contained either very small populations of fish or none at all. Lakes that had a pH greater than 5.0–5.5 contained normal fish populations. Lakes varied in pH because of rainfall patterns and local geology. In southern Sweden it was estimated that fish populations in more than 15,000 lakes are affected by acid precipitation. In southern Norway acidification of thousands of freshwater lakes and streams has affected fish populations in an area measuring 33,000 square kilometers (13,000 square miles). This situation obviously represents a major environmental problem.

Essentially all of the other biological components of the acidified lakes are affected. Among the more interesting effects are that the moss *Sphagnum* grows on the bottom of some of the acidified lakes. Commonly, *Sphagnum* encroaches on the surface of lakes, particularly around bog lakes, and it eventually may cover them. But in the acidified lakes *Sphagnum* covers the bottom and effectively minimizes the flux of nutrients from the sediments to the overlying water. As a result the water tends to become nutrient poor, highly acidic, and very clear. Few other life-forms may occur in the

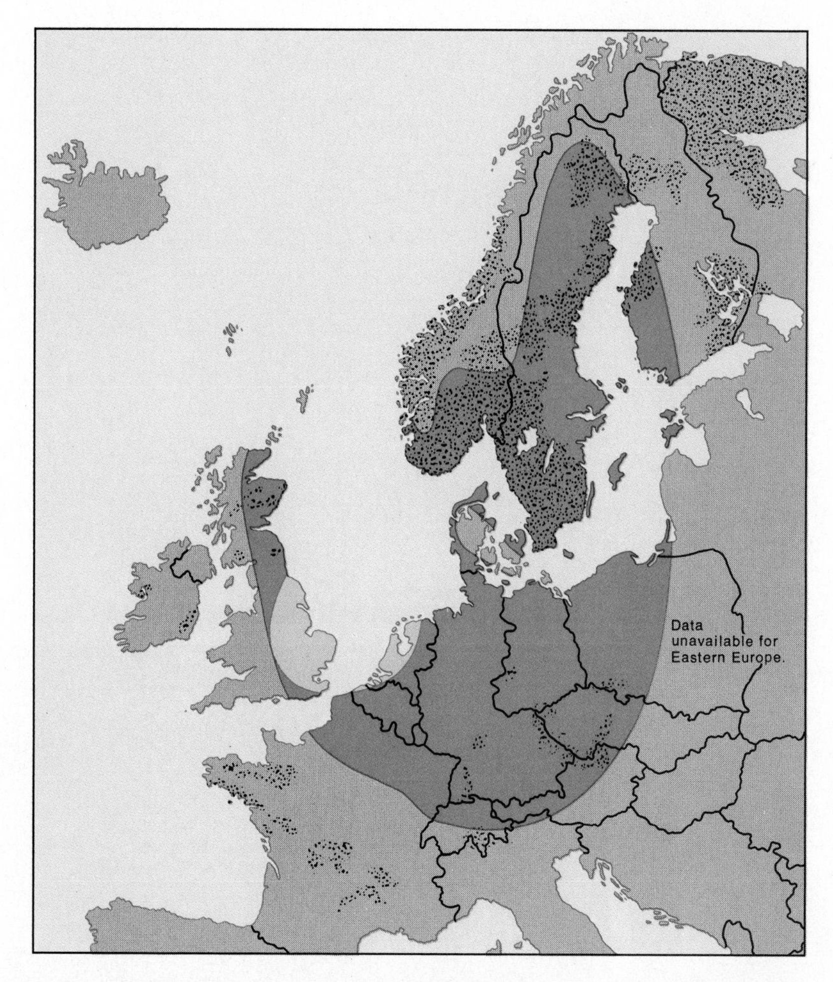

water column. Also, in some of the acidified lakes a mat of blue-green algae grows over the bottom. In places it looks like a carpet of felt, and it is very dense with so much structure that it can be picked up in large pieces.

These acidified lakes are indeed unusual. There was much concern during the early 1970s about Lake Erie being "dead." Lake Erie was never dead. The problem with Lake Erie was that it was too much alive! There was too much biological activity but not of a desirable type. In the Scandinavian lakes, however, the water column is crystal clear and by and large devoid of most organisms.

In the Adirondack Mountains of New York state a similar condition has developed. In 1930 only a very few Adirondack lakes had no fish. By 1980 more than 100 did not contain viable populations of fish, and 237 had pH values of less than 5.

Another aspect of the acid precipitation problem is that much of the acid accumulates during the winter as snow. When snow melts, because of the freezing and thawing process the first part of the meltwater may be much more acid than the remainder. Studies suggest that the first 10% of meltwater could contain about 50% of the chemicals accumulated in the snowpack. These proportions depend upon the environmental conditions during the melt and the rate of the melt. The effect of this meltwater on lakes and

Airplane spreads lime (right) on acidified lake in the Adirondack Mountains of New York. Effects of acid rain are revealed by the clear water of Chapel Lake in the Adirondacks in contrast with vegetation on the surface of a normal lake (below and below right). At the bottom left are normal young brown trout in water with a pH of 5.5; in the center, trout in water with a pH of 5.0 are impaired; and at the right in water of pH 4.5 the trout are fatally damaged.

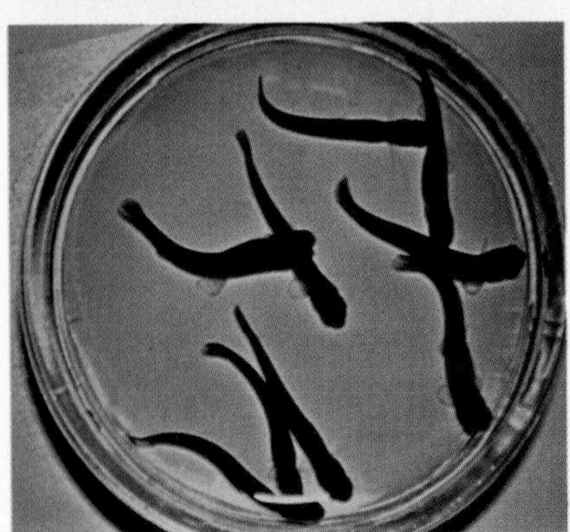

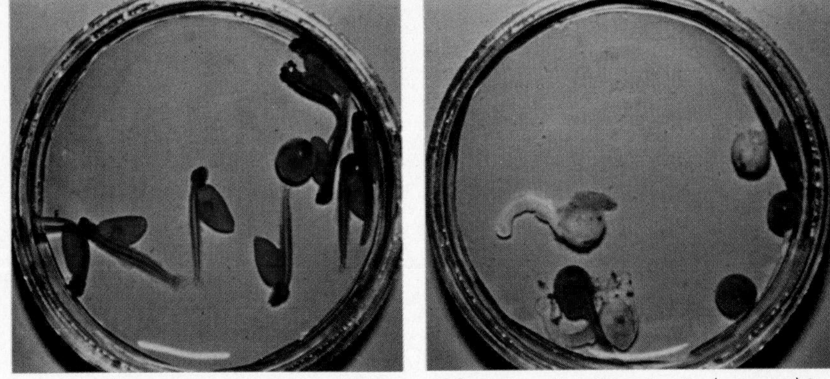

(Top) Dwight A. Webster, Cornell University; (center, left) © John Bova—Photo Researchers, Inc.; (center, right) Owen Franken—Stock, Boston; (bottom, left, center, and right) © Lars Overrein, Norway; photos provided courtesy of Ellis Cowling streams may be sizable. For example, in an ice-covered lake in southern Sweden the incoming snow that fell onto the lake had pH values ranging from 4.0 to 4.4. In January there was a thaw, and much of the snow on top of the ice melted and meltwater flowed into the lake. As a result the acidity of the surface water of the lake increased to about pH 3.3, far greater than the snow that had fallen on the lake. Very acid water was seen at depths of two and even three meters below the surface as the meltwater mixed into the lake. Thus, episodic events such as the melting of the snowpack can produce relatively low pH values in the aquatic environment for short periods of time.

A Norwegian scientist, Arne Henriksen, proposed an empirical model to predict the amount of acidification in lakes based on the calcium concentration of the lake water and the pH of precipitation. This model is an important development in understanding the effects of acid precipitation on aquatic ecosystems, and it provides a framework for asking critical questions about mechanisms of acidification of lakes. Detailed studies of entire watershed ecosystems can provide some of these important answers for landscapes affected by acid precipitation.

A recent and particularly important finding was that aluminum is leached from the soil in regions subjected to acid precipitation. Dissolved aluminum may be very toxic to fish and other aquatic organisms. However, organically complexed aluminum appears to be less toxic to fish than dissolved free (aquo) aluminum or aluminum complexed with inorganic ligands such as fluoride, sulfate, or hydroxide. Thus, at the same low pH value fish can live in naturally acid brown-water lakes (those high in dissolved organic matter) but not in clear lakes (low in dissolved organic matter). Both hydrogen ions and aluminum affect the regulation of body salts and deplete the blood plasma concentration of sodium and chloride in fish.

One remedial measure is to add lime to acidified lakes. This procedure has been attempted in Sweden, Norway, Canada, and the U.S. However, it is expensive and difficult to treat remote lakes, and the beneficial effects are only temporary.

It can be shown in the laboratory that acid precipitation has a variety of effects on terrestrial vegetation, such as leaching of nutrients from foliage, formation of spots of dead tissue, erosion of external surfaces, and inhibition of nitrogen fixation. Thus, it is significant that the lowest pH values for precipitation in the eastern U.S. occur during the summertime, because this is the time when plants are actively growing and presumably are most sensitive to stresses, such as acid precipitation. However, actual effects reflect a statistical problem. It does not rain every day, and all rains are not equally acid. Moreover, plants have periods in which they are more sensitive than at other times. If all three circumstances occur at the same time—if the plant is sensitive, if it rains, and if the rain is very acid—then there can be major effects on the yield, the growth, or the reproduction of the plant. But if all three of these do not occur simultaneously, then the effect may be very small. Thus it is difficult to predict effects in the natural situation. In addition there are several other factors that affect plants, including amount of solar radiation, availability of nutrients and water, ozone concen-

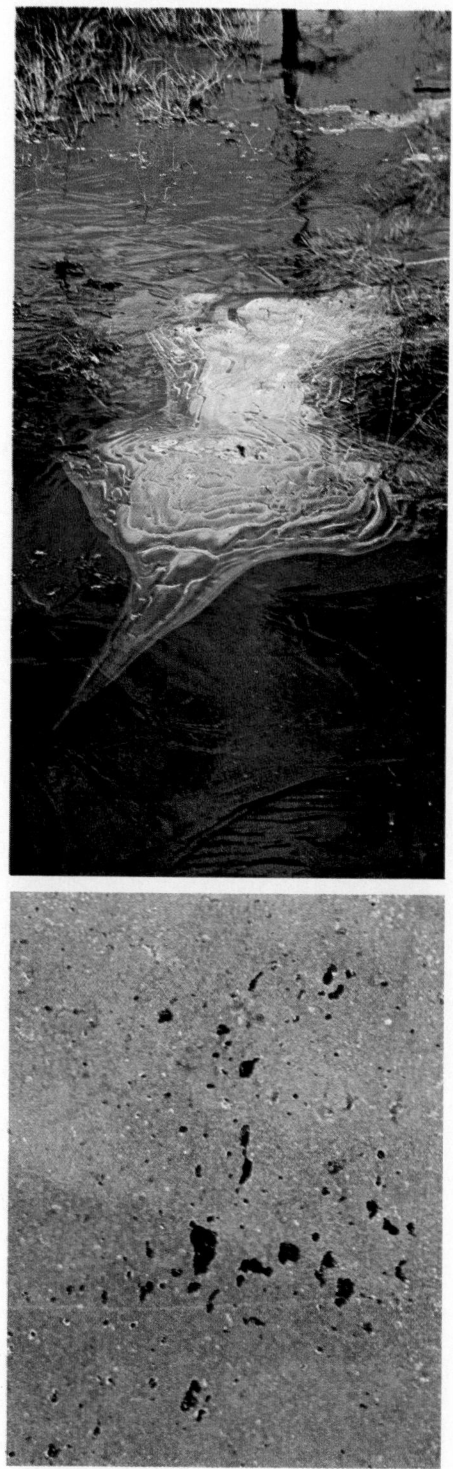

Smoke particles form on ice (above), an example of dry deposition of pollutants. When the ice melts (top), such particles increase the acidity of the water.

Photos, Corvallis Environmental Research Laboratory, U.S. Environmental Protection Agency

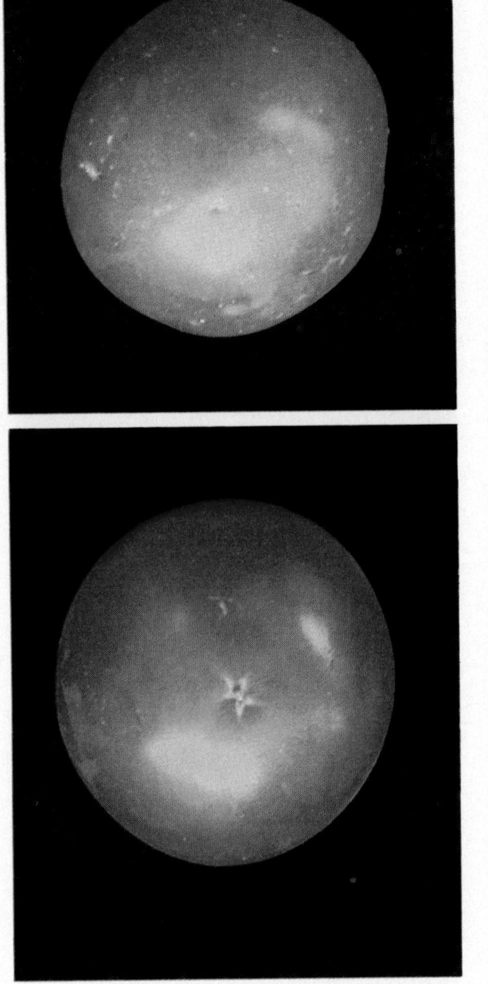

Plants exposed to sulfuric acid rain suffered damage that adversely affected their marketability. At the top are a tomato and spinach leaf that were subjected to rain with a pH of 3.0. The tomato and spinach below were exposed to rain with a pH of 5.6.

tration, and diseases. Because plants respond to the combined effects of all of these factors, it is difficult and probably unrealistic to try to separate their effects in the natural environment.

The effects of acid rain on soil are scarcely known. Clearly one effect is the increased leaching of aluminum and other metals. Concern has been raised about the contamination of drinking water supplies by these metals. In humid regions, however, most of the rain strikes vegetation surfaces and not the soil directly. As rain filters through the vegetation to the soil, its chemistry is greatly altered. Much research is needed to evaluate these changes and their effects fully.

A short-sighted, but widely practiced philosophy in pollution control is dilution, and tall smokestacks are often the mechanism for control of air pollution. However, tall stacks also allow emissions to travel farther, and because the emissions are in the atmosphere for longer periods they undergo more chemical transformation to acid substances. Thus, in the case of acid precipitation, someone is literally dumping "garbage" onto someone else's backyard, but it is usually not clear to the recipient who is doing it or what can be done to stop it. In this way air pollution standards in one political jurisdiction can have an effect on the natural resources of other areas. The problem has enormous political, economic, and ecological ramifications.

Courtesy, Gary E. Glass, U.S. EPA; photo, Jim Boeder

Chemist collects samples of snow in northern Minnesota for analysis of acid content.

Future prospects

As yet it is not possible to determine precisely the overall effect of acid precipitation on vegetation, soils, and a variety of other materials, but the effects on aquatic ecosystems are clear warnings. They raise a major question central to regulation in all environmental matters: when is a pollutant a serious problem and when should action be taken to regulate the pollutant and alleviate the problem? If a decision is delayed until all the facts are in, it may be too late to deal with the problem successfully. On the other hand, if action is taken incorrectly in haste, vast amounts of money and time may be wasted.

The technology is available or nearly available to reduce emissions of sulfur and nitrogen to the atmosphere. However, such procedures are expensive, and so the question narrows, as it often does, to one of economics. Is it worth the cost to remove the offending material at the smokestack, or is it better to distribute it in the environment? These and similar questions will plague policymakers during the next few years unless there is an effective and fair means of accounting for all costs involved in environmental problems.

Even more vexing is the fact that it is currently impossible to make quantitative predictions of the effect that reductions in emissions will have on the acidity of precipitation. There are too many unknowns in the process at present.

The cost of reducing emissions is enormous. On the other hand, the total cost of reduced forest and crop production, fishless lakes, and damage to structural materials may be even larger. At present, scientists lack the answers to these difficult questions. Clearly quantitative information on costs related to acid precipitation and its ecological effects is urgently needed.

FOR ADDITIONAL READING

L. S. Dochinger and T. A. Seliga (eds.), *Proceedings of the First International Symposium on Acid Precipitation and the Forest Ecosystem* (USDA, Forest Service General Technical Report NE–23, 1976).

T. C. Hutchinson and M. Havas (eds.), *Effects of Acid Precipitation on Terrestrial Ecosystems* (Plenum Press, 1980).

G. E. Likens, R. F. Wright, J. N. Galloway, and T. J. Butler, "Acid Rain," *Scientific American* (October 1979, pp. 43–51).

PLANT BREEDING TO FEED A HUNGRY WORLD

Agricultural productivity has increased greatly through the efforts of plant breeders. Future gains are expected to result from the use of genetic engineering.

by Garrison Wilkes

The population of the world in 1980 was approximately 4.5 billion, almost triple the number in 1900. To feed such numbers adequately has posed a formidable challenge for agricultural science. In recent years great successes have been achieved without which the current population could not be fed. Many of the most significant of these have centered on improvements in plant breeding.

There are three basic ways to increase the productivity of agriculture: to increase the land under cultivation; to improve growth conditions by better control of growth inputs, such as water, fertilizer, and protection from insects; and to breed plants for greater productivity. At present there is little land suitable for plant agriculture not now in production, and in many regions improved growth conditions have reached their upper limits faster than expected because of their dependence on fossil fuels (for gasoline and fertilizers), the price rise of which has been dramatic. Thus, plant breeding remains the area in which major improvements can still be made.

About 7,000 years ago humans first began to cultivate, independently in different parts of the world, a diversity of crop plants—maize (corn), beans, and squash in Mexico; wheat, barley, and peas in the Middle East; and rice, millets, and soybeans in the Far East. The earliest domesticated crops were probably not much more productive than their wild progenitors, but the act of cultivation was a radical break with the past. This restructuring of the food supply set in motion numerous forces, many not at the time consciously intended, that have directed the evolution of these crops. This was the first stage of plant breeding, or human control over crop plant evolution.

Agriculture was a subject for religious mythology among Mexico's Aztec Indians (opposite page, top). At the top right the corn plant flourishes in the years when the rain god Tlaloc is dominant and makes the earth rich and green; however, in the years when Xipe Totec rules (top left) the plant is small and rootless and is attacked by birds and animals. In early China (opposite page, bottom) at a rice paddy south of the Yangtze a farmer and his wife operate a chain pump with their legs and feet. At the right a boy drives a water buffalo to turn a larger water-pumping device.

***GARRISON WILKES** is an Associate Professor of Biology at the University of Massachusetts, Boston.*

(Overleaf) Webb Photos; (inset) NASA

Migration and acclimatization

The number of plant species that has historically fed the human population is only about 5,000. This small quantity is less than a fraction of 1% of the total world flora. As the human population has grown, it has depended increasingly on a shorter list containing the most productive plants. Today only about 150 plant species are important in meeting the caloric needs of humans.

Most domesticated plants, including all of the food plants, are the products of a long selection process by means of which people have produced a plant that is totally dependent upon their care for survival. This selection process is called domestication. In many cases these crops have been so genetically altered that they can no longer disperse their own seed or compete in a natural plant community; they have become dependent on growers' care because they cannot exist any longer in the wild. The process is a co-evolution because humans have expanded their population based on these domesticated crops and for them too there is no escape to the hunting and gathering of wild plants. Thus, the changes made in these crops by plant breeding have made people captives to this assured food supply.

Each of the basic food plants originated in a distinct geographic region. These regions are called Vavilov Centers of Crop Plant Gene Diversity after the Soviet geneticist and plant breeder Nikolay Ivanovich Vavilov, who first studied these regions where the crop originated or underwent a secondary rapid evolution following a domestication outside the center. When plotted on a world map, nine major and three minor centers that account for the vast majority of cultivated plant diversity if not origin can be identified. They are located in mountainous regions along the belt of the two tropics (Cancer and Capricorn) in areas long populated by agricultural peoples but isolated by steep terrain, arid land, or other natural barriers. In the 1920s Vavilov found these regions to be essentially untouched by the changing world and still supporting ancient agricultural customs. Much has changed in the last 50 years, but these regions are still the reservoirs from which many of the strains and valuable genes used by plant breeders have come. Each of them has been tapped for useful genes that changed some plant characteristic, conveyed increased resistance to fungal diseases or insect attack, or improved the durability or nutritional qualities of the harvest.

For many crops the truly wild plant from which the domesticated crop was derived no longer exists because of extensive agricultural disturbance and habitat destruction. In those cases the varieties closest to the ancestral form are the primitive land races still being grown in the regions of the world where the plant was originally domesticated or in regions where the crops had acclimatized long ago. These regions are the genetic resource areas to which the plant breeder turns for additional germ plasm in the plant improvement process.

Primitive varieties or land races are the largest depository of genes for a crop, but they are also the largest unknown because they are usually heterogenous and because little data exist for their morphological, biochemical, and genetic traits or for their responses to parasites or environmental stress. Generally these land races perform poorly under inputs of high fertilizers,

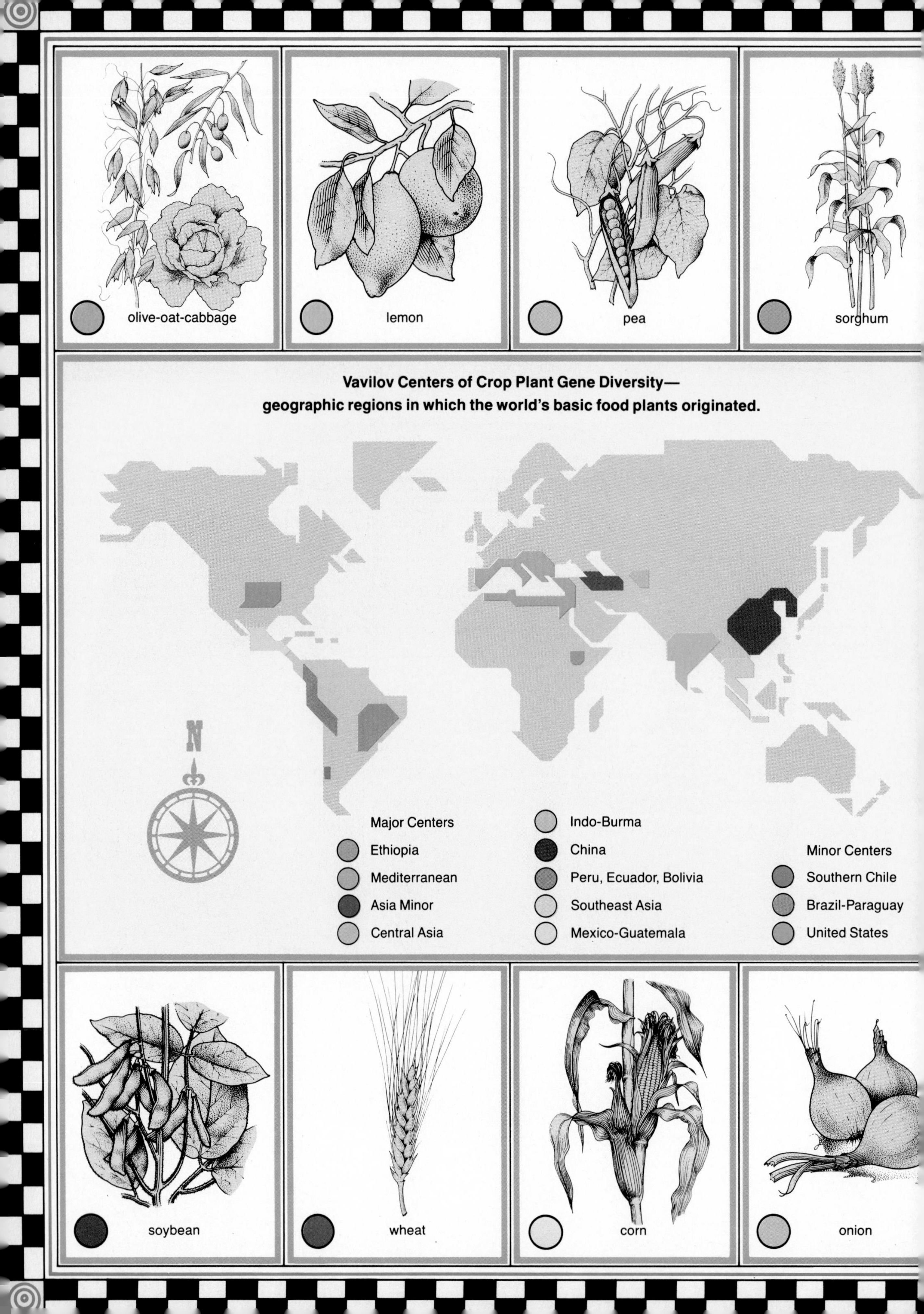
olive-oat-cabbage
lemon
pea
sorghum
Vavilov Centers of Crop Plant Gene Diversity—
geographic regions in which the world's basic food plants originated.
N
Major Centers
Ethiopia
Mediterranean
Asia Minor
Central Asia
Indo-Burma
China
Peru, Ecuador, Bolivia
Southeast Asia
Mexico-Guatemala
Minor Centers
Southern Chile
Brazil-Paraguay
United States
soybean
wheat
corn
onion

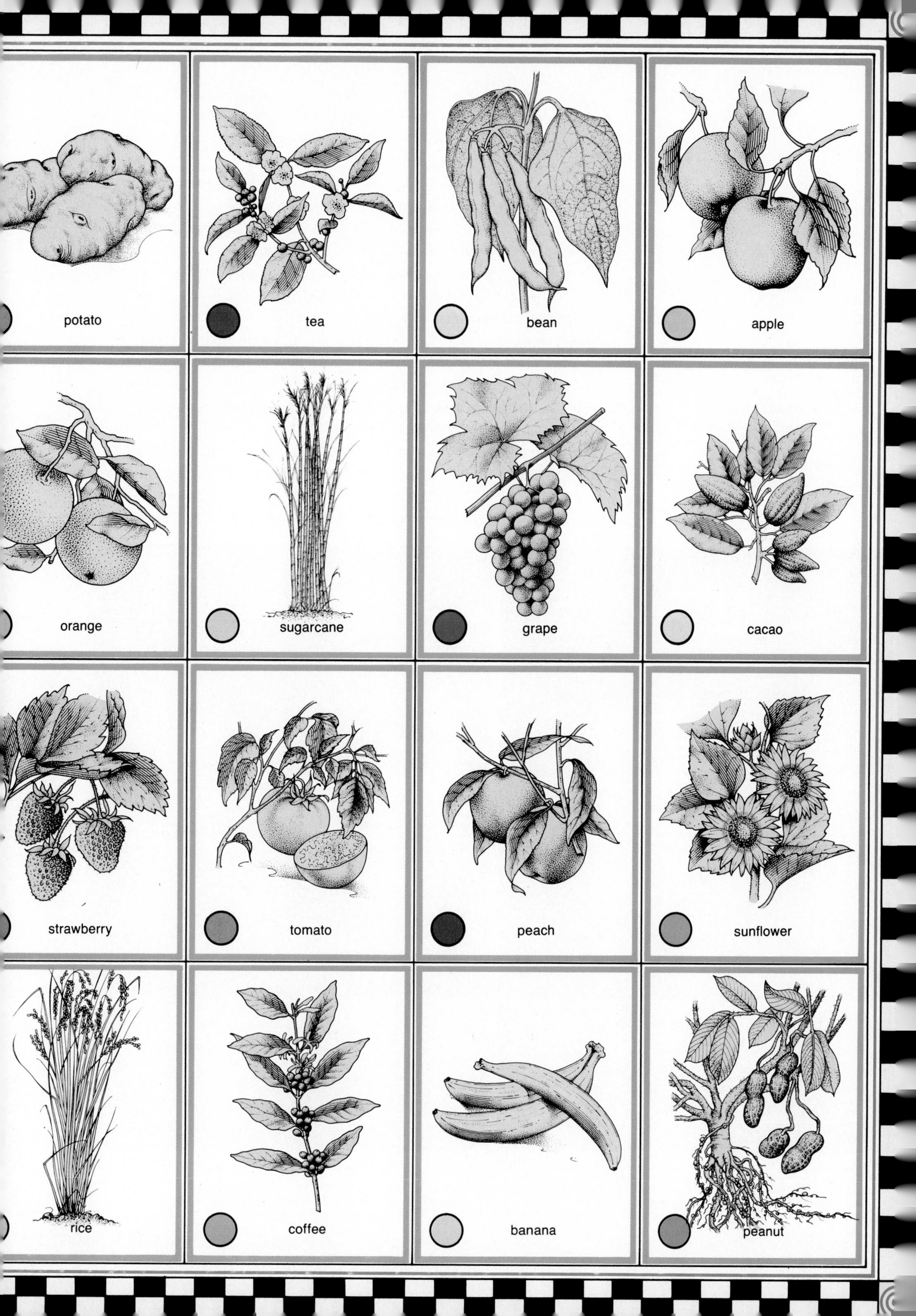
potato
tea
bean
apple
orange
sugarcane
grape
cacao
strawberry
tomato
peach
sunflower
rice
coffee
banana
peanut

water, and intensive cultivation because they represent the product of a long evolutionary history in an environment lacking those inputs. On the other hand there is a fairly wide variation in the ability of land races to withstand cold, drought, diseases, and insects.

The process of plant breeding is a dynamic one of genetic selection in response to changing diseases, parasites, agricultural techniques, and human use. In the agriculture of the United States the varieties in current use have generally undergone a rather rigorous selection process by plant breeders and are more or less homogeneous. These varieties possess a "highly tuned" set of genes but a considerably narrowed gene base over the native land races from which they were derived. These advanced varieties are the ones most widely and frequently used as parents in current breeding programs or in introducing a variety into an area of comparable climate.

A major goal of plant breeding is to develop varieties with increased resistance to insects and disease. Among the diseases and pests affecting corn are corn earworms (below) and maize smut (below right), a parasitic fungus.

The U.S. experience with agricultural productivity has been rich because of the worldwide genetic contributions to the nation's plant breeding and improvement process. The earliest settlers brought with them from their homeland the land races that they knew and grew them in the new land. Thus, barley from England grew beside barley from Germany in Massachusetts fields. Ship captains often brought back such cargoes as wheat from Calcutta and rice from China or Madagascar for a relative or friend

John H. Gerard

© Heather Angel

to try on his farm. The Spanish missions in the West introduced arid land crops completely foreign to farmers on the east coast. And later, with the large-scale immigration from central and southern Europe, new and distinct genetic diversity was added to the basic crop plants. The service that the immigrants rendered by bringing seed with them was to establish a broad genetic base for plant breeding and improvement.

The importance of the worldwide contribution of these settlers to U.S. agriculture and plant breeding is immediately apparent in looking at the geographic origins of this country's ten most valuable crops. Only corn is native to the U.S.

The most widely grown corn is a truly U.S. variety and is a product of the nation's westward expansion. This variety, Corn Belt Dent, is itself a hybrid race the two parents of which were widely cultivated by the Indians on the east coast of the U.S. One of the parents was Gourdseed, or Southern Dent, a many-rowed, thin ear borne on a heavy stalked plant, which was widely grown from Virginia south. The other was a New England Flint, an eight-rowed, slender ear with broad shallow kernels borne on a much tillered but not excessively tall plant, which was cultivated by the Indians and first encountered by New England settlers. (A tillered plant has sprouts extending from near its base.) Out of these two very dissimilar parents came the highly productive corn of American farms.

(Top) Oxford Scientific Films; (bottom, left and right) Webb Photos

Plant breeders are seeking to eradicate rust fungus in barley (top) and black rot in potatoes (above). At the left is wheat that has lodged—grown too fast and toppled over. Because plants that have lodged often rot on the ground breeders are working to eliminate this trait.

Walter Dawn

Recently bred descendants of primitive corn (above) include Shoepeg sweet corn (below) and Rainbow Flint (below right). Such varieties are noted for high productivity and resistance to diseases.

As the settlers pushed west through the Cumberland Gap, they took from Virginia the Gourdseed types with which they were familiar and planted them on the newly plowed lands. Sometimes the germination was poor, and the empty hills were replanted with the more rapidly maturing New England Flints. The two races both flowered at the same time, hybridized, and formed a unique American race, which is one of the most productive corns in the world.

Since the 1920s plant breeders have been making hybrid seed by using parents of good combining ability to produce a first-generation hybrid. This gene-engineered hybrid of an already hybrid race was a key element in the increased yields per acre between the 1920s and 1950s. The increases from the 1950s to the present have been mostly selection for types that respond favorably to large applications of nitrogen fertilizer and also for those that mature later.

But corn is the only American plant on the list; all the rest have come from somewhere else in the world. In regard to wheat, the second most important cereal crop in the U.S., two introductions from other countries (Marquis and Turkey) have figured prominently in the ancestry of the nation's present crop. Marquis, an outstanding bread wheat, was introduced into the U.S. in 1912 from Canada. It had been bred there from a cross of Red Fife (a wheat with ancestry traced back to Scotland, Germany, and Poland) and Calcutta (a hard red wheat from India).

Photos, Grant Heilman

Though Marquis wheat is a hybrid of two very dissimilar parents, the other major strain, Turkey, is an introduced land race from Russia. Throughout Europe the newly constructed Atchison, Topeka and Santa Fe Railway had posters extolling the virtues of settling the open lands along its right-of-way. In 1873 a band of Mennonites left Russia for farmland in Kansas, and with them they brought the seed of their crops. One of the crops was Turkey, the hard red winter wheat that they had grown in the Ukraine-Caucasus regions of Russia. Today most of the hard red winter wheat presently grown in the U.S. can be traced in large part to this introduction of the Mennonites.

Kenn Goldblatt—DPI

Bred varieties of wheat incorporate some characteristics of the primitive plant but lose others. The variety above retains the spikelike awns but has a shortened head; below, the head remains long but the awns are gone.

© A–Z Botanical Collection

The Green Revolution

The examples of wheat and corn described above represent the second phase of plant breeding. This phase involves the migration and acclimatization of crops throughout the world and often the hybridization between dissimilar forms far from their Vavilov Centers. The third phase of plant breeding and improvement began with the rediscovery of Gregor Mendel's classic experiments on the heredity of garden peas and the beginning of the science of genetics. For the first time the plant breeder had a clearer idea of how to proceed with crop improvement. A good example of this is the origin of the wheat and rice varieties that formed the Green Revolution, the development of high-yielding crops for the world's less developed nations.

Plant breeding for specific traits has been one of the most distinctive elements in the Green Revolution. The traditional land races used until recently in those regions did not make effective use of fertilizer; the fertilizer caused them to lodge (grow too fast and tall and thus topple over) and often rot on the ground. But the specially bred high-yielding rice and wheat varieties incorporated dwarfing genes that gave the plant short, stiff straw and enabled it to respond to fertilizer without lodging. Other genetic improvements included resistance to certain pests, insensitivity to day length in controlling the flowering process, and rapid maturation so that a second crop or double cropping could be practiced. These changes were relatively simple, but like the original domestication of plants their impact in feeding a hungry people was dramatic.

The Green Revolution began in 1943 when the Rockefeller Foundation, in collaboration with the Mexican Ministry of Agriculture, began a research program designed to increase the production of Mexico's basic crops: corn, beans, potatoes, and wheat. A broad-based breeding program was initiated with each of these crops. Because most wheat grown in Mexico was highly susceptible to stem rust, the initial breeding focused on increasing resistance to this fungus disease in existing Mexican varieties. The development of new wheats possessing an accumulation of rust-resistant genes and the improvement of agricultural practices, such as the use of fertilizers, pesticides, and better seedbed preparation, made a substantial impact on wheat yields. By 1957 Mexico had achieved self-sufficiency in wheat for its population. Over a ten-year period the average wheat yield rose from 740 kilograms (1,628 pounds) (1945–46) to 1,440 kilograms (3,168 pounds) per hectare in 1957–58. (One hectare equals 2.2 acres.) Then the yield response began to

Walter Dawn

W. H. Hodge—Peter Arnold Inc.

level off because the most productive fields lodged when nitrogen fertilizer was applied in quantities of 80 kilograms (176 pounds) per hectare or more. To increase the yields further more fertilizer-responsive wheat varieties would have to be bred.

The first extensive breeding program to develop semidwarf spring wheats was started by Norman Borlaug in 1954 when a wheat, Norin 10 × Brevor, was crossed with indigenous Mexican varieties. The dwarf wheat Norin × Brevor came from the U.S. and the dwarfing trait in Norin 10 from Japan. Following World War II an agricultural adviser to the U.S. Army of occupation in Japan had observed Japanese farmers growing short, stiff-strawed wheat varieties that remained erect under heavy fertilizer application. This variety, Norin 10, has an international ancestry. The dwarfing short-stature gene came from a Japanese wheat which in 1917 was crossed with Glassy Fultz, a selection of the American soft red winter wheat variety Fultz, at the Central Japanese Agricultural Experiment Station to produce Fultz-Daruma. This variety in turn was crossed with the American hard red winter variety Turkey at the Ehime Prefectural Agricultural Experiment Station in 1925 in an effort to produce rust-resistant, short-stemmed, early maturing varieties. Following seven cycles of selection by plant breeders Norin 10 was registered and released in 1935 for Japanese farmers (The Japanese word "norin" means "agriculture and forestry," and varieties officially released are so named and given a number.) The Norin 10 brought to the U.S. in 1946 was not adapted for direct planting in American fields but was introduced into breeding nurseries and released by Orville Vogel from Pullman, Washington, as the variety Gaines in 1962. Vogel had supplied the Norin 10 × Brevor cross to Borlaug.

Of the thousands of hybrid seeds containing the dwarfing gene grown in the Mexican program three plants were selected as showing promise. The selected progenies of these semidwarf spring wheats possessed the short stature of the Norin 10 and the disease resistance of the Mexican parents along with genes for an increased number of fertile florets per spikelet and an increased number of stalks per plant.

The introduction of Norin 10 genes into the Mexican program led to the development of the first short-statured and lodging-resistant spring wheat varieties, which first were grown by Mexican farmers in 1962. Thus, Mexican wheat yields, which had leveled off after 1958 because heavier fertilizer application was not possible, started rising again. With the use of these short-stature varieties yields as high as eight metric tons per hectare became common.

International diffusion of these varieties began almost immediately at the experimental level. India and Pakistan were involved in the program from an early date. The new Mexican wheats were first grown in India in 1962. By 1965 India had an order for 18,000 metric tons of Mexican wheat and Pakistan for 42,000 metric tons. This successful rapid transfer of plant-breeding technology halfway around the world broke the static yield potentials of those regions.

The Mexican varieties proved remarkably well adapted to India and Pakistan because in accelerating the Mexican wheat program two generations of the breeding material were grown each year at different climatic and day-length regimes. A valuable side effect of this system was to establish a plant relatively insensitive to day length. The normal winter crop was grown on the northeast coast of Sonora essentially at sea level. The summer crop was grown at high elevations (2,600 meters, 8,530 feet) in central Mexico near Toluca. The Toluca site has heavy rainfall and severe epidemics of both stem and stripe rust. Selection for broad disease resistance and the use of widely adapted varieties that were not bred to pure line standards meant that these semidwarf Mexican wheats—Penjamo 62, Lerma Rojo 64, Ciano 67, INIA 66, and Sonora 64—possessed a reservoir of genetic diversity that could be incorporated into the breeding programs of new host countries. The adaptation and use of those wheats that is now occurring in less developed nations is a continuing process of crossing them with indigenous varieties and of selection for local conditions. Good examples are Kalyansona and PV-18, both wheat releases of Punjab Agricultural University (Ludhiana) in the wheat belt of India. These two varieties were selected out of seed which Borlaug had sent from the Mexican program in 1963. The wheat variety PV-18 was developed from a cross of Penjamo (from Borlaug) and Gabo 55. The variety Kalyansona is a sister strain of PV-18 and resembles the latter in almost all the plant characteristics except that the grains are amber (the preferred color in Punjab bread wheats). Both varieties possess high-tillering capacity and a wide adaptability to the various climatic conditions of north India.

The break with the static yield potentials for rice in less developed countries was also based on plant breeding. The high-yield varieties of rice have

Webb Photos

David Mangurian

Varieties of rice developed for their productivity and resistance to disease, insects, and cold include Oryza sativa indica *(opposite page, top) and* Oryza sativa japonica *(opposite page, bottom). Both varieties respond well to fertilizers. Above, an agronomist shows the difference between the tall native criollo corn and the shorter but higher yielding H-309 variety at a government agricultural station in southern Mexico. At the left alternate rows of corn are detasseled to control the cross-pollination process by preventing inbreeding, which reduces the vigor of the plants.*

Breeders use a variety of methods to improve the productivity and hardiness of their plants. At the top a vacuum is used to remove anthers from a spikelet of rice in preparation for cross-pollination with another variety. Workers in Mexico (center) depollinate a new dwarf variety of bread wheat to prepare the plants for cross-fertilization; their goal is to achieve a higher-yielding strain. At the bottom rice is cross-pollinated at the International Rice Research Institute in the Philippines.

(Top) W. O. McIlrath—USDA-SEA-AR; (center) David Mangurian; (bottom) Bill Gillette—Stock, Boston a history of origin similar to that of wheat. In Taiwan in 1949 at the T'ai-chung District Agricultural Improvement Station a cross between the semidwarf Dee-geo-woo-gen, a Chinese variety of the *japonica* series, and a tall native *indica* variety was made. An outstanding selection from this cross, T'ai-chung Native I, was the first semidwarf, fertilizer-responsive *indica* variety developed through plant breeding; it was released in 1956. Prior to this time only dwarf, fertilizer-responsive *japonica*-type varieties with short sturdy straw and dark-green, thick, narrow, and somewhat erect leaves were known, primarily in Japan, Korea, and Taiwan. The problem with these varieties was that they were not adapted to the tropics. By 1961 T'ai-chung Native I was recording yields in excess of eight metric tons per hectare, and by 1964 about one-third of Taiwan had been planted with this variety.

In 1962 the International Rice Research Institute (IRRI), founded by the Ford and Rockefeller foundations in cooperation with the government of the Philippines, began research on rice production in the tropics, hoping to be as successful as the forerunner of the International Maize and Wheat Improvement Center had been in Mexico. From the start there was a recognition that one of the most important single reasons for the low yields of tropical rice was the lack of lodging-resistant varieties. The traditional tropical varieties are tall and leafy and lodge under heavy nitrogen application.

The IRRI geneticist, T. T. Chang, who came from Taiwan, was familiar with the breeding work then in progress in his homeland, and the rice varieties Dee-geo-woo-gen, I-geo-tse, and T'ai-chung Native I were used in the earliest crosses. The most successful crosses that first year (1962) were between Peta, a tall Indonesian *indica* type that had disease resistance and heavy tillering, and the Chinese Dee-geo-woo-gen. Less than three years later the first cross, IR-8, was released. This new variety responded favorably to fertilizer and produced its maximum yield of 9,477 kilograms (20,849 pounds) per hectare at a nitrogen application of 120 kilograms (264 pounds) per hectare. In 1980, IR-8 remained the preferred variety in some districts of the less developed countries. Current rice breeding has shifted to improved grain quality and resistance to disease, insects, and cold, resulting in such new strains as IR-20, IR-22, IR-28, and IR-42.

The key element in the Green Revolution was breeding for high-yielding varieties of wheat and rice. These varieties, combined with improved agricultural practices, have increased production dramatically, sometimes as much as tenfold, over broad areas of India, Pakistan, and Iran for wheat and of Indonesia, the Philippines, Sri Lanka, and Bangladesh for rice. Before the Green Revolution these nations had been part of a general pattern of yield stagnation in less developed countries. In fact, per capita grain production had actually decreased over the previous 25 years.

Future prospects

Clearly plant breeding will be a major factor in meeting the immediate challenge of optimizing agricultural productivity. The high-yielding varieties of wheat and rice that started the Green Revolution were the first of a series of plant breeding advances. Food production to support the existing human population must become more productive per unit land area, per increment of water, and per unit of fossil fuel input. Plants will have to be more efficient in converting sunlight into usable food energy, and at the same time be more resistant to the destructive inroads of insects, nematodes, and fungi and less dependent on high levels of soil fertility.

Many of the genes to help engineer these new varieties are still to be found by a systematic exploration and characterization of useful gene assemblages in land races. The kind of plant breeding presently practiced requires the backup of both national and international systems for maintaining as large a collection of crop-specific germ plasm as possible to be used for future breeding to meet the ever-changing pathogen, insect, and environmental variables. A successful new variety from plant breeding is always a transitory event because after millions of such plants are grown they become an organizing force for the evolution of predators. At such time a new resistant strain must be bred. And soon afterward this new variety will also be confronted by a new host-specific pathogen. Crop improvement is thus a continual evolutionary process.

Possibly some break in this cycle will be achieved by radical new techniques of plant breeding now on the threshold of achieving reality. Research is in progress for hybridizing unrelated plants by techniques such as somatic cell fusion as is currently being investigated in potatoes or for moving specific genetic information from one plant into a totally unrelated crop by means of a carrier such as a virus or by wide hybridization as is being explored in corn × sorghum crosses. By using these techniques scientists could, for example, transfer specific genes that increase photosynthetic efficiency in hot, humid summer weather from corn to wheat or the assemblage of genes that controls the fixation of nitrogen in the root nodules of legumes to the roots of cereals. Plant breeding of the latter type involves not only the change of the crop plant but also of the bacteria that will invade or be associated with the roots of this crop and thus establish an effective nitrogen-fixing association.

In conventional plant breeding there is always a delay of some three to ten years between the first genetic crosses and the release of the variety to farmers' fields. Therefore, it is doubtful that many of the new techniques will have much effect on production by the end of the decade, but certainly by the year 2000 they will have restructured agricultural productivity.

The challenge to supply enough food throughout the world is a clear and urgent one. The best backup any plant breeder could have to meet the changing productivity demands of the future is the largest pool of genetic variation possible. Currently much of this genetic variation is in peasant agriculture, but it is disappearing because of the inroads of genetically uniform seeds, such as the dwarf wheats. The process is a paradox in social and economic development: the product of technology (plant breeding for yield and uniformity) displaces the resource upon which the technology is based (the genetic diversity of locally adapted land races not yet altered by plant breeding).

Science
Year in Review

Science Year in Review Contents

Anthropology

A recent article in *Science* magazine detailed the unpreparedness of the United States for the events that brought revolutionary change to Iran. The U.S. embassy in Teheran was orphaned in the midst of change, armed only with ignorance and a false sense of security. Reasons for the failure were advanced; among them were that Iranians are skilled at concealing their true political beliefs, and U.S. scholarship on Iran had been severely stifled by financial cutbacks at universities and colleges.

During the time that officials from the U.S. embassy were held hostage in Iran, anthropologists with firsthand knowledge of that nation offered their opinions. Mary Catherine Bateson, who worked there from 1972 to 1979, suggested that the best strategy for securing the release of the hostages would be for the U.S. negotiators to do nothing. They should cease to negotiate, as if Iran did not exist. Based on her analysis of an Iranian technique for controlling hostilities in which the parties simply stop talking to each other, Bateson counseled: "In short, we should refrain from actions that tend to confirm our diabolic role in the Iranian morality play of good versus evil." U.S. negotiators, however, did not follow this advice.

U.S. officials were not the only ones unaware of the nature of the Iranian revolution. Few intellectuals and observers, inside or outside of Iran, anticipated the part that the Islamic religion would play, first in opposing the shah and then in forging the society that emerged after his departure. Members of the Iranian middle class themselves believed religion to be merely a "political vehicle" to topple the shah, according to a recent study by Michael Fischer. It soon became apparent, however, that the masses of Iranians took more than a "vehicular" approach to their support for the Islamic fundamentalism of the Ayatollah Ruhollah Khomeini.

As the shaky coalition that overthrew the shah itself began to unravel, anthropologists made their predictions about Iran's future. Some were also analyzing U.S. participation in the Iranian drama. In the wake of their new insights into Iranian and Islamic culture, anthropologists were working with greater resolve to create a theory of culture worthy of the complexities of human societies and histories.

Social transformations. Recently, Eric Wolf described anthropology as "a holding company of diverse interests." Although his assessment is essentially correct, the existence of diversity is not necessarily all bad. Two recent studies of social change in South America —one ecological and the other Marxist—illustrated through contrast the scope of current anthropological research.

Within the ecological framework anthropologists investigated the question of differential participation by remote Brazilian villages in external market economies. Research by Daniel Gross and his co-workers established a correlation between the difficulty of making a living by the traditional methods of slash-and-burn agriculture and the participation in an external marketing system for central Brazilian Indian groups. Applying a measure of the capacity of the environment to sustain slash-and-burn agriculture, the researchers demonstrated that the group confronting the most environmental resistance (that is, having the lowest caloric yield for their investment of time in gardening) showed the highest level of market involvement. The authors concluded that the variations in market involvement are a consequence of environmental pressures. They ruled out such factors as the allure of trade

Carl Frank—Photo Researchers, Inc.

Tin miners in Bolivia were among the subjects of a major anthropological study during the past year. Many of them are former peasants who believe that their new employment for wages represents a contract with the devil.

goods and the relative isolation of the groups, although the most isolated group, the Mekranoti, was also the least active in the external market. The question of differential participation within a particular group was also left for future investigation.

Anthropologists learned that participation in the marketplace was not the only way that the Indians of Brazil felt the impact of development in their part of the world. According to anthropologist Kalervo Oberg encounters with "civilization" left such groups as the Nambiquara of western Brazil hateful, distrustful, and despairing. As Brazilian society expanded and expressed interests in the mineral wealth contained in their tribal lands, these people faced increased pressure to overutilize the land that was still available to them. The American Anthropological Association joined with several other organizations in an appeal to the Inter-American Commission on Human Rights regarding what they considered to be violations of Brazilian Indians' human rights.

The human dimension of the transition from a precapitalist to a capitalist mode of production in Colombia and Bolivia was the focus of a study unveiled in 1980 by Michael Taussig. In *The Devil and Commodity Fetishism in South America* Taussig analyzed the beliefs about contracts with the devil held by sugarcane workers (Colombia) and tin miners (Bolivia) as aspects of their experience of the unnaturalness of life under the capitalistic order. The confusions, criticisms, and ambivalences of former peasants who have become wage laborers testified to the loss of a coherent vision of human relations under the new order. The devil contract, according to Taussig, represents the sense of loss that typically accompanies the self-employed peasant as he moves into the ranks of the wage-earning working class. It "is an image illuminating a culture's self-consciousness of the threat posed to its integrity."

These peasants believe that their new employment for wages represents a contract with the devil. Any earnings thus obtained are thought to be barren; the goods purchased with them will disintegrate, and for the sugarcane workers any cane cut under contract with the devil will never again sprout. Taussig's argument is strengthened by the fact that peasants working either their own lands or the lands of other peasants do not report a belief in devil contracts. Taussig concluded that the devil mediates between two worlds, one of which is rapidly dying, and that the belief symbolizes the pain and human costs of capitalism as seen from a precapitalist perspective.

Taussig compared the newcomers' views of the commodity system, who consider it the workings of the devil, with our own received notion of its naturalness. While some anthropologists questioned the "peasant knows best" approach by Taussig, most accorded respect for his approach to the study of societal transition through the analysis of ideology.

Human sexuality and evolutionary theory. The search for a general theory of human sexuality continued in 1980, and for many sociobiologically minded anthropologists it followed upon the publication of *The Evolution of Human Sexuality* in 1979 by Donald Symons of the University of California at Santa Barbara. Symons's book represents a synthesis of sociobiological thinking on human sexuality, building on the work of R. D. Alexander, R. L. Trivers, G. C. Williams, and E. O. Wilson, among others.

Symons argued that in the arena of sexual activity male and female natures are essentially different: "Men and woman differ in their sexual natures because throughout the immensely long hunting and gathering phase of human evolutionary history the sexual desires and dispositions that were adaptive for either sex were for the other tickets to reproduc-

Along with Bolivian tin miners many sugarcane workers in Colombia believe that working for wages is a contract with the devil. They are sure that any cane cut under such a contract will never sprout again.

Victor Englebert—Photo Researchers, Inc.

Culver Pictures, Inc.

A Turk with his harem in the late 19th century exemplifies the theory of human sexuality that male reproductive success is best accomplished through multiple sexual partners while female success is achieved through mating with just one male.

tive oblivion." Men and women pursue different reproductive strategies, but, since they require one another to reproduce, conflict and cross-purpose abound.

Specifically, according to Symons, male reproductive success is accomplished through multiple sexual partners, whereas female success is best accomplished through mating and bonding with just one male. Reproductive success is the ultimate and only goal of human sexuality, according to the theory. Following on this assumption, females then have to be more selective in their sexual unions than do men because females have only the delimited procreative potentials of their own bodies with which to work, whereas males can make use of virtually unlimited numbers of female bodies which will do the bulk of the reproductive work for them. Therefore, random matings, which form the heart of the male strategy, are to be avoided by females, who should seek the fittest possible male.

In the scramble for reproductive success among males, Symons saw a host of consequences: "Sexual selection favors calculated risk-taking in male-male competition, hence the evolution of large body size, strength, pugnacity, playfighting, weapons, color, ornament, and sexual salesmanship." Coyness, not competition, is selected for on the female side of the field. Males who have won their battles presumably copulate with a great variety of sexual partners.

Reproduction provides only part of the story of sexuality, according to critics of sociobiological approaches. They maintain that sexuality provides a wealth of meanings for human beings that cannot be condensed by an argument focused exclusively on reproductive success, devoid of social and cultural considerations. The sociobiological argument is framed by a Western philosophical tradition that supports and justifies the present state of affairs between the sexes as unalterable consequences of ultimate (evolutionary) causes. The critics warn of the serious need to separate biological origins of human traits from their subsequent operation in the evolved species.

The sociobiologists R. D. Alexander and K. M. Noonan anticipated such rejections of the reproductive success explanation for sexuality. They suggested that it is universally the case for individuals (male or female) to cite goals other than reproduction when thinking about their motivations for engaging in sexual intercourse.

Although Symons's data were drawn from a worldwide sample of ethnographic reports, his most convincing and consistent case for male supremacy was that offered in Western culture. Comparative research undertaken by Karen Sacks indicated that, contrary to Symons, femininity is not everywhere the same. Sacks argued that the views of other cultures within the sociobiological framework are reduced to reflections of industrial society and are not true portraits of the relationships between the sexes in those societies.

Commentators on Sacks's book, such as Stephen Gudeman, were dissatisfied with her treatment of kinship relations as being based exclusively on production. Thus, both Symons and Sacks investigated their topics from narrow perspectives. The positions of females vis-à-vis males in total social systems promised to be an important and controversial topic for research in the years to come.

—Lawrence E. Fisher

Archaeology

In the Book of Genesis it is recounted that after a great deluge that drowned the Earth, Noah's Ark, containing the stock of living beings from which the Earth was to be repopulated, came to rest "upon the mountains of Ararat." A number of attempts to identify the biblical Mt. Ararat and to locate the remains of the Ark have been made. Expeditions in 1955 and 1969 to the upper slopes of Agri Dagi, a volcanic complex in northeastern Turkey, brought back pieces of handworked wood said to have come from a glacier at approximately 4,200 m (1,380 ft) elevation. Fernand Navarra, a French industrialist, suggested that this wood, preserved in ice far above the timberline, was a relic of Noah's Ark itself.

Over the years since this discovery accounts have appeared in the press, and several tests to determine radiocarbon dates have been run on the wood samples by different laboratories. In a recent report R. E. Taylor of the University of California, Riverside, and Rainer Berger of the University of California, Los Angeles, collected and discussed the scattered dating evidence, amounting to six radiocarbon determinations on wood samples independently received by five laboratories (National Physical Laboratory, England; University of California, Los Angeles; University of California, Riverside; University of Pennsylvania; and Geochron Laboratories, Inc.).

Five of the six dates were identical, within the limits of normal experimental variation in radiocarbon measurement, placing the age of the wood most probably in the 7th to 8th century AD. The sixth date fell in the 3rd to 4th century AD. All were very much younger than the age of approximately 5,000 years advocated by proponents of the Ark theory.

Taylor and Berger refer to previous comments suggesting that organic contamination of the wood, or the direct formation of carbon-14 in the wood itself as a result of bombardment by solar protons unscreened by the thin atmosphere at 4,200 m, may have increased the radiocarbon content of the dated fragments, making them appear much younger than their true age. They point out, however, that any known sources of organic contamination would have been removed from the samples by normal pretreatment procedures. Furthermore, other research on the effects of solar bombardment at high altitudes shows that the maximum possible effect from this source would be only 0.5%. Indeed, radiocarbon dating of high-altitude bristlecone pine wood and low-altitude European oak wood of the same tree-ring ages suggests that the effect of altitude on carbon-14 levels is negligible.

In attempting to account for the presence of wood that is demonstrably ancient high on a mountain long identified with the biblical Mt. Ararat, Taylor and Berger refer to a well-established tradition within the Armenian Christian and Byzantine church communities of constructing shrines dedicated to important biblical traditions. Perhaps, they suggest, the wood comes from the remnant of a cenotaph erected by early Christians on what they believed to be the spot where Noah's Ark came to rest.

Electron microscopy. An exciting recent development in archaeological analysis techniques is the application of the scanning electron microscope to the study of butchering marks on ancient bones. Patricia Shipman of Johns Hopkins University and Richard Potts of Harvard University found shallow cut marks on fossil bones from Olduvai Gorge in Tanzania that may be as many as two million years old. The electron

The Temple of Isis, built on the shores of the Nile River in the third century BC, has been moved a few hundred yards from its original site to save it from the flood waters of the river caused by the Aswan Dam and restored to its former glory.

Michele Da Silva—Camera Press Ltd.

microscope technique allows researchers to distinguish between the smooth, U-shaped tracks made by the teeth of carnivores and the rougher tracks made by the jagged facets on the edge of a hand-held stone flake. The approach promises to expand significantly the archaeologist's perception of early human activity. One intriguing observation made by Shipman and Potts, of cut marks superimposed over carnivore tooth marks on the same bone, provides support for the theory that prehuman primates developed their taste for meat through the scavenging of carnivore kills.

Another promising new application of the electron microscope to archaeological research was reported by Patricia C. Anderson of the University of Bordeaux, France. Anderson studied the polishes and residues on the working edges of both archaeological and experimentally produced stone tools. As a result she was able to identify the kinds of organic materials on which the tools had been used.

Electron micrographs showed that the working edges of flint tools typically exhibited a gloss or polish, which, seen in cross section, penetrated below the surface of the stone itself. This suggested that the gloss forms when heat generated through the friction of tool use, in conjunction with the moisture and acidity of the material being worked, dissolves or "melts" the surface layer of the stone. An amorphous silica gel is apparently formed by the dissolution of the stone (flint is a cryptocrystalline silicate), and the gel forms a vitreous gloss over the working surface of the tool. Phytoliths, distinctive and virtually indestructible microscopic bodies that form in growing plants through the silicification or silica incrustation of plant cell walls, were often found adhering to or even sunken into the glossy buildup on the tool's working edge. Distinctive deposits of calcium carbonate, calcium oxalate, phosphorus, hydroxyapatite, and collagen could also be seen adhering to or embedded in the gloss. It is possible to identify these residues with broad categories of grasses, woody plants, bone, and certain other animal tissues and thus to infer how the tools were used.

Since the surface gloss and the minerals it contains are virtually indestructible, this form of analysis can yield information about the organic constituents of an ancient archaeological context, even in circumstances where only stone tools are preserved. One particularly intriguing result of Anderson's research was the discovery, on the edge of a convex scraper from the Middle Paleolithic Mousterian site of Combe-Grenal, in southwest France, of a phytolith representing the tribe of grasses to which wheat, oats, and barley belong. The Combe-Grenal site dates to a time far earlier than any previously known evidence for the human use of ce-

Archaeologists excavate carved jaguar figure (above) in a 20-story stone pyramid recently discovered in the jungles of Guatemala. The pyramid, measuring some 300 meters (1,000 feet) across the base, was built by the Mayan Indians in about 300 BC. Radar imagery taken from a flight over the Guatemalan rain forest (below) reveals canal systems dug by the Mayans between 250 BC and AD 900.

(Top) Courtesy, Brigham Young University; photo, Christopher Sherriff; (bottom) Jet Propulsion Laboratory real products, suggesting that exciting discoveries may await the archaeologist as this analytical technique becomes more fully developed.

Women's status in early societies. Some interesting insights into the status of women and the differentiation of social roles in Middle and Upper Paleolithic societies were reported in a comparative study by Francis B. Harrold of the University of Texas at Arlington. Harrold studied the sex, age, burial position, and grave furnishings of subjects from archaeological sites over a large part of Europe and Asia. Most of the 36 Middle Paleolithic subjects were early *Homo sapiens* of the Neanderthal variety, assignable to the interval 75,000–35,000 years ago. The preeminent characteristic of these burials was the strongly preferential treatment accorded to males. Among those interments for which sex could be determined, eight out of ten males were found with grave goods, but none of the seven females was so honored.

The Upper Paleolithic sample included 96 subjects, all of fully modern *Homo sapiens sapiens* type, from sites dated between 35,000 and 10,000 years ago. The data for these subjects indicated that significantly more males than females had been accorded formal burial, but among the burials themselves there were no discernible differences between the sexes in terms of grave furnishings. The strong social distinction between the sexes noted for Middle Paleolithic times, with females clearly assigned a subordinate position, was not found in the Upper Paleolithic sample.

Gilt-bronze lamp, dating from the second century BC, was among the rare and beautiful objects displayed in the exhibit "The Great Bronze Age of China."

Courtesy, The Metropolitan Museum of Art; photo, Wang Yugui

Other differences between the two samples were evident as well. The grave goods of Upper Paleolithic times were richer and more diverse than those of the Middle Paleolithic. Articles of symbolic or ornamental significance, such as necklaces of perforated teeth and ocher pigments, were much more common in the later burials than in the earlier ones. So were stone and bone artifacts and mollusk shells. Other indicators of increased diversity within the Upper Paleolithic group were a greater variety of burial positions and more multiple interments.

These correlations, established statistically, allowed Harrold to conclude that the status of women changed significantly from the Middle to the Upper Paleolithic. Further, they showed that, for both sexes, a significant expansion in the variety of forms of individual expression developed in Upper Paleolithic times. This probably indicates an expansion in the variety of individual social roles. These findings are congruent with and provide support for archaeological discoveries from other quarters, and they suggest that human societies during the Upper Paleolithic came to be characterized by larger social groups and more complex mechanisms of social integration than had existed during the Middle Paleolithic.

Ancient astronomy revisited. In recent years many of the great stone monuments of antiquity, particularly in Great Britain, have been interpreted as structures designed for making astronomical observations. Archaeoastronomers have projected the alignments of stones within such structures into the heavens in an attempt to determine whether the stones might have been used as sights over which to observe the positions of certain stars or planets on significant dates. Aubrey Burl, lecturer in the Hull (England) College of Higher Education, is skeptical of many of the claims for ancient scientific knowledge advanced by some archaeoastronomers. With a touch of irony, he notes that: "Over the last decade many claims have been made that early prehistoric Britain was the focus of a scientifically learned society whose mathematical, geometrical, and astronomical discoveries anticipated those of the Babylonians and Greeks. . . ."

In a recent study Burl outlined some of the reasons for his skepticism about what he feels are exaggerated claims for ancient science and offered an alternative explanation for a major class of stone monuments found in the north of Great Britain. Burl's proposed interpretation acknowledges a relationship between the monuments and astronomical phenomena, but of a much simpler kind than has been proposed by others—a relationship that is primarily symbolic rather than scientific.

Previous studies of certain stone circles in the Brit-

Courtesy, Aubrey Burl

Ancient stone circle in Scotland, built about 2250 BC, features a recumbent stone between the two tallest upright ones. Like other Scottish circles it is oriented to fit the path of the Moon as it travels across the sky low above the southern horizon, appearing to float just above the recumbent stone.

ish Isles have shown, in one case, a nice alignment of some of the monument's stones on minor movements of the Moon. Another monument was seen to contain stones that aligned on the position of the Sun at the time of the midsummer and midwinter sunsets. At other monuments alignments with the stars Antares, Capella, Deneb, and Pollux were noted. Burl points out, however, that there was little consistency or pattern among the various monuments regarding the astronomical phenomena for which alignments could be found. Furthermore, for many monuments no significant astronomical alignments could be found at all. Burl suggests that there was a note of "special pleading" in the archaeoastronomers' stress on certain stones as sights, to the exclusion of others.

To put this criticism in its baldest form, if a very large number of upright stones (from many monuments) are sighted across in various ways by an archaeoastronomer seeking alignments with stellar or planetary objects, a few such alignments will certainly be found. The alignments may well be fortuitous, however, and not intended by the people who built the monuments. If this is the case, then the stone structures are probably not the sophisticated "astronomical observatories" that some archaeoastronomers have suggested but must have some other explanation.

Burl restudied the layouts of 50 ancient stone monuments from northeastern Scotland, resurveying some of them with extremely accurate modern equipment. The monuments were of a type referred to as recumbent stone circles, consisting of rings of upright monoliths associated with a large recumbent or horizontal stone, placed between the two highest uprights. Compass bearings from true north across the circle to the midpoint of the recumbent stone showed that all these monuments were oriented within an arc between 155° SSE and 235° WSW.

No star or planet could be fitted consistently to the compass azimuths of the various monuments. However, the southerly orientation does fit the path of the Moon as it travels across the sky low above the southern horizon in the high latitudes (around 57°) of Scotland. Such an orientation would have produced a dramatic effect on nights when the full Moon, in its transit across the sky, would seem to onlookers within the stone circle to float just above the recumbent stone at the southern edge of the circle. Burl suggests that the ceremonies for which these monuments were constructed took place on such nights, when the activities around the recumbent altar stone were silhouetted by the brilliant Moon as it passed above, flanked by the highest monoliths of the circle on either side.

The challenge mounted by Burl is thought-provoking and is also appealing in its simplicity. As the field of archaeoastronomy matures, such challenges and reinterpretations will play an important part in testing and evaluating its claims and will thereby contribute to its establishment as an increasingly solid body of knowledge.

—C. Melvin Aikens

Architecture and civil engineering

During the past 15 years an entire way of thinking about architecture has been called into question. The debate has centered on a growing disenchantment with Modernism and the International Style and a renewed concern for historic precedent. While the belief re-

Hedrich-Blessing

Corporate headquarters building for AT&T in New York City, designed by Johnson/Burgee, posed the challenge of reconciling its granite exterior with a lightweight, flexible steel frame.

mained that architecture is a science that can be measured and quantified, the commitment to architecture as art with an overt interest in architectural form began to dominate.

The 1970s were marked by the creation of an intellectual climate that permitted an explosion of publications throughout the world, the formation of discussion groups, the exhibition of architecture and architectural drawings in galleries, and the appearance of architects and architecture in the popular press. This period of debate, however, generated much more passion and many more words than buildings. The debate and discussion of architecture continued into 1980 with two major events.

Probably the event that will have the longest impact resulted from a decision by the directors of the Biennale of Venice, Italy, to establish an exhibition dedicated specifically to architecture. Their intent was to provide a center for the elaboration and promotion of architectural debate and research and to hold an exhibit on architecture every two years in conjunction with exhibitions on other visual arts. Billed as "The First International Exhibition of Architecture," the theme of the 1980 architectural section was "The Presence of the Past." The exhibit, which opened in July, addressed the many changes taking place in architecture, particularly the importance of history as a source of form. Included was the work of 76 architects from throughout the world, chosen to display the many manifestations of the theme. Also featured were tributes to three architects: Philip Johnson (U.S.), Ignazio Gardella (Italy), and Mario Ridolfi (Italy). The tributes were intended to recognize the importance of their contributions to the creative reintegration of historic heredity into architecture. Of particular interest at the exhibit was the "Strada Novissima," a real street built with temporary materials inside an old building in Venice. Each one of 20 international architects designed a facade along the street expressing his or her own sense of form with reference to the theme of the Biennale, "The Presence of the Past."

The second event was an exhibit held in Chicago entitled "Late Entries to the 1922 Tribune Tower Competition." In 1922 an international competition was held to design a building for the *Chicago Tribune*, a newspaper in that city. Because the competition was held at the beginning of a period of architectural transition, it revealed a range of architectural thought by many young architects who had not previously designed a major building. Many of them, though relatively unknown at the time, were later to be recognized as important architects. The exhibit organizers believed that 1980 was also a period of transition and to hold the Tribune Tower competition again would afford the opportunity to see changes that had taken place. The drawings that were exhibited reinforced the growing interest in history, symbolism, and ornament.

While the two preceding events continued the architectural debate, 1980 was also a significant year as many of the promises of architectural change moved closer to reality in the form of major public buildings. As these buildings entered their construction phase, the relationship between architecture and civil engineering began to take on new meaning. Twenty years ago the structural system of a building was not only often expressed in the final form but was also very often the same thing. The civil engineer, particularly in large and high-rise buildings, dominated the form of the building. By 1980, however, this was not necessarily the case. Rather than influencing the shape of a building in order to optimize the structural system, the engineer now was more often optimizing the structure to achieve the particular shape desired by the architect and the owner.

The corporate headquarters building for American Telephone & Telegraph Co. by the architectural firm of Johnson/Burgee, under construction in New York City, is an excellent example of this new relationship between architecture and civil engineering. Johnson/

Burgee borrowed extensively from the Renaissance revivalists in creating a 660-ft-high granite and glass tower. Since the exterior was to be clad in rigid pieces of granite, structural engineer Leslie Robertson needed to reconcile the application of materials not used in some time with the kind of structural system—a lightweight, flexible steel frame—common today. Thus, Robertson was offered a challenge to combine current technology with historically evolved form, a practice that may become more common.

The most controversial building of the year was the Portland Public Service Building in Portland, Ore., by Michael Graves. Graves was awarded the contract for the building as a result of a competition. The controversy surfaced not only because of the design, which was considered outrageous by the "established" architects in the area, but also because of the architect. Graves's work generally reflects 18th-century Romantic-Classicism and employs forms unthinkable in Modern architecture. Critics argued that his buildings are two-dimensional paintings that are purely picturesque.

As reported in the August 1980, *Architectural Record,* "(the) Portland (building) clamors for attention; it is colorful, gregarious, outgoing. It proclaims that buildings serving the public ought to do so with verve and dash; they should excite the viewer, delight the visitor; encourage participation and inspire observation." The 17-story building, which is located between the classically styled city hall and county courthouse, has a classical skyscraper tripartite division of base, shaft, and capital. This division corresponds to the legs, torso, and head of the body, a classical metaphor Graves used instead of the "bodyless" glass box. The base is of a green color that makes a visual connection to the three-block-long park on which one side of the building faces. The shaft is a concrete box with small square windows. Its color is cream to lighten the apparent mass and to contrast with two terra-cotta pilasters which support a large keystone that caps the structure. Construction of the building began in July 1980.

Portland Public Service Building in Portland, Oregon, shown below in model form, created controversy because of its break with accepted Modern architecture.

Michael Graves

In Chicago two recent projects by Helmut Jahn of Murphy/Jahn Associates (formerly C. F. Murphy Associates) further illustrate the change in architecture. In the 500,000-sq ft (46,500-sq m) addition to the Chicago Board of Trade the formal character of Jahn's design is an abstraction of the original Art Deco building. The massing recalls the symmetry and tripartite elements of the original, but the highly articulated planes of the walls and roof are constructed using the current technology of a tight-fitting glass skin rather than the limestone of the old building. The result is something quite different from the original structure. Another project under construction by Jahn and C. F. Murphy worth noting is the One South Wacker office tower in Chicago. Again Jahn translates the forms of the 1920s into gray and silver reflective glass that is deployed decoratively onto the highly modeled facade.

Similar changes in architecture were also taking place in other parts of the world. For the 1980 Venice Biennale, Aldo Rossi of Italy was commissioned to design a floating théater. The finished structure has obvious references to the floating theaters used in Venice in the 16th century, but it also has been likened to certain types of Lombard farm structures that Rossi knew in his childhood. The 250-seat theater is constructed of iron tubes forming a structure similar to a scaffold. The scaffolding is welded at its base to a barge and is completely covered on the outside with yellow pine that is either clear-sealed or painted blue. In form, the theater is a tall rectangle that rises to an octagonal, pyramid roof. It is deceptively simple, classical, and symmetrical. Although small, it is monumental and complements the domes and towers of Venice.

The British architect, James Stirling, who received the Royal Gold Medal for Architecture from the Royal Institute of British Architects in 1980 and is noted for his "high-tech" work at the Engineering Laboratories at Leicester, recently turned to classicism as demonstrated by his Science Center of Berlin. A requirement for this building, on which construction began in 1980, was to preserve the facade and the larger part of the 19th-century Beaux Arts Law Courts building. Rather than house the Science Center in one large building,

James Stirling, Michael Wilford and Associates; photo, John Donat

Model of the Science Center of Berlin reveals the architect's concept of clustering around a central garden several buildings of varying, though not entirely dissimilar, form.

Stirling's solution was to cluster around a central garden several buildings each of differing, though not completely dissimilar, form. The form of each building relates to a familiar structural type and like the Law Courts building is symmetrically planned. The building base and window pattern are also highly influenced by the existing buildings. Stirling used pink and gray on alternating floors to form bands that visually unite this cluster into a single entity.

These and other steps that architects are now taking to renew contact with history may be the first acts in a long process of increased concern for aesthetics. The year 1980 has been significant in this process.

—Roger H. Clark

Astronomy

Among the major developments of the past year was an increase in knowledge of the solar system, particularly with regard to the planet Saturn. Also, a major astronomical facility, although in effective use while under construction, was dedicated. The first direct evidence of the cosmological distances of quasars was announced. Observations in shortwave radiation from above the Earth's atmosphere along with visible and infrared ground-based observations continued to improve astronomers' understanding of both nearby and distant objects.

Instrumentation. The Very Large Array (VLA) radio telescope was formally dedicated on Oct. 10, 1980, nearly a year ahead of schedule. The facility is located on the Plains of San Augustin near Socorro, N.M., a site chosen because it is a large, flat, high-altitude plateau surrounded by mountains that afford excellent protection against man-made radio interference.

The VLA consists of 27 interconnected radio dishes, each having a 25-m (82.2-ft) aperture. The dishes are arranged in a Y-configuration and can be moved on a standard-gauge railway track to four different sets of spacings along the arms of the Y. The lengths of the arms are 21 km (13 mi) in the southwest and southeast directions and 19 km (12 mi) in the north. Collectively, the receiving surface of the 27 dishes is equivalent to that of a single dish 122 m (400 ft) in diameter. More significant, however, is the angular resolving power of the array, which, in its largest arrangement, is equivalent to a dish having a diameter of 25 km (15.5 mi). The array, therefore, has a resolving power equivalent to that normally associated with large optical instruments, permitting the direct comparison of radio and optical data with like detail. The system is designed to operate at wavelengths of 21, 6, 2, and 1.3 cm.

A major component of the VLA is its computer system. Each dish in the array, under computer control, precisely tracks the object under study. The 27 antennas form 351 different interferometer pairs, which, as the Earth rotates, sweep out different paths in space. By storing the signals from each dish and combining them as required by the instantaneous spatial arrangements of the interferometer pairs, the computer synthesizes what would be seen by a continuous dish over the area swept by the array. This technique, pioneered at the Mullard Radio Astronomy Observatory of Cambridge University and known as aperture synthesis, allows the construction of spatial maps of the radio signals comparable to photographs taken with optical telescopes.

The dedication of the VLA was in many respects anticlimactic because it had already been used to collect significant observations over the past several years. Each of the dishes was assembled and put into operation before the next one was prepared. Even in partial assembly the array was a powerful tool. At full strength the VLA will undoubtedly have a significant and continuing impact upon astronomy.

Another device of note is the Acoustical-Optical Spectrograph, developed in Australia by the Commonwealth Scientific and Industrial Research Organization to study radio signals that change rapidly both in frequency and time. The amplified signal from a radio telescope is impressed upon a piezoelectric transducer (a device actuated by power from one system and supplying power in the same or another form to a second system) constructed of crystals that change size with the impressed electrical signal. This causes an ultrasonic sound wave to be generated in the crystalline substance of the transducer. The characteristics of this wave depend upon the frequencies present in the impressed radio signal. The contractions and rarefactions in the wave cause local changes in the index of refraction of the transducer which travel with the wave as it passes through the crystals of the transducer. The transducer thus behaves as a variable diffraction grating capable of causing deviations in a beam of light. The angle at which the deviations occur depends upon the spacings of the refraction changes, which, in turn, are related to the frequencies in the sound wave. The amount of light deviated, on the other hand, depends upon the strength of the refraction changes, which are proportional to the sound wave intensities. Consequently, when a laser beam is shone on the transducer, it generates a diffracted beam that reflects both the amplitude and frequency content of the original radio signal. If a lens is placed in the dispersed beams, the dispersed light corresponding to the various frequencies present can be measured by an array of photodiodes. In this fashion a simultaneous measurement of the spectrum of the radio signal, both frequency and amplitude, is generated.

The overall bandwidth of the spectrograph depends primarily upon the frequency response of the piezoelectric transducer. At present several thousand frequency channels, each the output of one photodiode, can be measured in a radio band more than 200 MHz wide. For strong signals the time resolution of the device is about one millisecond.

The Acoustical-Optical Spectrograph is particularly suited to the study of radio pulses from pulsars. It also is proving useful in studying scintillation (rapid changes in intensity) in radio signals induced by interstellar plasma and the Earth's ionosphere. Solar research is also benefiting from this instrument because of its capability of giving detailed information about radio scintillation caused by the solar corona.

Galileo revisited. One of the newest observations of Neptune turned out to be over 350 years old. Steven Albers, an amateur astronomer from Stewart Manor, N.Y., became intrigued with the possibility of seeing one planet occult another. Consequently, he wrote a computer program that predicted the occurrence of such events from the year 1557 to 2230. In all, 21 such possibilities were predicted. He published his results in the magazine *Sky and Telescope,* where they caught the eye of Charles T. Kowal of Palomar Observatory. According to Albers Neptune should have been occulted by Jupiter in January 1613, during the time when Galileo was known to have been making telescopic observations of Jupiter and its four bright satellites. Kowal was interested in the orbit of Neptune because it is

The Very Large Array radio telescope in New Mexico consists of 27 interconnected radio dishes, each of which (right) has a 25-meter (82.5-foot) aperture. The angular resolving power of the Y-shaped array is equivalent to that of a single dish having a diameter of 25 kilometers (15.5 miles).

Photos, National Radio Astronomy Observatory, operated by Associated Universities, Inc. under contract with the National Science Foundation, Charlottesville, Va.

not known with great precision, partly because the planet has yet to make one complete circuit around the Sun since its discovery in 1846. The only known prediscovery sighting of Neptune with sufficient accuracy for use in orbit determination was made in 1795 by the French astronomer Joseph Lalande while he was telescopically compiling a catalogue of star positions. Lalande failed to recognize Neptune as a planet, taking it instead as a field star, and his position for Neptune does not agree with that computed from the currently accepted orbit of the planet.

Encouraged by Albers's calculation of an occultation by Jupiter, Kowal and Stillman Drake, a historian of science, requested copies of the entries in Galileo's observing notebook. They found that during December 1612 and January 1613, Galileo had included a faint star for reference in his diagram of Jupiter and its Galilean satellites. He had also plotted the position of another object, a star now known as SAO 119234, which Kowal and Drake could identify from Galileo's surprisingly accurate records. Fortunately, on Jan. 28, 1613, Galileo gave the positions of both this star and the faint star relative to Jupiter in terms of the Jovian radius. The comparison of Galileo's record of SAO 119234 with that of his location for the unrecognized faint star convinced Kowal and Drake that the latter was Neptune and that Galileo's position for it was of high quality. But Galileo's position for Neptune varied by one minute of arc from the position predicted using the currently accepted orbit. The difference is large enough to cast doubt on the accuracy of the present orbit, and suggests that some unseen mass is disturbing the motion of Neptune.

Thus, Galileo's sighting is more than a mere historical oddity. Because of its precision, it may lead to the discovery of another planet that disturbs the motion of Neptune, much as the disturbance of the motion of Uranus led to the discovery of Neptune itself.

The solar system. Certainly the greatest excitement in the past year centered on the continued success of Voyager 1 as it rendezvoused with Saturn. The U.S. space probe passed within 124,200 km (77,000 mi) of the planet in November and was able to observe Saturn's rings both on their sunlit and opposite sides. Voyager discovered three new satellites of the planet, photographed the circulation pattern of Saturn's atmosphere, obtained extremely detailed pictures of the rings, measured the atmosphere of the satellite Titan, and sent back reasonably detailed pictures of six of Saturn's eight major satellites.

Features on Titan's surface were not revealed because of the clouds in its dense atmosphere. The

F-ring of Saturn (below left), photographed by Voyager 1 from 750,000 kilometers (470,000 miles), is a unique braided structure that defies a simple gravitational explanation. Photomosaic (below right) of Rhea, from 80,000 kilometers shows it to be the most heavily cratered of Saturn's moons.

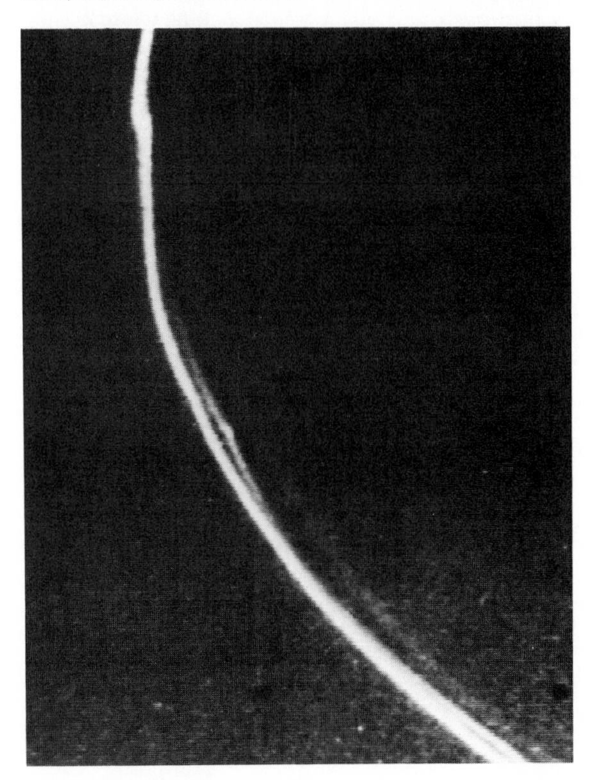

Photos, Jet Propulsion Laboratory

atmosphere was shown to be much denser than expected and to consist of about 99% nitrogen. The pressure at the surface may be as much as twice that of the Earth's atmosphere. The satellite Mimas displayed a huge crater, the deepest yet found in the solar system, which measures one-quarter of the satellite's diameter across. Rhea, with the best determined detail of all, is saturated with craters. Tethys revealed a large circular feature 180–200 km (110–125 mi) in diameter just opposite from a huge crack that may have been caused by shock waves induced by the collision that formed the circular feature. Dione is heavily cratered, with bright wisps emanating from one particularly bright basin. Iapetus is six times brighter on the side opposite to the one that leads in its orbital motion. Enceladus, while always more distant than 200,000 km during the encounter, does not have the craters that should easily have been seen at these distances. The satellites have densities within 10–20% that of water except for Dione, 40% greater, and Titan, nearly 100% greater. They, therefore, must contain large proportions of ice.

Voyager showed that Saturn's rings are much more complicated than anyone had expected. Of the three rings readily seen from the Earth only the A-ring, or outermost, appeared much as expected. Cassini's Division, the dark gap separating ring A from B, contains at least twenty ringlets. The B-ring on close examination comprises myriads of ringlets, somewhat resembling the grooves in a long-playing record. This ring also revealed radial features, or spokes, that co-rotated with the planet. Such motion was in contradiction to the normal Keplerian motion of the rings, which requires the inner parts to move faster than the outer parts. The spokes are thought to be composed of small electrically charged particles that are lifted from the plane of the rings and move in step with transitory disturbances in Saturn's rotating magnetic field.

The C-ring appears to be made up of icy boulders with an average size of about one meter. Voyager supplied an explanation for the narrow width of the F-ring, which exists beyond the A-ring. Two small satellites, one at the inner and one at the outer edge of the ring, appear to herd the ring particles, keeping them within their narrow confines. But Voyager also saw a feature of the F-ring that has yet to be explained: braided ringlets that seem to defy a simple gravitational interpretation.

Elsewhere in the solar system, the size of Pluto has again been revised downward. S. J. Arnold and A. Boksenberg of University College, London, and W. L. W. Sargent from Palomar Observatory used a speckle interferometer at the 200-in (5-m) telescope to measure the planet's diameter. Based on Pluto's reflecting characteristics, the measurements are consistent with a diameter between 3,000 and 3,600 km (1,900 and 2,200 mi). This means that Pluto is significantly smaller than the other planets and is close to the size of the Earth's Moon. Combining this size with the mass deduced earlier from Charon, Pluto's nearby satellite, gives another decided difference for Pluto; its density lies between 0.5 and 0.8 that of water, implying that it consists of frozen volatiles.

The Sun. The Solar Maximum Mission spacecraft was launched on Feb. 14, 1980, to be aloft while the Sun goes through its present period of maximum sunspot activity. The instrumentation aboard was designed to observe solar flares in the range from visible light through gamma rays. The major aim of the observations was to understand the mechanisms by which such flares occur. Another instrument aboard was the solar constant monitor, which had already detected daily variations in the Sun's radiant energy of as much as 0.4%, fluctuations completely undetectable from within the Earth's atmosphere.

Young stars. Martin Cohen and Leonard V. Kuhi completed an extensive survey of T-Tauri stars using the Lick Observatory three-meter telescope to make high-dispersion spectral scans of 500 of these objects. The infrared brightnesses of the stars were also measured using telescopes at the Mt. Lemmon and Kitt Peak observatories. T-Tauri stars are variable stars that can be identified by their irregular light variations, intense infrared emissions, bright spectral lines, and association with dark nebulae. They are believed to be stars in an early stage of formation, actually contracting from interstellar matter in the clouds with which they are associated. When a large interstellar cloud complex begins contracting, stellar formation appears to commence abruptly and continues at a more or less uniform rate.

From their observations Cohen and Kuhi concluded that these young stars typically have masses ranging from 0.2 to 3 solar masses, and that they vary in size from one to five times the solar diameter. They estimated that the age range of the stars they studied extends from 10,000 to 6,000,000 years.

Hot halos. Observations made with the International Ultraviolet Explorer (IUE) satellite led K. S. deBoer and B. D. Savage of the Washburn Observatory to conclude that the Milky Way Galaxy and the Magellanic Clouds are engulfed in large, hot halos, or coronas, of extremely rarefied gas. Their observations confirmed a prediction made by Lyman Spitzer at Princeton University in 1956 that such a halo should exist around our Galaxy. Using the IUE, deBoer and Savage were able to obtain high-dispersion ultraviolet spectra of bright stars in the Magellanic Clouds. The exposure times required were as long as five hours; this was possible only because of the favorable orbit of the satellite. In the spectral range from roughly 1200 to 2100 Å (angstroms; one Å equals 0.0000001 mm), they found a number of strong absorption lines that were due to interstellar matter. The lines appeared to be double, revealing components with small Doppler shifts that implied their origins in gas associated with our Galaxy and also showing other components with large Doppler shifts characteristic of the Magellanic Clouds. Triply ionized lines of carbon and silicon were found; these require temperatures of the order of 100,000 K. The gas density was found to be extremely low, about three particles per 10,000 cc.

The halo surrounding our Galaxy appears to extend outward about 50,000 light-years from the Sun. Similar halos around other galaxies may account for the absorption lines with small red shifts frequently seen in association with the highly red-shifted emission-line spectra of quasars. All that would be required for such a finding would be the chance interposition of a galaxy and its halo between the Earth and the quasar.

Multiple quasars. The quasar 0957 + 561, discovered in 1979, continued to receive attention during the past year. New observations indicated that this apparent double quasar is in fact single and only seems to be double because of the gravitational lens effect, in which light is deflected when it passes through a gravitational field generated by an intervening object. The system was observed by Peter Young, James E. Gunn, Jerome Kristian, J. Beverly Oke, and James A. Westphal with the 200-in (5-m) telescope at Palomar. They used a charge-coupled device (CCD) rather than the usual photographic plate. The CCD is a solid-state device that is capable of measuring the brightness of faint objects even in the presence of much brighter ones. Their results showed that the southern image of the quasar pair had an elongation of about one second of arc toward the north. The other member of the apparent pair had a circular shape. This discrepancy between two images that should appear identical in shape if they were double images of one object at first cast doubt on the gravitational lens explanation.

Independently, Alan N. Stockton, using the 2.2-m telescope of the Mauna Kea Observatory of the University of Hawaii, obtained a series of short-exposure photographs of 0957 + 561. By taking short exposures, he could select the photographs taken when the Earth's atmosphere was particularly steady and thereby obtain better resolution of detail than would occur in a longer exposure, which would be smeared during the more unsteady moments. His photographs showed a small fuzzy patch to the north of the southern image. Using an image processor, he combined five of his exposures to enhance the detail. Next he scaled the northern quasar image to the intensity of the southern image and subtracted it from the southern one with the image processor. He was left with an image of an elliptical galaxy. Thus, the distortion seen with the Palomar telescope and the fuzzy patch in Stockton's

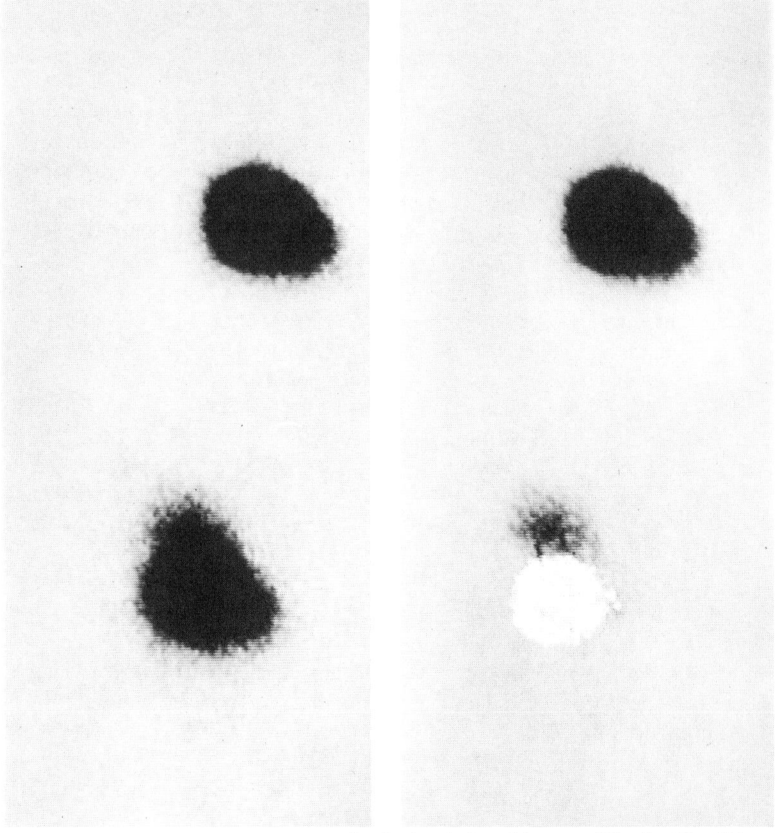

Quasar 0957 + 561 appears to be double in a digital superposition of five one-minute exposures (left panel). In the right panel the upper component has been scaled to the same brightness as the lower and then subtracted from it, revealing an intervening galaxy (white area). The gravitational lens effect, in which light is deflected when it passes through a gravitational field generated by an intervening object, caused the quasar to appear to be double.

Alan Stockton, Institute of Astronomy, University of Hawaii, Manoa pictures was an intervening galaxy. This was confirmed by the observers at Palomar when they examined the spectrum of the southern image. It showed the same quasar spectrum as the northern image but superposed was a spectrum of a galaxy. Both quasar images showed a red shift of 1.40, indicating extreme distance from the Earth according to the cosmological red-shift relation. But the faint superposed galaxy spectrum had a red shift of only 0.4 and, therefore, was only about half as distant as the quasar. The suspected galaxy was found by the Palomar team to be a slightly elongated giant elliptical one, and was further revealed to be only the brightest member of a cluster of galaxies situated in the vicinity.

Observers from the Massachusetts Institute of Technology, David H. Roberts, Perry E. Greenfield, and Bernard F. Burke, used the VLA to observe 0957 + 561, as they had a year earlier. This time they observed the quasar for 12 hours rather than 42 minutes as they had in 1979, thus obtaining a much more detailed radio map of the system than they had in the previous year. The new map showed the same five radio concentrations as before, but the two intense regions associated with the optical quasar images showed small features that strongly implied that the regions were duplicates of each other. The northern concentration, however, had a weak bridge of energy apparently joining it to the two large diffuse regions to the northeast. The southern concentration, on the other hand, had an extension that was centered within 0.02 arcsecond of the position of the elliptical galaxy image, seemingly confirming the gravitational lens.

The gravitational lens explanation was still in doubt, however, because conventional wisdom called for a point mass (a compact gravitational source massive enough to bend light) midway between the two quasar images, not off-center as the galaxy obviously was. Furthermore, the other radio structures in the radio data were still unexplained. The Palomar observers, however, reexamined the properties of a gravitational lens, but for the extended mass of a galaxy rather than the point mass of an object such as a black hole. They found that an extended-mass lens, such as the intervening elliptical galaxy, can produce one, two, three, or even more images, and not just the two expected from a point-mass lens. The number and displacement of the images formed by an extended mass could be not only multiple but asymmetrical as well, depending on the geometry of the source, the lens, and the Earth. In the case of 0957 + 561 the position of the light-deflecting galaxy is such that the southern quasar image is actually two images so close together that they cannot be resolved. The northern image is a third image formed by the gravitational lens. The two most intense of the remaining radio concentrations seen by the VLA are associated with the quasar but are far enough from the line joining the Earth and the gravitational lens that only one displaced image of each of them is seen. The weakest concentration of the five seen in the radio map is not associated with the quasar. Though this model was only one of several that were offered by the Palomar team using an extended mass as the gravitational lens, the proposal of Dennis Walsh, Robert F. Carswell, and Ray Weymann in 1979 that the appearance of 0957 + 561 resulted from such a lens was by now firmly established.

In the meantime Weymann, David W. Latham, and others in a continuing program of observing quasars with the multiple-mirror telescope in Arizona found that the quasar PG 1115 + 08 appears to be triple. All components show the same red shift. Although the intervening mass had not been identified at the year's end, this observation undoubtedly is the second example of a gravitational lens. The phenomenon of the gravitational lens, interesting enough in itself and as another confirmation of relativity, has a significant cosmological meaning. The quasar 0957 + 561 cannot be an object within our own Galaxy; it must be more distant than the remote galaxy which influences the radiation that the Earth receives from it. Thus, the red shift seen in the quasar spectrum must be cosmological, a result of the expansion of the universe. This is the first direct evidence that quasars must be extremely remote and highly luminous objects.

—W. M. Protheroe

Chemistry

Syntheses that favor production of one molecular configuration over its mirror image; novel polymers with silicon-to-silicon backbones; details about the energy states of molecules before, during, and after they chemically interact; and progress on several fronts in the search for alternate energy sources capped research in chemistry during the past year. In increasing evidence was the dependence of the chemist on computers to perform calculations, store and retrieve information, and supplement human intelligence.

Inorganic chemistry

During the year inorganic chemistry received considerable attention from research chemists around the world, both in academia and in industry. Although university work concentrated on fundamental problems of bonding, reactions, syntheses, and structures of inorganic compounds, more and more effort was being spent on problems related to either energy or health, in part because of the increased funding available for such projects. Fields of intensely active research in inorganic chemistry included bioinorganic chemistry, homogeneous catalysis, photochemistry, and solid-state chemistry.

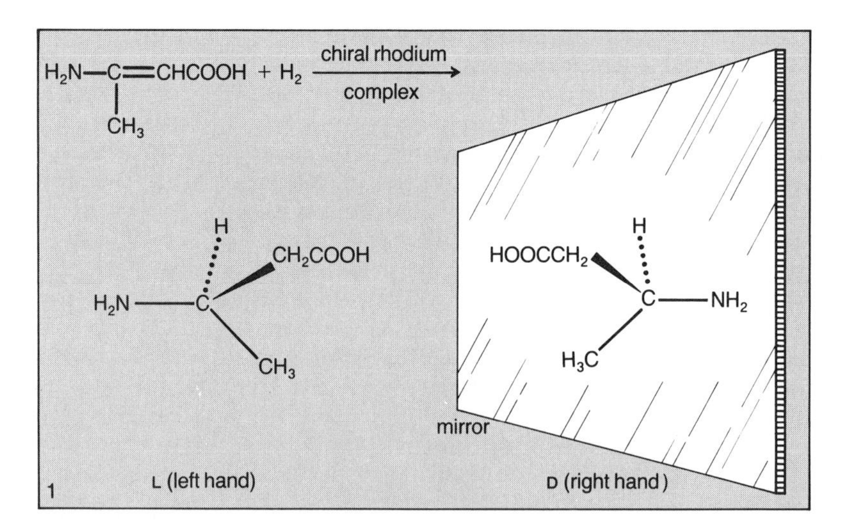

1

Stereoselectivity of homogeneous catalysts. Organometallic chemistry of the transition metals has become an important field of research for both inorganic and organic chemists. Among exciting aspects is its potential to provide homogeneous catalysts that selectively promote desired reactions and facilitate production of certain needed products. (A homogeneous catalyst is present in a reaction in the same phase as at least one of the reactants, usually as a gas or liquid.) For example, a particular rhodium phosphine complex is used as a homogeneous hydrogenation (hydrogen-adding) catalyst in manufacturing the amino acid L-dopa (dihydroxyphenyl-L-alanine), a drug used to treat Parkinson's disease, by the Monsanto Co., St. Louis, Mo. A similar catalyst may be used for producing L-phenylalanine, which is needed to make the sweetener aspartame, when U.S. government approval is given to G. D. Searle & Co., Skokie, Ill.

These examples are perhaps the most impressive achievement to date in selectivity of man-made catalysts, which rival the long-known selectivity of natural enzymes. The commercial promise of such man-made catalysts and the intriguing theoretical problems associated with asymmetric rhodium catalysts have drawn many chemists to explore the mechanisms of reaction and to design new catalysts. Brice Bosnich and his co-workers at the University of Toronto have had tremendous success in designing chiral rhodium complexes, which are highly stereoselective catalysts. (Chiral molecules exist in mirror-image forms that differ in the same way left and right hands differ.) The reactions catalyzed are the hydrogenation of olefinic amino acids to give saturated amino acids with a high percentage of either left- or right-handed (L or D) isomers. Such a reaction can be illustrated, in principle, with the simplest possible olefinic amino acid (*see* 1).

Normally such a reaction with an ordinary catalyst would yield a product composed of left- and right-handed isomers in equal amounts. What is wanted is a catalyst that is stereoselective and results in the formation of approximately 100% of the desired isomer. A given isomer may be more in demand than its counterpart because it is biologically active while the other is not. Such is true of L-dopa, which is an excellent drug for Parkinson's disease; D-dopa is not effective.

In the past year Jack Halpern and his student Albert S. C. Chan of the University of Chicago reported on their detailed investigation of the kinetics and mechanism of a particular hydrogenation reaction using a chiral rhodium complex as a catalyst. They used a complex discovered earlier by Bosnich, and they studied the several intermediate steps involved in hydrogenation of ethyl-α-acetamidocinnamate to the ethyl ester of *N*-acetyl-D-phenylalanine in methanol solution (2). All of the intermediate steps proved to be well-known reactions in transition-metal organometallic chemistry. The significant contribution made by Halpern was in pointing out that the intermediates detected by sophisticated modern techniques of probing such reaction mixtures would require the formation of the L product, whereas in fact the D isomer is produced. This seeming contradiction indicates that although the D-intermediate route is present in such low concentration as to escape detection, it is the more reactive of the two and provides a low-energy path for exclusive formation of the D product. This finding poses extremely significant questions in relation to current understanding of stereoselective reactions. It is now generally as-

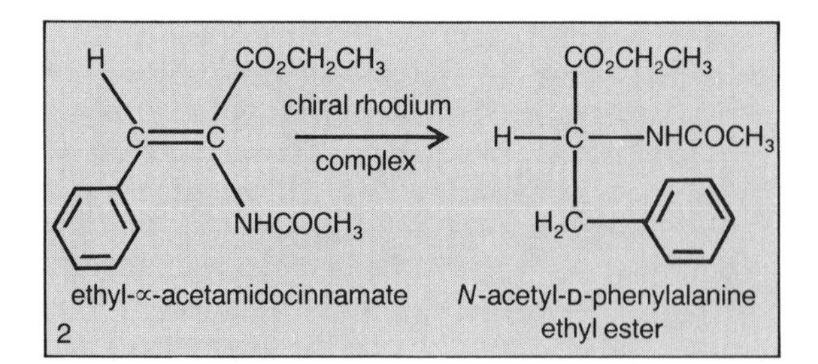

2

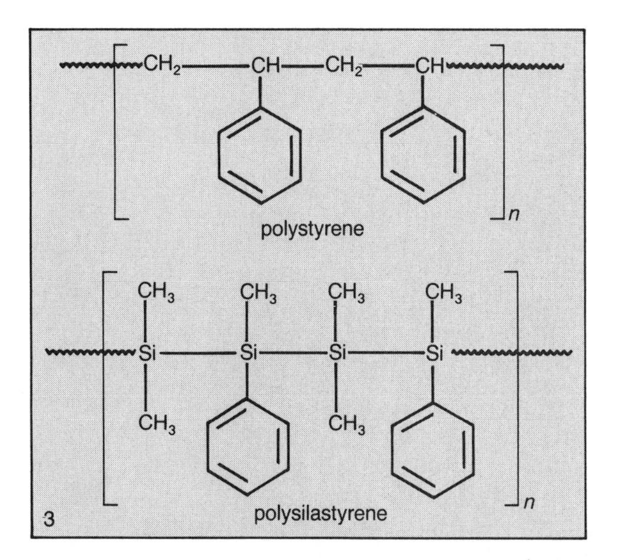

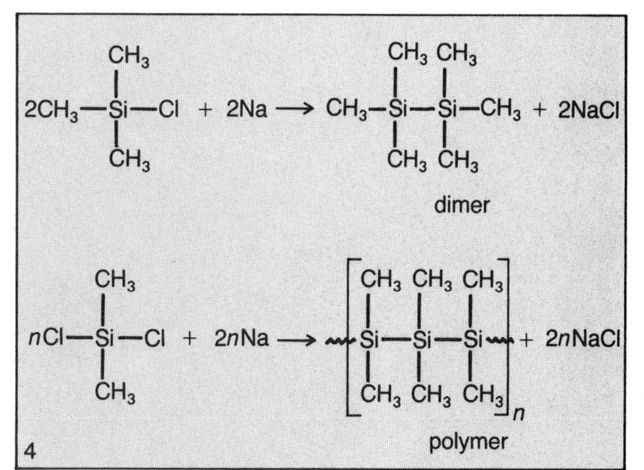

sumed that enzymes are very selective catalysts because of the "lock and key" concept attributed to the adducts of enzymes and their substrates; *i.e.*, the molecules they accept for catalysis. The results of Halpern bring this assumption into question, for the catalytic intermediates may not be these stable adducts.

Polysilastyrene. Polystyrene is a most versatile and useful polymer, produced in tonnage quantities by many chemical industries around the world. During the year Robert C. West and Lawrence D. David at the University of Wisconsin prepared a polymer with a formula resembling polystyrene, which they dubbed polysilastyrene. Polystyrene has a backbone made of a linear chain of carbon atoms, whereas polysilastyrene has a corresponding backbone of silicon (3).

The silicon polymer was prepared by the standard reaction of alkylchlorosilane with sodium, modified to give a polymer instead of a dimer (4). Early experiments using bulk chips of sodium gave a viscous liquid with an apparent molecular weight of about 8,000 daltons, but more finely dispersed sodium gave solid polymers of molecular weights in the range of 100,000–400,000 daltons. To yield polymers with different characteristics the ratio of methyl to phenyl groups can be varied, and other groups can be used.

The practical uses of polysilastyrene need investigation, but it is certain that such polymers will receive considerable attention in the years to come. There has long been an interest in making inorganic polymers with backbones of elements other than carbon. Previous efforts to do this had largely failed, for which reason this margin of success is so important. In contrast to other such inorganic polymers, polysilastyrene can be molded, drawn into fibers, and cast into films much like conventional organic polymers. (See *Applied chemistry*, below.)

Water photolysis. The world's most abundant chemical energy source is water. Its use as such will require the efficient, economical, large-scale splitting

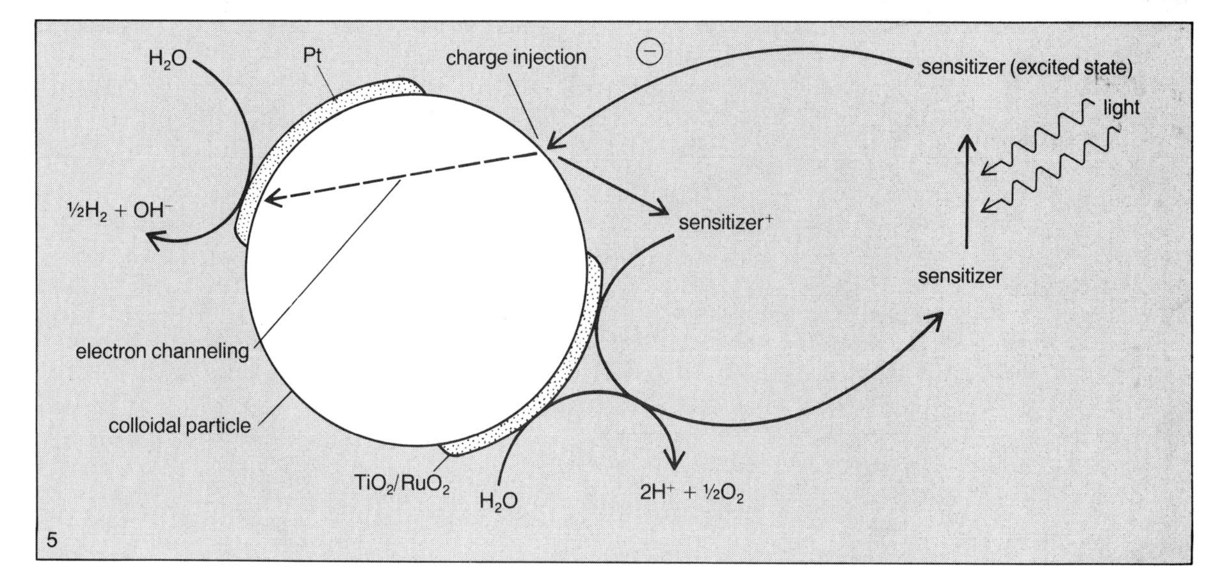

Courtesy, Oak Ridge National Laboratory

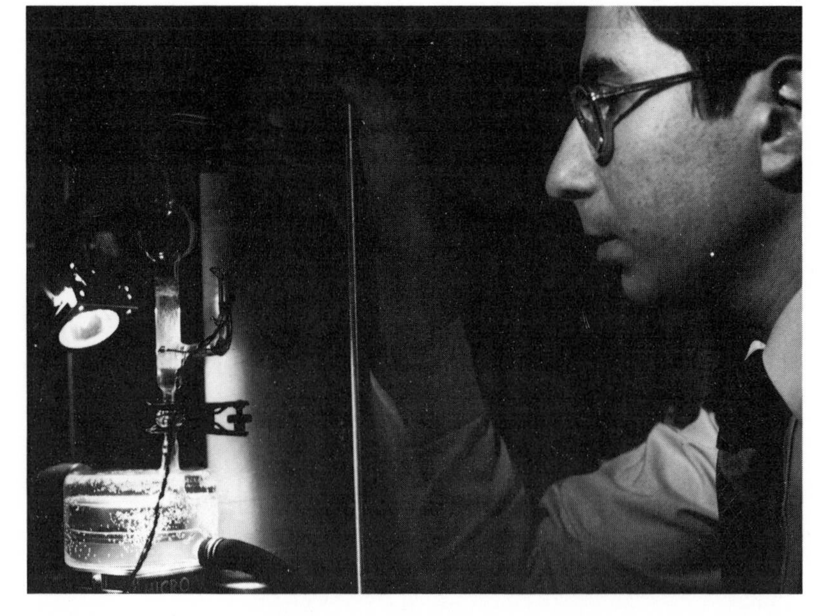

Scientist Elias Greenbaum of Oak Ridge (Tennessee) National Laboratory works with light-powered water-splitting system based on bioinorganic reactions involved in natural photosynthesis. The system uses isolated spinach chloroplasts, metalloproteins called ferredoxins, and the metal-containing enzyme hydrogenase to generate gaseous hydrogen and oxygen.

of water into hydrogen and oxygen. Hydrogen, the cleanest of fuels, could then be burned in air to release energy and to restore the water. This system lies at the heart of an energy plan for the future often called a hydrogen economy.

Water is a stable molecule, and in theory the same amount of energy is required to split it into hydrogen and oxygen as is returned from their combination. Consequently the conventional ways of splitting water by heat or electricity will use as much or more energy as is yielded from burning the hydrogen that is formed. It is not surprising, therefore, that many scientists in the world are trying desperately to harness solar energy to split water.

At the Swiss Federal Polytechnic Institute in Lausanne, Michael Grätzel and his associates recently developed a system that uses visible light to decompose water. Success was also reported for an ultraviolet-light system at the Institut le Bel of the Université Louis Pasteur in Strasbourg, France, by a team of scientists led by Jean-Marie Lehn. The basic concept used in the two laboratories, which were working independently, is that more than enough energy exists in sunlight to decompose water. This process does not happen spontaneously because there is needed a compound that efficiently absorbs the light energy (a sensitizer) and an appropriate surface to permit electron flow (a semiconductor).

One example of a system that has given encouraging preliminary results is diagrammed in (5). It consists of a semiconductor comprising colloidal microspheres of titanium dioxide (TiO_2) on the surface of which plati-

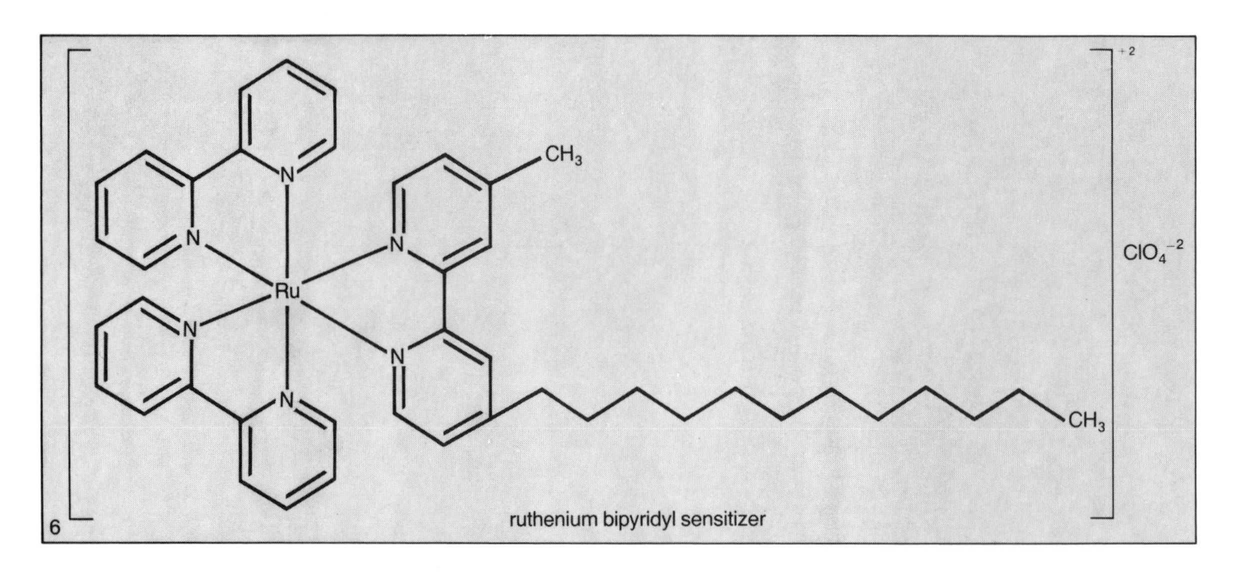

6

ruthenium bipyridyl sensitizer

num (Pt) and ruthenium dioxide (RuO_2) are co-deposited. The sensitizer (6) is a derivative of *tris*(2,2′-bipyridine)ruthenium(II), or $[Ru(bipy)_3]^{2+}$, in which an *n*-alkyl group is attached to one of the bipyridine moieties. When the ruthenium complex and the colloidal particles are dispersed in water and the mixture is exposed to visible light, both hydrogen and oxygen are evolved. As of early 1981 the rate of hydrogen generation was low, and the process was not efficient. Efficiency, however, depends on the way in which the semiconductor is prepared and on the nature of the sensitizer. This system is but the beginning, and chemists may be at the threshold of a process destined to bring to the world the ultimate solution of its energy problem. (See *Applied chemistry,* below.)

—Fred Basolo

Organic chemistry

Reflection on the past year's activity in organic chemistry points to continuing emphasis on synthesis, both in developing methods and in preparing specific compounds. Also noteworthy is the increasing use of computers in organic chemistry. This usage is manifold and reflects the organic chemist's entry into such nontraditional fields as X-ray crystallography, theoretical chemistry, information science, and mathematical constitutional chemistry.

Natural product synthesis. For several years research in organic chemistry has focused on synthesis of complex molecules of natural origin. The past year has been no exception. Some remarkable achievements were reported in the synthesis of natural products of medical interest, including compounds possessing antibiotic properties—*e.g.,* rifamycin S (*see* 1), thienamycin (2), lasalocid A (3), and monensin—as well as compounds with anticancer properties—*e.g.,* jatrophone (4), quassin (5), and maytansine.

One reason for developing synthetic routes to such products is to supplement their natural supply, as some are extremely scarce. Another reason is that chemical synthesis can be adapted to produce "unnatural" analogues, which in some cases may be more potent and less toxic than the natural compound. Usually, the more potent the drug the lower is the dose, and the lower the dose the fewer the toxic effects. A distressing illustration of this effect surfaced recently with reference to a synthetic analogue of morphinelike alkaloids that has many times their analgesic activity. The compound has been responsible for many overdose-related deaths among drug addicts unaware of its enhanced activity. This illicit drug, pushed as China White, has been identified as 3-methylfentanyl (6).

Structure determination of rare natural products is also a field in which synthesis plays a role. For instance, the biologically potent substance leukotriene C (7) is produced in the lungs during asthma attacks

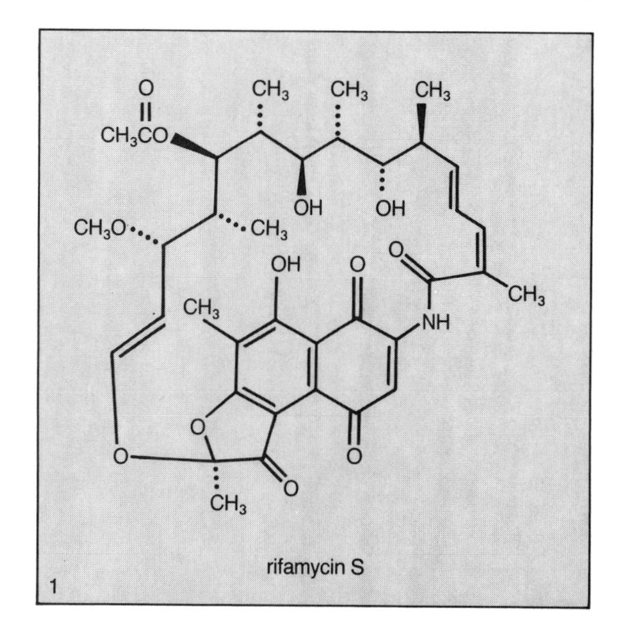

1 rifamycin S

but elucidation of its structure and mode of action has been hampered by its scarcity. Recently the compound was synthesized by Elias J. Corey and co-workers at Harvard University in collaboration with Swedish chemists Bengt I. Samuelsson and Sven G. Hammarström. Their success constitutes a proof of structure and a practical way of making the compound in quantity. Similarly, British chemists from the University of Southampton reported the isolation, structure, and synthesis of the sex attractant of the olive fly (8). This pest is widespread in the Mediterranean, and identification and synthesis of its sex attractant is an important step in the direction of its eventual control.

Asymmetric synthesis. A glance at the structures of the compounds whose syntheses are mentioned herein shows the presence of many functional groups and numerous centers of chirality, in which spatial arrangement of the atoms (the stereochemistry) is absolutely specific. It is crucial in synthesis that the reactions used to build the structure produce the correct stereochemistry at each chiral center and that other

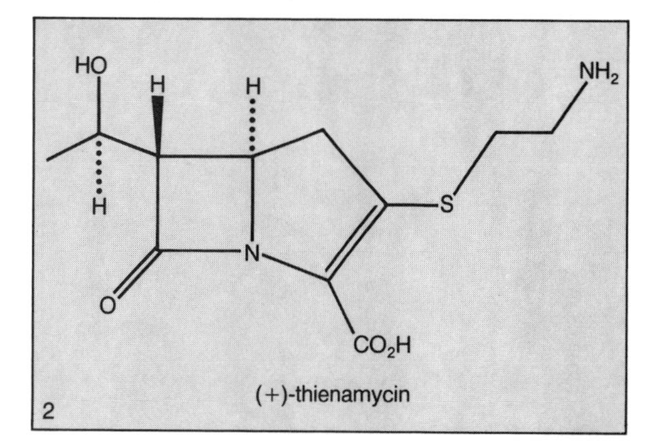

2 (+)-thienamycin

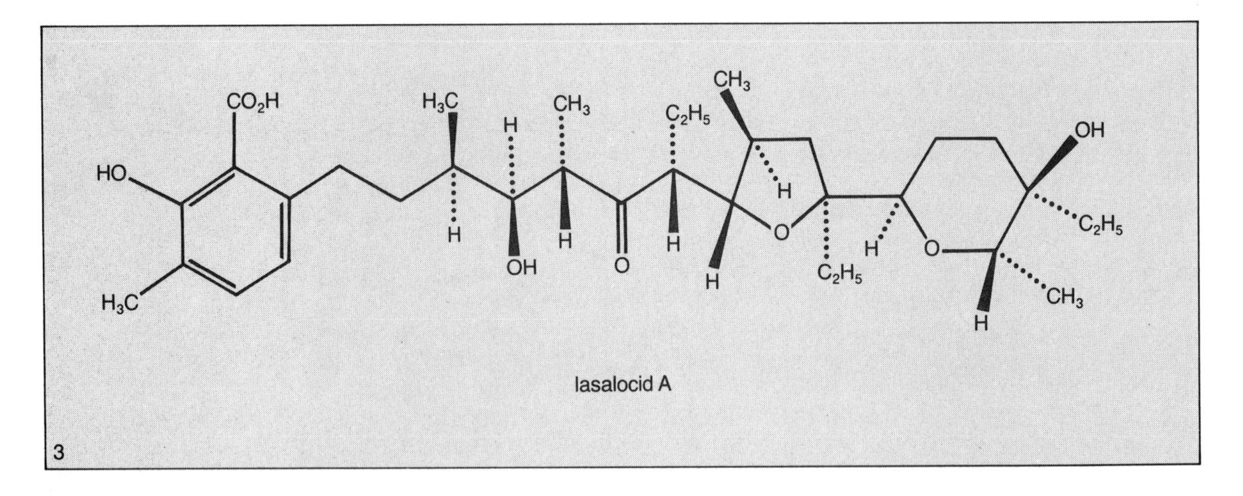

lasalocid A

3

parts of the molecule remain unchanged in the process. In recent years chemists have made great progress in designing stereocontrolled or asymmetric syntheses.

Progress is due to a combination of ingenious approaches to the basic objective of forming one chiral form (one of two nonsuperimposable mirror-image forms, or enantiomers) in preference to the other. One approach is to start with readily available natural compounds that already have chirality of known configuration and to build the target structure from this base. For example, Corey's synthesis of leukotriene C starts with the sugar ribose, which is plentiful and of known absolute configuration. By building the structure from ribose in a stereocontrolled way, the absolute configuration in the final product is unambiguous. Another strategy is to separate, or resolve, the chiral forms at some stage in the synthesis and to complete the synthesis with the enantiomer of the correct configuration. The resolution step is more difficult than it may appear and is analogous to the task of sorting through countless pairs of gloves to separate the right-hand gloves from the left. Nonetheless, efficient resolutions have been achieved using chromatographic methods, such as in the resolutions reported recently by William H. Pirkle of the University of Illinois using high-pressure liquid chromatography.

Most of the activity in asymmetric synthesis, however, has been and continues to be in the development of enantioselective reactions; that is, reactions that are controlled to produce one enantiomer in preference to the other. Advances in the degree of stereoselectivity achieved have been impressive, and prospects for the near future are that chemists will be able to tailor reagents, catalysts, solvents, and physical conditions to achieve virtually complete stereocontrol in chemical reactions. In designing stereocontrolled reactions, it helps to know the mechanism. Thus, mechanism studies provide valuable insight into the nature of the stereoregulating step. Recent studies by Jack Halpern of the University of Chicago on the mechanism of asymmetric hydrogenation using chiral rhodium catalysts have provided this kind of insight (see *Inorganic*

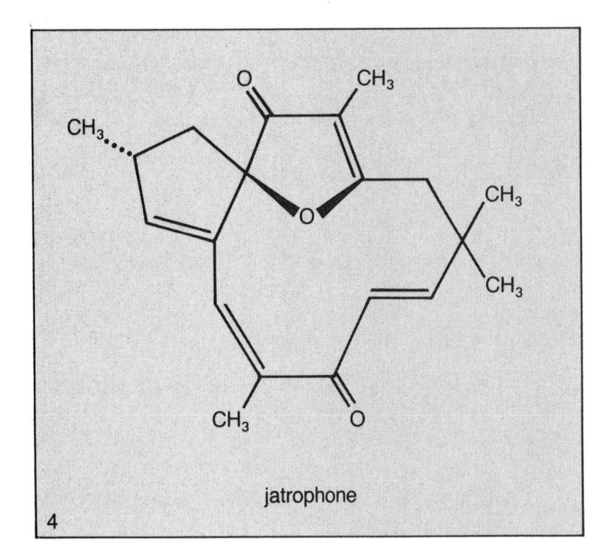

jatrophone

4

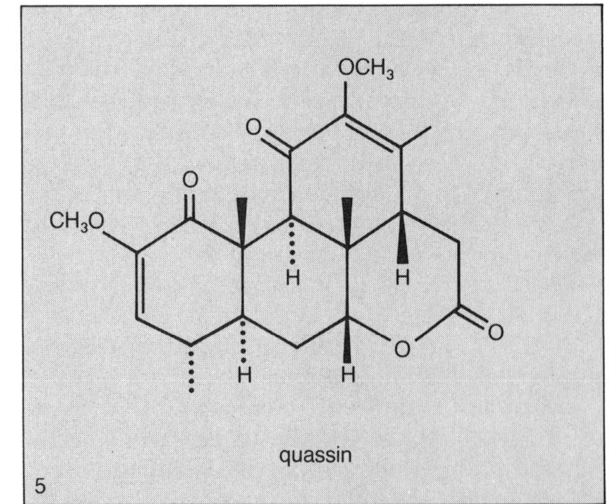

quassin

5

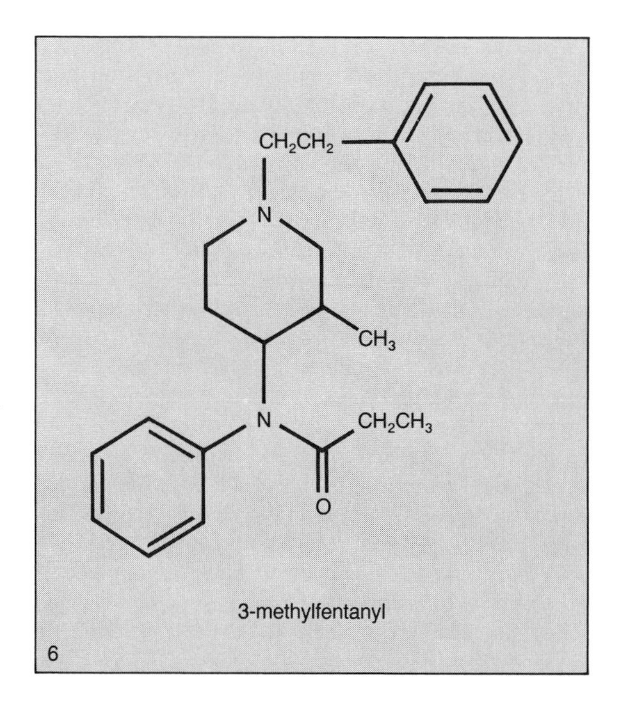

6 3-methylfentanyl

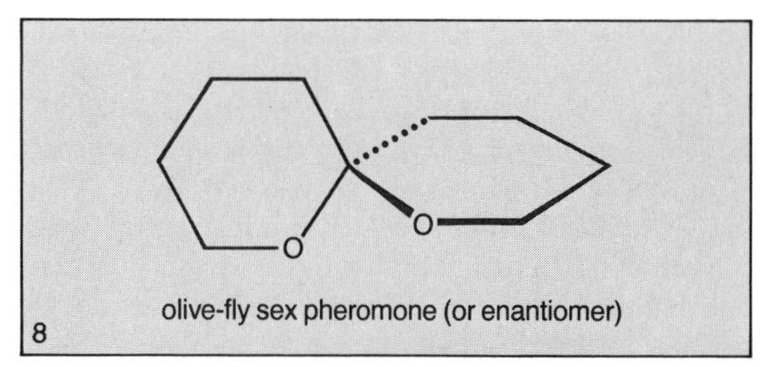

8 olive-fly sex pheromone (or enantiomer)

chemistry, above), and chances are good that improved stereocontrol will result from rational catalyst design and choice of reaction conditions indicated by the mechanism studies.

Use of computers. Computers are essential tools in chemistry today, and their increasing importance to organic chemistry warrants special mention. The three main functions of computers in chemistry are to facilitate computation, to store and retrieve information, and to augment human intelligence. The computational function is well known, and the recent advances in theoretical organic chemistry have been made possible because of advances in computer technology. Major objectives of theory include the calculation of the preferred geometries of molecules, their energies, and their spectroscopic and chemical properties. Calculations of some or all of these properties, based on quantum mechanical molecular orbital theory or molecular mechanics theory, are becoming increasingly reliable such that theory now supplements experiment and is a powerful predictive tool.

As one example, circumstances under which the reactive species known as carbenes (R_2C:, in which the two dots represent unbonded valence electrons) behave as electron donors as opposed to electron acceptors have been predicted by theory (Kendall N. Houk of Louisiana State University) and verified by experiment (Robert A. Moss of Rutgers University in New Jersey). Medicinal chemistry and pharmacology also benefit from quantum mechanical calculations that are designed to clarify the mechanisms of drug action. The results in turn suggest the design of better drugs.

Computers have become essential for the storage and retrieval of chemical information. Chemical Abstracts Service (CAS) of the American Chemical Society maintains the largest file of chemical information in the world, and access to this file is critically important to science in general. In 1980 CAS announced registration of the five millionth compound (most of which are organic) mentioned in the literature since 1965, when

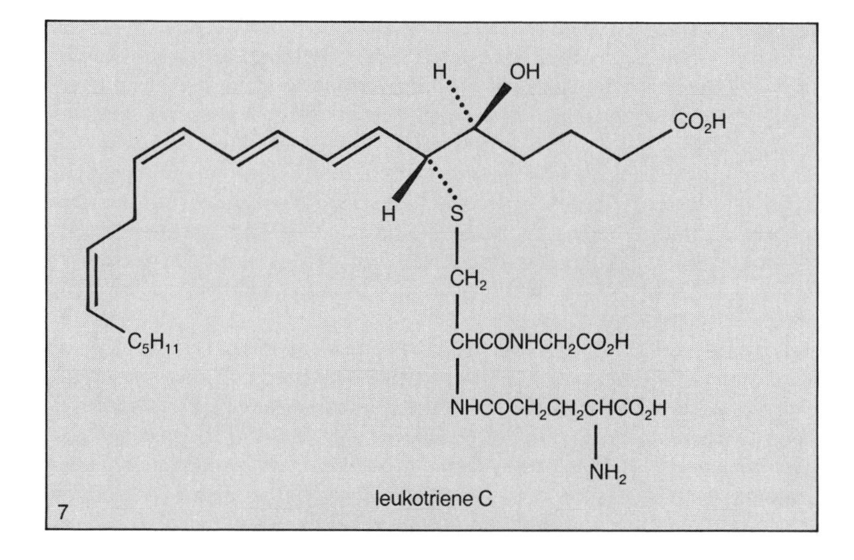

7 leukotriene C

the file of chemical substances was first established. CAS also announced the start of a new service that provides access to the information in the chemical registry file. Availability of this file for direct structure and substructure searching has long been sought but has not been possible until now because of the sheer size and complexity of the file. The task of searching the entire file is handled by groups of minicomputers working in parallel. This arrangement keeps search time to a minimum by permitting each computer to search separate parts of the file concurrently. As the file becomes larger, a search will not become longer; it will just take more minicomputers to do the job. Although the service, called CAS ONLINE, is just starting up, there is little doubt that it will become of great value to industry and research.

The use of computers to solve chemical problems through deductive reasoning and decision making lies in the realm of artificial intelligence. Programming computers to think logically about organic chemistry was reported by chemists at Purdue University, West Lafayette, Ind., and the Technical University of Munich in West Germany. These programs depend on classifying and documenting chemical structure in terms of a mathematical model, a matrix representing bonds between atoms. By combining a matrix for the starting materials in a hypothetical reaction with a reaction matrix describing bond formation and bond breakage, the computer can generate a matrix representing bond relationships in the hypothetical products. In this way, one can generate all conceivable reactions. Intelligent control of the number of solutions is achieved by biasing the reaction matrix in some way, as the Munich group did by imposing the restriction that the reactions proceed with the minimum structural change, which is a well-known heuristic principle of chemistry. The utility of these programs remains to be demonstrated, but their potential use lies in their ability to predict new chemical reactions, to design syntheses of complex molecules, and to elaborate reaction mechanisms.

—Marjorie C. Caserio

Physical chemistry

It may have been the start of a new decade, but physical chemists were loath to give up a research tool that rose to prominence in the 1970s—the laser. Combustion, comprising complicated and rapid chain reactions in flames, is a case in point. Understanding what happens in a flame has largely eluded chemists, and as a result such important processes as the explosive burning of an air-gasoline mixture in the engine of an automobile are not well enough understood to permit researchers to use other than empirical approaches in designing internal combustion engines and other machines that involve burning flames.

Combustion research. In 1980 the U.S. Department of Energy opened a new combustion research facility at Sandia National Laboratories in Livermore, Calif. The $10 million complex has two high-powered infrared and visible-light lasers (two additional lasers to be added by 1982) whose beams are carried by an overhead optical-distribution system to several different laboratories in the facility. Each laboratory has its own minicomputer, and each of these is also connected to a central computer. The combustion facility is open to outside users as well as staff researchers.

A variety of laser techniques was being used to analyze combustion processes in working engines. In one experiment under way, Sandia researcher Sheridan Johnson was measuring the fuel-air ratio 360 times during each combustion cycle at 19 points along the axis of an engine's cylinder. Johnson used a technique called Raman scattering. The amount of scattered light is a measure of the fuel concentration at the point at which the light is collected.

Reaction kinetics. A very popular use of lasers in physical chemistry research during the 1970s was in studying the kinetics of chemical reactions in a way that could be directly correlated with theoretical calculations. Molecules are characterized by a quantum mechanical energy state that is defined by three types of motion within the molecule. The electrons that form the chemical bonds between atoms in a molecule have energies that depend on the size and shape of the orbits of the electrons. Also, the atoms in a molecule do not maintain a fixed separation from one another but actually vibrate back and forth like balls held together by springs. These vibrations have characteristic energies that increase with the violence of the motion. Finally the molecules, or parts of them, can rotate, and the more rapid the rotation the higher the energy associated with the molecule.

The probability that one molecule will react with another during a collision depends on the energies of each of these three types of motion, that is, on the energy state of the molecule. In a reaction vessel each molecule is in a different energy state, so that the overall rate of a reaction is an average over the rates for molecules in the individual states. Yet theorists calculate the reaction rates of molecules that are in particular energy states. The use of lasers provides a way for experimentalists to do so-called state-to-state chemistry either by putting the reacting molecules into preselected energy states or by detecting the states of product molecules.

In 1980 Arunava Gupta, David Perry, and Richard Zare of Stanford University used two lasers to simultaneously select the vibrational states of reacting molecules and detect the energy states of the products. By this means they were able to verify a decade-old prediction of John Polanyi and Wing Hing Wong of the University of Toronto. The Canadian chemists had ar-

A. T. Winfree, The Institute for Natural Philosophy

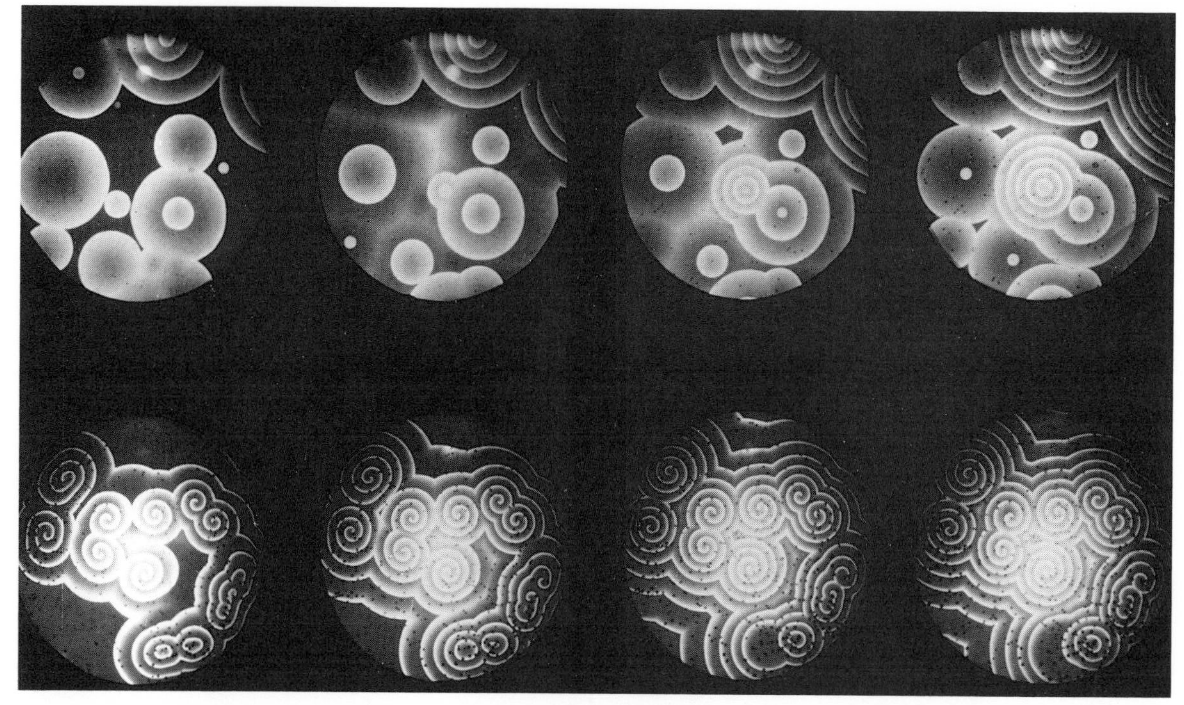

Circular "bull's-eye" waves (top sequence) and spiral waves (bottom sequence) in a special acidic reagent were filmed in studies of oscillating chemical reactions. The spiral waves, of a class yet to be fully described mathematically, may provide clues about heartbeat irregularities in humans.

gued in 1969 that in certain cases the most effective means of promoting a reaction would be to put the reacting molecules into energetic vibrational states, whereas in other cases it would be more fruitful to give the reactants higher kinetic energies so that they collided more violently.

The Stanford chemists studied the reaction between hydrogen fluoride gas in a reaction cell and a beam of strontium atoms. An infrared laser boosted the hydrogen fluoride reactant into an energetic vibrational state. When this was done, the reaction rate was found to be greatly enhanced compared with the effect of increasing the kinetic energy of the beam of strontium atoms by an amount equal to the vibrational energy.

Lasers are by no means essential for state-to-state chemistry, as an experiment by Randall Sparks, Carl Hayden, Kosuke Shobatake, and Daniel Neumark of the University of California at Berkeley demonstrated. The California researchers examined the reaction between hydrogen molecules and fluorine atoms in a crossed molecular-beam apparatus. Molecular beams are thin streams of gaseous material emanating into an evacuated chamber from orifices that separate the chamber from the sources of reactant molecules. The molecules collide and react only where the two beams cross. A mass spectrometer determines the velocity and direction of the product molecules, and from this information it is possible to calculate the vibrational energy of the product.

In their experiment the investigators found that when the kinetic energy of the fluorine and hydrogen reactants exceeded a certain value, a new quantum mechanical effect called a dynamical resonance came into play. The resonance tended to produce hydrogen fluoride molecules only in an excited vibrational state. A second effect of the resonance, which had earlier been predicted by theorist Robert Wyatt of the University of Texas at Austin, is that the reaction can take place even when the reactants do not collide head-on. Ordinarily such head-on collisions are most effective in overcoming the energy barrier that tends to inhibit the reaction.

Laser-induced chemistry. Ultimately, chemists would like to use lasers to control chemical reactions. One idea is that by putting all the energy of a laser into exciting specific vibrations within the reactant molecules, it may be possible to choose which molecular bonds are broken and thereby determine in advance what the product molecules will be. In conventional reactions driven by heating the reactants, all vibrational excitations occur, and the chemist has to make do with what nature provides in the way of a spectrum of broken bonds and product molecules.

A problem that has plagued so-called laser-induced chemistry is that the energy in a laser-excited vibration rapidly drains away into other vibrations, so that the laser simply becomes an elegant furnace for heating the reactants. The key issue is whether the laser can

Courtesy, IBM Corp.

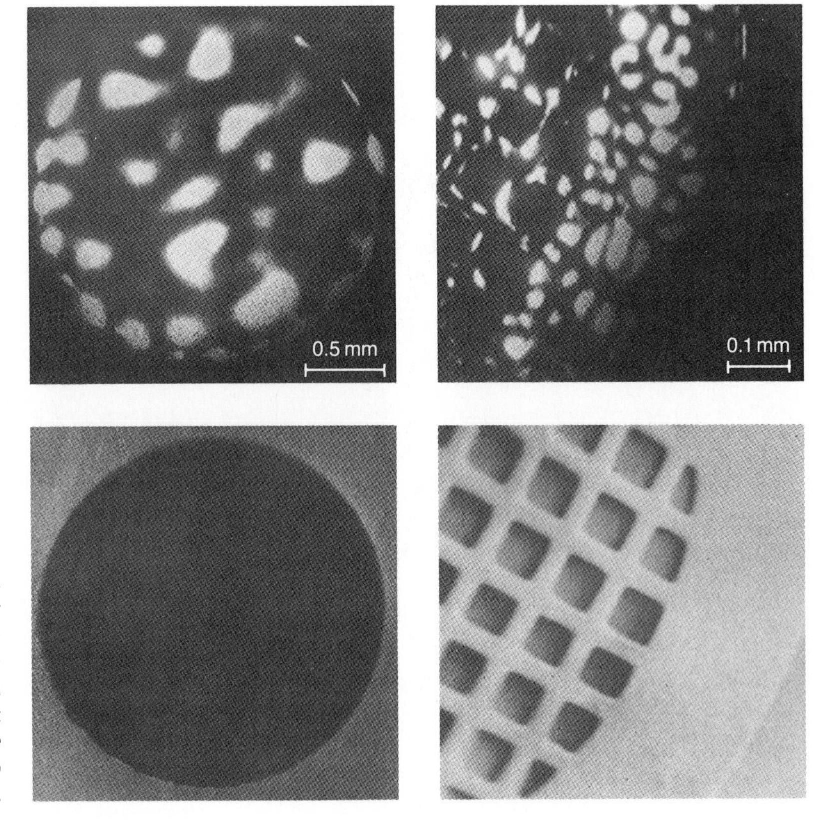

Electron-acoustic images of an aluminum rod in an invar alloy matrix (top left) and a copper grid on a copper plug (top right) are compared with their scanning electron micrographs (bottom). Contrast differences in acoustic microscopy, a comparatively new technique, reflect differences in thermal and elastic properties, rather than topographic differences, between adjacent regions of the material under study. Ordinarily a laser is used to generate an ultrasonic pulse in the specimen, but a recent refinement—used for the top photos—employs an electron beam, resulting in improved image resolution. A piezoelectric transducer in contact with the specimen detects the pulses, and its output is used to synthesize the final images.

deposit all its energy and break a bond before this energy redistribution process can start. One way of overcoming the problem would be to find molecules with vibrations that do not communicate well with other vibrations, so that energy transfer takes place slowly.

In 1980 Ahmed Zewail and J. W. Perry of the California Institute of Technology showed that the strategy may be workable in some cases. The Caltech scientists investigated durene, or tetramethyl benzene, which is a ring-shaped molecule with two kinds of carbon-hydrogen bonds. Twelve of the 14 hydrogen atoms in the ring are part of methyl (CH_3) groups, whereas the remaining two hydrogen atoms are attached to carbon atoms in the ring. By comparing the absorption spectrum (absorption of light as a function of its wavelength) of ordinary durene with that of durene in which the heavy hydrogen isotope deuterium replaced the hydrogen that is not in the methyl groups, the investigators were able to deduce that the methyl groups formed a semi-isolated "subsystem" and that energy deposited in the vibration of the methyl group was not rapidly transferred to the rest of the durene molecule. The experiment is regarded as just one of many that will be needed to fully explore the issue of laser-induced chemistry.

An alternative approach to laser chemistry is not to use a laser to vibrationally excite the reactant molecules themselves but to affect the very short-lived entity that exists as an intermediate or transition state between the breaking up of reactant molecules in a collision and the formation of the products—and thereby to control the course of the reaction. In 1980 Philip Brooks and his colleagues at Rice University, Houston, Texas, demonstrated at least the possibility of a laser's influencing the transition state in the reaction between potassium atoms and mercury dibromide in a crossed molecular-beam experiment. The researchers illuminated the interaction region where the beams crossed with laser light of a wavelength that could not be absorbed by either the reactants or the products (potassium bromide and mercury bromide). Nonetheless, when the laser was turned on, they observed a weak fluorescence from the reacting mixture. The interpretation is that laser light was absorbed by the transition state and, therefore, that one of the products, mercury bromide, was created in an electronically excited state. The fluorescence was emitted when the excited mercury bromide decayed from its high-energy quantum state to a lower energy state.

Zeolite structure. Laser chemistry is obviously one of the leading research topics in physical chemistry, but other kinds of investigations do take place as well. One of these in 1980 was the use of high-resolution electron microscopy to examine the structure of zeolites. Zeolites are silicates of metals like sodium, potas-

Courtesy, J. M. Thomas, University of Cambridge

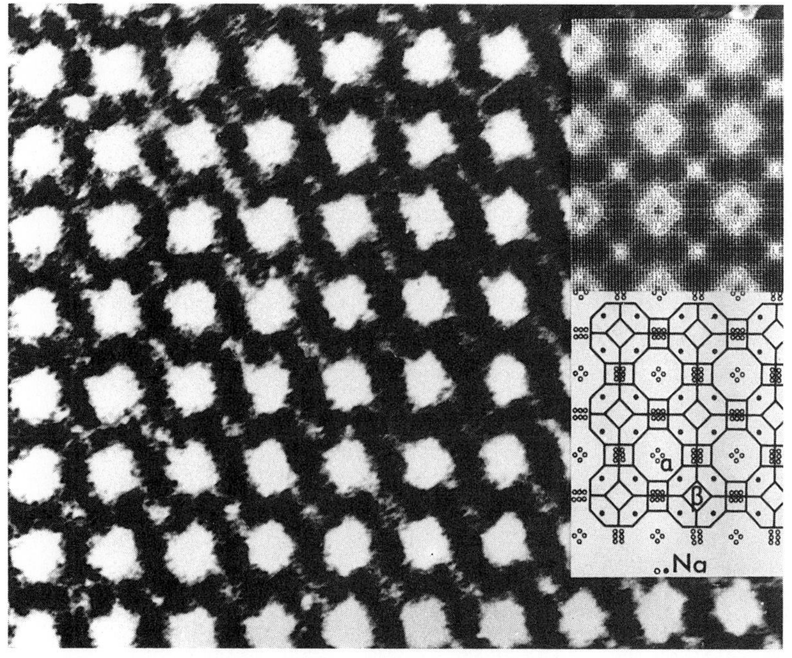

High-resolution electron micrograph of a synthetic zeolite reveals pattern of channels that give the material the properties of a molecular sieve. Distance between centers of neighboring channels is 12.3 angstroms. The image matches well with a computer-simulated image (top inset) derived from crystallographic studies. In an interpretative drawing (bottom inset) two types of channels in the aluminosilicate framework are labeled α and β; black dots and small circles represent sodium atoms (Na).

sium, or calcium which have certain peculiar structures that make these materials of practical use. The structures are of two types. The first is a three-dimensional network of open channels, and the second is a three-dimensional cagelike lattice. These structures allow the zeolites to select molecules of a given size or shape for various purposes. One important application comes in the petroleum industry, which is in increasing need of efficient catalysts that help transform raw feedstock into gasoline and other fuels or chemicals. Zeolites can act as catalysts, entities that promote chemical reactions but are not changed themselves in the process.

In 1980 Les Bursill of the University of Melbourne, Australia, and Elizabeth Lodge and John Thomas of the University of Cambridge used high-resolution electron microscopy to image one of these zeolites directly, an aluminosilicate with a channel structure. Among features of the material that the investigators resolved were defects in the crystal lattice called dislocations and the transformation of the crystalline material into a glassy state under the heating action of the electron beam in the microscope. The researchers believed that further improvement in resolution from the three to five angstroms presently achievable to two angstroms (one angstrom is a hundred-millionth of a centimeter) will allow them to see ions actually sitting in the open channels, a configuration whose existence X-ray crystallographers heatedly debate. It is knowledge of this type that chemists expect will allow them to tailor the structure of zeolites to achieve each of their possible applications.

—Arthur L. Robinson

Applied chemistry

Once again, use of alternative sources of energy was the focus of much of the past year's efforts in applied chemistry. Other important developments included the discovery of new silicon-based polymers and of a technique to improve the manufacture of glass fibers for light-wave communications.

Biomass energy. That many countries are showing increasing interest in obtaining larger fractions of their energy from biomass (trees, crops, manure, seaweed, algae, and urban waste), a previously ignored alternative to oil, was underscored by the World Congress and Exposition on Bio-Energy, held April 21–24, 1980, in Atlanta, Ga., and attended by 1,700 scientists, businessmen, and policymakers from 76 countries. Although presently biomass constitutes only a small percentage of the world's energy supply, several countries —Brazil, Sweden, China, and the United States, in particular—have begun ambitious biomass programs.

Brazil, a nation with a major successful investment in energy derived from grains, by the end of 1980 was running 330,000 automobiles on a water-alcohol mixture, which replaced 10% of its previous oil supply. This number was expected to double in five years, with an eventual goal of total replacement. In 1980 Brazil planned to build 250,000 alcohol-burning cars and to convert another 70,000–80,000 existing automobiles to use it. Brazil's success, however, may not be easily achieved elsewhere. The country possesses plentiful crops of surgarcane and cassava, from which the fuel is produced; the fuel mixture requires a low engine-vapor temperature, which is not possible in a colder climate; and the government was supporting the development primarily to provide jobs and to reduce the country's dependence on foreign oil.

Sweden, which has no coal, oil, or natural gas, began a biomass program to move away from its 70% dependence on foreign oil. Because a large portion of the country is heavily forested, Sweden planned to shift to wood as a prime energy source by growing fast-rotation trees that can be harvested every three to five years. Wood is also the best source of biomass energy in the U.S.; about five million tons of dried wood residue lie on the nation's forest floors, enough to supply 7% of its energy needs. In many developing countries trees were being cut down for cooking fuel, resulting in depletion of forests and consequent soil erosion.

In 1980 China had about 7½ million installations to produce methane, the chief constituent of natural gas, from biomass wastes. In southern China replacing wood with methane produced from human and animal wastes was making reforestation programs feasible. Anaerobic digestion of agricultural wastes to produce heat was used by about 30 million Chinese peasants.

According to Thomas Stelson, assistant secretary of energy for conservation and solar applications, biomass provided 2.5% of the total energy supply of the U.S., a figure that could be doubled by 1990 and doubled again by the year 2000. A more optimistic report by the Office of Technology Assessment concluded that the U.S. could obtain 20% of its energy needs from biomass by the end of this century. In that country at least 600–700 different research and development projects in biologically derived energy were under way, but virtually the only biomass energy development to stir the public imagination has been the production of ethanol from grain. According to S. David Freeman, chairman of the Tennessee Valley Authority, "corn-based alcohol may be good business for the farm lobby, but it can be very expensive for the rest of us. Breaking the OPEC habit by digging into our breadbasket poses the grave risk of driving up the price of food in a hungry world."

Fermentation of biomass to ethanol received the greatest share of federal funding in the U.S., while liquefaction of wastes and of biomass was still in the research stage. Pyrolysis (thermal decomposition) of wastes was the most widely used process in biomass conversion programs (the largest such plant was designed by Monsanto and operates in Baltimore, Md.), but research and development on nonthermal biomass conversion increased in the past year largely because of the high energy requirements of pyrolytic conversion and the high cost of industrial ethanol produced from petrochemical feedstocks.

One nonpyrolytic process for ethanol production based on bioconversion of cellulose was shown to be technically feasible by Gulf Science & Technology Co., Pittsburgh, Pa. Since the process uses waste materials (*e.g.*, municipal wastes and solid pulp-mill discards) as a feedstock, there is no competition with potential food substances or with petrochemical feedstocks. According to Gulf's George H. Emert, 2,000 tons of cellulosic wastes per day will yield 570,000 l (150,000 gal) per day of 95% ethanol. However, because forest and agricultural residues contain 10–40% hemicellulose, which cannot be converted by this process, enzyme techniques that can hydrolyze both cellulose and hemicellulose were being investigated. (Hydrolysis is a decomposition process in which chemical bonds within molecules are split and the severed ends "patched" with the elements of water.)

Research chemist John F. Cooper at Lawrence Livermore National Laboratory in California works with aluminum-air fuel cell being developed as a rapidly refuelable power source for electric cars. An aluminum alloy plate serves as both fuel and anode. Air and water pumped through the cell react with the plate to produce electricity and a form of aluminum hydroxide, which can be recycled to extract the aluminum. To repower the cell, the expended plate is removed and a fresh one added.

Courtesy, Lawrence Livermore National Laboratory; photo, Don Gonzalez

One bioconversion system, developed by Robert P. Chambers, S. Veeraraghavan, and Y. Y. Lee of Auburn (Ala.) University, involves the acid-catalyzed hydrolysis of hemicellulose obtained from natural lignocellulose on a packed bed at 150° C (300° F) with 0.2% sulfuric acid. The products—five-carbon sugars that are not good substrates for fermentation by the usual microorganisms—are converted to 73% 2,3-butanediol, 10% ethanol, and the rest acetic acid and acetoin by a bacterium recently isolated from decaying wood.

At the University of Connecticut at Storrs, Donald W. Sundstrom and co-workers reported that cellulose of woody and herbaceous plants can be broken down more selectively with an enzyme mixture of endoglucanase, cellobiohydrolase, and β-glucosidase than with conventional acid hydrolysis. To deplete the accumulation of cellobiose, which inhibits the reaction, the group immobilized β-glucosidase on alumina and improved the yield of glucose in the enzymatic hydrolysis of cellulosic materials. A process for the enzymatic hydrolysis of bagasse (plant residue) from sugarcane under development at the University of Queensland, Australia, uses two enzymes, invertase and cellulase, for complete hydrolysis.

Coal gasification. During the past year development of underground coal gasification continued. If successful, commercial gasification could quadruple the proved resources of U.S. coal. Results from the Hoe Creek No. 3 underground coal gasification experiments, carried out in northeast Wyoming and sponsored by the Department of Energy (DOE) and the Gas Research Institute in Chicago and directed by Douglas R. Stephens and others of the Lawrence Livermore Laboratory in California, demonstrated the feasibility of in-situ conversion of the large low-grade coal reserves found in many western states. Current experiments tested the Uniwell method, which involves sinking a single, vertical well into a coal seam and cementing in two or more concentric pipes. The central pipe feeds oxidant into a burn cavity, and product gases issue from the outer pipes. This technique should prevent massive subsidence caused by removal of large quantities of coal in an underground burn.

An in-situ burn of Texas lignite (an incompletely formed coal) at Tennessee Colony, Texas, was carried out by Texas Utilities Co. and its consultant, Basic Resources Corp., and used technology licensed from the U.S.S.R. It showed that the vast Texas lignite deposits can produce gas economically with minimal effects on the environment. According to Clarence W. Garrard, Jr., vice-president of Basic Resources, the tests demonstrated the possibility of conducting a reverse burn in a thin, wet lignite seam and that water infiltration plays an important role in the Tennessee Colony region. Because the tests gave no evidence of a pollution plume in the groundwater surrounding the burn cavity, the researchers believed that bacteria and clays in the region had adsorbed, absorbed, or consumed any resulting pollutants, a hypothesis being tested at the University of Texas. They estimated that a low-BTU gas (85–100 BTU per standard cubic foot) can be produced from Texas lignite at less cost than the present deregulated market price for heat from natural gas. Underground coal gasification can extract energy without extracting the coal, and Soviet technologists have been conducting burns for at least 40 years.

Other coal gasification tests carried out during the year included those at Westinghouse Electric Corp., Waltz Mill, Pa. (production of gas rated at 300 BTU per standard cubic foot from lignite); Elgin-Butler Brick Co., near Elgin, Texas (conversion of 140,000 tons of lignite per year to fuel gas at a cost of $3.50 per million BTU); and Avco-Everett Research Laboratory's two-stage, high-throughput, entrained-flow reactor that processes all kinds of coal without special accommodations. Bechtel Power Corp. conducted preliminary studies for a plant in Fall River, Mass., to gasify coal to medium-BTU gas for combined-cycle electrical generation and for producing methanol.

In research sponsored by DOE for the flash pyrolysis of coal that does not require high pressures or a large supply of hydrogen, researchers at the Occidental Research Corp. were studying a process in which coal is heated rapidly at atmospheric pressure in an oxygen-free chamber filled with an inert gas. The coal decomposes to a carbon-rich char and hydrogen-rich gases and liquids. In a coal liquefaction process developed by the Dow Chemical Co., Pittsburgh, Pa., No. 8 seam coal containing 4% sulfur (a poor candidate for the process but cheap and plentiful) is crushed, dried, and slurried with a process-derived oil containing an ammonium molybdate catalyst. The two-stage flash process produces the lowest-cost energy of any current liquefaction process: $3.13 per million BTU.

Synthesis-gas conversions. In 1980 processes involving catalytic compounds of ruthenium, a transition metal, for hydrogenation of (addition of hydrogen to) carbon monoxide and synthesis of commercially important petrochemicals from synthesis gas—a mixture of carbon monoxide and hydrogen—were developed in the laboratories of three major U.S. petrochemical companies. At Mobil Research & Development Corp., Tracy J. Huang and co-workers created a heterogeneous (two-phase) ruthenium catalyst system on an intermediate-pore zeolite (a metal silicate with a channel-forming lattice structure) that converts synthesis gas to high-octane gasoline in a single step. Although the ruthenium-catalyzed reduction of carbon monoxide has been studied extensively for 30 years, the reactions show poor hydrocarbon selectivity and produce no aromatic compounds. Mobil's catalyst, however, converts at least 95% of the synthesis gas to a product mixture that boils in the gasoline range and is rich enough to give an octane number of 80–104.

B. Duane Dombeck and colleagues of Union Carbide's research laboratory in South Charleston, W.Va., converted synthesis gas to ethylene glycol, a widely used and versatile organic compound, using a homogeneous (one-phase) ruthenium catalyst (an acetic acid solution of ruthenium dodecacarbonyl) and a pressure of 340 atmospheres (5,000 psi) or below. Similarly, John F. Knifton of Texaco produced carboxylic acids from synthesis gas using a homogeneous ruthenium catalyst (any of several ruthenium compounds and an iodine-containing promoter) at a temperature of 200° C (390° F), and a pressure of 270 atmospheres (3,970 psi). In this process a small carboxylic acid such as acetic acid can be converted to the aliphatic carboxylic acid that is three carbon atoms longer, and because acetic acid itself can be made from synthesis gas, the process makes possible the synthesis of higher-molecular-weight carboxylic acids exclusively from coal-derived synthesis gas without any petroleum-derived reactants.

Solar energy. According to a statement made on May 15, 1980, by Robert R. Ferber, manager of Solar Photovoltaics Technology Development at the National Aeronautics and Space Administration's Jet Propulsion Laboratory, Pasadena, Calif., there is enough space on south-facing roofs across the U.S. for solar power cells to generate all of the nation's electrical energy. Ferber expected photovoltaic systems to be commercially competitive within six years.

Describing it as a breakthrough in thin-film, solar-cell technology, researchers at Boeing Aerospace Co., under contract to DOE's Solar Energy Research Institute, developed a copper indium selenide/cadmium sulfide cell with light-to-electricity conversion efficiency of 9.4%, close to the 10% goal for thin-film cells set by DOE's National Photovoltaic Program and a great improvement over the 6.7% efficiency attainable from such cells less than a year earlier. The cell is an inexpensive substrate, alumina, coated with semiconductor films that are applied by vapor deposition.

T. W. Fraser Russell, director of the University of Delaware's Institute of Energy Conversion, announced development of a cadmium sulfide/copper sulfide solar cell with a conversion efficiency of 9.2%, which was expected to be raised to 12%. Another cell, a cadmium-zinc sulfide/copper sulfide version, being developed at the University of Delaware had an efficiency of 10.2%, which was expected to be raised to 16%. The principal advantage of such polycrystalline sulfide cells over more efficient single-crystal cells (*e.g.*, silicon wafer cells with conversion efficiencies of 15%) is low cost. Total cost of the latter cell is estimated at 50 cents per watt, well below DOE's goal of 70 cents per watt.

In 1980 processes for the photolysis (decomposition by sunlight) of water to produce hydrogen and oxygen came closer to commercial application. At the Swiss Federal Polytechnic Institute at Lausanne, Michael Grätzel, Enrico Borgarello, John Kiwi, Ezio Pelizzetti, and Mario Visca devised a system "capable of efficiently decomposing water into hydrogen and oxygen under visible light illumination" with "astonishingly high" yields by use of a cofunctional oxidation-reduction catalyst—a combination of platinum metal and ruthenium dioxide, each codeposited on a common carrier, colloidal titanium dioxide. The sensitizer is a derivative of *tris*(2,2′-bipyridine)ruthenium(II), in which an *n*-alkyl group is attached to one of the bipyridine molecules. The system can generate as much as 300 ml (ten fluid ounces) of hydrogen per liter (1.1 qt) of water per hour if oxygen generation is inhibited.

Jean-Marie Lehn, Jean-Pierre Sauvage, and Raymond Ziessel of the Institut le Bel of the Université Louis Pasteur in Strasbourg, France, reported that "deposition of a suitable catalytic metal on a semiconductor [*e.g.*, rhodium on strontium titanate] yields an efficient catalyst for water photolysis under irradiation with ultraviolet light." One drawback is that the system operates only with ultraviolet light (only about 5% of solar energy), but preliminary experiments with the catalyst doped with trivalent chromium or iron showed promise of a broader spectral response.

In a related development F. T. Wagner and Gabor A. Samorjai of the University of California, Berkeley, showed that oxide semiconductor surfaces can serve as the anode for photochemical production of hydrogen from water even when no platinum electrode is present to serve as the cathode. They found that the illuminated surface of strontium titanate crystals can assist the decomposition of water in the presence of a deliquescent (water-attracting) compound such as sodium hydroxide at a rate of 20–100 monolayers (layers of single molecules) of water per hour.

Harry B. Gray of the California Institute of Technology at Pasadena recently reported that it may be possible to increase dramatically the 4% efficiency of one of his earlier discoveries, a dimeric rhodium complex containing four 1,3-diisocyanopropane molecules used for photolysis of water (see *1979 Yearbook of Science and the Future* Year in Review: *Chemistry: Applied chemistry*). He was able to add electrons needed in the reaction to the complex with an efficiency of 90%, which may produce hydrogen with an equally high efficiency.

Peter O'D. Offenhartz and co-workers at EIC Laboratories, Newton, Mass., were developing an innovative chemical heat-pump system that uses solar energy to heat a home in winter, cool it in summer, and provide hot water all year round. The heat-exchange fluid from commercially available solar collectors circulates in copper tubes through a bed of calcium chloride methanolate pellets. The fluid's heat releases the complexed methanol as vapor, which, after being heated to 120°–130° C (248°–266° F), enters an air-cooled condenser in which the vapor transfers some of its

Courtesy, Bell Laboratories

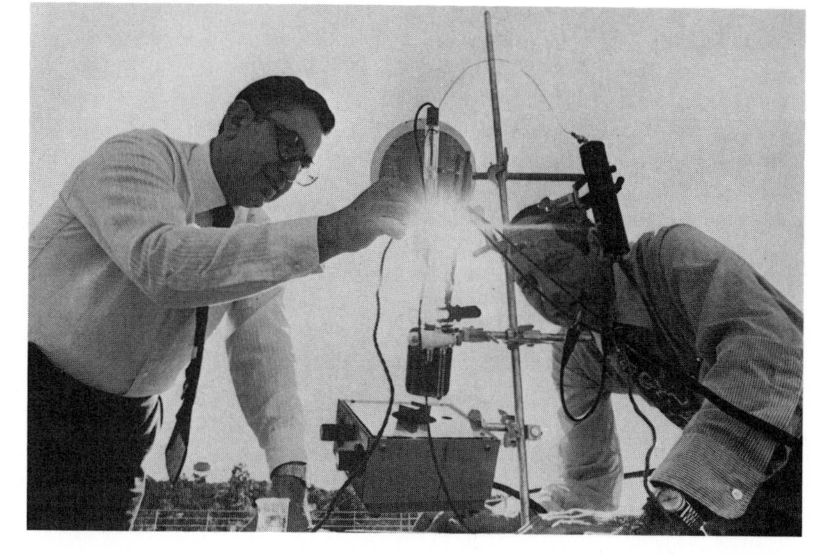

Bell Laboratories investigators Adam Heller (left) and Barry Miller test a new liquid-junction solar cell that demonstrates an 11.5% conversion efficiency, which is competitive with some commercially produced solid-state single-crystal cells. Its p-type-semiconductor cathode, an indium phosphide crystal in contact with an acidic solution of vanadium, functions as the light-sensitive electrode. Sunlight striking the cathode drives a flow of electrons toward the crystal surface. This action acts to protect the electrode from degradation, a serious problem in cells in which the anode is the light-sensitive electrode.

heat to air. Cooled to 40°–50° C (104°–122° F) the vapor condenses, heating the air to 40°–50° C, which is hot enough to be used for space heating in winter. The condensed methanol flows from the condenser to the evaporator, where it picks up heat from a stream of warmed ambient air and evaporates, a process that cools the air stream, providing an air-conditioning effect for summer use. The cold methanol vapor then enters the absorber, in which it reacts with a bed of dry calcium chloride to reform the methanolate complex.

A new battery. A zinc/chlorine battery system for motor vehicles and for electric utilities was reported by Energy Development Associates, Madison Heights, Mich., a subsidiary of Gulf & Western Industries. The battery operates by reaction of chlorine at one graphite electrode and zinc plated onto another, producing zinc chloride. In recharging, zinc is replated onto one electrode, while chlorine is regenerated at the other. G&W President David N. Judelson estimated that the 545-kg (1,200-lb) automotive unit would cost $3,000 in mass production; in a small converted car, *e.g.*, a Volkswagen Rabbit, it would carry four passengers 150 mi (240 km) at 55 mph (88.5/km/hr) on a single charge for 2.3 cents per mile, compared with 6.5 cents per mile for gasoline.

Silicon polymers. In an attempt to prepare cyclic compounds useful for organosilicon syntheses, Robert C. West and graduate research assistant Lawrence D. David at the University of Wisconsin, Madison, unexpectedly obtained a new family of silicon polymers with many properties markedly different from those of other known polymers. The reaction involves condensing mixtures of dimethyldichlorosilane and phenylmethyldichlorosilane in toluene with an excess of metallic sodium. Although early experiments using bulk sodium chips gave a viscous liquid with a molecular weight of about 8,000 daltons, use of finely dispersed sodium yielded solid polymers with molecular weights of 100,000–400,000 daltons. The new polymers are polysilanes with backbones of connected silicon atoms and are not to be confused with silicones, which have chains of alternating silicon and oxygen atoms. The proportion of phenyl and methyl groups attached to the backbones can be varied to yield polymers with different characteristics. The one under most intense study has one phenyl group for every two silicon atoms. Because the structure is analogous to the carbon-chain polymer polystyrene, West and David called it polysilastyrene.

Since polysilastyrene can be drawn into fibers, it may be possible to use it to make silicon carbide in fiber form. Ceramics investigator K. S. Mazdiyasni of the Air Force Materials Laboratory at Wright Patterson Air Force Base near Dayton, Ohio, used it to strengthen silicon nitride ceramics used for high-temperature-resistant gas-turbine engine parts and rocket nozzles. Electrons can move easily from one silicon atom to another in the polysilanes, just as in silicon transistors and photocells, and films of polysilastyrene become semiconducting when doped with antimony pentafluoride. According to polymer chemist Ryuk Yu at Wisconsin, there is a good possibility that polysilanes can be developed to yield useful elastomers.

Meal and chemicals from an alga. Mordhay Avron of the Weizmann Institute in Israel and Ami Ben-Amotz of the Israel Oceanographic and Limnological Research Co. identified a red, single-celled alga, *Dunaliella bardawil*, that thrives in saline water, produces 40% of its dry weight as glycerol (glycerin), and can accumulate as much as 8% beta-carotene (a source of vitamin A), which can be extracted to leave a high-protein (70%-protein) meal similar to that obtained from soybeans. Since *D. bardawil*, unlike most algae, has no cell walls, the meal is readily digestible by live-

Adapted from "Performance of Photovoltaic Cells Improved," CHEMICAL & ENGINEERING NEWS, vol. 58, no. 40, p. 37, October 6, 1980

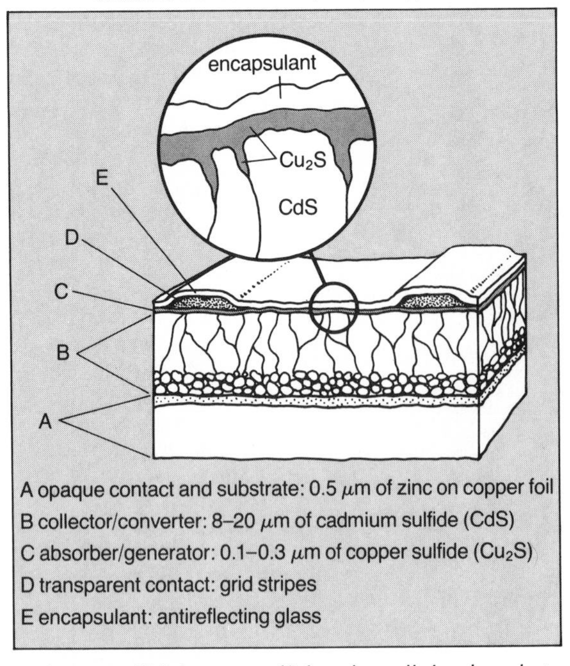

Cadmium sulfide/copper sulfide solar cell developed at the University of Delaware has five functional layers beginning with a copper-foil substrate and topped with an antireflecting encapsulant made of two layers of glass. The grid stripes are for circuit control.

stock and fish. The alga can be raised on barren tracts in semiarid regions containing seawater and brackish water unsuitable for ordinary agriculture. Koor Foods of Haifa, Israel, would develop the alga commercially, which eventually may serve as a source of biochemicals such as chlorophyll, fatty acids, steroids, vitamins, and enzymes.

Optical fiber process. Jim Fleming, John B. MacChesney, and Paul B. O'Connor of Bell Laboratories, Murray Hill, N.J., devised a method to make light-wave communications by means of glass fibers more competitive economically with alternative transmission media, such as metal cable and microwaves. The researchers found a technique to speed up fabrication of glass preforms from which glass fibers are made. Because preform production is an important part of fiber fabrication, increasing the production rate should lower the cost of fiber manufacture substantially.

The technique uses a plasma and is a variation of the modified chemical-vapor deposition process developed by MacChesney and O'Connor and patented during the summer of 1980. Silicon chlorides or hydrides are passed as vapors together with oxygen and doping agents through a fused quartz tube. A furnace two centimeters (three-fourths of an inch) long travels back and forth over the outside of the tube, which is rotated at 100 rpm, heating the tube to 1,200°–1,600° C (2,200°–2,900° F). The resulting reaction deposits oxides on the inner surface. These melt to form a new glass layer with each passage of the furnace. The tube collapses to solid rod when furnace temperatures are raised to 1,900° C (3,450° F), and the rod stock is drawn into optical fibers 1.25 mm (0.0491 in) in diameter. In the plasma-assisted process, instead of heating the vapors with only an external source, they are heated within a zone that includes a plasma formed and maintained inside the tube by electromagnetic energy produced by an induction coil. The technique speeds up the fabrication process while maintaining the desired transmission properties of the fiber.

—George B. Kauffman

See also Feature Article: THE PROMISE OF SYNTHETIC FUELS.

Earth sciences

Major events in the Earth sciences during the past year included the volcanic eruptions of Mt. St. Helens in Washington state and strong earthquakes in Algeria and Italy. Investigations into the causes and effects of these disasters occupied many scientists. Other research efforts focused on such subjects as atmospheric aerosols, acid rain, the safe dispersal of toxic wastes, the use of sound waves to explore for oil, and hydrothermal vents in the deep ocean.

Atmospheric sciences

During the past year there was interesting and productive research on a broad spectrum of topics. Of particular note were studies of the gaseous and particulate constituents of the atmosphere and their effects on weather and climate, the nature and consequences of "acid rain," and the properties and processes of the upper atmosphere.

Atmospheric aerosols. Minute solid and liquid particles are present in large numbers in the atmosphere. They arise from many sources—land surfaces scoured by the wind, smokestacks and tail pipes, sea spray, and volcanic eruptions. Some form in the air as a result of gas-to-particle conversion. Over the last few years the origin and fate of aerosols and their effects on weather and climate have been receiving increased attention.

A large fraction of the atmospheric particles having diameters of about 0.1 μm (micrometer) are composed, at least in part, of sulfate compounds. There is widespread agreement that they occur as a result of the direct conversion of sulfur dioxide gas to aerosols consisting of substances such as ammonium sulfate. In sufficient numbers these particles can influence the radiation balance of the Earth. In addition, the sulfate aerosols are important in cloud and precipitation processes, because they act as condensation nuclei around which cloud droplets form. The more numerous the cloud droplets, the smaller they will be for a given quantity of available water and the less likely will be the growth of raindrops taking place through the process of coagulation.

The series of eruptions of Mt. St. Helens in Washington had as one of its consequences the injection into the atmosphere of large quantities of gaseous and particulate matter. According to Thomas J. Casadevall of the U.S. Geological Survey and his colleagues, the sulfur dioxide emissions have been highly variable from day to day. Average emissions were about 1,300 metric tons per day in August 1980 and decreased more or less gradually to about 290 tons per day in March 1981. These levels are much smaller than the man-made emissions of sulfur dioxide, amounting to about 75,000 metric tons per day in the United States.

As noted recently by Sean Twomey at the University of Arizona, because the absorption and reflection of solar radiation and the outward loss of infrared radiation from the Earth depend in part on cloud droplet concentration, condensation nuclei indirectly influence the radiation balance of the atmosphere. In this way they affect weather and climate processes.

In calculating the absorption and scattering of radiation by atmospheric aerosols, the common practice is to use a theory derived by Gustav Mie in 1908. It can be applied exactly when the scatterers are spheres. Since particles in the air may be far from spherical, there has been some concern about the validity of the use of the Mie theory. In particular, questions have been raised as to whether it gives an adequate estimate of the Earth's global albedo, the fraction of the incident solar energy reflected back to space.

James Pollack and Jeffrey Cuzzi at the Ames Research Center of the U.S. National Aeronautics and Space Administration (NASA) formulated an approximate method for calculating the scattering of electromagnetic radiation by randomly oriented nonspherical particles of sizes comparable to the wavelength of the incident radiation. They found that irregularly shaped particles in the lower atmosphere have greater reflectivity than do spherical particles of comparable mass and composition.

Acid rain. Sulfuric and nitric acid particles in the atmosphere serve as cloud condensation nuclei and

Below, volcanic ash cloud caused by the eruption of Mt. St. Helens hovers over Richland, Washington, 130 miles to the east, on May 19, 1980. The ash killed many insects, such as the bee at right, and was deposited on plants throughout the area (bottom right).

Wide World

(Top) R. D. Akre; (bottom) Bill Dean

lead to the acidification of rain and snow. Measurements in many parts of the world, particularly downwind of heavily industrialized regions, have revealed appreciable increases in the acidity of some lakes—by factors as large as 50 in the Adirondack Mountains region—over the last 50 years.

It is clear that various types of vegetation and marine life are affected adversely by acid precipitation, in some cases to a very serious extent. Certain stones, marbles, and metals can be damaged by sulfuric and other acids. An interesting research program on these effects, currently in progress through the support of the U.S. Environmental Protection Agency (EPA), involves the study of the condition of marble headstones and markers in National Cemeteries in various parts of the U.S. The headstones are of similar materials and shapes and exist in large numbers. Initial investigations involve the study of the condition of headstones of various ages in about a dozen cemeteries in three distinctly different climatic regions. They are in the far western, northeastern, and Appalachian areas.

In June 1980 the federal government enacted Public Law 96-294, which included as Title VII, "Acid Precipitation Program and Carbon Dioxide Study." It assigned national responsibility for the conduct of a ten-year research program on acid rain to a task force having a joint chairmanship composed of the chief officials of the U.S. Department of Agriculture, the EPA, and the U.S. National Oceanic and Atmospheric Administration (NOAA).

On the international scene the U.S. and Canada in August 1980 agreed to start formal negotiations having as their objective a reduction in the acidity of rain. In late 1979, 33 nations, including the U.S., Canada, the Soviet Union, and Western European countries, agreed to limit air pollution, particularly under circumstances in which the emissions of one nation are likely to damage another. *See* Feature Article: THE NOT SO GENTLE RAIN.

Carbon dioxide. The concentration of carbon dioxide (CO_2) in the atmosphere continued to increase, as it had throughout the century. The increase, currently amounting to about one part per million per year, is primarily a result of the combustion of fossil fuels. This combustion releases annually about five billion tons of CO_2 into the air. There has been considerable debate over the contribution of worldwide deforestation to increased levels of atmospheric CO_2. The burning of forests, the slash-and-burn agriculture that replaces them, and the oxidation of exposed forest humus all liberate the carbon contained in those materials into the atmosphere in the form of CO_2. Deforestation also indirectly contributes to the CO_2 budget by destroying the vegetation that can remove CO_2 from the air by photosynthesis. In the early 1970s George Woodwell of the U.S. Marine Biological Laboratory at Woods Hole, Mass., estimated that the biological contribution to atmospheric CO_2 through such practices was probably between four and eight billion tons per year. Such a quantity seemed excessive to some experts who were studying the capacity of the world's oceans to absorb atmospheric CO_2 and who could not reconcile the figures for such a huge release of CO_2 with the observed buildup in the atmosphere and the oceans. Wallace Broecker at Columbia University in New York City calculated that the oceans could not absorb more than about 2.5 billion tons of CO_2 per year and thus concluded that the biospheric contribution to atmospheric CO_2 could not exceed one billion tons per year.

During the past year Woodwell reexamined the effects of deforestation and estimated that the 1970 CO_2 release was probably between two and four billion tons, about half of the earlier figures. On the other hand Paul Crutzen and Wolfgang Seiler at the Max Planck Institute for Chemistry in Mainz concluded that the land can be either a small sink or small source of CO_2. They proposed that the charcoal remains from forest clearing by fire could store from 0.5 to 1.7 billion tons of carbon per year.

Mathematical models. Weather forecasting moved from an art to a science when high-speed electronic computers made it possible to solve the set of equations that describes the atmosphere. These equations specify, as a function of time, changes in pressure, air motions, and temperature, and the conservation of energy, mass, and water substance. When the state of the atmosphere and the properties of the underlying surface are known, the equations make it possible to calculate future conditions of the atmosphere. Because of the complex nature of the equations, it is necessary to use numerical methods and high-speed computers in order to obtain solutions in reasonable periods of time.

Since the late 1940s many scientists have been involved in the development of mathematical models of the general circulation of the atmosphere. Such models have been used to evaluate the effects on the global atmosphere of increases of CO_2 and aerosols and to simulate the state of the atmosphere many centuries ago. General circulation models are employed by the national weather organizations in various countries for making weather predictions.

In the 1960s atmospheric scientists began the development of mathematical models of hurricanes. These models involve equations similar to those used to describe the general circulation of the atmosphere, but information is needed in smaller units of distance. In hurricane models it is essential to take into account the roles of convective clouds—cumulus and cumulonimbus—in the vertical transport of energy, moisture, and momentum.

Increasing attention is being devoted to the mathematical modeling of mesoscale systems—those having diameters of several hundred kilometers—in middle latitudes. It is widely recognized that in order to account for and predict rain and snow more accurately it is necessary to understand the interaction of clouds and the mesoscale environment in which they occur. Recently, James Fritsch and Charles Chappel at the Atmospheric Physics and Chemistry Laboratory of the NOAA developed a numerical model for predicting mesoscale pressure systems in which convective clouds play an important role.

Climate. The Earth's climate—its changes in the past and how it is likely to change in the future—received much attention during the past year. Based on studies of geological data and analyses of isotopes, climatologists began to reconstruct the temperature history of the atmosphere over the distant past. A particularly large amount of information is available about the climate over the last few centuries. This comes from direct instrumental measurements and from such indirect indicators as the records of lake levels, extent of sea ice and glaciers, the width and isotopic composition of tree rings, and the history of harvests.

The long-period changes of climate accounting for the ice ages have been ascribed to the drift of continents and variations in the Earth's orbit around the Sun. There still is no satisfactory theory or set of theories that can account for climatic fluctuations over periods of decades or centuries. An increasing number of scientists were working on the formulation of such theories.

The well-recognized, steady increase of atmospheric CO_2 is expected, according to most experts, to lead to a warming of the lower atmosphere. Scientists at the Geophysical Fluid Dynamics Program at Princeton University (N.J.) and elsewhere tentatively concluded that a doubling of the CO_2 (to about 650 ppm) would lead to a temperature increase averaging 2° to 3° C (3.6° to 5.4° F), with perhaps twice as much warming at high latitudes. These results were obtained by means of mathematical models of the general-circulation atmosphere. In 1980 Roland A. Madden and Veerabhadran Ramanathan of the National Center for Atmospheric Research reported that they were still unable to verify the predicted warming of the atmosphere and surmised, as had others, that the thermal inertia of the oceans is likely to delay the warming expected to occur as a result of the effects of CO_2. They reiterated concerns that the current atmospheric models do not yet adequately incorporate interactions between the atmosphere and the oceans.

The consequences of a globally averaged warming on the total environment are not known with any degree of confidence, but they are expected to be important. It seems reasonable, as a number of scientists have concluded, that there would be a melting of landlocked ice and a breaking off into the ocean of Arctic and Antarctic ice. These processes could lead to potentially disastrous rises of sea level. Climatologists at the University of East Anglia in England, on the basis of studies of climatological data, concluded that worldwide warming would be accompanied by drier conditions over much of the United States, Europe, and the grain belt of the U.S.S.R. If this were to occur, the food-growing capacity of the world would be appreciably reduced.

When particles in the stratosphere are sufficiently numerous, as was the case following the massive Mt. Agung volcanic eruption in 1963, the Earth's surface can, as a result, cool a few tenths of a degree Celsius. The period of reduced temperature can continue for a year or two after the eruption and be associated with significant weather anomalies. It appears that to date the eruption of Mt. St. Helens has not injected enough material into the atmosphere to cause a detectable decrease in surface air temperatures.

Upper atmosphere research. Powerful radar sets, balloons, rockets, and satellites were used during the past year to measure the properties of the upper atmosphere, that region above about ten kilometers (six miles). Sounding balloons usually cannot get above about 50 km (30 mi), and satellites are not effective below about 150 km (90 mi). Thus the middle region of the upper atmosphere, from about 50 to 150 km, can be observed directly only by rockets, though it can be sensed remotely from satellites, aircraft, and the ground. In 1981 there were four major radar systems capable of measuring constituents, temperatures, and motions in this middle region. Equipment was located in Massachusetts, Puerto Rico, and Peru along a line

Aggregate-type dust particle is typical of extraterrestrial particles found in the stratosphere. When such particles are numerous, they cause cooling at the Earth's surface.

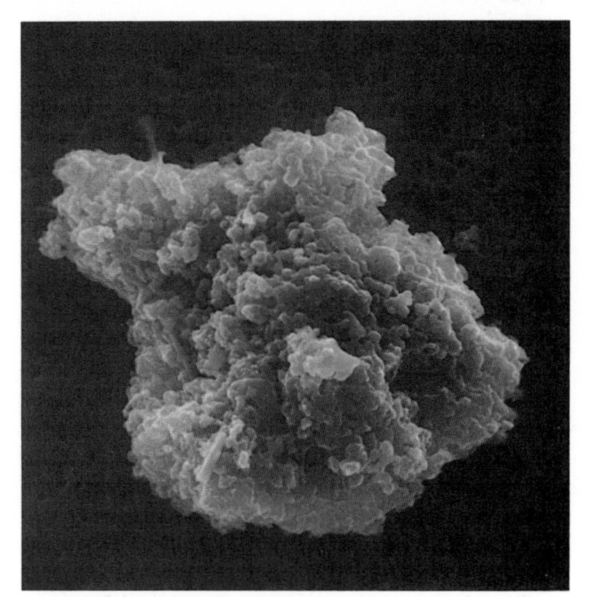

D. E. Brownlee, California Institute of Technology

roughly north–south. The fourth radar was in Alaska, but efforts were being made to move it to Greenland. This would allow time-resolved data to be obtained along a nearly pole-to-pole cross section.

A series of satellites in the NASA Explorer program has been involved in the study of the Earth's environment out to great distances. Atmospheric Explorer spacecraft have been used to study how solar energy in the ultraviolet region produces ionospheric and ozone layers. A new Dynamics Explorer program is planning the launch of two satellites that will investigate the interactions of the solar wind with the Earth's magnetosphere. The program will also include research on the coupling of the magnetosphere with the ionosphere.

The highly interactive nature of the physical, electrical, and chemical properties of the upper atmosphere makes it difficult to formulate comprehensive theories until much more is known. As of 1981 most research effort was aimed at observing and describing the state of the environment and understanding the interactions between the various parts of the system.

—Louis J. Battan

Geological sciences

During 1980 the geological profession continued to be strongly influenced by the economic climate, particularly as it related to energy. The demand for geologists qualified to participate in this complex and difficult enterprise was higher than ever. But despite the priority given to economic considerations, the more purely theoretical disciplines of geology continued to flourish.

Geology and geochemistry. There were no profound theoretical breakthroughs during 1980, but impressive contributions to knowledge were made within the context of the dramatic theoretical advances of the 1960s and 1970s. The man who is generally credited with initiating the "revolution" in geology by proposing the hypothesis of continental drift was honored by an international symposium in Berlin in February. The 100th anniversary of the birth of Alfred Wegener was commemorated by sessions devoted to plate tectonics, continental drift, polar research, paleoclimatology, and paleogeography, all of which disciplines figured importantly in his research.

Mt. St. Helens. Geology, being a historical discipline, is as influenced by the occurrence of natural events as it is by the development of theories. The most dramatic event of geological significance to occur during 1980 was the eruption of Mt. St. Helens in southwestern Washington state. This peak, the youngest of the five large volcanic mountains in Washington, had been dormant since 1857. There is no question that this event assumed great significance for Americans simply because it occurred in their own country and resulted in great destruction and a significant loss of human life. In 1980, as in almost any year, there were other volcanic eruptions occurring throughout the world. Significant volcanic activity was reported from Sicily, the Philippines, Japan, Indonesia, and Chile.

The scientific significance of the Mt. St. Helens eruption should not, however, be minimized. The Cascade Range in general and Mt. St. Helens in particular had been a subject of interest to geologists for more than a century, and geological investigations in the region had intensified in recent years. Studies of the present eruptive episode can, therefore, be viewed within the context of a considerable body of knowledge accumulated over an extended period of time.

Once the first steam and ash ventings began on March 27, 1980, many agencies and individuals immediately started intensive investigations. Among the institutions involved in this effort to collect and interpret data were the United States Geological Survey (USGS), the United States Forest Service, and the Washington State Department of Natural Resources. A large number of individual geologists from government agencies, colleges and universities, and private industry were also involved in the study of various aspects of the eruption.

In a report by Dwight R. Crandell and Donal R. Mullineaux published by the USGS 1978, Mt. St. Helens was identified as the most active volcano in the Cascade Range during the past 4,500 years. Crandell and Mullineaux reported that, although the present visible cone was formed during the relatively brief period of 1,000 years, activity was initiated about 1,900 years ago with the formation of extensive lava flows. About 450 years ago ash and pumice was spread over a wide area in an explosive eruption, and at that time the growing peak was crowned by a lava plug several hundred feet in height. In a major eruption that began about 180 years ago ash was distributed over much the same area covered in the 1980 eruption. Intermittent activity continued between 1831 and 1857.

The beginning of the present phase of activity may be marked by the occurrence of a 4.1-magnitude earthquake whose focus was about five kilometers (three miles) beneath the peak on the afternoon of March 20, 1980. During the following week many earthquakes with magnitudes between 2.8 and 4.4 were recorded in the immediate vicinity of the peak. Some of these tremors resulted in minor avalanching. On March 27 a plume of steam and ash appeared and quickly rose above the crater. Steam venting and the eruption of ash occurred frequently until early in April, when activity began to decrease. Then in the period from May 7 to May 14 activity increased, with columns of steam and ash rising to an elevation of 4,000 m (13,000 ft). Eruptive activity again decreased between May 15 and May 17.

Two distinct craters that had formed early in the present episode had merged by early April into a single

(Top, left) © 1980, Vern Hodgson; (top, right) James Mason—Black Star; (center) courtesy, General Electric Corporate Research and Development; (bottom) UPI

Mt. St. Helens in Washington at the beginning (top, left) and during (top, right) the explosive eruption of May 18, 1980. At the center left is a photomicrograph of a sample of volcanic ash from the eruption; the zoned structure probably results from crystal growth during varying pressures and temperatures. A growing lava dome (bottom) is seen in the crater atop Mt. St. Helens on October 22, 1980.

crater 500 m long, 300 m wide, and 250 m deep. Mudflows, identified by Crandell and Mullineaux as a principal hazard in their 1978 report, began to occur with the onset of warmer weather in late March. Especially interesting and significant from the standpoint of the prediction of volcanic activity was the recording of harmonic seismic tremors beginning on April 1. This pattern of seismic activity was thought by seismologists and volcanologists to result from the movement of magma (molten rock).

Continuous monitoring of changes in the configuration of the peak led to the confirmation on April 23 of a bulge on the north face. The bulge apparently resulted from the rise of gas-charged magma. This development, together with extensive faulting on the north side of the peak, was recognized as a possible precursor of catastrophic activity, and travel north of the peak was sharply curtailed.

On the morning of May 18 a gigantic landslide, apparently triggered by a 5.0-magnitude earthquake, slid down the north face of the mountain and released pressure over the developing bulge. The result was a nearly horizontal blast of incandescent gas and ash with an initial velocity estimated at 50 m per second and which within minutes devastated an area of approximately 500 sq km (200 sq mi). The landslide plunged down the north flanks of the peak into Spirit Lake 8 km (5 mi) to the northeast and more than 20 km (12 mi) north down the North Fork of the Toutle River, filling the river valley to a depth of 60 m. The landslide, together with mudflows that occurred after the major eruption, blocked streams and rivers and caused extensive flooding in the drainages north of the mountain.

The catastrophic horizontal blast dissipated within a few minutes. A vertical column of ash and steam that rose to an elevation of 13–19 km then developed. Ash carried northeastward on strong prevailing winds resulted in extensive ash falls. Collapse and ejection resulted in the removal of 400 m from the north slope of the peak. On June 14 magma appeared as a dome in the crater and rose to a height of 30 m.

Intermittent eruptive activity, some of it resulting in a wide dispersal of ash, occurred after the eruption of May 18. Volcanologists were in some disagreement as to whether another catastrophic eruption of Mt. St. Helens was to be expected in the near future. The consensus seemed to be that such an event was unlikely. Geologists agreed, on the other hand, that some activity was to be expected for years or even decades.

A vast amount of data was collected at Mt. St. Helens, and it was expected to be a long time before volcanologists would be able to assimilate all of it. It can confidently be expected, however, that the information gathered will contribute to the understanding of volcanism in the Cascade Range and, more generally, of the tectonic setting of western North America.

The hypothesis of plate tectonics, which has played a significant role in contemporary geological explanations, was invoked in the explanation of Cascade volcanism. According to this hypothesis the Earth's crust consists of a number of more or less rigid plates that are moved with relation to one another by thermal convection currents in the Earth's upper mantle. The small Juan De Fuca plate, which lies off the coast of northern California, Oregon, Washington, and southern British Columbia, is being driven toward the much larger North American plate and is being forced beneath it. In the process rock material is dragged tens of kilometers beneath the axis of the Cascade Range, where it melts to form magmas. It is the rise of these magmas that initiates episodes of volcanism. A principal reason for believing that volcanic activity will recur in the region is that there is every reason to suppose that this process is continuing.

Geothermal resources and oil shale. The Cascade Range has been the subject of study not only from a purely scientific point of view but from an economic standpoint as well. A three-day conference on the tectonics, volcanology, and geothermal potential of the range, which was held at the USGS regional office in Menlo Park, Calif., in February, nearly coincided with the renewed volcanic activity in the range. Earth scientists attending the conference concluded that existing data suggest promising geothermal resources in the area, but that new regional and detailed mapping and the drilling of deep exploratory holes are needed before a more conclusive evaluation can be made. In *California Geology* C. T. Higgins reported that the California Division of Mines and Geology was undertaking a statewide survey to locate low- and moderate-temperature geothermal resources, those involving fluid temperatures between 90° C (195° F) and 150° C (300 °F). Resources in this temperature range are employed mainly in the direct heating and cooling of buildings, in the heating of domestic water supplies, and in various industrial applications. The division was compiling data on all known and newly discovered thermal springs and wells in the state. In an attempt to extend the scope of geothermal energy applications, the feasibility of extracting energy directly from magmatic bodies was being studied. In a review of investigations undertaken by the Sandia Laboratories, the National Magma Energy Advisory Panel recommended that research be continued. Methods for finding and mapping magmatic bodies and for improved drilling techniques were areas designated for continued study.

Attempts to devise an economically feasible and environmentally acceptable method of producing petroleum from oil shales continued during 1980. A panel of the U.S. National Research Council cautioned that surface mining of oil shales and tar sands could have a long-lasting and severe impact upon the environment. Of particular concern to the panel was the production of pollutants. Devising economically feasible methods for removing pollutants from the vast quantities of water that have been used in the processing of oil shales continued to present difficulties.

In the meantime, several test projects aimed at the production of oil and gas from oil shales were undertaken. Using a process similar to one developed by Texaco Inc. and Raytheon Co., Colorado Synfuels Co. obtained access to land in southwestern Wyoming for the experimental production of oil from oil shale by the use of microwave technology. In this process a microwave generator is introduced directly into the shales through drill holes in order to separate oil from shale. In a demonstration plant in northwestern Utah the Ramex Synthetic Fuels Corp. of Salt Lake City was attempting to produce gas by introducing heat-transfer equipment into oil shale bodies by means of single drill holes. This procedure, which proved viable in the laboratory, showed promise in the field, according to Ramex president D. H. Nelson.

Toxic wastes. A goal of high national priority involving geologists was the development of environmentally acceptable means of disposing of toxic wastes. Since 1976 the U.S. government has sought through the National Waste Terminal Storage Program to find secure sites for the storage of high-level radioactive waste. Geological conditions such as the permeability of repository rocks, hydrological parameters of the region, and the risk of tectonic activity are crucial factors to be considered in the selection of suitable sites. Field studies to determine these factors and others were being carried out in Washington, Nevada, Utah, Louisiana, Mississippi, and Texas.

The salt domes of the Gulf Coastal Plain have long been considered promising sites for the storage of radioactive waste. In 1963 geological studies led to the identification of 36 domes as potentially acceptable. Since 1978 intensive investigation, including the drilling of 34 deep exploratory holes to test hydrological conditions, resulted in the elimination of all but seven of these as potentially suitable disposal sites. In the Paradox Basin of Utah and in the Permian Basin in Texas bedded salt deposits were being investigated as possible waste depositories. Salt has for many years been regarded as the most suitable matrix for the storage of highly toxic materials, but other rock types were also being considered. At the Nevada Nuclear Test Site, tuffaceous rocks formed by the welding of volcanic ejecta appeared to offer some promise as waste repositories, while the feasibility of disposal in the basalts of the Pasco Basin in Washington was also being considered.

An intriguing possibility, discussed by W. S. Fyfe in an article published in *Episodes*, was that canisters of radioactive material be introduced into oceanic trenches that are adjacent to subduction zones (regions where one crustal plate is thrust under another). At a subduction rate of ten centimeters (four inches) per year, the canisters would reach a depth of one kilometer in 10,000 years.

Tibetan Plateau. The participation of China in the international geological community during the past several years presented Western geologists with their first access in a number of years to areas of great geological interest. A multidisciplinary symposium on the Tibetan Plateau was held in Beijing (Peking) from May 25 to June 1, followed by a 14-day field trip to south-central Tibet sponsored by the Academia Sinica. Of special interest to the Earth scientists in attendance was the fact that the Tibetan Plateau is perhaps the fastest rising land mass in the world. The movement of the rigid plate underlying the subcontinent of India beneath the Asian plate was causing the plateau to rise at a rate estimated at 5 mm (0.2 in) per year. This rapid rise has resulted not only in significant geological changes during the past one million years but also in changes in the fauna and flora of the region during the same period.

Growing scientific cooperation between the United States and China was evident in an agreement signed by H. William Menard, director of the USGS, and representatives of China's Ministry of Geology. The two countries agreed on plans for cooperative Earth science projects, with joint research on earthquakes to begin immediately.

—David B. Kitts

Geophysics. In the years ahead enormous effort will go into the search for new petroleum reserves. Although they are not the kind of developments that capture headlines, important progress has been made during the last decade in the oil-exploration industry in acquiring a detailed three-dimensional picture of the Earth's structure at depth. The fate of peoples and the course of history may well turn on the results of the search for oil, and from a human perspective this is probably the most significant development in geophysics of our time.

A potentially important advance announced during the year was in the area of earthquake prediction. A Japanese scientist proposed that a simple relation exists between the amount of slip that occurred in the last great earthquake on a given fault and the time until the next one. Preliminary investigations were encouraging, suggesting that it may someday be possible to issue long-range forecasts to within a decade or less of when a great earthquake will occur. Two disastrous earthquakes that struck in the Mediterranean region during the year served as grim reminders that a workable scheme for prediction will come none too soon.

Seismic imaging in oil exploration. Most oil and gas deposits form originally on the bottoms of shallow seas from decomposition products of organic material. They become incorporated into the sedimentary deposits that continually are being deposited on the seafloors.

Courtesy, Sandia National Laboratories

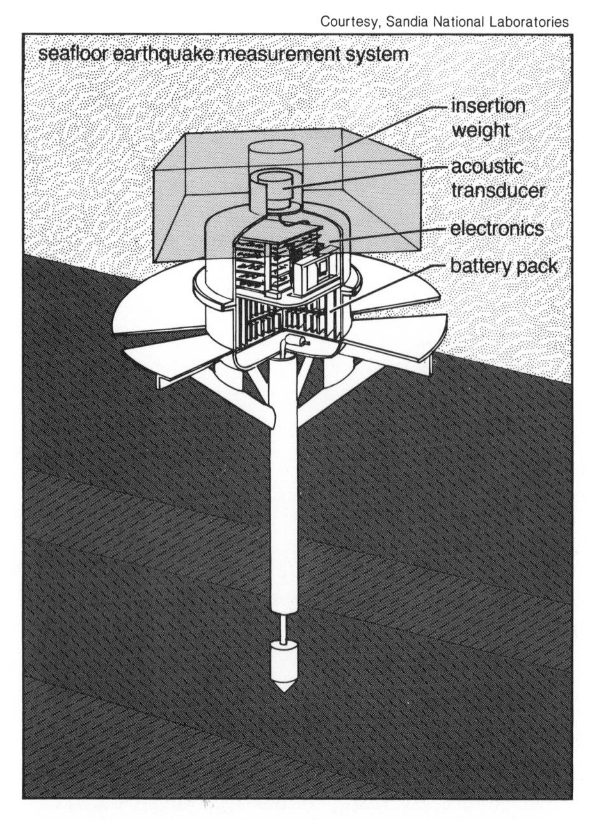

Consisting of a pressure vessel atop a seven-foot probe, the above device may aid in the recovery of offshore oil by measuring the response of the ocean floor to earthquakes, thereby helping design safe oil rigs.

Over a long period of time these sediments consolidate into rocks. When conditions are favorable, the incorporated hydrocarbons, along with water, become trapped within more porous rocks such as sandstone. Because they are less dense than water, the hydrocarbons tend to migrate upward and concentrate in structural or stratigraphic traps. It is this tendency to concentrate that has made them important as an energy source, and it is the identification of the characteristic subsurface features that act as natural underground traps—faults and anticlines in particular—that is the principal goal of exploration geophysics.

The principal tools of the exploration geophysicist are sound waves generated at the Earth's surface by dynamite, large thumpers mounted on trucks, or air guns that produce a sound pulse by the explosive release of compressed air. These waves propagate downward in the Earth until they encounter a change in the compressibility, rigidity, or density of the rocks, at which point some fraction (usually small) of the energy is reflected back toward the surface. These reflections are recorded at the surface on geophones, small electromagnetic devices that detect changes in pressure or velocity. The signals from these geophones are transmitted along wires or by radio to a central collection point, where they are recorded on magnetic tape.

Because of the large number of factors—economic, scientific, and political—that go into decisions about where to do exploratory drilling, it is difficult to assess objectively the contribution of any one factor, such as geophysics, in the discovery of new oil fields. Given the low success rate, with only one in seven wildcat wells paying off, luck certainly plays a large role. But drilling is an extremely expensive business, and even a small improvement in the success rate pays large dividends.

There is some evidence that success rates did improve in the continental United States in the 1970s, especially in the search for small fields. Two important developments introduced in the 1960s in the petroleum exploration industry may well have played a part in this improvement. First was the ability to record more seismic traces; 100 channels were in routine use by the end of the decade. The second was common-depth-point processing, which allowed data from many different source pulses to be combined by computers in a way that enhanced detail at depth while suppressing noise introduced by near-surface features.

The principal difficulty in the interpretation of seismic data lies in separating the desired signal from the noise—the reflections, refractions, and diffractions—from all the other interfaces. The major problem is that the upper crust of the Earth is usually quite heterogeneous; faults, folds, and lateral changes in rock type are pervasive features of most sedimentary basins. This complexity makes it essential to sample the reflected waves at as many points as possible. During the 1970s progress accelerated in field techniques and recording hardware, and by the end of the decade some exploration companies were advertising the capacity to record up to 1,000 separate data channels. These channels are sampled up to 1,000 times per second, resulting in data rates of up to a million points per second. However, these data must first be unscrambled to obtain a picture of the configuration and properties of the rocks at depth.

The problem is analogous to the formation in the human visual cortex of a three-dimensional image of the world about one, except that in this case it is the pattern of chemical activity induced by the reflected photons on the retina of the eye that provides the information. The process of constructing an image in one's mind involves a prodigious feat of data processing. Substantial progress has been made in the last decade toward the development of a comparable capacity to resolve a "seismic image" of the Earth's crust.

Among the important computational developments was the use of equations from the field of optics to remove distortions caused by the irregular geometry of the boundaries between different rock units. Another development was the application of linear inverse theory to correct for the highly variable effects of the near-surface layer of soil and badly weathered rock. A

Courtesy, General Electric Research and Development Center

Scientists record concentrations of radon in the Earth's crust at sites in New York state and Alaska. Variations in the concentrations can be related to stresses deep within the Earth and thus may be useful in predicting earthquakes.

third was the use of powerful digital filtering techniques to remove multiples, or echoes, in the data. These are a serious problem in data collected along the continental margins, where energy reverberates in the near-surface layers and thus obscures reflections from deeper horizons.

Of course, their eyes tell people much more than just the shape of the world around them; on the basis of color, texture, and tone, they infer the actual material composition of the objects viewed. Similarly, in the exploration industry great progress was being made in interpreting the amplitudes and frequency content, as well as the arrival times, of reflected seismic waves. Estimates of compressional velocities have been used routinely for some time in the interpretation of seismic data; shear velocities and absorption properties were beginning to be used to estimate not only which rocks are present but their porosity and fluid content as well. In one exciting development oil company scientists discovered that one particular type of high-amplitude reflection (a "bright spot") is often associated with the presence of a pocket of natural gas. The implications of this discovery are awesome, for if means can be found to extend this technique to the direct detection of oil as well, it could reduce enormously the costs of oil exploration.

Probably the most important advance in petroleum exploration, however, is a result of putting together all the above-mentioned developments in recording, processing, and interpreting massive amounts of seismic data. The possibility now exists of deploying seismometers in two-dimensional arrays and making real three-dimensional interpretations. Until recently, exploration seismologists were limited by economic, logistical, and computational considerations to deploying long linear arrays of geophones, and interpreting them as if the crust varied in just two dimensions. The removal of this limitation will almost certainly have a profound effect on the search for oil in the decades ahead.

Earthquake prediction. An international meeting in May 1980 at New Paltz, N.Y., gave earthquake seismologists from throughout the world a chance to review progress during the last few years in earthquake prediction. While few would claim that final solutions are at hand, a consensus of the scientists present was that slow but significant progress was being made.

One important new result was described by Kunihiko Shimazaki, a seismologist from the Earthquake Research Institute in Tokyo. He found that the time between two successive great earthquakes would be proportional to the rate of strain accumulation. Most observations had suggested that the scatter about the average inter-event time was about 25%, making estimates of the time of occurrence too uncertain to be of any practical value. Shimazaki's work, however, suggests that the scatter in the observations is due to variation in slip from event to event, and that if one can estimate the slip in the last event, the time to the next event can be estimated much more precisely. This optimistic result was expected to be the focus of a great deal of attention and work in the years ahead.

William Ellsworth of the U.S. Geological Survey reported on preliminary application of this idea to the San Andreas Fault in central California. This portion of the North American-Pacific plate boundary last failed in 1906 in the great San Francisco earthquake. For that event two estimates of the slip are available, one based on surface offsets across the fault and

Francolon/Siccoli—Gamma/Liaison

Earthquake in Algeria destroyed about 80% of the city of El Asnam, causing some 3,000 deaths and leaving more than 200,000 people homeless.

another on changes in angle between geodetic monuments. When combined with measurements of the plate-motion rate in this area, both measurements suggest that another great earthquake should not be expected for about another 50 years.

Less comforting, however, are the results of a similar calculation for the San Andreas Fault in southern California. About 120 years have passed since the last great earthquake in this region, and detailed geologic investigations suggest that along the portion of the San Andreas Fault closest to the metropolitan Los Angeles area another great earthquake is possible at any time. The annual probability of occurrence is about one in 50, but some scientists believe that the sequence of geophysical anomalies that has been observed in southern California in the last few years—abrupt changes in vertical and horizontal strain, seismicity, and radon emissions from wells—are sufficient to increase this probability to an annual risk of one in 20. The task faced by geophysicists is to further refine this probability and provide some measure of the uncertainty in the calculation.

Rescue workers explore ruins in San Lorenzo, Italy. A November 1980 earthquake destroyed 100 towns in southern Italy and killed about 3,000 people.

Keystone

Destructive earthquakes. On October 10 a magnitude-7.5 earthquake occurred near El Asnam, near the Mediterranean coast in Algeria. Almost the entire city was destroyed, as was much of the surrounding area. Approximately 3,000 people lost their lives. Just six weeks later a magnitude-6.9 event struck southern Italy. In the mountains east of Naples 100 towns and villages were destroyed and some 3,000 people died.

These earthquakes were reminders of the enormous human good that could be accomplished by a workable scheme for short-term earthquake prediction. A large fraction of the world's onshore earthquakes occur along the great mountain belts that stretch from the western Mediterranean to India. Tragically, throughout much of that region unreinforced masonry construction is the rule, and even moderate earthquakes often exact a heavy toll. Economic and social considerations make it unlikely that people will be living in more earthquake-resistant houses in those areas in the near future. The only way to reduce the terrible loss of life that occurs is to find some way to issue short-term warnings that will allow people to get out of their homes before they collapse.

—Allan G. Lindh

Hydrological sciences

Hydrologists and oceanographers during the year pursued research on such subjects as the general circulation of the ocean, the geology and geophysics of the seafloor, the structure and distribution of hydrothermal vents in the deep ocean, and the effect of the eruption of Mt. St. Helens on the hydrology of the surrounding region.

Hydrology. The May 18, 1980, eruption of Mt. St. Helens was of interest to hydrologists as well as to volcanologists and petrologists. The major lateral northward blast on that day devastated an area of some 500 sq km (200 sq mi); rock, ice, and organic debris together with material directly from the volcano covered the upper Toutle River drainage basin to depths of up to 180 m. The debris flow traveled more than 20 km (12 mi) down the North Fork of the Toutle River. Floods heavily charged with sediment continued down this river into the Cowlitz River (raising its bed five meters) and the Columbia River, blocking many tributaries of the latter and halting navigation on it. The Swift Reservoir received approximately 11,000 ac-ft of water, mud, and debris from the Pine and Smith creeks and Muddy River, which drain the eastern and southern flanks of the mountain.

The eruption dramatically altered the hydrology of the surrounding area by reducing the capacity of major river channels to convey water and by altering the rainfall-runoff relationships in the affected watersheds. Within 36 hours of the eruption the U.S. Geological Survey (USGS) began studying its hydrological impacts. The work, which continued into 1981, included installation of new data collection stations, flood hazard assessment, and water quality impacts.

Many data collection stations were destroyed by the eruption, and new data had to be collected to help in redefining the hydrological relationships. New stations were installed on affected streams to monitor discharge, sediment transport, and selected chemical and biological quality characteristics. Monitoring of rivers, lakes, and reservoirs for flood warning was also taking place. Such information is vital for carrying out studies to define quantitatively the current and long-term effects of the eruption on the hydrology of the region and for future flood warning, resource planning, and management.

One initial concern was the potential flooding from possible failure of the debris pile damming Spirit Lake. Data from previous flood studies and from gauging stations along with geological maps were used to simulate the debris dam failure. The potential flood release was simulated for the post-eruption river characteristics, and hazard assessments were made for downstream areas. Besides Spirit Lake, similar analyses were made for the reservoir system on Lewis River and for a small debris reservoir on Maratta River. This small reservoir later failed, and the recorded peak discharge at Highway 99 was within 7% of that predicted by the model.

A continuing concern related to changes in infiltration characteristics of the ash-covered watersheds and subsequent changes in rainfall-runoff relationships. The May 18 blast totally killed aboveground plant cover within 60,000 ha (150,000 ac) north to northwest of the volcano, changing the water balance relationship. The volcanic ash sealed the macro-pore infiltration system found in forest hydrological settings and reduced infiltration rates to 5–10% of pre-eruption levels.

In 1981 a USGS infiltration and erosion study was

Keystone

Steel gates arrive aboard a barge in the Thames River at Woolwich, London. They will be positioned there between the huge concrete piers in an effort to protect London from flooding by the Thames.

being carried out in the East Fork Shultz Creek drainage area on selected plots approximately 24 m × 6 m in size. A portable rainfall generator was being used, and variations in infiltration and erosion rates with changes in volcanic ash characteristics, hillslope, rainfall intensity, and antecedent ash moisture conditions were being measured. Two permanent plots furnished with precipitation, discharge, and sediment sampling equipment were being established to measure changes in infiltration and erosion through time under natural conditions. Infiltration studies on eastern Washington and northern Idaho cropland soils affected by ash fallout were also being undertaken.

Results from these and related studies were to be used as inputs into a hydrologic model for flood runoff and sediment transport prediction. The latter is especially important because sediments carried off the steeper parts of the watersheds may be deposited into the Columbia River and its major tributaries, reducing their conveyance capacities and increasing flooding potentials. In response to this possibility the U.S. Army Corps of Engineers dredged river channels extensively to increase flood-carrying capacities.

Data collected during heavy rainfall in late December 1980 appeared to indicate that less sediment will be deposited in the tributaries than previously believed. Furthermore, some previously deposited sediments may become eroded and washed down the Columbia River by the higher runoff flows associated with heavy rainfall.

In conjunction with the USGS the University of Washington performed an initial assessment of changes in the rainfall-runoff response related to ash deposition. The U.S. National Weather Service River Forecast model was initially applied to the Cispus River, a tributary of the upper Cowlitz. The mathematical simulation model was first calibrated for pre-eruption conditions using historical data. Subsequent tests were made using the limited number of post-eruption rainfall events to determine the extent of basin changes and to better predict future flood hazards.

Potentially adverse effects on water quality could have resulted from the eruption. The USGS immediately initiated a statewide surface water sampling program in Washington to determine the nature and magnitude of the volcanically induced variations. Streams to the east of Mt. St. Helens received major ash fallout. Chemical variations were pronounced but short-lived (two to three days). Streams draining to the south (tributaries of the Columbia River) showed few observable changes in water chemistry. Streams flowing west, however, had elevated levels of chloride and sulfate anions for up to one month after the eruption.

The hydrologic effects from the Mt. St. Helens eruption were expected to persist in some watersheds for many years. Because of the destruction of vegetal cover, erosion processes changing watershed and river characteristics will persist. These may adversely affect the lower Columbia River, and if so their economic impact would be substantial: increased potential for flooding, the necessity of large expenditures for sediment dredging, reduced navigation on the Columbia because of sediment deposition, and decreased salmon catches due to fish kills from high levels of turbidity caused by the volcanic ash.

—Eric F. Wood

Oceanography. *Hydrothermal vents.* In late 1979 and early 1980 the first reports of one of the major discoveries of the decade in oceanography were reported to the scientific community. Scientists from the United States, France, and Mexico were all involved in establishing the existence of major hydrothermal vents, or very hot springs, at the floor of the deep ocean. The vents were found on the East Pacific Rise crest

Sewage sludge is converted to soil conditioner at a wastewater treatment plant near Washington, D.C. The use of sludge as fertilizer could help preserve water quality.

Robert C. Bjork, USDA, Science and Education Administration

near Baja California at approximately 21° N and 109° W, at a depth of about 2,600 m (8,500 ft).

The plate tectonics model of the Earth's crust had identified the rise crests as regions of the Earth where geological processes are most likely to be evident. And, in fact, the discoveries yielded new evidence on the creation of oceanic crust, on metal chemistry, on hydrothermal effects, and on major types of biological communities dwelling on the ocean bottom.

The observations were carried out with the Deep Tow (a remote camera and seismic system) and the research vessels "Melville" and "New Horizon" of the Scripps Institution of Oceanography, the "Angus" (a camera and temperature sensor sled) and the submersible "Alvin" of the Woods Hole Oceanographic Institution, and the French submersible "Cyana."

The major results revealed that the vents jet out water and dark particulate matter at temperatures approaching 380° C (715° F); the water is kept from boiling only by the very high pressures at the seafloor. The hottest waters were found to issue from mineralized chimneys that are blackened by sulfide precipitates. As sites of actively forming massive sulfide mineral deposits and other active chemical reactions, these hydrothermal springs have an important effect on the overall chemistry of the ocean. For example, they appear to be responsible for removing much of the magnesium from the abyssal ocean.

Cooler springs are also found near the hot vents, similar to those already observed 320 km (200 mi) northeast of the Galápagos Islands at the Galápagos spreading center by scientists in 1976; these springs are clear to milky and, like the hot vents, support exotic bottom-dwelling communities of giant tube worms, clams, and crabs. They emit water from the seafloor with temperatures of about 20° C (68° F).

The discovery of the high-temperature vents raises interesting questions for future biological research. The vents are apparently the first places where water temperatures well in excess of the normal boiling point at one atmosphere have been found in open contact with the biosphere. In these dark, hot regions the biological communities appear to feed by chemosynthesis rather than by photosynthesis. The chemosynthetic bacteria use hydrogen sulfide as an energy alternative to the Sun. Growing in waters at temperatures above 100° C (212° F), they could be the food-chain base for the observed community of mussels, clams, crabs, tube worms, and other animals, some of which are bright red due to the presence of hemoglobin. Most of the

Submersible "Alvin" (left) took part during the last year in explorations of hydrothermal vents on the deep ocean floor. A picture taken from "Alvin" (right) shows a vent at a depth of about 2,500 meters (8,250 feet) on the East Pacific Rise. The temperature of the vent water is about 350° C (660° F).

Vicky Cullen, WHOI

John Edmond, MIT

Photos, courtesy, Royal Society

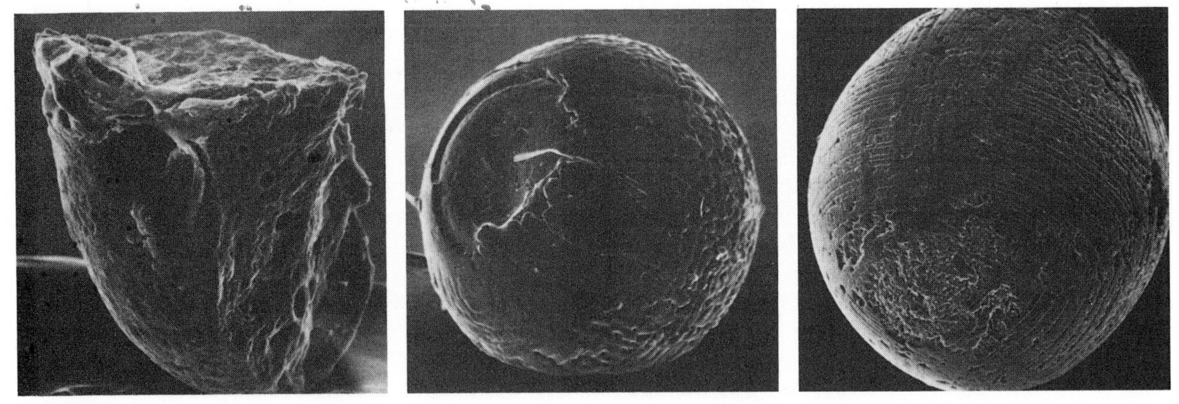

Spherules found in deep-sea sediments are believed to be chips thrown off by colliding asteroids. At the left is a chip from a meteorite; the differences between it and the iron (center) and stony (right) spherules indicate that the latter did not originate in meteorites.

animals found in these regions appear to be unique and may be representatives of previously unknown animal families. These communities must have unusual dispersal systems that allow them to populate various areas of the world rift system as hydrothermal vents evolve and die out.

In these studies the submersible "Alvin" was used for the first time for geophysical research. The gravity and seismic measurements made on the bottom yielded the first at-the-site determinations of seismic velocity and density in shallow crust. Further analysis was expected to place important limits on the depth of hydrothermal activity and fissuring of the crust, and also on the extent of the deep magma chambers below. The magnetic measurements were expected to yield understanding of how marine magnetic anomalies are generated by sources deep below the crust.

Climate research. The World Climate Research Program was established on Jan. 1, 1980, by the World Meteorological Organization and the International Council of Scientific Unions. This program identified as an immediate priority the need for research on the controlling effect of the dynamics and thermodynamics of the oceans on the global cycles of heat, water, and chemicals (especially carbon) in the Earth's climate system.

Studies in 1980 of the ocean circulation, heat budget, and mixing of the air-sea interaction contributed to understanding the role of the ocean in shaping the Earth's climate. The heat transported across the latitude of 25° N in the North Atlantic was estimated from oceanographic measurements to be about 1.1×10^{15} watts northward; this figure agreed with the value obtained from charts of energy exchange between ocean and atmosphere but was about 50% smaller than the values derived from satellite radiation measurements. This major discrepancy will have to be resolved by further studies being planned under the World Climate Research Program.

Ocean circulation. General ocean circulation experiments remained an area of intense interest by oceanographers. In 1980 studies using moored and drifting buoys, temperature and conductivity probes, and devices that measure profiles of velocity were undertaken in oceans throughout the world.

Studies on the role of mesoscale eddies in the general circulation of the ocean continued. These medium-sized fluctuations in the ocean currents contain much of the ocean's kinetic energy. They can extend from the surface to the bottom, are several hundred kilometers in diameter, and rotate in either a clockwise or counterclockwise direction.

The effect of eddies on heat flux was emphasized by calculations from several years of data in the Drake Passage, a strait connecting the Atlantic and Pacific south of the southern tip of South America. These show a much stronger eddy heat flux than had been expected. The extrapolation of these results to the circumpolar region suggests a major eddy contribution to net heat and salt flux in high southern latitudes. During 1980 special emphasis was also given to the design of studies of coastal circulations.

The remote sensing techniques of satellite measurements made great strides during the past year. Satellite altimetry, which measures the height of the ocean surface, was used to reveal features in the Gulf Stream, the Kuroshio Current, and eddy motions in the North Atlantic. Sea surface temperature measurements from satellites were used for many purposes; among them were monitoring the Somali Current in the Indian Ocean during the summer monsoon, following the California Current, and observing the boundary between the shelf and slope water off Nova Scotia. Wind stress was shown to be calculable from satellite data, a major step for providing a global coverage of this important driving force at the ocean surface.

Seafloor studies. In the general area of the geology and geophysics of the seafloor the International Program of Ocean Drilling continued its successful program of deep sampling with the drilling ship "Glomar Challenger." The first expeditions of the year were to the South Atlantic Ocean to collect samples for studying the history of the Falkland Plateau and the Rio Grande Rise and ancient ocean circulations. This work also included coring with the newly developed hydraulic piston coring device on the Mid-Atlantic Ridge. Because of the undisturbed nature of the core retrieved by this method, expanded correlations could be made between geological age determinations and geomagnetic field phenomena.

Later expeditions took the "Glomar Challenger" to the Walvis Ridge and then north to the Blake-Bahama Basin and Outer Ridge, and the Florida Straits. In these expeditions scientists from Austria, Canada, West Germany, France, Japan, the United Kingdom, and the United States were involved, revealing the strong international flavor of the project.

The "Glomar Challenger" program was planned to continue into the 1980s with strong international support from scientists and the various governments that had supported the international phase thus far: France, West Germany, Japan, the U.K., the U.S.S.R., and the U.S. At the same time planning began on a new, complementary program that would focus on drilling in the margins of the ocean, an area that had not yet been studied much. The technical difficulties of drilling in those regions required a larger ship than the "Glomar Challenger"; the U.S. ship "Glomar Explorer" was contemplated for the project.

—D. James Baker, Jr.

Electronics and information sciences

The use of computers in communications networks and of optical fibers for transmission systems continued to undergo further development and expansion during the past year. A microprocessor that can execute two million instructions per second was introduced, and the first high-speed digital communications satellite was launched.

Communications systems

During the year worldwide attention in the communications field was drawn to the potential of home information services. The British Prestel system expanded its commercial operation, while in France a far-reaching plan was intended to put millions of low-cost communications terminals into French homes. The world's telecommunications networks continued their evolution toward all-digital plants, and fiber-optics links began regular operation in several locations, displacing the copper cables that have carried telecommunications traffic for the last century. In the United States the Bell System announced commitments to fiber systems in the Northeast and in a proposed transatlantic cable.

In computer communications the year was preoccupied with attention to local data networks, with a system called Ethernet attracting international interest. Several electronic mail systems began operation, and the satellite for the Satellite Business Systems network was placed into orbit. None of these technological achievements seemed to have as much impact on the future evolution of communications in the United States, however, as did the attempts to restructure the Bell System.

Home information services. Videotex systems link the home television receiver, via the telephone network, to a computer data base. Users select information they wish to receive from a list of choices appearing on the TV screen. Alphanumeric characters and graphic elements (forming low-resolution pictures) transmitted from the data base are displayed on the screen. The telecommunications systems, in consort with a host of information providers, are thus able to bring into homes information on entertainment, travel, news, weather, sports, shopping, and a great variety of other subjects. The potential of this new medium received much publicity during the past year, and a number of countries began trials of videotex systems.

In the U.K. the Prestel system, which had begun operation in 1979, was considerably expanded in 1980. The installation program aimed for access to the ser-

Dark square central area on the raised ceramic module is a silicon chip on which has been made an optical receiver that could result in more economical use of data-communications links between computers. A fiber-optic cable extends from the module.

Courtesy, IBM

Courtesy, Northern Telecom Inc.

The Visual Ear allows deaf and speech-impaired persons to communicate by telephone. With the phone headset in place as shown a user dials the desired number. On a Visual Ear at the other end a blinking light alerts the receiving party to an incoming call. The two then type messages back and forth on the keyboards; the words appear above the keyboard on a moving display.

vice for about 60% of telephone owners in 1981. Customer acceptance in the U.K. was closely watched, but the high costs of the specially equipped television sets and of the per-page access capability prevented significant penetration of the general consumer market. Instead closed user groups of business Prestel customers with common informational needs, such as travel agents, were the chief supporters of the service.

In France the Télématique program was launched in March 1980. The program encompassed a videotex service, a low-cost facsimile terminal, a telewriter terminal to transmit graphics and handwriting, and an electronic directory service. The latter was probably the most significant, since the intention was to provide free of charge to telephone customers an information display terminal based on a cathode-ray tube in lieu of a telephone directory. Tests of the electronic directory service were scheduled to begin in 1981 with service to about 250,000 customers in northwestern France.

The Canadian teletex system, Telidon, featuring a graphics capability to draw cartoon-quality pictures on the television screen, was operational in several areas of Canada by early 1981. In the United States AT&T joined forces with a newspaper publisher, Knight-Ridder, for a trial of a videotex service in Coral Gables, Fla. The trial began in mid-1980 with 30 home terminals installed in separate residences. The terminals were moved approximately every three weeks to enlarge the customer base. About 15,000 pages of information were accessible to the Coral Gables customers. Although AT&T had scheduled further tests of videotex and electronic directory service for Austin, Texas, in 1981, these tests were being vigorously opposed by newspaper publishers who feared the loss of direct access to information consumers that would occur if an electronic medium were controlled by a regulated monopoly such as AT&T.

Optical fiber transmission systems. Continued improvements in light-wave technology for signal transmissions made optical transmission increasingly attractive to augment the traditional approach that used electrical signals over copper wires or coaxial cables. Optical fibers offered the advantages of smaller size, much higher information capacity, and immunity to electromagnetic interference; they also required fewer repeater stations along the transmission route. Experimental use of optical fiber transmission between telephone central offices began in 1977. In 1980 the first optical transmission system using standard installation techniques began regular service in the Bell System, connecting three central offices in Atlanta, Ga. The new system, designated FT-3, carried 44.7 million bits (0s and 1s) per second of information on each hair-thin fiber. In the optical cable 12 fibers are embedded between two strips of plastic in a flat ribbon, and as many as 12 ribbons are stacked in the cable. A 12-ribbon, 144-fiber cable enclosed in a sheath fabricated of polyethylene and wire is one-half inch in diameter and carries the equivalent of more than 40,000 voice channels.

In addition to many applications of FT-3 within metropolitan areas the Bell System announced early in 1980 plans to utilize optical technology for intercity communications. This would begin with a 983-km (611-mi) intercity fiber system in the northeastern U.S. linking Boston, New York City, and Washington, D.C. The Bell Northeast Corridor Network will link 23 digital switching centers along the route. The New

York-to-Washington portion of the network was scheduled to begin service in 1983.

First-generation optical transmission systems operate at a wavelength of 0.82 micrometers (in the near infrared). Operation at a longer wavelength of 1.3 micrometers offers the prospect of significantly less signal attenuation.

Another significant optical development was taking place in Bell Laboratories, where an undersea optical cable system was being planned. In the proposed transatlantic cable indium-gallium-arsenide-phosphide lasers operating at a wavelength of 1.3 micrometers input light to the fibers. Each section of the cable has one operating laser and one or more standbys to enable an estimated mean time between failure of eight years. Repeating stations were being placed at intervals of about 30 km (18½ mi). As compared with previous cables the lighter weight of the optical design, its higher capacity, and the longer distance between repeaters (9 km in existing cables) are especially important in the undersea application.

Data communications. The amount of data communications, representing traffic between computers, data terminals, and other digital equipment, continued to grow strongly during recent months. While progress on some of the anticipated nationwide data networks (notably the XTEN network of Xerox Corp. and AT&T's Advanced Communications Service) was slower than had been originally anticipated, much of the interest in the field centered on various proposals for local data network designs. In mid-1980 Xerox Corp. announced its Ethernet system, which would be used for local interconnection of its office equipment. Xerox was joined by Digital Equipment Corp. and Intel Corp. in pressing for adoption of the Ethernet standard. (For additional information on Ethernet, see *Computers and computer science*, below.)

Electronic mail systems also received considerable attention in 1980. In the U.S. and Western Europe postal agencies were developing electronic systems to transmit documents and data. The U.S. Postal Service planned a system called E-COM (Electronic Computer-Originated Mail) in which customers could enter messages and address lists through data terminals into postal computers, where they would be printed and routed through the postal system. In June 1980 an international electronic facsimile system, Intelpost, began operations between Canada and the U.K.

Several U.S. corporations also instituted public electronic mail services in 1980. GTE Telenet, a subsidiary of General Telephone and Electronics Corp., began a service called Telemail, while Tymnet, Inc., started tests of a system in Los Angeles and San Francisco. In the Telemail system users may store information within the network itself for later access. A user's "electronic mailbox" can be accessed from any point in the country using a data terminal. An alternative to these

Courtesy, Bell Laboratories

The Digital Signal Processor chip, smaller than a telephone pushbutton, contains some 45,000 transistors and can perform more than 1,000,000 calculations per second. Voice recognition is one expected use.

systems was electronic voice storage, which was being provided in a system developed by the 3M Co. AT&T also developed an electronic voice storage system that digitally encodes and stores for later delivery voice messages at an electronic switching office.

Communications satellites. Communications satellites continued to be launched at a steady rate. One of the newest belonged to Satellite Business Systems, formed by IBM, Aetna Life and Casualty Co., and Communications Satellite Corp. (Comsat). The SBS service was inaugurated in early 1981 to provide data communication, facsimile, and video conferencing services directly from rooftop-mounted 5.5- and 7-m antennas. The service was initially aimed at large businesses where the conglomeration of data requirements would be sufficient to support the costs of individual Earth stations. Like other new satellites SBS used an important new multiplexing method called Time Division Multiple Access, in which the individual time slots in the digital stream are shared on a need basis.

At Bell Laboratories an integrated circuit fabricated in 1980 should play an important role in satellite voice transmission. The echo canceller chip, containing about 35,000 transistors on a 7.95-mm × 9.04-mm surface, removes echoes from communications circuits by effectively canceling them out. In satellite circuits the roundtrip delay to orbit positions results in annoying voice echoes. These echoes must be eliminated to achieve voice communication quality comparable to that achieved in terrestrial communication systems. (For additional information on satellites, see *Satellite systems*, below).

Restructuring the Bell System. In 1980 the U.S. Federal Communications Commission, which had set out in its Computer Inquiry II to define the elusive boundary between regulated telecommunications and unregulated data processing, issued a decision. A clear boundary was impossible to define, but in order to increase competition the FCC continued "basic services" under regulation while deregulating customer premise equipment and so-called "enhanced services." The Bell System could operate in the deregulated area only through a fully separate subsidiary. The situation was confused, however, by the existence of a 1956 consent decree that restricts Bell to furnishing common carrier communication services.

In late 1980 Bell announced an internal realignment in anticipation of its restructuring and entry into the unregulated market. Every action was closely watched as the whole fabric of U.S. communication policy seemed to be reforming along new lines, spurred by the prospects of huge new data processing/communications markets.

—Robert W. Lucky

Computers and computer science

John Mauchly, one of the inventors of the electronic computer, died on Jan. 8, 1980. At its national conference in October 1980 in Nashville, Tenn., the Association for Computing Machinery, the largest and most important professional organization in the computer field, presented each registrant with a medallion carrying the likenesses of Mauchly and John Presper Eckert in order to honor the two men who were responsible for designing and building the first large-scale electronic computer, ENIAC, and who were the founders of the modern computer industry.

As in the case of most inventions, many others made significant contributions. There were earlier experimenters with electronic computation, but their goals were much less ambitious and their accomplishments quite limited. It was Mauchly, in a memo written in 1942 at the Moore School of Electrical Engineering at the University of Pennsylvania in Philadelphia, who first proposed the construction of a large general-purpose digital electronic computing machine. Along with Eckert he was most responsible for its successful completion in 1946. (*See* Year in Review: SCIENTISTS OF THE YEAR: *Obituaries*.)

Computer networks and communications. Local and long-distance communications systems involving computers continued to gain in importance. The powerful minicomputers and the inexpensive microcomputers that had become available in recent years made it possible to distribute computing power throughout organizations in which computers were used. A few years ago the typical computer installation in business, government, or education would feature one or two large central computers and perhaps a few separate small computers for special projects. By 1981 the typical major installation was a network of interconnected communicating computers.

There are networks of all kinds, from limited local installations within a single building or office complex to large national and international interconnections. Several events that occurred in 1980 illustrated the wide scope of network activity. One was the announcement by Xerox Corp., Digital Equipment Corp., and Intel Corp. that they were jointly offering a communication system called Ethernet, a network designed to provide efficient and economical high-speed communication among computers and associated devices separated by less than 2.5 km (1.5 mi). Another was an announcement by Satellite Business Systems in the fall of 1980 that it had successfully launched its first satellite dedicated to a major new data communication service. Satellite Business Systems is a joint venture of International Business Machines (IBM) Corp., Aetna Life and Casualty Co., and Communications Satellite Corp. Initially at least, it planned to sell its services to business organizations with large-volume, long-distance data communication needs.

Much data communication is done on lines leased from telephone companies. The cost of such lines depends on the distance traversed, and the full cost is incurred even if the amount of data transmitted is small and the line sits idle most of the time. One way to make such transmission lines less expensive is to use satellites to provide the communication links. Another is to provide for effective sharing of the lines. A practical method introduced in recent years to permit such sharing is known as packet switching. It was developed and tested in the late 1960s and early 1970s in connection with Arpanet, a computer network.

In a packet switching system each sending computer transmits fixed-length packets of data. Each packet carries source and destination addresses plus checking and sequencing information. Packets from many different senders can be intermixed and transmitted over high-speed data links to the other computers in the network. Each receiving computer selects all of the packets addressed to it and reconstructs its incoming messages.

A number of public packet switching networks have been established that provide data communication to computer installations throughout the world. One example is Telenet, a company originally established by people from the Arpanet project and now a part of General Telephone and Electronics Corp. A typical customer pays a fixed monthly charge to rent a minicomputer that provides an interface with the network. Beyond that the customer pays an hourly charge for the time during which a communication channel is used plus another charge that depends on the number of packets transmitted. These last charges are independent of the distance between the sending and the receiving computers.

Even in the simple case in which a line exists between the sending and the receiving computers, it is necessary to establish agreed-upon protocols, *i.e.*, rules and conventions that will be observed to make sure that the data received are identical to the data sent. Physical protocols deal with the nature and the timing of the electrical signals on the line. In addition, there are conventions concerning packaging of the data, control characters and their meaning, and how to distinguish control information from data. It is usually assumed that errors may occur when data are transmitted over communication lines. Checking information is usually included in such transmissions, and there are protocols about the nature and position of checking information and about how retransmissions can be requested and completed.

Communications in and between networks of computers that may involve computers made by different manufacturers or that may have sensitive data moving across international boundaries can require the existence of a number of different levels of protocol. This is an area that has been studied by the International Standards Organization, which proposed a model of a computer communication system that involves seven distinct layers of protocol.

Major computer manufacturers were anxious to move rapidly toward the implementation of computer networks. Ideally, everyone should wait until communication standards are established, but in practice it has been necessary for companies that want to be leaders in the field to set up internal standards and protocols. A number of mutually incompatible network architectures already exists. As of 1981 it was not clear

Scanning electron micrograph (below left) reveals circuitry of an experimental system that uses Josephson tunnel junctions, the fastest known low-power switches. These devices, one of which is visible as a small circle in the center of the micrograph, consist of two superconducting electrodes separated, sandwich-style, by a layer of insulator so thin that currents of electron pairs can pass from one to the other up to some maximum supercurrent. Two scientists (below right) show a display on the system's oscilloscope of an ultrafast electrical signal. The system measures such signals with unprecedented resolution and sensitivity.

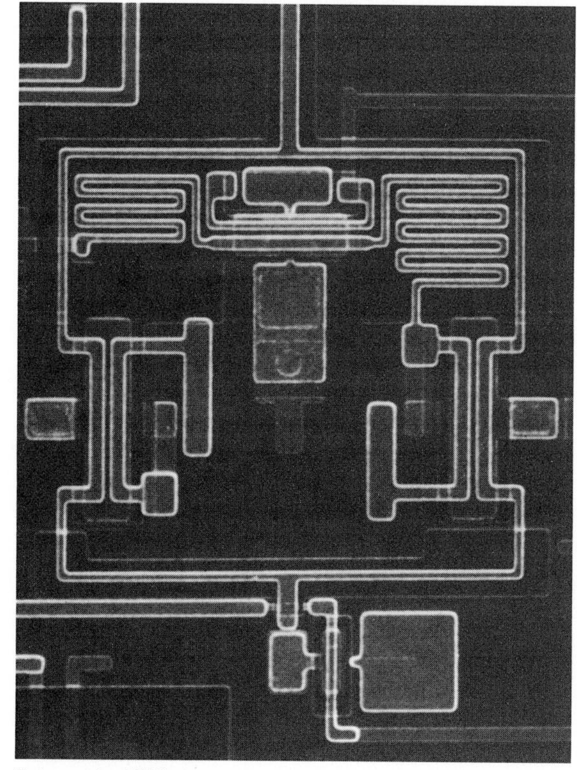

Photos, courtesy, IBM Corp.

Courtesy, IBM Corp.

Displayed on the computer terminal is a digital sound spectrogram of Alan Cole (right) saying "John saw one example of speech from several thousand runs." This represents one step of a process in which a computer transcribes ordinary speech, composed of sentences drawn from a 1,000-word vocabulary and read at a normal speaking pace, into printed form with an accuracy of 91%.

how the problems raised by such incompatibility will be resolved.

Packet switching is an area in which a great deal of progress has been made toward the setting and implementation of international standards. Most computer manufacturers planned to provide hardware and software interfaces to packet switching networks that are compatible with the protocols prescribed in a recommendation developed by the Consultative Committee on International Telegraph and Telephone (CCITT). This recommendation is widely known by its CCITT serial number, X.25, and is expected to become an official international standard. It is a technical document of more than a hundred pages that was published in 1976 and revised in 1980.

The Ethernet system mentioned earlier is a relatively simple implementation of a local computer network that uses a common coaxial cable as its transmission medium. Each sending node on the network is able to create packets that are transmitted to all receiving nodes. Each of the latter can select and read those packets addressed to it. A node that needs to send a packet waits until there is no traffic on the cable and then starts sending. A collision occurs when two or more nodes try to send data at the same time. Automatic collision detection circuits indicate to those nodes that the colliding transmissions are invalid. Each node involved in a collision then stops transmitting and tries again after an interval selected at random. Because the line speed is ten million bits per second the transmission time for each packet is usually short, and collisions will thus not be a serious problem if the volume of data is not very large.

Computer network and communication systems are not the exclusive province of users of large and sophisticated computer systems. Users of microcomputers can install inexpensive communications equipment that connects them to the public telephone system. For example, Edunet, an educational computer network organization, developed an interface that permits owners of the popular Apple microcomputer to gain access to public packet switching networks through any telephone. A home computer banking system featuring the Radio Shack TRS-80 microcomputer was announced by a bank in Tennessee during 1980. AT&T announced late in 1980 that it planned to test a computer-based system in Austin, Texas, in which customers use home telephone lines to access information that can be displayed on a television screen. Systems of this kind are already in use in some countries in Europe and will probably spread throughout the U.S. in the 1980s.

Very-large-scale integration. Most people are now aware of the small silicon chips that have revolutionized electronics and computers and have introduced computer technology into almost every area of human activity. Many, perhaps most, 1981 model automobiles use microcomputer controls based on silicon chips to improve engine performance. Chips that make it possible to simulate the human voice have been used to create devices that "talk" to their users. Chips are used in the many electronic toys now available and are also employed in home appliances.

The typical integrated-circuit chip contains a number of interconnected transistors on a rectangle of silicon that measures less than a quarter of an inch along each side. As the technology has developed, it has become possible to put more and more transistors on a single chip. By the late 1960s the phrase "large-scale integration" (LSI) had come into general use to characterize the chips with hundreds and even thousands of transistors that could then be produced.

Chips that consist of regular arrays of circuits, such as those found in random-access memories, are the easiest to design and fabricate, and so LSI found its initial applications in the production of memory chips. By 1980 64K-bit memory chips—chips that can store 65,536 bits of information—were used in many of the new computers.

The use of LSI in the implementation of the logic functions of large computers developed more slowly. This was partly due to the fact that the early LSI circuits were relatively slow, but another major reason was the difficulty and the high cost associated with the design and fabrication of special-purpose LSI chips.

The design of high-performance logic chips that are used in computers is a complicated, difficult, and time-consuming process. A great deal of expensive computer and computer graphics equipment is used in the design phase, and considerable amouts of time on the largest and fastest computers are required during the checkout. The design usually must undergo a number of tests because it is important to eliminate all errors before large-scale production is begun.

In 1979–81 major computer manufacturers announced new computer models that make extensive use of LSI. Typical chips used in these computers contain between 1,000 and 3,000 transistors. Because of the time required to develop new circuits for a whole line of computers, it can be expected that this technology will be in use through a large part of the 1980s.

Several of the new computers that were introduced in 1979–81 use a "gate array" technology that was first proposed about ten years earlier. The chip starts out as a regular array of logic circuits known as gates. Thus, all of the logic chips are identical up to a point. They are then differentiated in the final design and production steps, during which the interconnections among the gates are established. While these chips are not as efficient in the use of space as those that are fully customized, their much lower design cost makes it possible to use large-scale integration in computers in which the expense of using custom LSI design technology would be prohibitive.

Recent advances in the semiconductor industry have made it possible to put hundreds of thousands of transistors on a single chip. Before the end of the 1980s it seems likely that it will be possible to increase this number to millions. The development of this very-large-scale integration technology (VLSI) provides almost unlimited potential for creative computer design, but the complexity and the costs associated with such designs make it difficult to realize that potential. If one extrapolates design costs based on custom design of complex LSI chips, it seems that design costs may become astronomical. A reasonable $25–$50 per gate design cost could result in a $2.5 million–$5 million design cost for a projected 100,000-gate (300,000–400,000 transistor) VLSI chip. New methods and new ideas are needed to bring design costs down to a reasonable level.

Computer scientists continued to search for practical ways in which to use VLSI to do things that cannot be done with the earlier technologies. They also looked for ways to teach VLSI technology to computer scientists and computer engineers. Recent progress in both these areas was encouraging. New structured design methods and cheaper, more powerful minicomputers made it possible to do research in a university laboratory environment. Opening up the field of VLSI technology to large numbers of new people at universities throughout the world may make it possible to realize some of the potential of the technology more rapidly than seemed possible just a few years ago.

In the last two years courses in VLSI design have been introduced at a large number of universities. Most of them were derived from a course developed by Carver Mead at the California Institute of Technology. A number of interesting experimental VLSI chips designed by students in these courses were reported in technical periodicals during 1980.

—Saul Rosen

Courtesy, AT&T Co.

After requests for information are typed and relayed to a computer, the desired data are displayed on home TV screens, as in the Viewtron system.

Electronics

Underlying the astonishing progress made in electronics over the past decade is the integrated circuit (IC). There are essentially two kinds of IC's: the monolithic and the hybrid. In the monolithic variety literally thou-

Courtesy, IBM Corp.

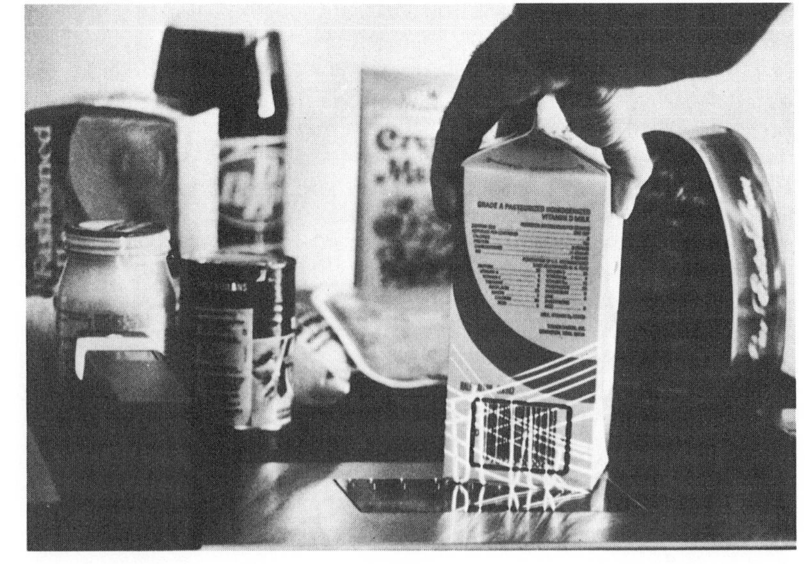

Scanner developed by IBM uses holography to read information on packages such as the milk carton shown at the right. When a package bearing the appropriate marking nears the scanning window, a laser light "wraps around" the item. The code label can be read by the scanner whether it is upside down, skewed, or right side up. Such a scanner can process a 22-character label in less than a second.

sands of transistors and resistors are embedded in a thin (about 100 micrometers thick, a micrometer equaling one-millionth of a meter) chip of silicon measuring about 0.2 × 0.2 cm. For very-large-scale integration (VLSI) tens of thousands of transistors are contained in a silicon chip measuring typically less than 0.6 × 0.6 cm. The microcomputer, often referred to as a computer on a chip, is an example of a VLSI circuit. Invading the areas of consumer, industrial, and military applications, the microcomputer has wrought a revolution in electronics.

The transistor is the heart and soul of an IC. Like its predecessor, the vacuum tube, a transistor can amplify weak signals, whether they be sound or the 1s and 0s a computer processes. A resistor is a device that impedes the flow of current. These two components constitute the basic elements of an amplifier, a microcomputer, and a host of other ICs.

The tooling-up costs for the development of a monolithic IC can be enormous, millions of dollars for a very-large-scale model. This huge outlay of money can only be justified if the IC has a potentially large market such as is the case for the microcomputer. For smaller production quantities or special applications, such as amplifiers operating at microwave frequencies, the hybrid IC is the answer.

In the hybrid, resistors and conductors are deposited on a ceramic chip, and transistors, monolithic ICs, and other components are externally connected to the circuit. There are two methods that are used to deposit the resistor and conductor patterns on a chip. In thick-film technology the patterns are printed on the chip, in a manner similar to silk-screen printing. In thin-film technology the patterns are usually evaporated onto the chip.

Manufacturers of ICs continued to try to scale down the sizes of transistors and resistors in a silicon chip. This reduction in size has two salutary effects: an increase in packing density (more devices per chip) and an increase in operating speed.

Microcomputer advances. In 1980 Intel Corp. announced the development of a 32-bit three-chip microprocessor that is able to execute two million instructions per second. (A bit in a computer system is the smallest increment of usable data, expressed in machine language as a 0 or a 1; a 32-bit microprocessor can handle that many bits simultaneously.) The computer power of the Intel unit matches the central processing unit (CPU) of the IBM 370/158 mainframe computer. Previous generations of microprocessors could handle 8- or 16-bit words only. A 32-bit capability provides greater precision and faster operation than 16- or 8-bit microprocessors.

Multiplication and division, when programmed using software, takes a much longer time to execute on a computer than addition or subtraction. Using specially designed IC's, however, Rockwell International announced an 8-bit by 8-bit multiplier that is capable of yielding a product in slightly more than five nanoseconds (a nanosecond is one-billionth of a second). Intel developed a special processor that can multiply, divide, and extract square roots. Containing 65,000 transistors, the chip can handle 16-, 32-, or 64-bit data. Multiplication can be done in 16 microseconds, and division and the extraction of a square root require 35 microseconds.

An essential element in a microcomputer is its memory. There are two IC types used: the read-write, or random-access memory (RAM) and the read-only memory (ROM). The data in a RAM can be changed during the execution of a program. Data in a ROM, however, cannot be changed. A ROM may contain such information as a program or look-up tables which, for a given application, are fixed.

Prior to 1980 the maximum storage capacity of RAM's and ROM's were 64 kilobits (kb), a kilobit equaling 1,024 bits. In 1980 experimental 256 kb RAM's and 512 kb ROM's were introduced by the Nippon Electric Co., Ltd., of Japan. These high-density IC memories were expected to become commercially available in a few years.

In 1980 Radio Shack introduced a pocket-size computer that can be programmed in the software language BASIC. Weighing 170 g (6 oz), the BASIC instructions can be loaded into a cassette or stored in the computer's memory. Earlier in the year Matsushita Electric Corp. of America introduced a hand-held computer that can be connected to a host of peripheral devices. Its many features included a language translator and an electronic memo pad capable of storing up to 500 characters.

To protect programmers from having their programs appropriated by others, the U.S. Congress passed the Computer Software Copyright Act in November 1980. The protection afforded by this law should result in the availability of a greater variety of software packages.

Amorphous silicon solar cells. Silicon used in the manufacture of transistors, IC's, and other solid-state devices is in pure crystalline form. Crystalline silicon, which is obtained from noncrystalline, or amorphous, silicon, is characterized by a well-defined, repeated arrangement of atoms. To achieve this a number of costly processing steps are required to transform amorphous to crystalline silicon.

In the past few years several firms have begun to make solar cells from amorphous silicon. This material yields cells that are less costly than solar cells made from crystalline silicon. Cells fabricated by RCA, a pioneer in the field, have yielded a conversion efficiency of 6.3%. (Conversion efficiency is the percentage of incident sunlight, when directly overhead the cell, that is converted into electrical energy.) Although amorphous silicon cells are less efficient than crystalline silicon cells, some experts believe that when a conversion efficiency of 10% is reached the amorphous cell will become economically viable. (*See* Feature Article: A NEW WORLD OF GLASSY SEMICONDUCTORS.)

Speech synthesizer mimics the sounds heard by the human ear. A speech processor chip (SPC) connected to a ROM digitizes and reproduces speech wave patterns.

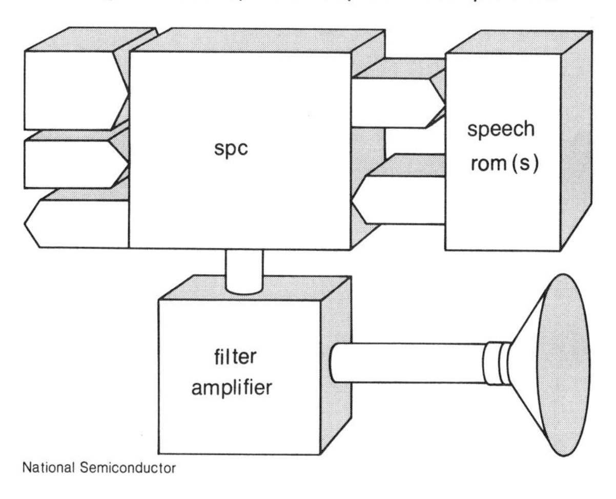

National Semiconductor

Talking machines. "You are running out of gas" and "The engine needs a tune-up" are examples of commands having a humanlike quality that may be emanating from the dashboard of one's car in the future. Or the smoke detector in an office will announce "fire" and then direct people to the nearest exit. These and many other applications stem from the availability of IC's that can synthesize speech.

Three basic methods are used in speech synthesis. In one method, phoneme coding, speech sounds and the rules for their combination are stored in a computer memory. With suitable programming a microprocessor blends the sounds to form speech that is intelligible, although it has a robotlike quality. The second method, developed by Texas Instruments Inc. for its Speak and Spell series of toys, stores basic sound and computer instructions in a memory that imitates the way sounds are produced.

In 1980 National Semiconductor Corp. introduced a method that mimics what the human ear hears rather than what the vocal chords produce. Speech elements that are heard are stored in a ROM. Connected to the newly developed speech processor chip, the speech wave patterns contained in the ROM are digitized (converted to 1s and 0s) and then compressed. National Semiconductor claims that any original voice in any language can be reproduced in a clear and lifelike manner.

Electronic home banking. On a limited basis electronic home banking was introduced in October 1980 in Knoxville, Tenn. The backers of the experiment included the United American Service Corp., Radio Shack, and CompuServe Inc., a subsidiary of H & R Block, Inc. The participating bank was the United American Bank in Knoxville.

The estimated cost for the service was between \$15 and \$25 per month. Each of the bank's customers was equipped with Radio Shack's TRS-80 Color Computer, which could be connected to a television set and a telephone. With this equipment customers could pay bills, apply for loans, and be kept current on the balance in their checking account. In addition, the user had available a comprehensive news service.

Entertainment electronics. The integrated circuit continued to increase its impact on the home television set. RCA developed an IC version of a comb filter that separates the intensity and color components of a TV signal. The result was improved picture quality and sharpness. Zenith Radio Corp. introduced a Space Phone that permits a viewer to respond to telephone calls through the television set. When the phone rings,

University of Utah

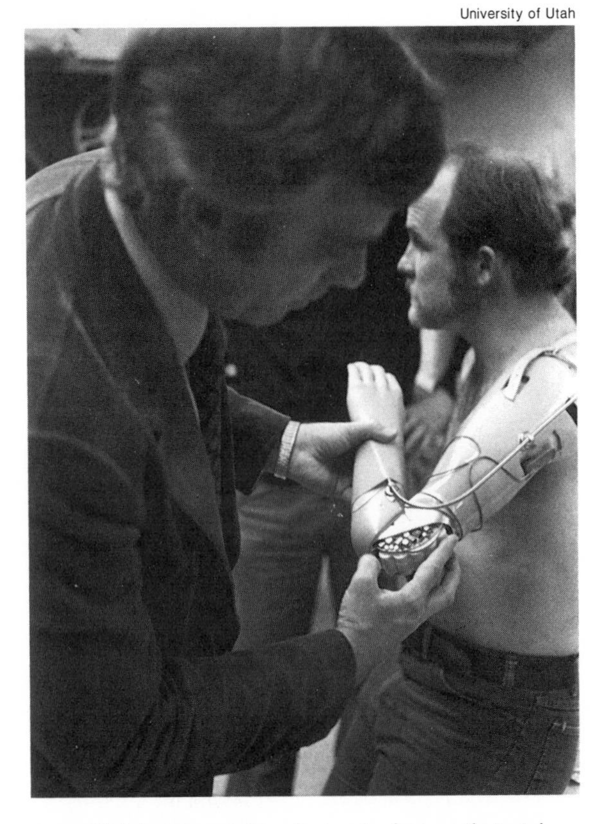

An artificial arm contains electronic devices that pick up electrical signals from the brain to muscles in the amputee's limb remnant and then translate these signals into appropriate arm movements.

a button is pressed on the remote control unit. This causes the sound of the television show to fade, permitting the viewer to have a telephone conversation through the set while viewing the picture only.

Activity in video cassette recorders and videodiscs increased in 1980 with the availability on the market of many units from RCA, Sony Corp., North American Philips Corp., and others. One serious problem with this segment of the industry was the lack of a universal standard. For example, Philips used a laser-based system while RCA offered a groove-type videodisc. The same chaotic situation existed in regard to video cassette recorders.

A standard for AM stereo was finally approved by the U.S. Federal Communications Commission in April 1980. The system developed by Magnavox Co. won over those submitted by four other competing firms. Owing to additional circuitry, AM stereo receivers will be more expensive than the conventional monaural set. The conventional set, however, will be able to reproduce a stereo signal monaurally.

Using chips that synthesize speech, a number of interesting toys and games made their bow in 1980. A doll called Baby Soft Sounds that has a vocabulary of 16 words was introduced by Fisher-Price Toys division of Quaker Oats Co. An electronic bridge game, the Voice Bridge Challenger, was offered by Fidelity Electronics Ltd. of Miami, Fla. In this game the electronic unit announces bids in bridge jargon.

A firm in Los Angeles, Bambino Inc., marketed some interesting electronic sports games in 1980. In their Super Football Game, for example, a player's movements are shown on a multicolored display. In addition, the microcomputer contained in the unit displays the score, yards required for a first down, and time remaining in a quarter.

Medical electronics. Programmable heart pacemakers were introduced by Medtronic, Inc., of Minneapolis, Minn., in 1980. Containing an IC chip that allows a physician to change the pacemaker's functions in order to meet a patient's changing medical condition, the unit, called the Spectrax, is 10 mm (0.4 in) thick and weighs 45 g (1.5 oz).

Owing to the microprocessor, considerable progress was made during 1980 in the improvement of prosthetic devices. At the University of Utah, for example, a bionic arm was developed that responds to muscle commands. The commands are sensed by existing skin on the shoulder and arm regions.

At the Massachusetts Institute of Technology a programmable artificial leg was developed. Replacing the hinges found in a conventional mechanical knee, a motor-brake arrangement is used that provides sufficient power to allow its user to climb a flight of stairs. In addition, by the placement of a sensor on the good leg of a one-legged person information on its motion can be transmitted to the microprocessor contained in the artificial leg. Properly programmed, the microprocessor can help to achieve a natural gait for both legs.

—Arthur H. Seidman

Information systems and services

The first White House Conference on Library and Information Services in 1980 presented its final report to U.S. Pres. Jimmy Carter. The conference was convened at a time when information was expanding rapidly, when people's need for information was increasing, and when science and technology were testing new and better ways for providing information—a time when the United States and all of the industrialized world were entering the Information Age.

The conference delegates considered the role of the library and other information services in providing improved access to information both at home and abroad. They approved a total of 64 resolutions. Five of the most significant called for: (1) the elimination of all barriers to free and full access to information and the provision of adequate library services to special user groups such as children, the aged, racial and ethnic minorities, and the physically and emotionally handicapped; (2) the encouragement of cooperative efforts among public educational agencies, appropriate private nonprofit organizations, and libraries to plan and implement literacy programs for adults and out-of-school youths; (3) the strengthening of personnel development and training programs for librarians and information specialists, and the improvement of library and information services through research and the application of new technologies; (4) the development of a national policy requiring government agencies at all levels to work together and provide, to the maximum extent possible, the new and existing library and information services while protecting the privacy of all segments of society; and (5) the establishment of an office of library and information services within the U.S. Department of Education, directed by an assistant secretary responsible for administering all contracts and grants programs related to library and information services. All of the conference results and resolutions, together with the president's recommendations, were to be sent to Congress for action.

U.S. information systems. In response to a growing concern of the long-term costs of presidential libraries, it was suggested by the head of the General Services Administration that instead of having individual libraries a central library complex be built to accommodate the papers of the next six U.S. presidents. The proposed structure would consist of a cluster of buildings, including a common visitors' center; six presidential units, each housing the historical papers and memorabilia pertaining to an individual president; and common storage, laboratory, and maintenance facilities. This plan would assure that presidential records and related valuable materials would be effectively preserved for future generations while curbing excessive governmental costs.

The Lister Hill National Center for Biomedical Communications has, as one of its aims, the rapid transfer of new medical research information to health practitioners. To further this goal, knowledge bases on viral hepatitis, peptic ulcers, and human genetics were being developed. (A "knowledge base" is an integrated medical information system in which current biomedical literature in specialized areas is identified, selected, reviewed, condensed, and synthesized by experts, and then stored in a computer retrieval system for use by physicians and other health professionals.) As of 1981 the hepatitis knowledge base had been in use for several years, while the ulcer and human genetics knowledge bases were still under development. The human genetics knowledge base was using new optical videodisc technology for storing both visual and textual materials and was thus providing a new dimension to information storage and retrieval system design.

The National Center on Child Abuse and Neglect is a unit in the U.S. Department of Health and Human Services authorized to engage in activities designed to identify, treat, and prevent child abuse and neglect. In

Courtesy, RCA

Disk coated with a thin layer of tellurium can hold 100 billion bits of information, the entire contents of a multivolume encyclopedia. This is 10–20 times the usual capacity of current magnetic storage disks.

the center's clearinghouse was a computerized data base containing: (1) bibliographic data on journal articles, books, reports, and other publications; (2) descriptions of ongoing research projects; (3) descriptions of service programs for abused children and their parents; (4) descriptions of films, videotapes, audio cassettes, and other items dealing with child abuse and neglect; and (5) excerpts from current state laws, including relevant welfare, criminal, and juvenile court codes. The products and services of the clearinghouse, including on-line information searches and various publications and directories, were available at no charge to the user and could be obtained by written or telephone request.

Providing information on the availability, cost, and use of approximately 10,000 aids and devices for handicapped people was a data base and search service named ABLEDATA. It was being developed by the National Rehabilitation Information Center together with the California Department of Rehabilitation and the Rehabilitation Engineering Demonstration Units.

To assist corrections officers in obtaining quick individualized responses to requests for information, the National Institute of Corrections (NIC) implemented a program to coordinate and expand existing criminal justice information systems. It was designed to provide

Courtesy, Bell Laboratories

Graphic display reveals computation involved in computer speech research for such applications as an airline reservation system.

general reference services on criminal justice and correctional activities, information on volunteer programs in corrections, information on correctional staff development and training, and special clearinghouse services to disseminate information to attorneys representing correctional agencies and to sheriffs and jail administrators. These information services were available, through NIC, cost-free to corrections officers working in community-based programs or in jails and prisons.

A comprehensive collection of writings on the history of women was compiled and recorded on microfilm by Research Publications Inc. in Woodbridge, Conn. The collection, obtained from nine separate archival sources, consisted of approximately 10,000 books, pamphlets, and periodicals, 800 photographs, and 80,000 manuscript pages, for a total of approximately three million pages. It included material on such themes as women's role in the U.S. crusade for social justice, the social and intellectual history of women since 1975, and the role of women in the westward movement. Each microfilm image was assigned a sequential reference number for easy identification. A subject index was available, and search was facilitated by means of a rapid-scan target on the film.

The Neighborhood Information Sharing Exchange, a component of the Department of Housing and Urban Development, supports community groups in their efforts to revitalize their neighborhoods and enables local leaders, city officials, business and philanthropic organizations, and concerned citizens to share solutions to community problems. Priority items selected for initial emphasis included housing rehabilitation, energy conservation, community revitalization, and neighborhood planning.

The U.S. Department of Energy's Office of Oil Imports provides information on all shipments of oil and petroleum products into the United States. This information is published monthly by the American Petroleum Institute. It is also available as a computer-searchable data base called IMPORTS, which stores information on the date of each shipment, commodity type, importing company, quantity, destination, fee, etc. Special retrieval languages facilitate searching and analysis.

A Nuclear Waste Material Characterization Center was established during the year by the Department of Energy. Located at the Battelle Pacific Northwest Laboratories in Richland, Wash., it was established to help scientists and laboratories obtain and compare information about nuclear waste materials.

International information systems. UNESCO launched a preliminary study for planning an international information system relating to new and renewable energy resources, such as solar, geothermal, wind, and ocean energies. Included in the plan were studies of information needs and sources, obstacles to the flow of information, and suggestions for overcoming those obstacles. The goal was to arrive at a mechanism for improving information flow in this field, perhaps by creating an operating information system.

The UNESCO General Information Programme and the International Federation for Documentation during the year sponsored a clearinghouse for information education and training materials at the University of Maryland. Their aim was to facilitate and improve the training of information specialists. The clearinghouse collected, organized, announced, and distributed copies of teaching and training materials, including reading lists, course outlines, bibliographies, and notes developed by faculty members.

The Museum of the Jewish Diaspora, located at Tel

Aviv University in Israel maintains, on permanent exhibit, a data base on Diaspora Jewry dating back to the Middle Ages. (The Diaspora was the settling of scattered Jewish communities outside Palestine.) The three main files contain data on communities, personalities, and historical events. The museum has its own computer and terminals for the public to use. A visitor who wishes to obtain information about a particular Jewish community or person types the name in either Hebrew or English at one of the typewriter-display terminals. Because this is an interactive retrieval system, if the name cannot be found in the data base, possibly because it was spelled incorrectly, the screen will display a list of similar names from which the visitor may choose or may try again. Displayed information can be read on the screen or taken away as a printed copy. In addition to using the console to retrieve information, the visitor can play numerous educational games that test knowledge of Judaism and Jewish history.

FRANCIS (French Retrieval Automated Network for Current Information in Social and Human Sciences) is an automated retrieval system covering philosophy, sociology, anthropology, archaeology, art, education, religion, and many other subjects. The magnetic tape files contain more than 500,000 bibliographic references and grow by approximately 70,000 references per year. Though the Centre de Documentation Sciences Humaines is located in Paris, the files can be searched on-line in many countries in Europe and also in Canada.

Information science research. Purdue University (West Lafayette, Ind.) received a grant to develop error-correcting procedures for use in automated language-processing systems. Errors, often called noise or distortions, occur in all systems of communication and information retrieval. The proposed research utilizes formal or abstract languages for expressing syntax or grammatical structure in order to determine whether a given sequence of words is a possible sentence, and, if so, what the structure of that sentence is. This approach exploits an analogy between the structure of languages and the domain of pattern recognition to identify natural language patterns.

Researchers at the University of Illinois were developing mathematical models of human behavior in information generation and retrieval. The specific tasks being studied included techniques of text editing, searching, and browsing. The long-range goal of this research was to develop analytical procedures for designing self-modifying computer systems that can adapt to the needs of particular users of the information system. The relationship between information and market performance was being explored at the U.S. National Bureau of Economic Research. The study focused on the effect of imperfect information on the competitive marketplace.

Syracuse University was studying ways of improving the performance of information storage and retrieval systems by investigating the effect of different information representations, such as controlled or free index vocabularies, classification codes, or word stems derived from the title and/or abstract of a document. Preliminary results indicated that although the user may expect to retrieve approximately the same number of relevant articles with different representations, the actual articles retrieved may not be the same. By isolating and controlling specific components of an information retrieval system, the investigator should be able to determine the impact of different information representations on the retrieval characteristics of the system and suggest ways of improving the design and structure of such systems.

—Harold Borko

Satellite systems

Applications satellites are Earth-orbiting satellites that utilize their vantage points in space for economic benefit and military purposes rather than scientific research. There are three basic classes of applications satellite systems: communications, Earth observation, and navigation. Users are individual and groups of nations, and private industrial concerns.

Satellite Business Systems (SBS) satellite undergoes testing at Hughes Aircraft Co. Launched in November, the SBS is the first high-speed digital communications satellite, relaying data at 480 million bits per second.

Courtesy, Hughes Aircraft Co.

Courtesy, TRW Inc.

The U.S. Navy FleetSatCom spacecraft is placed in a space environment for preflight testing. The satellite was launched in October to help provide global tactical communications.

The U.S. and the Soviet Union in 1980–81 continued to dominate such activities because of their large booster rockets, which they also used to launch satellites for other nations. Yet France, Japan, and China, as well as the European Space Agency (ESA), continued to develop their own space launch capabilities. Eleven ESA member nations planned future use of the French-developed booster (Ariane) in competition with the U.S. McDonnell Douglas Delta launch vehicle or the piloted space shuttle.

The space shuttle, after many technical delays, made its first, eminently successful manned flight in April 1981. The Ariane second developmental flight in May 1980 suffered engine failure, delaying schedule of the third flight until at least June 1981.

Communications satellites. Of all types of applications satellites the one that has exhibited the greatest activity in growth and economic value is the communications satellite. The rapid growth of national and international communications during the 1970s resulted primarily from the use of satellites. The International Telecommunications Satellite organization (Intelsat), a consortium of 105 nations, of which the Communications Satellite Corp. (Comsat) is the U.S. member, continued to grow in size and capability. During the year Honduras, Niger, and Guinea became Intelsat members. Global transmissions of telephone, television, facsimile, and digital data are provided by spacecraft in geostationary orbit. (A geostationary, or geosynchronous, orbit is at an altitude of 35,900 km [22,300 mi] above the Equator. At that height a satellite travels at the same angular velocity as the surface of the rotating Earth and thus remains at a constant point above the Earth. Three such satellites can provide global coverage except at the highest latitudes.) At the beginning of 1981 Intelsat had five satellites in operation stationed over the Atlantic, Pacific, and Indian oceans plus one or more standbys at each location.

During the 1980 Winter Olympics Intelsat transmitted more than 400 hours of television coverage by satellite. In the summer some 1,000 hours of the Moscow Olympics were relayed similarly around the world. The first of the improved Intelsat 5 satellites was launched December 6. The largest communications satellite ever, weighing 1,950 kg (4,300 lb), its capacity is 12,000 simultaneous telephone calls plus two television channels. This is double the number of telephone circuits in the previous Intelsat 4A satellites. Four more Intelsat 5 satellites were scheduled to be launched by the U.S. National Aeronautics and Space Administration (NASA) during 1981. Testifying to the projected growth of the market, three Intelsat 5A's with increased capability were ordered. Lloyd's of London agreed to insure the Intelsat 5 satellites for $65 million "against failure during launch and for 180 days thereafter" and for $500 million "against injury or damage to third parties."

In November NASA launched the first high-speed digital communications satellite for Satellite Business Systems (SBS). Owned by Comsat, IBM, and Aetna Life and Casualty, SBS will relay data point-to-point in the U.S. at the rate of 480 million bits of data per second. By comparison, a typical landline telephone circuit can move only 9,600 bits per second. Electronic equipment can receive the satellite signals and print high-quality copy at the rate of 70 pages per minute.

In December Comsat filed an application with the Federal Communications Commission for a new subsidiary, Satellite Television Corp. (STC). Permission was requested to broadcast three channels of pay television without commercials directly to the houses of individual subscribers. STC would be competitive with existing cable TV systems, particularly in rural and sparsely populated areas. Homes of subscribers would have a small rooftop antenna, and, if approved, STC would usher in the world's first satellite direct broadcast system.

American Telephone and Telegraph Co. (AT&T) ordered three new communications satellites from Hughes Aircraft Co.; the craft are designated Telstar 3. One of the design features is an increased life ex-

(Left) Courtesy, Hughes Aircraft Co.; (right) courtesy, NASA

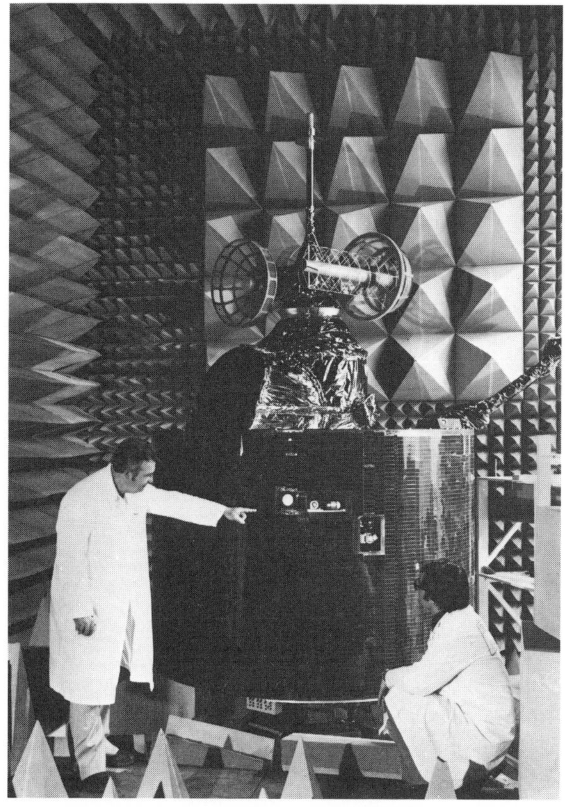

GOES meteorological satellite (left) is inspected prior to launch. Among the photographs taken by the satellite from its orbit at 35,600 kilometers (22,300 miles) was one (right) showing Hurricane Allen in the Gulf of Mexico and Hurricane Isis off the U.S. Pacific coast.

pectancy of ten years instead of the usual seven for most satellites.

Comsat completed installation of a 13-m (29-ft) receiver station at Susupe in the northern Mariana Islands. Previous telephone communications to those isolated islands had been by high-frequency radio circuits, often subject to delays and interference due to atmospheric disturbances. Satellite communications not only provided improved telephone service but also made available telex data, facsimile, and television to the islands.

In Indonesia the utilization of the two Palapa communications satellites to link the nation's many hundreds of islands continued. By 1981 Indonesia's national telecommunications company, Perumtel, had more than 115 ground terminals. Twenty more were ordered for installation during 1981.

In 1980 Intelpost, International Electronic Post via satellite, began operations. Initial facsimile services were between New York City, Washington, D.C., Toronto, and London. Services to Bern, Switz., and Amsterdam were scheduled for 1981. The U.S. Postal Service charged $5 a page, which was transmitted in less than ten seconds. Operations were on a 24-hour basis.

On the military side NASA launched the fourth U.S. Navy FleetSatCom into geosynchronous orbit, completing the system to provide global tactical communications. These satellites, weighing 1,880 kg (4,140 lb), have 23 channels in the ultra-high and super-high frequency bands. In addition to the U.S. Navy, service was supplied to the U.S. Air Force and other Defense Department users. Instant communications were thereby constantly available to surface ships, aircraft, and small ground-mobile forces. A fifth FleetSatCom was to be launched in 1981 as an in-orbit spare.

Earth observation satellites. This category of applications satellites consists of three major types: meteorological (weather), Earth resources, and military reconnaissance.

Weather satellites. On May 29 an improved TIROS-N weather satellite was launched by NASA for the National Oceanic and Atmospheric Administration (NOAA), but it did not achieve proper orbit because the Atlas booster rocket failed. Nevertheless, the existing TIROS-N and NOAA 6 satellites, in polar orbit at 870 km (540 mi) altitude, view every part of the Earth twice during each 24-hour period. Meteorologists, oceanographers, and hydrologists are thus provided continuous data on solar energy, the atmosphere, surface sea ice, and water and weather conditions. Of particular value to weathermen were satellite sensors that provide 24-hour imagery of cloud distribution and movement, ice and snow distribution, and sea-surface temperatures. In addition, this series of satellites obtains temperature profiles of the atmosphere along with its water vapor and total ozone content. In 1981 more than 120 nations received such information directly as the satellites passed overhead. An additional estimated 1,000 amateur and educational groups received such imagery using small receivers.

Two Geostationary Operational Environmental Satellites (GOES) in Equatorial orbit provided continuous views of U.S. weather. U.S. GOES satellites transmitted imagery each half hour around the clock. These pictures were seen daily on television news and weather reports. The ESA Meteosat, in geosynchronous orbit so as to cover European weather, operated well for two years before the imagery circuit failed in late 1979. Meteosat 2 was scheduled for launch in 1981. The Japanese GMS satellite similarly obtained weather data in the Far East.

Apart from short- and long-range weather forecasts one kind of information of economic value is that concerning sea temperatures, derived from GOES infrared sensors. Commercial fishermen are informed of likely locations of schools of fish. In the eastern Gulf of Mexico the Gulf Loop Current, moving at a rate of three-and-a-half knots, is tracked by measuring seawater temperature differences. Commercial shipping may obtain an updated map of this current and thus sail with the current or avoid it completely, resulting in shorter travel time and fuel savings.

During the year the Nimbus experimental satellite was used to track the migration of polar bears that had been fitted with a small transmitter beacon. This tracking technique was also used on a 96-kg (212-lb) loggerhead turtle, an endangered species. The signal, beamed every four days, was relayed by Nimbus to NASA's Goddard Space Flight Center, where the path of the turtle was plotted from Oct. 16, 1979, until June 15, 1980.

Earth resources satellites. The two U.S. polar orbiting Earth observation satellites, Landsats 2 and 3, continued to provide useful data despite some operational problems. Monitoring and mapping the Earth's surface with multispectral imagery, the Landsats provided information that in 1980 was being used by 37 states.

In 1980 a multi-agency program, AgRISTARS, was initiated to determine the usefulness, cost, and extent to which aerospace remote sensing data can be useful in crop forecasting and management systems. NASA also began a three-year program with the Pennsylvania Department of Environmental Resources. Using Landsat data, the project planned to study the damage to Pennsylvania forests by the gypsy moth caterpillar. The spectral images of known defoliated forests throughout the state were to be mapped to detect the damage caused by this insect. The results of this program were to be used to inform other states in the eastern U.S. of techniques that can monitor the southwestward spread of gypsy moth blight.

In addition to three Landsat ground stations in the U.S., nine foreign stations now receive data from these satellites. During 1980 stations became operational in Japan, India, Australia, Argentina, and South Africa, joining stations in Canada, Brazil, Italy, and Sweden. Agreements were signed with China and South Africa in 1980 calling for establishment of facilities in those countries for direct access to Landsat data.

During 1980 U.S., Canadian, and French agencies participated in a project to evaluate a Satellite-Aided Search and Rescue System (SARSAT). They formally confirmed an understanding with the Soviet Union regarding cooperation between SARSAT and a similar compatible Soviet system. Norway asked to join the SARSAT experiment, and Brazil, Italy, Japan, and Sweden were considering joining the project. Bulgaria, East Germany, and Poland were considering whether or not to join the Soviet system. The presently approved arrangements, which utilized both Soviet and U.S. satellites, were scheduled to begin undergoing tests in 1982.

Military reconnaissance satellites. The United States and the Soviet Union continued launching reconnaissance satellites to observe and remotely sense the other nation's military movements, electronic transmissions, nuclear explosions, and ballistic missile launchings. During the conflict between Iraq and Iran in September the U.S.S.R. launched film-return military reconnaissance spacecraft. Such spacecraft usually remain in orbit for two weeks. However, during the Arab-Israeli war in 1973 similar spacecraft were directed to return their photographs to the Earth after only six days. It is recognized by both powers that such capabilities are valuable for strategic defense.

From 1960 to 1976 the U.S. was reported to have utilized film-return spacecraft for reconnaissance missions. For the past several years the U.S. Air Force has developed long-duration, digital-image spacecraft that remain aloft for two or more years. In a manner similar to Landsat processing, images of surface targets are transmitted to U.S. receiving stations on each pass.

In April the U.S.S.R. apparently launched the sixth in a series of antisatellite weapons tests. In such tests a target space vehicle is launched, followed several days later by another to rendezvous with the first. Such a rendezvous could be for purposes of satellite inspection or destruction. The U.S. was reported to be conducting research on laser-beam weapons that might be the basis for an antiballistic missile defense system.

Navigation satellites. The fifth in the series of developmental Navstar global positioning satellites was launched in February 1980. Key elements of these satellites are cesium and rubidium atomic clocks. The U.S. Navy Transit navigation satellite system continued to be operational for Naval units as well as for commercial vessels equipped with Doppler radar and related computer systems. Two U.S. Navy Transit satellites, which were termed Nova, were scheduled for launch in 1981.

—F. C. Durant III

Energy

U.S. Pres. Jimmy Carter achieved the enactment of a comprehensive energy program with the passage of the Energy Security Act, signed into law June 30, 1980. However, during the first several weeks of the administration of Pres. Ronald Reagan several actions seemed to point toward a fundamental change in direction from that incorporated in the 1980 legislation. On the international front the Organization of Petroleum Exporting Countries (OPEC) continued to be unable to agree on a pricing policy; also, the Iran-Iraq war interrupted petroleum supplies from those two countries, but there continued to be a glut of oil on the world markets. World consumption of petroleum did not rise, as recession and conservation efforts continued to play roles in petroleum demand. The year was marked by rapid price increases within the United States.

Energy Security Act. The Energy Security Act of 1980 contains eight titles, each addressing an area of energy concern. Title I deals with synthetic fuel; title II with biomass energy and alcohol fuels; title III with the preparation of energy targets; title IV with incentives for development of renewable energy; title V with solar energy and energy conservation; title VI with geothermal energy; title VII with acid precipitation and carbon dioxide studies; and title VIII with the strategic petroleum reserve. The thrusts of the titles dealing with specific energy sources are to establish mechanisms and incentives for developing these energy sources.

By far the most important title of the Energy Security Act is that dealing with synthetic fuels from coal, oil shale, and tar sands. The legislation established a specific goal of achieving commercial synthetic production of 500,000 bbl of crude oil per day by 1987 and 2,000,000 bbl of crude oil per day by 1992. Second, the Defense Production Act amendments of 1980 extended coverage to synthetic fuel projects under the terms of that act for the increase of capacity for synthetic fuels. This amendment places the U.S. Department of Energy in a position of responsibility for the administration of those sections of the Defense Production Act of 1950 and requires an annual report upon their actions by the president.

The third and most important section of the Act is the establishment of the United States Synthetic Fuels Corporation. With an initial budget authority of $20 billion, the corporation was established to achieve the national synthetic fuel production goal mentioned above. It was to be managed by a seven-member board of directors serving seven-year terms, of which no more than four could be members of the same political party. The chairman of the corporation, who would also be a member of the board, would be the chief executive officer.

To achieve the production goals the corporation has the authority to provide financial incentives for the development of synthetic fuels plants. It can commit no more than 15% of the total obligation authority to a single project or person, the objective of this provision being to assure that a diversity of technologies will

Computed tomography X-ray scanner (below left), borrowed from the medical profession, provides scientists with their first look inside a piece of coal that is being heated and gasified. The X-ray images (below right) show how the coal expands and melts (upper right to lower left) as it is heated.

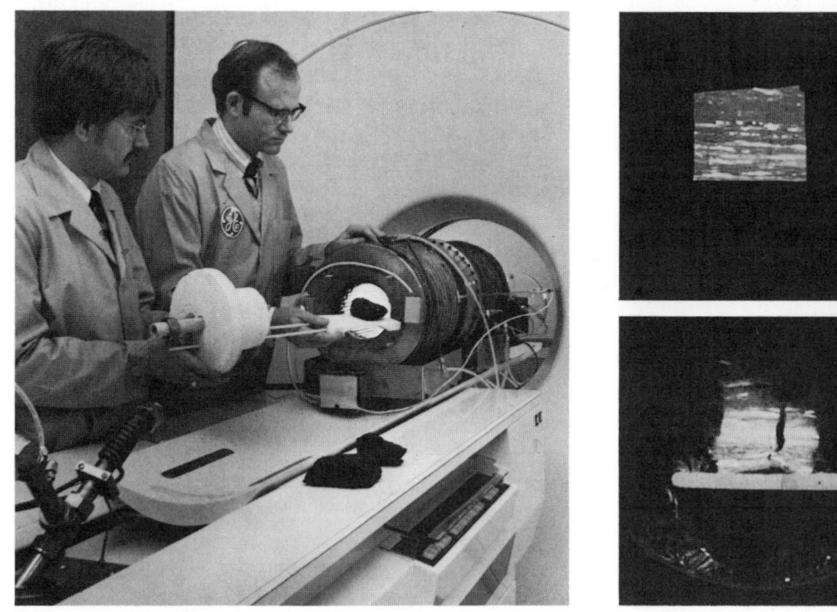

Photos, courtesy, General Electric Research and Development Center

be funded. Second, the corporation must also give due consideration to the promotion of a competitive industry and to competition in general. Third, the corporation is required to use competitive bids and to consider the cost in terms of dollars per barrel of oil equivalent as a criterion for offering incentives. Finally, the legislation does not permit the corporation to leverage its authorized funds; that is, the full dollar amount of the incentive offered must be appropriated in advance and held in the corporation as a contingency. Within these constraints the corporation can offer two kinds of incentives: first, market or price mechanisms and second, mechanisms to reduce the cost of capital and to increase access to capital.

The market and price mechanisms include price guarantees and purchase agreements. Such actions would encourage a market for synfuels under prices that would help cover the cost of production and provide incentives for firms to enter into production. They would have the effect of reducing uncertainty as to market price and as to future prices of competing fuels. The incentives for reducing capital costs and increasing access to capital involve the use of long-term guarantees, direct loans, and joint ventures. The corporation cannot guarantee a loan above 75% of the initial total estimated project cost, but additional guarantees may be made if project costs exceed estimated costs. With respect to loans the corporation is authorized to lend up to 75% of the initial capital costs to a synfuels project.

Finally, the corporation has the authority to enter into joint ventures with producers under certain circumstances. Given the circumstances, the corporation can provide up to 60% of the funding for a module of a synfuels plant, that is, for a portion of the plant that needs to be tested to lay the foundation for a full-scale project at the same site. In a joint venture the corporation, while providing part of the funding, is permitted to construct and operate the module so long as the amount of funding does not exceed 60% of the total costs of the module.

President Carter nominated John Sawhill to be the chairman of the Synfuels Corporation, and he also nominated persons for the other six board positions. The Congress did not act on these nominations before it adjourned at the end of October for the 1980 elections. The president then gave Sawhill and four other directors interim appointments so that the corporation could be organized. After he took office in January 1981, President Reagan withdrew the nominations of Sawhill and the other board members. Reagan indicated in his budget message to the Congress that he proposed to reduce sharply the funding commitments made by the Department of Energy to synthetic fuels projects and to transfer those projects from the Department of Energy to the Synfuels Corporation. Thus, the future shape and direction of the synthetic fuels program under the Energy Security Act was being reexamined by the government. (*See* Feature Article: THE PROMISE OF SYNTHETIC FUELS.)

Title II of the Energy Security Act establishes a biomass energy and alcohol fuels program. This title calls for incentives for the development of biomass energy through insured loans, loan guarantees, price guarantees, purchase agreements, and, in the case of municipal waste energy projects, construction loans. It calls for the establishment within the Department of Energy of an office of alcohol fuels, an Office of Energy from Municipal Waste, and an extension of the authorities of the Forest Service with respect to the utilization of national forest wood for energy development projects.

Title III of the act calls for the establishment of energy targets with provisions for congressional consideration of these targets. Title IV deals with renewable energy incentives. It provides for coordinated dissemination of information about renewable energy resources and conservation, the establishment of life-cycle energy costing for federal buildings, and incentives dealing with such items as energy self-sufficiency, photovoltaic technology, and small-scale hydropower.

Title V establishes a Solar Energy and Energy Conservation Bank. This bank will provide financing for the building of solar energy systems and for residential and commercial energy conservation improvements. The bank is also authorized to purchase loans and advances for credit to provide secondary financing of such loans. The title also establishes separate utility programs, residential energy efficiency programs, commercial building and multifamily development programs, weatherization programs, and energy auditor training and certification.

Title VI deals with geothermal energy, with loans for reservoir confirmation, reservoir insurance, and other aspects of geothermal development. Title VII, dealing with acid precipitation and carbon dioxide, is of a different nature. Acid rain has emerged as a major problem in the using of fossil fuels, and the carbon dioxide balance of the atmosphere poses a long-term limit to the emission of carbon dioxide from combustion to the atmosphere. This title is aimed at fostering an in-depth study of these two problems to provide guidance for future policy. (*See* Feature Article: THE NOT SO GENTLE RAIN.) Title VIII directs the president of the U.S. to resume filling strategic petroleum reserves and provides for several kinds of oil from federal lands to be deposited in that reserve.

Energy prices. The price escalation in gasoline and residual heating oil that began when President Carter initiated a phased decontrol of prices, announced in April 1979, accelerated during 1980. The price of gasoline for all types stood at $1.11 per gallon in January 1980 but increased to $1.247 cents per gallon by July. After July, because of a glut in gasoline supply, the

Courtesy, Sandia National Laboratories

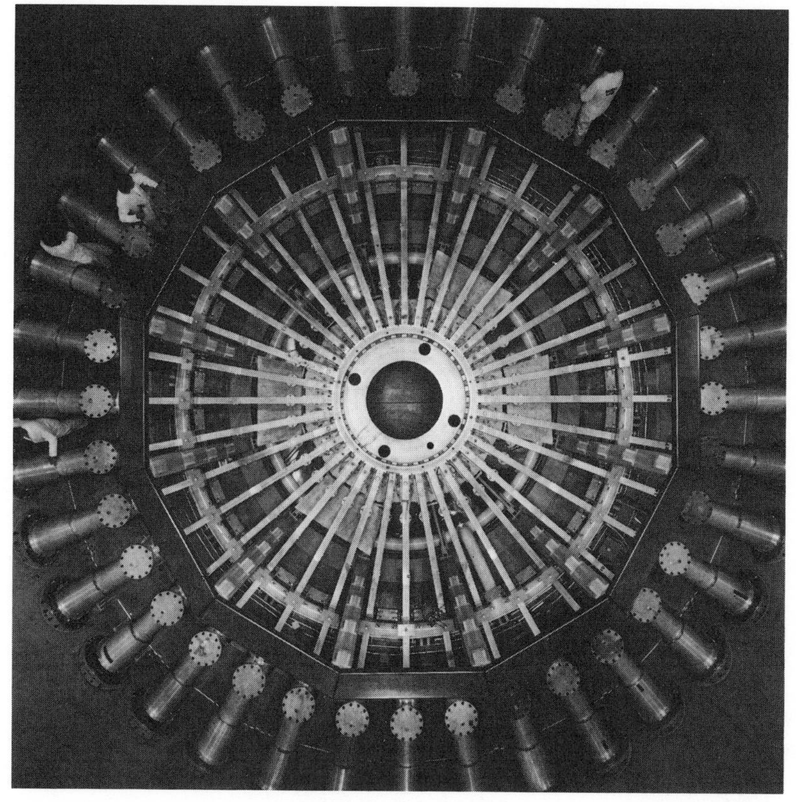

Overhead view shows central section of a particle beam fusion accelerator being built at Sandia National Laboratories. The 36 transmission lines will carry beams of particles to impact on a central hub where the fuel (heavy forms of hydrogen) will be located.

price declined to $1.231 per gallon by December. Residential heating oil started out the year at 90.8 cents per gallon and by November had reached $1.012 per gallon. The average residential heating charge per thousand cubic feet of natural gas increased substantially during the year, rising from $3.549 in January to $4.206 by October. The average retail electricity price also rose, from 4.19 cents per kilowatt-hour in January to 4.95 cents by October.

These price trends were sharply influenced by the decision in late January 1981 of President Reagan to dismantle all price controls on domestic crude oil and oil products. Although as of early 1981 official information concerning the effects on prices was not available, the gasoline prices reported in the *Oil and Gas Journal* for the week ending Feb. 23, 1981, indicated that the average price for unleaded regular gasoline was approaching $1.40 per gallon.

With respect to crude oil the weighted international average price increased sharply after January 1980. In that month the weighted price was $28.55 per barrel, and it rose steadily through the year with a sharp increase in January 1981 to $35.63. Responding to the phased price decontrol initiated by President Carter, the domestic average price for crude oil at the well rose throughout the year. In January 1980 the average stood at $17.86 per barrel, and it had increased by October to $23.08.

International developments. Oil production, as estimated by the *Oil and Gas Journal,* declined on a worldwide basis by 4.8% during 1980. However, this decline was not evenly distributed by area. While the Asian Pacific region was down 4.9% and the Middle East declined 15.4%, Europe increased production by 7.7% and the Western Hemisphere by 2.9%. The major declines in the Middle East were in Iran and Iraq as a result of their war, with Iranian production falling almost 60% for the year and that of Iraq about 25%. These declines were offset in part, however, by a substantial increase in production from Saudi Arabia, by far the largest producer in the Middle East. Production there rose 4% for the year.

During 1980 OPEC could not agree upon a common pricing policy. The meetings of the organization were put under great strain by the Iran-Iraq war, and moderate members kept warning about the danger of price increases in the face of a general oil glut. Nevertheless, prices increased during the year, and clearly OPEC had as an objective price increases somewhat in excess of the rates of inflation.

Reagan's energy policies. Although as of early 1981 President Reagan had not yet issued a comprehensive statement on energy policy, the thrust of his administration's approach to energy was contained in his economic report to Congress delivered Feb. 18, 1981. His program could be summarized under four major sub-

Courtesy, Gulf Canada

Gulf Canada Square building in Calgary, Alberta, receives all of its heat from the Sun, human body warmth, lights, and equipment; there is no other internal heating system.

jects: dismantling of regulations; increased domestic production through financial incentives; increased access to federal energy resources; and sharply reduced federal involvement in the development of alternative energy sources.

The major step for dismantling regulations was the ending of oil price controls, which had been due to expire in September 1981. Concurrent with the abandonment of these controls, production allocation and price controls on gasoline at the retail level were also abandoned. In addition to this major step the secretary of energy also announced that the national energy efficiency standards for major household appliances would not be issued pending a thorough review. The secretary also withdrew proposed standby energy conservation measures, such as a compressed work week, vehicle use stickers, and subsidized employer-based commuter and travel measures.

In a significant but little-noticed action the director of the Office of Management and Budget revoked the Department of Energy's clearance to collect industrial energy consumption data. President Reagan stated that the information requests were needlessly detailed and unduly burdensome and that his action would preclude the federal government from expanding its regulatory programs.

Increased domestic production of conventional fuels, including nuclear power, was clearly a goal of the Reagan administration. Oil price decontrol earlier than scheduled was one major step in this direction. Under the phased decontrol announced by President Carter, drilling for oil and gas in the continental United States rose dramatically, and every week set a new record for number of drills in operation. In addition, Reagan proposed a new system in which businesses could more quickly write off the costs of their investments. However, the administration did not call for modification or elimination of the windfall profits tax on crude oil production.

The new secretary of the interior, James Watt, made it clear that his objective was to increase access to the federal lands, including the outer continental shelf, for the production of energy and mineral resources. Several actions were taken pursuant to this objective. Watt reinstated a number of tracts that had been eliminated from the proposed California outer continental shelf sale, and he announced an acceleration in leasing oil lands off the coast of Alaska. Watt also announced the reduction of the federal involvement in enforcing the surface mine acts, with responsibilities being turned over to the states for administration of the surface mining regulations.

The new budget proposals included sharp reduc-

Table I. U.S. Energy Production, Consumption, Imports, and Exports, 1973–1980
10^{15} BTU

Year	Production	Consumption	Imports	Exports
1973	62.4	74.6	14.7	2.1
1974	61.2	72.7	14.4	2.2
1975	60.1	70.7	14.1	2.4
1976	60.1	74.5	16.8	2.2
1977	60.3	76.4	20.1	2.1
1978	61.2	78.1	19.3	2.0
1979	63.6	79.0	19.7	2.9
1980*	64.6	76.1	15.8	3.7

*1980 estimated on ten-month basis.

Source: U.S. Department of Energy.

Table II. U.S. Energy Production by Source, 1973–1980
10^{15} BTU

Year	Coal	Crude oil and natural gas liquids	Natural gas	Hydroelectric power	Nuclear electric power	Other	Total
1973	14.4	22.1	22.2	2.9	0.9	0.1	*62.5
1974	14.5	21.0	21.2	3.2	1.3	0.1	*61.2
1975	15.2	20.1	19.6	3.2	1.8	0.1	60.0
1976	15.9	19.6	19.5	3.0	2.0	0.1	60.1
1977	15.8	19.8	19.6	2.3	2.7	0.1	60.2
1978	15.0	20.7	19.5	3.0	3.0	0.1	*61.0
1979	17.5	20.4	19.9	3.0	2.8	0.1	63.2
1980†	18.7	20.5	19.7	2.9	2.7	0.1	64.6

*Totals may not equal yearly sum due to independent rounding.
†1980 estimated on ten-month basis.

Source: U.S. Department of Energy.

tions of the budget authority and obligations in the area of alternative energy sources. For 1981 the proposed reduction in budget authority for the Department of Energy amounted to $1.9 billion. Of this, $545 million were in the synthetic fuels program, $745 million in the alcohol from biomass subsidy, and $254 million from energy conservation. These reductions were substantially increased for fiscal 1982. The commercialization activities of the Department of Energy in regard to synthetic fuels were to be terminated; the focus on federal solar power activities was to be away from demonstration and commercial efforts and toward long-range research and development projects that are too risky for private firms to undertake; the energy supply programs dealing with geothermal, energy storage, electric energy systems, energy impact assistance and environmental studies, uranium resource assessments, and hydropower were to be reduced by a third; the technology development, regulation and information, and financial assistance to state and local governments for energy conservation was to be reduced; and the alcohol fuel subsidies program and the solar and energy conservation bank were to be eliminated.

Production and consumption. Energy production, consumption, imports, and exports for the U.S. since the beginning of the OPEC embargo are shown in Table I. As can be seen significant changes began in 1977. Domestic production of energy, stabilized at about 60 quadrillion BTU's for the period 1975–77, showed a continued increase after that date. In 1980 production was 7% higher than that recorded in 1977, and with the greatly increased activity in oil and gas exploration the total can be expected to increase more rapidly in the next few years. Consumption, on the other hand, increased through 1979 but declined 4% in 1980. The decline in imports, which reached their peak in 1977, was more dramatic. Imports in 1980 were 19% lower than in 1979. On a net import basis—imports minus exports—the decline was even more striking, 39% since 1977 and 37% in the last year alone.

The increase in U.S. domestic energy production, as seen from Table II, was almost entirely in coal and crude oil. The increase in coal not only represented a rise in U.S. consumption but also a rise in coal exports. Crude oil production grew slowly from its 1976 low, while natural gas production remained essentially stable after 1975. Nuclear electric power declined from its 1978 high and in 1980 was at about its 1977 rate.

The rising price of crude oil was reflected in crude oil consumption in the major industrial countries of the West. In 1980, with the exception of Italy, every country showed a decline in crude oil consumption, as indicated in Table III.

The decline in nuclear power generation in the U.S. was reflected throughout the non-Communist countries with the exception of France and Japan. Total world electric generation rose by an estimated 6% in 1980, or about 35 billion gross kilowatt-hours. The increases in generation for France of approximately 16 billion gross kilowatt-hours and for Japan of about 21 billion gross kilowatt-hours accounted for the increase. The data on nuclear electric generation are summarized in Table IV.

Table III. Crude Oil Consumption for Selected Major Non-Communist Countries, 1973–1980
000,000 bbl per day

Year	Japan	West Germany	France	United Kingdom	Canada	Italy
1973	5.0	2.7	2.2	2.0	1.6	1.5
1974	4.9	2.4	2.1	1.9	1.6	1.5
1975	4.6	2.3	1.9	1.6	1.6	1.5
1976	4.8	2.5	2.1	1.6	1.7	1.5
1977	5.0	2.5	2.0	1.6	1.7	1.5
1978	5.0	2.6	2.0	1.7	1.7	1.5
1979	5.2	2.7	2.1	1.7	1.8	1.5
1980*	4.8	2.4	1.9	1.5	1.7	1.6

*1980 estimated on partial-year basis.
Source: U.S. Central Intelligence Agency.

Table IV. Nuclear Electricity Generation for Selected Countries, 1973–1980
000,000,000 gross kilowatt-hours

Year	U.S.	France	Japan	West Germany	Canada	United Kingdom
1973	88.0	11.6	9.4	11.9	18.3	28.0
1974	104.5	14.7	18.1	12.0	15.4	34.0
1975	181.8	18.3	22.2	21.7	13.2	30.5
1976	201.6	15.8	36.8	24.5	18.0	36.8
1977	263.2	17.9	28.1	35.8	26.8	38.1
1978	292.7	30.5	53.2	35.8	32.9	36.6
1979	270.6	39.9	62.0	42.2	38.4	38.5
1980*	264.0	56.4	83.8	41.1	39.9	35.8

*1980 estimated on ten-month basis.

Source: U.S. Department of Energy.

Future prospects. The last year of the Carter administration witnessed the enactment of the president's energy program, culminating in the Energy Security Act of 1980. The strategy of the Carter administration, involving an emphasis on conservation, incentives for the developing of alternative sources for energy, and heavy taxation of conventional oil resources, was finally put in place. However, President Reagan has signaled major changes in emphasis toward increasing production from domestic sources of conventional fuels, a reduction in forced conservation through regulation, and a strengthening of the incentives to develop alternative sources of energy.

While both administrations moved toward market determination of energy prices, it is clear that Reagan places much more emphasis on developments in the private sector than on developments under government programs. The impact of this strategy on the energy position of the U.S. and the rest of the world remains to be determined. One significant fact is that even under the Carter administration OPEC's share of world production was declining, dropping from 50% in 1979 to 45% in 1980. If the new government policies stimulate U.S. production and consumption continues to drop because of higher prices, then perhaps the role of OPEC will continue to decline.

—William A. Vogely

Environment

The most important events in 1980 from the standpoint of environmental science were the release of *The Global 2000 Report to the President* by the U.S. government, a series of new revelations concerning the significance of water and the national railroad system as limiting factors in the future, new research results on problems relating to radioactive waste storage, and some theoretical breakthroughs of major significance.

The Global 2000 Report. On July 23, 1980, the long-awaited *Global 2000 Report* was released. This document discussed the results of the most massive project to assess the future of the environment yet undertaken. A small staff directed the operation under the joint management of the U.S. President's Council on Environmental Quality and the U.S. Department of State. This group was advised by a large number of senior professionals within the government, a core group of seven external to the government, and many others within and outside the government.

The document was enlightening in two respects. On the one hand, if one read all the technical material carefully, it provided the most comprehensive overview of the fate of mankind and the planet yet available. On the other hand, it was an extraordinarily frank and revealing portrait of the shortcomings of the U.S. federal government (or any other large organization) in providing an overview of the future. One example illustrated the type of overview provided by the document. About the time that shortages in availability of fossil fuels will be acute (1990–2000) and biomass as a source of fuel could appear as one solution, those parts of the world that already rely heavily on biomass as a source of fuel will have severely depleted their forests. This point is illustrative of the way in which a global perspective brings a more realistic assessment of supply and demand balances in the future.

The reader is given many reasons to be uneasy about the forecasting ability of governments. Two problems show up again and again: fragmentation and politicization of the forecasting activity. Several comments dramatize the fragmentation problem. The document concluded that the executive agencies of the U.S. government are not now capable of providing the president with internally consistent projections of world trends in population, resources, and the environment for the next two decades. The study team found that while the various departments and agencies of the government maintained many complex computer models for use in forecasting, each of them was entirely independent of the others. The lack of contact between the groups of experts working on these separate models was particularly revealing. The document reported that, with one or two exceptions at the most, none of the agency experts had ever met one another prior to the Global 2000 project and none knew anything about the others' calculation procedures—although on occasion they were required to make use of the projections developed by the other departments or agencies.

Also, forecasting by the government was revealed to be a political rather than a scientific activity. "Forecasts" by government agencies are not objectively derived products of a scientific activity. Rather, computer-generated projections by a typical government agency are the products of political activity in which a staff works backward from a politically desirable answer to a computer program that will generate that answer and then presents the answer as if it had some relation to objective reality. As the document explains, often an agency finds it helpful to use advanced analytic techniques as weapons in the adversary process of initiating, justifying, and defending its programs. Three examples described in the document reveal that this rather harsh criticism is realistic. The government's overall food production model assumes constant increases in world fish production, yet world fish catches have not increased significantly for the last decade. The trend in total world food production as projected from 1978 onward is unrealistic relative to the actual trend from 1970 to 1978. And, finally, the government's overall model assumes that the real price of oil will increase by 65% on the international market from 1975 to 1990; however, it had already increased by 72% in the first 5 years of that 15-year period.

One valuable feature of the *Global 2000 Report* was the demonstration, using computer simulation models, that the more a model mimics the real-world linkages between various sectors of national or world economies, the less optimistic the projections become. In other words, one only grasps the true severity of various problems by exploring the ways in which the impacts of various environmental limits are intensified because of competing demands from different sectors.

Water. The document *Environmental Outlook 1980* by the U.S. Environmental Protection Agency left no doubt that by the end of the century conflicts over water in the Rocky Mountain region will be intense. In that region water consumption for energy and manufacturing uses is expected to increase about thirteenfold. Approximately 90% of these increases are projected to come from water-intensive energy-related activities, such as oil-shale mining. Throughout the region farm and ranch lands and their associated water rights are being sold to energy companies.

Two other aspects of this situation give it a much larger significance. First, it is agriculture, not energy, that allows the United States to maintain anything like balance in its merchandise trade with other countries. The health of the U.S. dollar is critically sensitive to any downturn in the nation's ability to export food. Second, at the time the EPA document was prepared (it was published in August 1980) it was not known that the MX missile system would also place extraordinary demands on water in dry western states; the environmental impact statement concerning that program was only released on Dec. 18, 1980. That statement reveals

Keystone

Divers wearing breathing apparatus come ashore in Great Britain near the Isle of Wight. One carries a cylindrical container suspected of holding toxic chemicals. More than 2,000 such containers were washed into the ocean in November 1979 when the Greek freighter "Aeolian Sky" sank off the south coast of Britain.

Wide World

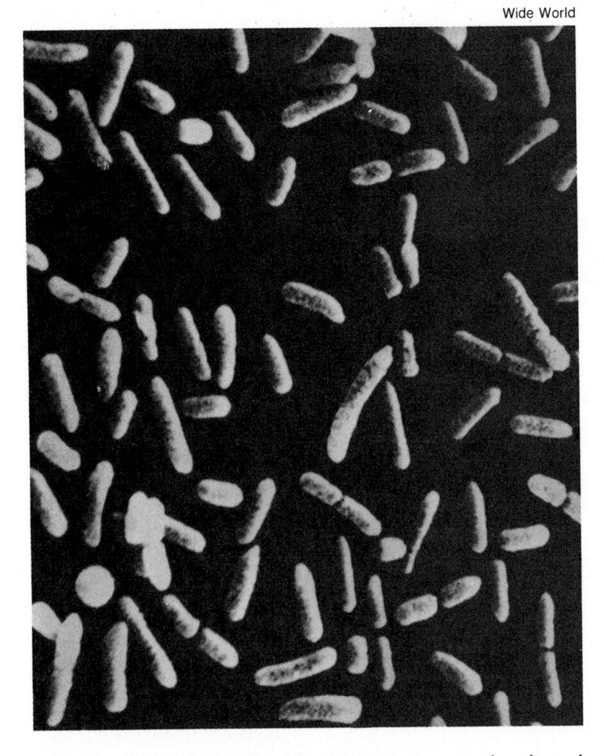

Bacteria, shown magnified 28,000 times, were developed by the General Electric Co. to fight oil spills by converting hydrocarbons into carbon dioxide and proteins. The U.S. Supreme Court ruled in 1980 that a patent could be granted on the microbe.

that water needs for missile system construction may significantly reduce groundwater levels in some states and that water may have to be brought in from other regions. At the least, it stated, the U.S. Air Force would have to buy water rights from other users.

The significance of this water demand had already been noted in a U.S. General Accounting Office document released in late February. It asserted that the annual water usage of the MX missile system would vary from 35 million to 10 billion gallons, leveling off after full deployment to 4 billion gallons. As of early 1981 no federal agency had attempted to reconcile the conflicting water demands of Western agriculture, prospective energy developments, and the MX missile system. Yet already there were signs that water was being depleted in the region of the country that must provide the water for these three competing demands. One source is the Ogallala aquifer, which lies under parts of the high plains states from Texas to Wyoming. The production of corn in semiarid western Kansas, for example, is made possible with heavy irrigation using water drawn from this aquifer. Wells that had been pumping water at 500 gallons a minute were down to 300 by the summer of 1980. West-central Kansas towns were facing the expensive prospect of going farther afield for purchase of water rights.

Thus, it would appear to be desirable to have integrated national planning at the highest level of government in order to allocate scarce water resources rationally over competing sectors. Otherwise, the U.S. could wind up in a curious situation in which it achieved a net improvement in international trade in coal at the cost of an even larger net loss in international trade in corn and wheat.

Railroads as a bottleneck. Railroads are not a natural environmental factor, but in the future it seems likely that they will gradually become an important man-made limitation to economic development in the U.S. The problem is vastly inadequate investment in the railroad system over the last several decades. A new document, the *National Energy Transportation Study*, revealed that the nation's railroad system as it now exists will not have sufficient capacity to move predicted coal traffic in 1990, particularly in the West.

One cause of this bottleneck is the expected great increase in western coal traffic, predicted to be sixfold from 1975 to 1990. The situation may well lead to competition between coal and agriculture for the use of the western railroads, although this problem was not mentioned in the study. Again, the importance of planning that considers all the competing demands of different sectors can been seen.

Storage of radioactive waste. A great deal of new information became available during the year on storage of radioactive waste. Finally, a study of the 1957–58 Soviet nuclear accident was released, which was as definitive as it could be without access to secret Soviet documents. The report, by John R. Trabalka, L. Dean Eyman, and Stanley I. Auerbach, drew on more than 120 documents, mostly from the Soviet open scientific literature. While it is still not absolutely certain what happened, a plausible explanation for the incident is the use of now-obsolete techniques of waste storage and isotope separation. The extent of the accident can be seen mostly clearly by examining maps of the area. Comparisons of high-resolution maps of the contaminated area before (1936–54) and after the accident (1973 and 1974) showed that about 30 names of small communities had been deleted from the latter. Radioactive contamination appears to have resulted in resettlement of the human population from an area of 100–1,000 sq km (40–400 sq mi).

One would hope that there is now a much better understanding of the physics and chemistry of radioactive waste storage containment so that such accidents would not occur again. However, in September two new studies indicated that previously unknown types of phenomena can occur when radioactive waste is stored and that these phenomena have implications for the lifetimes of any storage containers.

The first report, by J. C. Dran, M. Maurette, and J. C. Petit, considered the impact of radiation, chemical reactions, and the environment on degradation of glass,

Courtesy, EPRI and SRI International

DIAL, differential absorption lidar, is equipped with two laser beams to measure air pollutants over large areas and in three dimensions. The device sweeps beams of ultraviolet laser light across a plume of emissions from an industrial stack to determine the concentrations of the gases present. It can survey an area as far as three kilometers (two miles) from the emission source.

ceramic, or cement storage containers used to enclose radioactive wastes. They were concerned with the damage to the containment vessel that would result from an accumulation of very short particle tracks, the paths of alpha particle "recoils." That is, over storage periods of about 100,000 years, actinide atoms trapped in the storage material would undergo radioactive decay and emit alpha recoils that would penetrate the wall of the vessel. As these microscopic tracks accumulated, the wall of the vessel would in effect become etched and would be rendered increasingly susceptible to chemical reactions.

The etching rates under various conditions can be measured by using electron microscopy. The authors discovered that the etching rates increased by more than 20 times when the alpha recoil damage paths started overlapping. Furthermore, changes in the chemical environment also affected the etching rate. To illustrate, a high salt concentration in the environment increased the etching rate by a factor of ten. The significance of this finding is that it raises questions about the advisability of storing radioactive wastes in deep salt mines, into which highly saline waters could accidentally seep. These authors concluded that their electron microscope technique is extremely sensitive and could be used to identify types of materials that would be highly resistant to combined radioactive-chemical degradation in natural environments of the type in which radioactive wastes are expected to be stored safely for very long periods.

The second report, by E. H. Hirsch, also considered mechanisms by which materials such as glass can be rendered susceptible to chemical attack and breakdown when stored for long periods in natural environments. For example, sodium can be detached from sodium silicate glass by heat and also by irradiation. After that has happened, a cyclically repeated corrosion reaction is set in motion. Surprisingly, given the apparently chemically inert nature of smooth glass, corrosion processes form products that are not only water-soluble but also are readily detachable from the bulk of the glass. The author was able to point out technical omissions in previous research of this type and also showed that the processes described would be speeded up in the presence of high temperatures. This was not reassuring to those who might be uneasy in any case about storage of nuclear waste, because in the massive waste glass blocks used for radioactive material storage self-heating could produce temperatures of several hundred degrees Celsius during the first 100 years of storage.

Both of these reports leave the impression that it will be possible to identify materials that could be used to house radioactive waste safely. However, they both also leave the impression that previous research in this area has not detected important phenomena that must be understood thoroughly because of the great implications of any possible accidents. Such implications cannot be overstressed, given the knowledge from the Soviet experience that accidents are not only possible but have occurred, and on an extraordinary scale.

Theoretical breakthroughs. A paper by Karl Butzer developed a fascinating theoretical breakthrough by combining facts and theories from a large number of different fields, including ecology, history, anthropology, archaeology, sociology, and organization theory (from operations research). Butzer was stimulated by his researches in Egyptology to wonder if civilizations are organisms or systems. One observation was perhaps paramount in stimulating his line of reasoning. He noticed that in societies that have not become highly complex, a graph of their population fluctuations plotted against time reveals over many centuries a steady-state equilibrium pattern. That is, while the population would grow for a time to a peak and then decline because of famine or disease, the pattern of fluctuations could be visualized over a long period as a fairly regular one about a single imagined long-term steady state. However, when one examines the long-term pattern of fluctuations in population size for more complex societies, it can be seen that such fluctuations occur for some decades or centuries about one equilibrium, and then fairly suddenly there appears to be a jump to a new equilibrium level about which fluctuations occur.

Butzer explained that these jumps from one equilibrium level to another coincide with the adoption of new adaptive strategies. If the jump is to a higher level, the amplitude of population fluctuations declines. However, if the society encounters some negative social or environmental input, populations decline to a lower equilibrium, about which a new, wider-amplitude pattern of fluctuations is established.

Butzer linked two previously separate bodies of theory to explain these jumps in equilibrium level. On the one hand he used the notion of trophic levels in biological communities, a concept that originated in ecology. This notion invokes the image of a pyramid of feeding organisms, with a wide bottom layer of plants, a second level of herbivores that feed upon the plants, a third layer of carnivores that feed upon the herbivores, and successively higher layers of carnivores that feed on smaller carnivores. Butzer then pointed out that one can in the same way view an efficient social hierarchy, in which the bottom level of people consists of agricultural workers who support middle-level echelons of bureaucrats and, at the top, a ruling elite. Then he linked to that concept an idea developed by operations research analysts (such as Sir Stafford Beer) to describe how the shape of organization structures affects the information flow up and down in such structures and, thus, the efficiency of response of hierarchically organized institutions to new challenges. He thereby pointed the way to a single body of theory that explains how energy and information, both moving between layers and components in hierarchically organized structures, affect the dynamic behavior of those structures.

Butzer also examined a number of simple models of sociopolitical hierarchies to determine how their dynamic behavior (fluctuations through time) should be affected by the number of levels in the hierarchy and by the width of the base of the pyramid of hierarchical levels relative to the width at the top. If one envisions a very primitive society with no formal leadership, all the people are on one level, information flows horizontally but not vertically, and the society changes slowly. If a society has some leadership but all in one level (no "middle-level management"), there is some vertical information flow and increased dynamism. If a society has a relatively broad bottom level and a three-level hierarchical structure, energy and information flow is more efficient than in the previous cases, and each hierarchical level, as well as the whole system is in a dynamic steady state. However, if a society has a top-heavy vertical structure, there is impeded information flow from top to bottom or bottom to top, increased energy expenditure for system maintenance, excessive demands on the bottom (producing) level, and a metastable equilibrium (jumps in equilibrium level). A society with, for example, seven levels of organization is organized and integrated so as to exploit new opportunities rapidly. It is also vulnerable to fairly rapid disintegration in the case of an external shock. Butzer then applied these notions to an interpretation of 4,500 years of Egyptian history, seeking key variables that could be used to account for the dynamic behavior of Egypt as a system.

One important variable is progressive social pathology resulting from a top-heavy and metastable sociopolitical pyramid. This involves progressive overexploitation of the masses by a growing unproductive elite, with resulting social disequilibrium and eventual political and economic collapse. In terms of the model the base is too narrow given the superstructure, and also the superstructure has too many layers of bureaucracy to be supported by the base. Butzer found this situation in both the Old and New Kingdoms of ancient Egypt as well as during the Roman and Byzantine periods in Egypt. These ideas are particularly credible because they correspond to those put forth in the last few years by anthropologists studying quite different phenomena, such as the collapse of the classic Mayan civilization of Mexico and Guatemala.

Butzer also discovered that the quality of leadership is a key variable. Thus, the Egyptian pharaoh Ramses III marshaled sufficient support to beat off powerful foreign invaders who had destroyed the other kingdoms of the eastern Mediterranean world. He still managed to maintain internal order during the early years of growing food shortage. Other important variables were found to be foreign intervention and ecological stress produced by food shortages (failure of the Nile River system in Egypt).

Butzer argued that, over the long term, complex, "steep-sloped" systems are not stable. Because of the large number of components of such a system and the complexity of the information processing upon which it depends, an unexpected coincidence of troubles related to two or more of the four key variables just mentioned can trigger a catastrophic train of mutually reinforcing events to which the system is not able to respond.

During 1980 a group in the Environmental Sciences Division of the Oak Ridge (Tenn.) National Laboratory compared the number of trophic levels found in actual communities of plants and animals with the number found in model mathematical systems. Stuart Pimm found that the real-world systems had fewer trophic

G. F. Reynolds, The University of Reading, U.K.

Seasweep was developed in the U.K. to clear oil slicks and floating debris from water surfaces. The paddle wheel on the catamaran draws surface water over a small floating weir and into a floating confinement tank that has an outlet set low. Any oil and debris taken into the tank rises to the top and can be skimmed or pumped off, freeing from pollutants the water that continuously leaves the tank through the outlet.

levels than one would expect from populations of model systems. The real-world systems were simple because simplicity promotes long-term stability, and stability is a prerequisite for sustained survival.

A study by Donald L. DeAngelis appeared to be of considerable significance for future research because, like the paper by Butzer, it linked theories not previously considered in association and also provided a powerful new conceptual model that explained a wide variety of phenomena. Ecologists have become interested in a characteristic of the dynamics of multispecies communities described as "resilience." This measure expresses the resistance of communities or ecosystems to external shocks, or perturbations and the speed with which they return to equilibrium status after a perturbation. The faster a perturbed system returns from an initial displacement to its equilibrium point, the greater is its resilience.

Ecologists have been trying to identify the properties of communities and ecosystems that determine their resilience. DeAngelis used a combination of mathematical analysis and computer gaming to show that the resilience of systems decreases as the mean transit time of energy or matter through the system increases.

If the conceptual model of DeAngelis is combined with that of Butzer, one has the beginnings of a powerful set of ideas that can explain phenomena in a wide variety of fields, from natural plant and animal systems through harvested systems to the dynamics of human society. The following examples suggest some of the applications.

Ecologists have discovered that when different types of ecosystems are compared, some, such as tundra or cold-temperate bogs, recovered from perturbations very slowly, while others, such as ponds, recovered rapidly. The explanation, based on the concepts of Butzer and DeAngelis, is that the former have a low rate of energy input per unit of living tissue in the system, whereas the latter have a high rate.

Coral reefs and rain forests are examples of systems with very little throughflow of nutrients that come from outside; essential mineral elements are tightly cycled within the system. That is, if an organism dies, there is a tendency for it to be eaten by another organism within the system so that the minerals in its body keep recycling through the system. However, if systems of this type are perturbed, recovery may be very slow because there is so little input of minerals from outside.

Two other examples are more speculative but illustrate the type of phenomena on which this body of theory has a bearing. The first considers a commercially exploited resource, such as anchovies or sardines, that is eaten by one or more species of fish predators. A problem associated with sardines and anchovies is that while they sometimes are numerous enough to support very large commercial fisheries, their populations can decline suddenly, leading to a collapse of important industries. This happened recently with sardines off the coast of California and anchovies off the coast of Peru.

In attempting to explain these sudden declines DeAngelis pointed out that in such prey-predator systems an increase in the energy input (or matter input) to the system could be a destabilizing influence. The response of the predator population to an increased input of matter or energy could be to increase their overall numbers. In such a circumstance the prey (anchovies or sardines) would become increasingly predator-controlled and there would be less of it available for commercial fishermen.

A final example concerns the comparative dynamics of nations and civilizations. The curious reader of DeAngelis's paper might ask: "If DeAngelis is right, then how is it that the economy of a country such as Switzerland, which is more energy-efficient than the United States, is more stable than that of the United States, which has had wide-amplitude swings in inflation, interest, and unemployment rates in the last decade?" The answer is that Switzerland might well have a low resilience to external perturbations. However, like other countries of this type, such as Japan and Sweden, it has maintained stability in recent decades by avoiding situations that could lead to such perturbations (for example, military conflicts with other nations). Thus, there are two ways in which systems can be stable. One is to have a high rate of flux of energy and material through the system, as with a pond or the United States. If a perturbation impinges on such a system from the outside, the resilience is high and it takes a short time to return to the stable state. The other way is to have a low rate of nutrient or material flux, as with Switzerland, Japan, a coral reef, or a tropical rain forest; such a system requires avoidance of external perturbations.

It should be noted that this discussion has dealt with only an embryonic, preliminary theory, and the authors of it caution that more thorough work needs to be done. However, one can already see how this theory could be extended. If the work of the ecologists on matter and energy flow in systems were to be combined with the work of such anthropologists as Butzer who are considering the flows of information and control, and if these in turn were to be linked with the work of organization analysts such as Sir Stafford Beer, a powerful theory of systems would result.

—Kenneth E. F. Watt

Food and agriculture

During 1980 world food production increased 1.2%, only partially offsetting the 2.5% decrease in 1979. Although production rose in South and Central America, it declined in most African countries and in southern Asia. The greatest food needs were in the sub-Saharan African regions and southern Asia, but certain Central American, Caribbean, and South American countries also required assistance. In 1980 the U.S. Department of Agriculture (USDA) published "Global Food Assessment—1980," presenting data on food production for 79 low- and middle-income countries. In the U.S., 1980 was marked by continuing increases in the costs of planting, irrigating, harvesting, processing, transporting, storing, and cooking food.

Agriculture

U.S. agricultural exports reached a record $40.5 billion in 1980, according to the USDA, more than 25% above the 1979 value of $32 billion. This figure also represented a $23 billion surplus of agricultural exports over imports. This record is especially surprising in light of the boycott of U.S. goods by Iran (a $500 million market for U.S. farm products in 1979) and the partial suspension of trade with the Soviets. The Soviets responded by buying substantially larger grain supplies from other nations, particularly Argentina. U.S. trade increased with most of Latin America, Japan, and Eastern Europe. Hot, dry weather lowered production in Mexico, causing that country to increase its imports of U.S. grain. For the first time China became the leading buyer of U.S. wheat. Some disappointing harvests in other nations also contributed to the rise in U.S. farm exports.

It has been estimated that to feed the world's population as much food will have to be produced in the next 30 years as has been grown from the beginnings of agriculture to the present. By the year 2000 the current world population may exceed five billion. Feeding them will require an increase in annual food-grain production from the current 1.3 billion metric tons to about 2 billion. The goal is achievable. Today, the average U.S. farmer provides enough food for 65 people—46 in the U.S. and 19 abroad. Forty years ago a U.S. farmer produced only enough food for 11 people.

Integrated pest management. The integrated management of pests combines the use of chemicals, genetic principles, and biologic methods with farming techniques to achieve effective, economical pest suppression with minimum adverse effects on nontarget organisms and the environment. The idea has gained acceptance throughout the U.S. and other countries in recent years. It recognizes that an ongoing war began millions of years ago between humankind and certain species of insects, weeds, pathogens, nematodes, rodents, and other pests that compete for crops, gnaw at dwellings, infest domestic animals, or threaten health. Although the soldiers have not changed greatly during that time, the tactics and weapons have become more sophisticated. This holds true for both sides. Some pests that are sprayed with a pesticide have more resistance to it than others; with successive generations, a strain can inherit this resistance.

Several approaches to pest management have been used. In many insect species both sexes emit and respond to chemicals called pheromones, which attract the opposite sex. By building a trap containing a pheromone attractant, the density of insects of a given species can be surveyed. In the fruit orchards of western Colorado this approach has reduced the sprayings of peaches and apples from approximately 16 per season to as few as 2.

Another technique involves the breeding habits of a specific pest. For example, large numbers of screwworm flies can be sterilized artificially by irradiation or chemical sterilants. Subsequently released into an area inhabited by wild populations, sterile males will mate with wild females. Since the female screwworm fly mates only once before laying her eggs, fewer screwworm offspring will be produced. By repeating this procedure for several consecutive generations, it is possible to annihilate the wild population. This method was used to eradicate the screwworm fly from Florida and the melon fly and oriental fruit fly from the small island of Rota, near Guam.

A research scientist at North Carolina State University at Raleigh is attempting to cross domestic tomatoes with wild ones native to South and Central America. The leaves of the wild varieties are covered with glandular hairs that secrete a sticky substance, trapping small insects that land on the plant. The pests are paralyzed or even fatally poisoned in as little as 20 minutes. This built-in resistance can be relatively inexpensive, biomedically safe, and compatible with other control methods.

In recent years, the grasshopper population on the U.S. Great Plains has posed a major threat to small-grain crops. Spray programs and related efforts have brought the situation under partial control. One factor to be considered, however, is that grasshoppers control mites, and if all grasshoppers were eradicated, the mite problem in corn could become much more serious. Thus, as one pest is controlled, its relationship with other pests needs to be weighed carefully.

Mechanization research. During 1981 a trial was scheduled to take place that could have far-reaching effects on both consumers and producers of food, as well as on those responsible for conducting research on agricultural mechanization. The suit was filed in 1979 against the University of California by the federally funded California Rural Legal Assistance organization on behalf of a number of farm workers and a small farmers' organization. The plaintiffs charged that public funds were being used to finance research projects that could eliminate jobs, dispossess small farmers, and destroy a way of life, all in violation of the purposes of a land-grant university such as the University of California. Proponents of such research argue that certain farm chores can be done faster and more economically through mechanization.

Mechanical tomato harvester was developed by the University of California and Blackwelder Co. Introduced during the 1960s, such machines now harvest approximately 90% of the tomatoes in the U.S. used for catsup, sauce, and juice. In 1979 a lawsuit filed against the University of California on behalf of farm workers and small farmers charged that public funds were being used by the university to finance projects such as the harvester that could eliminate jobs and dispossess farmers.

J. B. Kendrick, Jr., University of California, Berkeley

The most celebrated issue in the suit involves the tomato harvester. The harvester was adopted in California in 1964 near the end of the federal bracero program, which had permitted Mexicans to work as farm laborers in the U.S. Researchers had stepped up efforts to develop a practical harvester, and a tomato variety was bred especially for machine harvesting. (For successful machine harvesting, tomatoes must ripen at the same time, detach easily from the vine, and withstand rough handling.)

Ninety percent of the tomatoes used for catsup, sauce, and juice are now machine harvested. California's acreage planted in these varieties of tomatoes has doubled, and the state provides 85% of the processed tomatoes grown in the U.S. (Most fresh-market tomatoes are grown in Mexico because of the costs associated with hand picking.) In comparison, very little okra is raised in the United States because no mechanized harvester has been developed to offset increased labor costs. Instead, okra is imported from Central America.

The University of California defends its position by pointing out that it helped save the processed-tomato industry. Proponents of mechanization research claim that it increases the efficiency of food and fiber production, reduces risks at harvest time, and reduces peak labor demands. Proponents also point out that mechanization reduces dependence on migratory labor, provides stable, year-round job opportunities, and removes the drudgery from farm work, particularly the stoop labor of hand harvesting. They warn that if mechanization is not encouraged, food prices will rise at a substantially faster rate. Humanists question whether migratory farm labor is a good social enterprise or whether it should be viewed as a social problem, since such migrant workers are among the poorest of the poor. It has been suggested that retraining programs be offered and agricultural laborers made eligible for unemployment insurance benefits.

Tomorrow's farmer. Over the last half century agricultural production has come to demand an ever higher level of expertise. Farmers and ranchers no longer can afford to make management decisions solely on the basis of "guesstimates." Now communications technology is being developed to help the agricommunity make decisions more quickly and efficiently.

The Kentucky Extension Service, in cooperation with the USDA and the National Weather Service, has designed and tested a new home information delivery system. One hundred Kentucky farmers in each of two counties were receiving weather, market, agricultural production, and home economics information via telephone lines from a microcomputer located in the county agent's office. The Green Thumb Box receives, stores in memory, and displays information on the screen of a standard television set. From a computer "menu" of up to 999 agricultural subjects, the Green Thumb system provides information on research findings, insect reports, and more. Localized information such as weather forecasts and frequently updated reports on marketing and weather conditions allow farmers to translate information into action immediately.

In Montana, Wyoming, the Dakotas, and Nebraska a related program called AGNET has been serving for five years as a management tool for agriculture. This program is based on the belief that computer systems should be built to service the "noncomputer" person's needs. Its program menu includes, among other items, land-purchasing models, simulation of grain-drying systems, irrigation scheduling, financial budgeting, economic analysis of feeder livestock performance, and a general accounting and bookkeeping system. As of 1980 AGNET had users in 27 states. Banks, farm equipment manufacturers, feed companies, and Extension Service offices have learned to take advantage of this simple system. Like Green Thumb, AGNET is reached through a typewriter-like device, but it differs from Green Thumb in that it uses a full range of

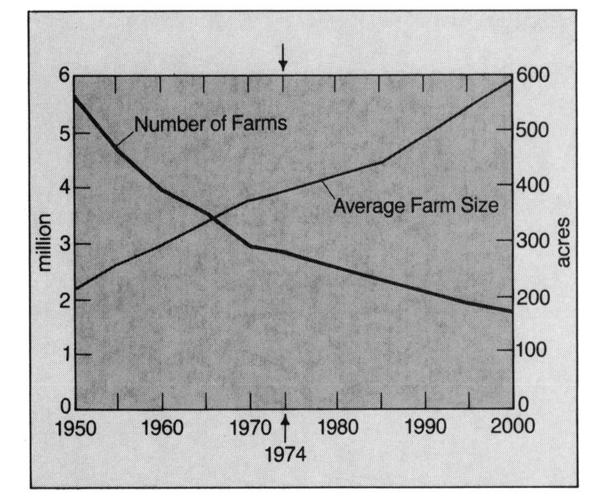

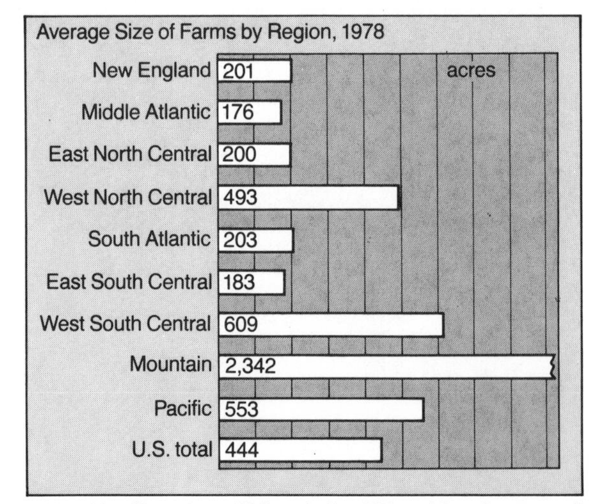

Adapted from "Fewer, Larger U.S. Farms by Year 2000—and Some Consequences," Thomas McDonald and George Coffman, USDA INFORMATION BULLETIN no. 439, pp. 3–4, October, 1980

alphanumeric keys, whereas Green Thumb uses only a telephone keyboard system. In both cases portable terminals can be used wherever an electrical outlet and telephone are available. The farmers and ranchers of the 1990s and the 21st century will use centralized computers to simulate alternatives in management and marketing systems, making it possible to do a businesslike analysis of management options.

Technology also has been developed that will allow a farmer to use a photoelectric sensor to monitor the ripening of cornfields, permitting, for example, computerized applications of water and chemicals at the optimum times. The principal obstacles to implementing such technology are economic, not technical.

Farms in the year 2000. The number of farms in the U.S. has declined steadily since about 1935. There were about 2.5 million farms in the U.S. in 1980. The USDA Economics and Statistics Service projects that, if the trend that began in 1950 continues, there may be only 1.8 million farms in the country by the year 2000. Meanwhile, the average acreage of a farm will continue to increase as small farms disappear and those that remain become larger. The number of farms with fewer than 100 ac (100 ac = 40.5 ha) is expected to drop by more than 500,000 by the turn of the century. Average acreage, however, is not the most useful measure of farm production because farms differ widely according to type and geographic area. For example, 1,000 ac might constitute an average wheat farm while 100 ac would be a large fruit orchard. Similarly, a ranch of fewer than 2,000 ac on the U.S. Great Plains might be very inefficient, while beef/cow operations in the southeastern U.S. may average only 250 ac. The relationship of farm size to location occurs because most regions specialize in certain commodities.

The USDA's projections were based on a 7.5% annual inflation rate. If farm price inflation in the next 20 years is less than 7.5% per year, there will be fewer large farms and more small farms by the year 2000 than the projection suggests. The statistics prompted then U.S. Secretary of Agriculture Bob Bergland to conduct ten public meetings at which farmers and ranchers discussed the future of U.S. agriculture. Generally, they felt that the increase in the relative size of farms was inevitable and insisted that the federal government should not arbitrarily restrict the growth of farms and ranches.

Food animal production in the future. Predictions are that artificial insemination will change markedly in the future. Whole herds of heifers will have their estrus cycles synchronized, allowing timed breeding. Some even predict that sexing of semen—that is, selecting only those sperm that will produce a certain sex in the calf—will be practiced with 80% success. Pregnancy rates will be increased substantially, and a number of endocrinological techniques for regulating reproductive functions of cattle will become available. In

Photos, Robert Kainer and David Lueker, Colorado State University

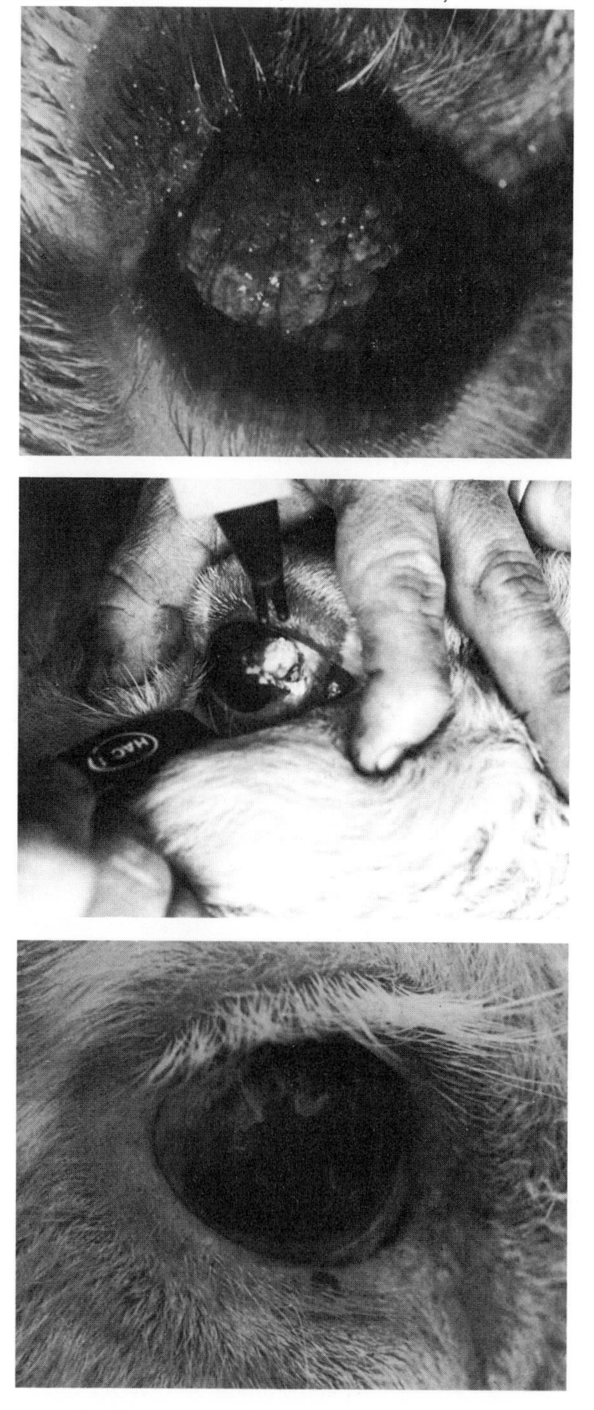

Eye tumors are a major problem for cattle, but a heat probe therapy developed in Colorado arrested 80% of them after a single treatment and 91% after two. At the top is a tumor before treatment. In the center a veterinarian begins to work on the eye with the heat probe, and at the bottom is the tumor-free eye two weeks later. The probe also proved successful in treating ringworm in cats and dogs.

addition to estrus synchronization these changes will include inducing puberty and shortening the time between delivery of one calf and breeding for another. The number of cattle produced by superovulation of the dam and transfer of fertilized eggs to an incubator cow is expected to rise from 20,000 in 1980 to perhaps 500,000 by 1990. There will be an increase in the number of twins in cattle, and "test tube" fertilization of cattle will become more widespread.

Techniques to effect cloning of meat-producing animals have been tested successfully on laboratory animals. The techniques are based on the knowledge that every nucleated cell in a mammal has the same genetic instructions. However, only a part of the genetic material is active at any given time in any given cell, so a full understanding of the control factors related to genetic activity is necessary for successful cloning. One advantage of cloning would be that ideal experimental animals could be reproduced exactly because the cloned animals would be genetically identical to the original. Cloning would also make it possible to build a breeding herd with consistent characteristics.

Other current research. One of the major problems of cattle is eye tumors. Investigators at Colorado State University arrested nearly 80% of the tumors with a single heat probe treatment developed by the Hach Chemical Co. After a second treatment, 91% of the tumors regressed completely.

Scientists at the University of Hawaii Agricultural Experiment Station have studied 85 varieties of taro. One of the basic crops of the Pacific Basin, this plant can be grown as either a flooded or an upland crop. Its roots provide poi, a staple in Polynesian diets. The leaf can be cooked and the stalk replanted. Taro has considerable potential as a cash crop. The Chinese use it to make chips and other snack foods, and European processors use taro roots to manufacture plastics.

Wheat breeders have had enormous successes in the last few years. The Vona variety has yielded more than 100 bu per ac on irrigated land in southeastern Colorado. Under dryland conditions in mid-Oklahoma it has yielded slightly fewer than 100 bu per ac. This is remarkable, considering that average wheat production in the U.S. for 1975 was 30.6 bu per ac.

The Arid Lands Research Center of the University of Arizona is growing plants that produce a petroleum-based product. One of these plants, commonly referred to as the gopher plant and scientifically known as *Euphorbia lathyrus*, produces a latex rich in hydrocarbons. This can be extracted with a solvent and converted to compounds similar to those made from petroleum. Botanists have suggested that such a plant could produce the equivalent of ten barrels of oil per acre. At $40 a barrel, the OPEC (Organization of Petroleum Exporting Countries) price early in 1981, this crop would be competitive with irrigated corn on the western U.S. Great Plains. Geneticists believe it might be possible to breed new varieties that could produce 25 bbl per ac. In limited experiments the plant did not seem to require irrigation, at least in areas that receive 35 to 38 cm (14 or 15 in) of rain per year.

—John Patrick Jordan

Nutrition

If health is understood to be a composite phenomenon consisting of a number of personal and environmental factors, then management of health calls for understanding as many of these factors as possible. In the past, educators such as Henry C. Sherman and Elmer V. McCollum have provided guides to assist people in choosing healthful foods. Not until the 1940s, however, did the U.S. government take an active role with the publication of "Food for Fitness," the Basic Seven Food Groups, and other guides. The Food and Nutrition Board (FNB) of the National Academy of Sciences published the Recommended Dietary Allowances (RDA), which, with frequent revisions by experts, is still the standard for planning diets to meet the nutritional needs of specific groups of healthy people.

In May 1980 the FNB released a carefully worded report, "Toward Healthful Diets," which elicited surprise, criticism, and approval from representatives of various government agencies, medical organizations, and other professional groups. Robert E. Olson, chairman of the task force that prepared the report, stated that its points of agreement with previous guides are greater than the areas of disagreement. Of the seven dietary goals promoted by the USDA and the U.S. Department of Health and Human Services (HHS; formerly the Department of Health, Education, and Welfare), four are endorsed by "Toward Healthful Diets": they included eating a variety of food; watching one's weight; reducing dietary salt; and, if one is a drinker of alcohol, drinking moderately. Not endorsed were those dealing with reduction in fat, saturated fat, and cholesterol; intake of complex carbohydrates and fiber; and reduction in sugar. Olson defended the findings on the grounds of (1) the force of epidemiological data; (2) lack of unanimity among scientists outside the FNB; (3) lack of unanimity among expert committees in other nations on the issue of modifying dietary intake of cholesterol; and (4) the significance of spontaneous changes in the mortality rates of coronary heart disease in other countries.

Food choices differ as much as people themselves and are strongly influenced by environmental background. Advice about diets varies even among qualified experts. Nevertheless, many impassioned lay persons feel they have a mission to persuade others to adopt their pet diets. The furor caused by "Toward Healthful Diets" was explained by George V. Mann of Vanderbilt University (Nashville, Tenn.) in terms of several factors: the pride and prejudice of advocates who support

Jean-Claude Lejeune

Adequate exercise and diet were urged as a desirable coordinated approach to health and physical fitness by the American Dietetic Association in 1980.

dietary restrictions on fats and sugars; bureaucratic advocacy of certain dietary strategies by such agencies as the National Heart, Lung, and Blood Institute, the USDA, HHS, and even Congress, all of which control research grants; and commercial greed for high profits from foods promoted as especially healthful, such as products low in saturated fat. It should be noted that while diet/heart policy has been profitable to some segments of the food industry, others—for example, producers of eggs, milk, and meats—feel threatened by the public's loss of faith in their products.

Stephen B. Hulley and others at the University of California questioned whether the retrospective finding of high-serum triglycerides in patients suffering from myocardial infarction is the cause or the result of the disease. It is possible that heart disease and high triglyceride levels are both caused by a third factor. It was recommended that active intervention not be pursued until there is persuasive evidence that it will be beneficial and not merely harmless. This applies to behavioral as well as medical intervention, since attempts to modify a person's life-style may have adverse social and psychological effects.

Though dietary restrictions may not prevent heart disease in all healthy people, J. Michael McGinnis of HHS considers prevention as today's greatest dietary challenge. He bases his contention on three points: "(1) Prevention works. (2) Stronger efforts are needed. (3) Diet provides one of the most critical tools and challenges for prevention." Although some experts are skeptical about the likelihood of changing people's food habits, many persons are eager for reliable, practical suggestions about the wise choice and use of food. Scientific data about diet improvement must be presented in the form of everyday food-buying guides, cooking suggestions, and eating plans. Nutrition is one of the strategies in disease prevention, and diet plays a vital role in the prevention of health problems.

In May 1980 the American Dietetic Association published a statement strongly urging a coordinated approach to health and physical fitness involving both adequate diet and exercise. The key to weight maintenance or loss is a combination of dietary modification and regular aerobic exercise. A diet that supplies the RDA meets the everyday nutritional needs of healthy people. Habits of nutritionally balanced eating and physical fitness established in childhood will be maintained throughout life.

The dietitian is caught in the middle in the welter of medical-scientific-government guidelines and recommendations, according to Esther Winterfeldt of Oklahoma State University, president of the American Dietetic Association. The job of the dietitian is to tell the confused public what to eat and to translate the intricacies of scientific data into everyday meals while convincing the consumer of the benefits to be derived from proper dietary practices.

Changes in dietary practices. It has been estimated that 50% of Americans eat out daily, in sharp contrast to the 25% who ate out daily in the 1960s. One of the major factors in this changing pattern has been the school lunch program, initiated in the 1930s. Much has been written about the pros and cons of the school lunch. Is it only a "filling station" for empty stomachs or is it part of the educational system with planned learning experiences for all levels? Both aspects have their advocates. A child cannot learn when he is hungry, but he can learn by seeing and eating balanced meals, exploring new foods, and observing others at mealtime.

Today greater stress is being placed on nutrition education programs that use the lunch program as a learning laboratory. "Does school lunch supply nutrients more adequately than other lunches?" was the question pursued by S. M. Howe and Allene G. Vaden of Kansas State University in a study of two groups of high-school students. The group that ate school lunches had better total diet and lunch ratings than the group that ate elsewhere. The latter group had diets adequate only in protein.

To encourage participation in the school lunch, al-

Bill Grimes—Black Star

In a recent study high-school students eating school lunches were found to have a better total diet than those eating elsewhere. Diets for the latter group were adequate only in protein.

ternative lunch patterns were offered, including free-choice or a la carte. Free choice resulted in greater participation, less waste, greater meal satisfaction, and lower cost. Nutritional information of interest to the diner posted near the food, such as information on which foods are conducive to weight loss, influenced choices: skim milk sales rose and dessert sales decreased. Knowledge about how food can help to fulfill a goal can influence eating.

A survey of families eating at fast-food restaurants showed that nearly all families eat there 1.32 times weekly. The meals eaten supplied adequate protein, high sodium, and 20 to 30% of the RDA for thiamin, riboflavin, ascorbic acid, and calcium when the food was chosen carefully. Fast food failed to supply adequate vitamin A, biotin, folacin, pantothenic acid, iron, or copper.

Food and diet supplements. Americans are estimated to spend more than $1.5 billion annually for supplements to their food supply. Apparently the consumer feels insecure about the quality of the food he is eating or about his food choices, even though the nation is considered to be among the most affluent, healthiest, and best fed on Earth. Nutritionists have questioned whether consumers separate food from nutrition. Generally, supplements are chosen and consumed without regard to signs of dietary or biologic need. One-third of American households buy vitamin supplements without obtaining any data on need or amount. Surveys show that affluent families, which have the best diets, are most likely to buy vitamins.

News stories about the possible effects of additives, preservatives, and the like cause people to distrust the food supply and to suspect that it causes disease. The consumer then looks for more reliable food and grasps at items that carry such labels as "natural," "organic," "untreated," "fresh," or "raw." Indecisive reports by government agencies, researchers, and other purported experts add further doubt about the quality of available food: Is it safe? Is it good for you? Would a supplement help?

The protein quality of diet supplements and meal replacements was analyzed by M. L. Marable and associates of the Virginia Polytechnic Institute and compared with information on the labels and with egg proteins. Predigested liquid protein products with collagen listed as the major ingredient were notably lower in protein quality than other products or than eggs. The protein quality of the other products was comparable to that of eggs but the cost was much higher, even though the major ingredient was nonfat dried milk, soy protein, or skim milk. These products had no advantage over regular meals in protein quality or as reducing aids. Numerous other studies confirmed that diet supplements are seldom prescribed for medical reasons, are nonessential for people's needs, are very expensive, and fail to supply necessary nutrients. When medical needs are involved, a specific nutrient or compound is prescribed, not a hodgepodge of questionable composition.

—Mina W. Lamb

Life sciences

Interactions among viruses and bacteria and between microorganisms and the cells of plants and animals permeated much of the recent research in the life sciences. Investigators offered a new treatment for the fungus-blighted American chestnut; strengthened their cases for viral agents as the cause of multiple sclerosis, some forms of arthritis, and juvenile-onset diabetes; and speculated that the survival in nature of the bacterium responsible for Legionnaires' disease is aided by the presence of photosynthetic cyanobacteria. Molecular biologists and microbiologists continued their efforts to understand the genetic nature of bacterial resistance to drugs and other toxic substances and even suggested a treatment for humans based on bacteria that offer resistance to toxic mercury compounds.

Symbiotic associations between tree roots and fungi were explored, as were ways to benefit agriculture by transferring the technology of bacterial nitrogen fixation to tropical legumes. Other highlights included the first successful transfer of a foreign gene into living organisms and the discovery of a semantic communication system among monkeys in the wild.

Botany

Some of the past year's interesting developments in botany involve the possible comeback of the American chestnut, plant root symbioses, seagrass biology, pollen tube growth, and plant-animal relationships.

Another chance for the chestnut. The saga of the American chestnut (*Castanea dentata*) is one of the saddest in the history of American deciduous forests. Early settlers encountered these abundant, stately trees in forests from Maine to Mississippi, where they soon proved to be an immensely useful source of materials for construction, furniture, and tanning. Their nuts were delightful eating for humans, and the trees sustained countless woodland creatures. By the end of the 1940s nearly all were gone, killed by a blight-causing fungus, *Endothia parasitica,* which probably arrived on seedlings imported from the Orient and which was first observed at the New York Zoological Gardens in 1904. Fortunately, the root systems of the American chestnut continue to sprout, giving time for both the tree and researcher to find a good ending to this saga.

Early conventional efforts to control the fungus met with futility. By 1915 spraying and sanitation practices had ceased, allowing the unhindered spread of the blight through the full range of the chestnut. The nearest thing to preservation was accomplished by hybridization with Oriental varieties to produce immune but far less impressive trees. Recently some encouraging developments from Europe offered an approach that could bring back the American chestnut.

Chestnut blight spread to the European chestnut *(C. sativa)* in the late 1930s and became a serious con-

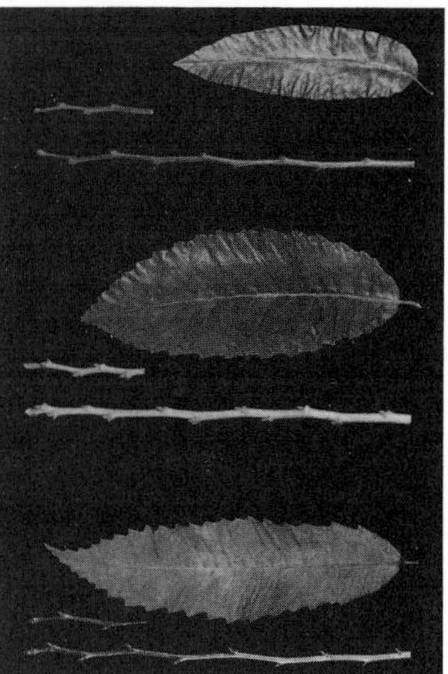

At the turn of the century massive American chestnuts rose from forest floors over much of the U.S., including the Great Smoky Mountains of North Carolina (left). By the end of the 1940s they had been leveled by a blight-causing fungus. Leaves and twigs of the Japanese, Chinese, and American chestnuts (above, top to bottom) are compared.

(Left) Forest History Society; (right) Connecticut Agricultural Experiment Station cern to those who were raising these smaller trees in orchards. Again conventional control methods were frustrated, but in the 1950s the spread of blight seemed to slow on its own in Italy.

During the 1960s Jean Grente, a French pathologist, collected samples of fungus cankers from diseased but surviving trees and discovered a weak strain of the blight fungus that apparently could be resisted by the trees. When this hypovirulent strain is grown with the virulent strain, they fuse and the latter also becomes hypovirulent. One current speculation is that a viruslike agent causes hypovirulence. Exploiting the transmissibility of hypovirulence, Grente developed a treatment method whereby a small number of cankers per hectare (10,000 sq m; one hectare equals 2.47 acres) are inoculated in an orchard. Complete healing is expected to take ten years.

There has been mixed enthusiasm over such developments in the U.S. Some hypovirulent strains may have been detected in some naturally healing trees in Michigan. As encouraging as this might be, however, the success of European techniques has not been duplicated in the U.S. because hypovirulent strains have not spread rapidly. American chestnuts have not responded to induced hypovirulent fungus treatments because of their low resistance to virulent strains. Current U.S. efforts involve the development of more resistant trees, which survive long enough to respond to treatment, and of methods for promoting the dispersal of the viruslike agents of hypovirulence.

Underground relationships. The complex below-ground relationships among plants, bacteria, and fungi have become increasingly more apparent to botanists and ecologists in recent years. As these symbiotic relationships are revealed, they serve to bring understanding to distribution patterns of certain land plants and promise to the use of such relationships in agricultural and other settings. For example, an immense amount of attention has been given to the use of nitrogen fixation to improve yields of agricultural crops. For years harnessing this phenomenon merely has meant growing crops known to support nodule-forming, nitrogen-fixing bacteria in their roots in order to build up soil nitrogen content. By contrast, some current efforts go so far as to seek to transfer the genes necessary for nitrogen fixation directly into crop plants. (See *Microbiology,* below.)

Perhaps less spectacular but no less interesting is research into other kinds of below-ground symbiotic relationships. One of these involves mycorrhiza, an intimate relationship between certain soil fungi and the roots of vascular plants. Although such relationships have been known to exist for many years, few specifics have been discovered about their function. It has been thought that the fungi may be instrumental in mineral procurement for the plant and that the plant may be a carbon source for the fungus.

During the past year three researchers from the University of Sheffield, England—J. A. Duddridge, A. Malibari, and D. J. Read—reported that mycorrhizae are involved in water uptake for the host plant. They raised a mycorrhiza-forming fungus, *Suillus bovinus,* in a flask containing a peat-vermiculite soil. After the unassociated fungus became established, seedlings of Scotch pine *(Pinus sylvestris)* were transferred to the flasks to allow mycorrhiza establishment. The association proved to be of the ectotrophic type, in which the fungus forms a mantle over the root tip but never penetrates root cells. The seedlings were then transferred to an apparatus in which water labeled with tritium, a radioactive isotope of hydrogen, could be added in such a way as to be in contact with regions of unmodified fungus but not with roots or mycorrhizae. Samples of all parts were examined for radioactivity in an attempt to discover a pathway for the water from its source to the needles of the seedlings.

The investigators concluded that the fungus absorbed the water and passed it to the plant roots through the mycorrhizae. The plants then transported the water through usual channels to the needles. Further investigation revealed that the fungus actually specializes into central hyphae (filaments of fungus) of comparatively large diameter surrounded by narrower, denser hyphae. This arrangement provides vessel-like structures to conduct water. Studies on plants under water stress showed that they are more likely to survive if they have mycorrhizae than if they do not.

Another below-ground symbiosis involves growth-promoting bacteria called rhizobacteria. One of these, *Pseudomonas fluorescens-putida,* has been used as a seed inoculant on potato, sugar beet, and radish to increase yield as much as 144% in field tests. A group of researchers from the University of California set out to determine why growth was enhanced. In experiments involving *Pseudomonas* and another bacterial species, *Erwinia carotovora,* they were able to show that the former effectively deprives *Erwinia* of iron in the root zone. Since *Erwinia* causes such diseases as potato soft rot and seed-piece decay, plant growth would be enhanced by their control in this way.

Pollination and fertilization. The regularly accepted pattern for flowering-plant pollination and fertilization portrays pollen landing on the sticky tip, or stigma, of a pistil (the female reproductive organ) of the same or different flower in the same species of plant. A number of pollen grains germinate tubes that penetrate the cuticle of the stigma and grow down through the length of its supporting stalk, or style, eventually reaching the ovules within the ovary. It is the ovule that contains the egg cell, which acts as the female gamete. The pollen tube contains two sperm nuclei, one of which acts as the male gamete by uniting with the egg cell nucleus.

Conventionally it is supposed that the sperm nucleus

Scott D. Russell, University of Alberta

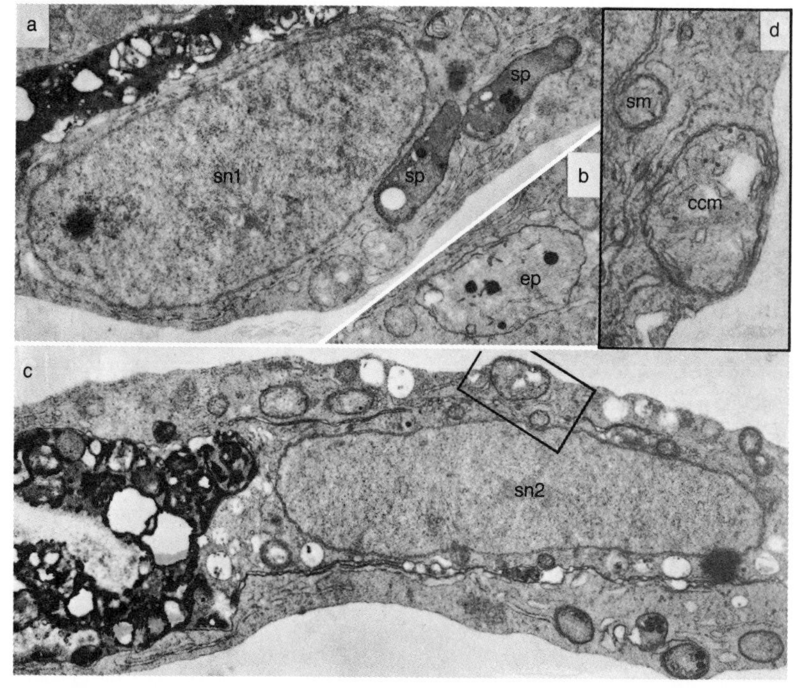

Recent microscopic studies of the embryo sac of Plumbago zeylanica *uncovered the presence of gene-carrying cytoplasmic organelles from male sperm in egg and central-cell cytoplasm shortly after gamete fusion. In a, a sperm nucleus (sn1) and two sperm plastids (sp) appear in the egg cytoplasm; for comparison, a larger, more elliptical egg plasmid is shown in b. In c, a second sperm nucleus (sn2) appears in the central-cell cytoplasm. In d, which is an enlargement of the box in c, a sperm mitochondrion (sm) lies near a central-cell mitochondrion (ccm). During fertilization the pollen tube carries two sperm nuclei: one fuses with the egg nucleus to produce the plant embryo, while the second fuses with nuclei of the central cell to produce the endosperm, the tissue that surrounds and nourishes the embryo in the seed.*

with its genetic content is the only contribution of the male parent. Any hereditary material found in the cytoplasm, such as is possessed by intracellular structures known as plastids (*e.g.*, chloroplasts) and mitochondria, would then be passed on only in female gametes. In this way cytoplasmic inheritance, as it is called, is thought to be maternal to organisms with small sperm and large eggs. Scott D. Russell of the University of Alberta reported a case in which male plastids, and thus male cytoplasmic genes, are passed on with the sperm nuclei. He was able to identify the presence of plastids from the male plant in both the fertilized egg cell and other fertilization products in *Plumbago zeylanica*. Although further work is needed to demonstrate the survival and genetic contribution of such male inheritance, Russell's work lent evidence to the existence of biparental cytoplasmic inheritance.

On another front three researchers showed that the rate of pollen-tube growth is correlated with such qualities as kernel weight and seedling vitality of the resultant generation. Working with corn *(Zea mays)*, Ercole Ottaviano and Mirella Sari-Gorla of the University of Milan, Italy, and David L. Mulcahy of the University of Massachusetts found that the longer the style, the more likely ovules will be fertilized by male nuclei of faster-growing pollen tubes. Because the styles of corn-ear pistils are quite long (they constitute the material commonly called corn silk) and increase in length for ovules nearer the base of the ear, kernels formed near the base show proportionately more traits carried by pollen with fast-growing tubes than that with slow-growing tubes in mixed batches. The researchers used the trait of colored versus uncolored aleurone (protein that contributes to kernel color) to demonstrate this phenomenon. They felt that faster growth in pollen tubes may be genetically related to plant traits, such as seedling vitality, and may be used to screen potential lines for hybrid crops. Also, because pollen-tube growth is partly supported by the style, tube-style compatibility would favor genetic combinations giving rise to vigorously growing seedlings.

Underwater reproduction. Interest in tropical seagrasses has grown since the early 1970s. These are flowering plants, growing in 1–20 m (about 3–65 ft) of water, that might be expected to require some variation in flowering and reproductive physiology from those of terrestrial plants. Many terrestrial plants flower according to a photoperiod (length of day and night) and have flower parts and pollen that are modified for dispersal by winds or animals. It is of interest to find what common factors favor flowering and pollen dispersal in underwater plants.

Some scattered observations were reported on flowering in the seagrasses *Halophila engelmannii* and *Thalassia testudinum* (turtle grass). There was indication that a temperature-photoperiod relationship exists for induction of flowering. A fairly large-scale study, reported during the year, compared field observation on manatee grass (*Syringodium* species) in numerous sites in the Gulf of Mexico and Caribbean Sea with laboratory studies. Calvin McMillan of the University of Texas grew collected plants in synthetic seawater in which temperature and day length were varied. He found that flowering is most likely to be induced at

Calvin McMillan, University of Texas at Austin

Turtle grass (Thalassia testudinum) *cultured in warm synthetic seawater was induced to flower after being kept under continuous light for more than a year and then supplied with water reduced to 24° C (75° F).*

temperatures of 22°–24° C (72°–75° F), which correspond to winter lows; this behavior explains why flowering is most frequent between January and June in the Caribbean. He also found that a day length shortened to 11 hours inhibited flowering; however, days are never this short in the Caribbean. A third factor, nutrient conditions, may also affect flowering.

Pollination adaptations of another seagrass, the Australian sea nymph (*Amphibolis antarctica*), were reported by another group of investigators. J. M. Pettitt of the British Museum (Natural History) and C. A. McConchie, S. C. Ducker, and R. B. Knox of the University of Melbourne collected plants off the coast of Victoria, Australia, and grew them in seawater in the laboratory. From their studies they noted a number of modifications for underwater pollination: pollen grains differ from those of most terrestrial plants by having no resistant outer layer (exine), by being flexible and as much as five millimeters (a fifth of an inch) in length, and by having a density similar to seawater. All of these contribute to the ability of pollen to be carried away from anthers (pollen-forming structures) by seawater currents. When pollen grains contact pistil stigmas they are bound to the surface by a waterproof adhesive, and germination is induced. The resulting pollen tube does not emerge from a preformed aperture, as with many terrestrial plants. Instead, an aperture seems to be produced relative to the position of the pollen grain on the stigma.

Herbivory. One contribution to the growing insight into relationships between plants and the animals that eat them was made by Phyllis D. Coley of the University of Chicago, who observed rates of insect grazing on the leaves of 27 species of trees on Barro Colorado Island, Panama. She measured total leaf area and the area of holes made by insect herbivores on both young and old leaves and reported the following relationships. Grazing rates were much higher on the mature leaves of fast-growing pioneer tree species than on those of slower growing persistent tree species. There was little difference between the damage done to young leaves on these two kinds of trees, but there was a moderate difference between that done to young and old leaves on the pioneers. These observations are consistent with the theory that some antiherbivore defenses are produced in later successional trees.

Melvin I. Dyer of Colorado State University reported positive effect of herbivory. He treated grain sorghum seedlings with mouse submaxillary-gland epidermal growth factor (mEGF) to simulate the effect of mouse saliva on plants that are clipped by mice. He showed that mEGF stimulates plant shoot growth for a brief time right after its addition. Dyer suggested that, because EGF is found in the salivary glands of several mammals and may even exist in insects, it may regulate plant productivity during herbivory and be part of the coevolution of plants and herbivores.

Seed dispersal by ants. Seed dispersal by ants in temperate regions has been found far more common than might be expected. Small-seeded species such as violets may benefit owing to a reduction in competition and seed predation and may experience certain advantages in seedling survival by virtue of ant dispersal. Two investigators, David C. Culver and Andrew J. Beattie of Northwestern University, Evanston, Ill., reported from experiments on two species of violets in England that ants modify seeds and deposit them in their nests or in refuse piles outside nests. Seeds thus modified are scarified (seed coats are scratched), and elaiosomes (fat bodies attractive to ants) are removed. In many plant species these effects generally result in enhanced germination. Germination in or near ant nests results in clumping of seedlings, which has uncertain advantages, but the nest environment itself has the positive effects of increased seedling emergence and survival because of higher nutrient levels. Because violets reproduce asexually to a large extent, increased survival of seedlings from sexually produced seed may be a way to maintain genetic diversity in such populations, according to the investigators.

Beattie and another Northwestern researcher, C. C. Horvitz, reported on what they felt to be the first well-documented case of seed dispersal by ants for wet tropical rain forest herbs. In their studies in Mexico they noted that two species of *Calathea* have seeds attached to a fleshy aril that is attractive to several species of carnivorous ants. The ants behave toward the seeds as they do toward prey by carrying the seeds to their nests, removing the edible aril, and burying the seed in refuse piles. Benefits from such treatment to germination and seedling survival were expected to be found and were being studied in the field.

—Albert J. Smith

Microbiology

Judging by research papers presented at scientific meetings and by publications in the scientific literature, microbiology during the past year contributed substantially to such concerns as the improvement of environmental quality, world energy production, food production and preservation, and human nutrition. Microbiological research also offered insights into cancer, sexually transmitted diseases, hospital-related infections, fungal diseases, allergies, new antiviral drugs, the potential use of recombinant DNA to develop influenza vaccines, and the potential of interferon as an antiviral and possible antitumor agent.

Bacterial diseases. A hitherto unrecognized disease called toxic shock syndrome (TSS), discovered in 1978, came to prominence in 1980 after many cases were reported. A severe illness, TSS begins with sudden high fever, vomiting, diarrhea, muscular pain, and a sunburnlike rash; it is followed by a drop in blood pressure and in severe cases by shock and death. The disease primarily strikes menstruating young women, although male cases and female cases not associated with menstruation have been reported. Epidemiologic and microbiological studies at the U.S. Center for Disease Control, Atlanta, Ga., established the association of both the bacterium *Staphylococcus aureus* and tampons in the development of TSS in menstruating women. *S. aureus* also has been isolated from sites of infection of the skin, bone, and lung from male cases and nonmenstrual female cases.

Unusual rectangular bacterium, collected from a saturated brine pool, was described in 1980. The virtual absence of internal pressure in the organism apparently allows its cell wall architecture to determine its shape.

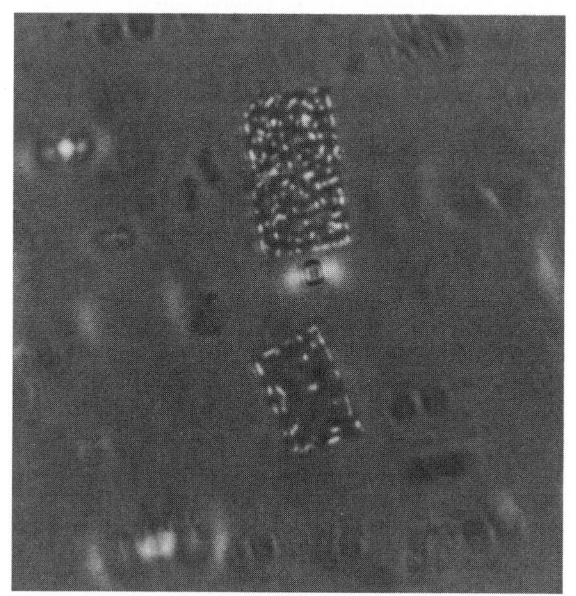

A. E. Walsby, University College of North Wales

S. aureus is a common cause of infections, but TSS seems to represent a new and severe disease manifestation. According to current knowledge, there is a significant association between tampons and TSS in menstruating women. Apparently *S. aureus* grows in the menstrual fluids absorbed by tampons and produces a toxin (perhaps more than one) that causes TSS. A California microbiologist reported discovery of a new toxin produced by *S. aureus*, but this claim has not yet been proved to everyone's satisfaction. (*See* Year in Review: MEDICAL SCIENCES: *General medicine*.)

From evidence developed over the past few years the bacterium that causes Legionnaires' disease seems to be quite common. It appears to be an organism that lives in water, and thus it is part of the human environment. Moreover, because the disease is generally not fatal and because it resembles pneumonia in many respects, it went unrecognized as a separate entity until recently. Curiously, in order to grow the bacterium in the laboratory, a rich medium is required; by contrast, its natural watery habitats are generally nutrient-poor. According to a scientist at the U.S. Department of Energy's Savannah River nuclear facility in South Carolina, an explanation for this paradox is that in nature the organism grows in association with cyanobacteria (blue-green algae) and that its nutrients, all or in part, come from this partner. It was also found that the organism can tolerate a wide variety of conditions with respect to temperature and chemical components in its watery medium.

That there was an outbreak of anthrax in the spring of 1979 near Sverdlovsk in the Soviet Union is certain. What is less certain is the cause of the outbreak. The anthrax bacillus has potential as a biological warfare agent, and there exists a facility on the outskirts of Sverdlovsk that U.S. intelligence agencies have long suspected of being a biological warfare installation. The Soviets reported that the outbreak of anthrax was of the common gastric variety, caused by eating meat from infected livestock. U.S. intelligence officials, however, believed that they had enough evidence to suggest that the outbreak was due instead to the inhalation of windborne anthrax spores following an accident in a germ warfare plant. Out of hundreds of cases of anthrax at Sverdlovsk estimates of deaths ranged from 40 to 1,000, an unusually high number to be blamed on contaminated meat.

New disease agents. The search for new disease-producing agents continued, particularly those that could be linked with known diseases of uncertain cause. Recently scientists from Colorado reported the isolation of two viruses from autopsy tissues of multiple sclerosis patients. The viruses were identified as members of a group (coronaviruses) that cause a disease resembling multiple sclerosis in animals. It is still premature, however, to claim that coronaviruses play a role in the multiple sclerosis disease process.

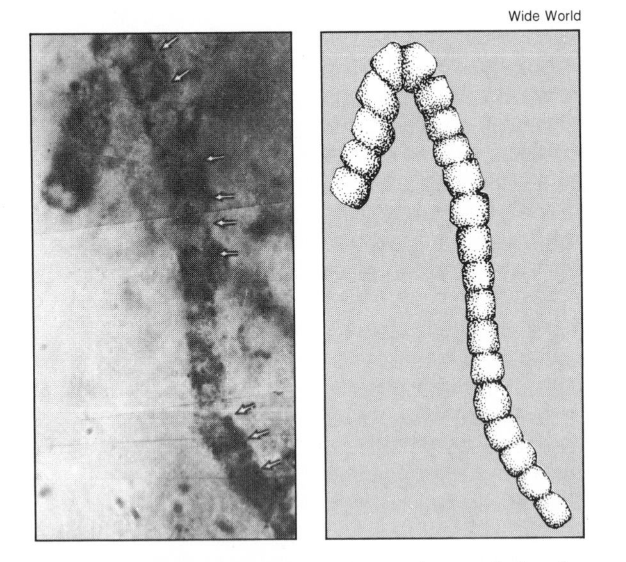
Wide World

Highly magnified thin section of Australian rock dated at 3.5 billion years contains what some scientists believe to be bacterial cells, the oldest known fossil organisms. Arrows indicate apparent cell walls within the structure.

That viruses can initiate chronic degenerative diseases has received considerable experimental support in recent years. During the year researchers isolated a virus, belonging to the retrovirus group, from an adult goat with chronic arthritis. In the kids of healthy goats this virus caused arthritic lesions similar to those in the chronically arthritic adult. Moreover, the virus could be reisolated from the experimentally induced lesions. These results provided a very strong link between viral infections and at least some forms of arthritic disease.

The possibility that viruses might be one of the causes of juvenile-onset diabetes in humans is supported by its abrupt onset, its seasonal incidence, the presence of inflammatory cells in the pancreas, and the destruction of insulin-producing pancreatic cells. Investigators at the U.S. National Institute of Dental Research, Bethesda, Md., were studying a virus that produces a diabeteslike disease in mice, one that resembles juvenile-onset diabetes in many respects. This animal model system may point the way toward an understanding of human juvenile diabetes.

Researchers in Buffalo identified *Capnocytophaga ochracea* as a predominant microorganism in periodontitis, an inflammation of the gums, in patients with juvenile-onset diabetes. A substantial number of patients with juvenile-onset diabetes also suffer from periodontitis. Knowledge of the causative organism in periodontitis is important because generalized effects of the periodontal infection may lead to a serious diabetic crisis. Identifying the predominant organism in the disease paves the way toward its control by such procedures as immunization and treatment with appropriate drugs.

In 1976 researchers in California discovered that the bacterium responsible for botulism *(Clostridium botulinum)* could grow and produce the toxin botulin in babies' intestines. Subsequent investigations led to the suggestion that infant botulism has been responsible for some cases of sudden infant death syndrome, or crib death. Botulism bacteria produce a toxin that is one of the most poisonous substances known. Formerly it was thought that botulism occurred exclusively as a result of eating food contaminated with *C. botulinum* and its toxin. This organism, which is widespread in nature, does not grow in intestines of adults or older infants. Most recently scientists in Wisconsin found that the bacterium can grow and produce toxin in the large intestine of infant mice 7–13 days old, but not in younger or older mice. Other bacteria normally found in the intestines of mice older than 13 days seem to exert a protective effect and prevent botulism infections. By extrapolation, it is thought that human infants probably lack such protective bacteria initially but acquire them as they grow older.

Hepatitis vaccine. Efforts to better understand and prevent hepatitis A (infectious hepatitis) and hepatitis B (serum hepatitis) have been hampered by the inability of researchers to cultivate the causative viruses. In the 1960s, however, U.S. scientists observed a viruslike particle in the blood of patients with hepatitis B, a substance later shown to be the protein coat of the hepatitis B virus. Recently scientists perfected a means of isolating this substance from blood for use as a vaccine to immunize patients against hepatitis B. In addition, there is evidence that this vaccine might provide protection against a form of liver cancer called primary hepatocellular carcinoma, because it is thought that infection with hepatitis B precedes this form of cancer. The first large clinical trial of the vaccine in the U.S., enlisting more than a thousand volunteers, was a solid success. Upon final approval by federal authorities, the vaccine will be used first on high-risk patients because of its scarcity and expense.

In a cooperative research effort scientists from West Germany and Switzerland developed a method of cultivating the hepatitis A virus. This achievement should lead to techniques for producing a vaccine against hepatitis A.

Plasmids and bacterial resistance. Many bacteria transiently harbor small, cytoplasmic circular DNA molecules, known as plasmids, that are capable of self-replication. An important recent development in understanding bacterial infections is that genetic determinants for certain disease-producing characteristics of bacteria, such as factors that destroy red blood cells or produce toxins and adhesives, can be carried by plasmids. Most recently a worker in the U.S. reported on the presence of a plasmid in a highly infectious bacterium that attacks fish. The plasmid mediates a novel infectious mechanism, that of an efficient iron-capturing system. On losing the plasmid the bacterium

From JOURNAL OF BACTERIOLOGY, David White, Jeanette A. Johnson, and Karen Stephens, vol. 144, no. 1, pp. 400–405, October, 1980

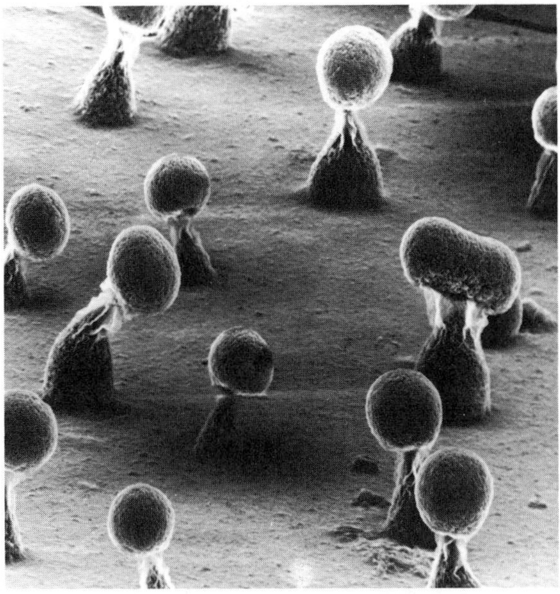

From FEMS MICROBIOLOGY LETTERS 9, Karen Stephens and David White, pp. 189–192, 1980

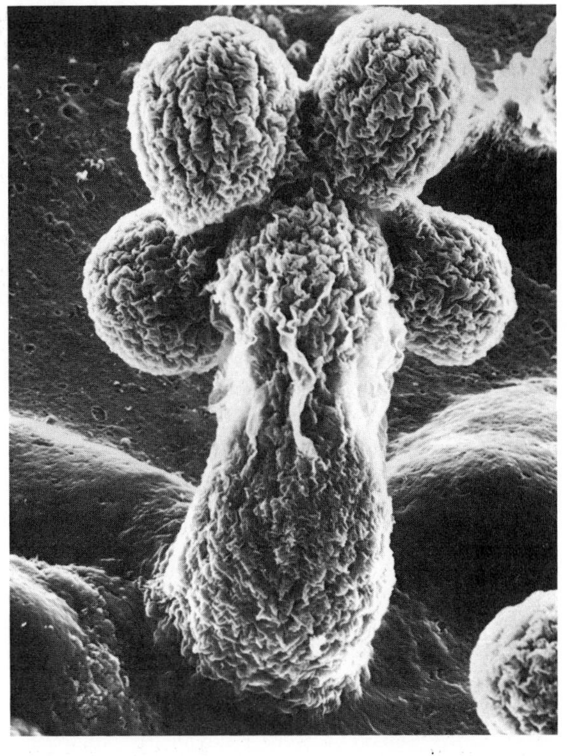

Ability of the myxobacteria to form fruiting bodies (above and right) is unique among procaryotes. Studies of the effects of chemicals and radiation on their intercellular cooperation is shedding light on how individual bacterial cells communicate.

loses its ability to infect fish and to grow under conditions of restricted iron content. Thus in this instance, as in many other cases, the ability of an organism to cause infection frequently depends upon the ability of the invading organism to compete successfully with the tissues of the host for iron.

Some plasmids also confer bacterial resistance to drugs and heavy metals. One such plasmid confers bacterial resistance to tetracycline, an antibiotic. It had been thought that the plasmid made its host bacteria unable to transport tetracycline across the bacterial surface membrane and into the cell. According to workers in Boston, this appears not to be the case. Instead, tetracycline-resistant bacteria were reported to transport tetracycline out of the bacterial cells as rapidly as it was taken into the cells, a novel form of protection. (For additional information on drug resistance in bacteria, see *Molecular biology,* below.)

Bacteria as well as humans are poisoned by such heavy metals as mercury. Some bacterial-resistance plasmids, however, confer mercury resistance. There is a possibility that mercury-resistant bacteria may one day be used to treat victims of mercury poisoning because these bacteria can chemically alter mercury compounds to less toxic forms. In a cooperative venture scientists from the U.S. and the U.K. were testing the feasibility of establishing mercury-resistant bacteria in the intestines of rats to determine whether the rats will become less susceptible to mercury poisoning. If successful, it may be possible to introduce mercury-resistant bacteria into the intestinal tracts of persons who are victims of mercury poisoning or whose occupations expose them to mercury.

Agricultural and environmental applications. Bacterial nitrogen fixation continued to be of interest. Recently an International Network of Legume Inoculation Trials was initiated, primarily to determine whether inoculating tropical leguminous plants with their appropriate nitrogen-fixing bacterial partners will increase crop yield without dependence on expensive chemical nitrogen fertilizers. Although legume inoculation has long been in wide use in temperate zones, a scarcity of basic and applied research has slowed the transfer of bacterial nitrogen-fixation technology to tropical legumes. Thus, the newly established Network is a means of promoting increased agricultural productivity throughout the tropics by encouraging appreciation of the benefits of the legume-bacterial interaction and by strengthening the capability of scientists in dealing with legume production.

A new species of *Rhizobium* bacteria that fixes nitrogen in association with mesquite, a desert shrub, was discovered by a scientist from California. This bacterium thrives under such adverse environmental conditions as high salt concentrations, acid soils, and drought. In association with desirable species of mesquite plants, the new bacterial species may be useful in a program to stabilize desert sands that continue to encroach on more valuable land.

Nitrogen-fixing root-associated bacteria also were

found in northern temperate and subalpine forest plants, and nitrogen fixation by cyanobacteria in high-temperature thermal springs was also observed. Thus, bacterial nitrogen fixation appears to be much more widespread than formerly thought.

Fungi grow in association with roots of trees in all soils and temperature zones. These fungal rhizomorphs absorb and transport phosphates to the plant roots. Recent research suggests that this association can be harnessed to achieve more economical use of costly superphosphate fertilizers and better exploitation of cheaper, less soluble rock phosphate. In related work scientists in England showed that fungal rhizomorphs also absorb and transport water to the tree root. In fact, these rhizomorphs appear to represent functional extensions of the root system. Such findings have important implications for the planting of trees on land reclaimed from strip mining or spent oil shale. The trees are unlikely to survive unless appropriate symbiotic rhizomorphs are also present.

In 1980 scientists were studying the microbial breakdown of constituents that might be encountered in waste water from coal gasification. The ultimate goal is an understanding of the metabolic dynamics that control the biodegradation of this complex organic waste. Other scientists were studying the microbial breakdown of oil-shale retort water to determine the biodegradability of this source of complex organic compounds. Still other investigators were probing such matters as the effect of shale retort water on nitrogen-fixing bacteria and microbial activity in soils overlying spent oil shale. It is commendable that such studies are under way before oil shale and coal gasification technologies are fully implemented.

—Robert G. Eagon

Molecular biology

During the intense debate in the mid-1970s over the safety and propriety of recombinant DNA experiments, proponents of the experiments offered as a principal argument the potential benefits of recombinant DNA: genetically engineered bacterial strains that produce insulin, growth hormone, interferon, viral antigens for use as vaccines, essential amino acids, vitamins, and useful enzymes. With the passage of time recombinant DNA experiments have been widely perceived as safe. The hazards that they were alleged to pose have not materialized, even in view of a burdensome and restrictive surveillance procedure unique in the history of scientific research. On the contrary attention has shifted so far from potential hazards to potential benefits that a new class of professional specialist has emerged—the investment counselor who evaluates the financial prospects of the dozens of new companies engaged in recombinant DNA research for profit.

Large fermenter at Genentech, Inc., of San Francisco provides an environment that encourages rapid multiplication of genetically engineered bacteria. The commercial company has programmed bacteria to make a number of human proteins of medical importance, including interferon and insulin.

Dan McCoy—Rainbow

The emergence of such companies has raised a new set of questions for discussion, some of which are particularly poignant for the financially troubled universities in whose laboratories the basic research leading to the formation of the companies was carried out. Should universities participate directly in genetic engineering companies, for example, as investors? What contractual arrangements for carrying out company research in university laboratories are fair and proper? How aggressive should a university be in pursuing patents and licenses for inventions that are made in its laboratories?

The scientific underpinning of these companies has been described in Year in Review: LIFE SCIENCES: *Molecular biology* since 1976. Recent developments include fine-tuning for optimum performance of bacterial factories and extension of the general recombinant DNA procedures to eucaryotic cells.

Improving bacterial performance. In the construction of a bacterial strain for maximum production of a desirable protein, several factors must be accommodated in the engineering design: (1) The gene for the target protein should be transcribed into RNA abundantly. (2) The RNA should be translated into protein efficiently. (3) The protein should be stable in the bacterium, or it should be secreted and stable outside the cell, or both.

Abundant transcription has been solved most often by fusing, with suitable recombinant DNA techniques, a strong promoter to the target gene. A promoter is a short stretch of DNA containing a sequence of 20–40 nucleotide pairs to which the enzyme RNA polymerase binds and near which the enzyme initiates the synthesis of messenger RNA. Promoters for different bacterial genes differ slightly in nucleotide sequence and perhaps also in structure. They differ greatly in strength; *i.e.*, the frequency with which RNA polymerase binds and initiates transcription of RNA. The difference between a strong and a weak promoter can be at least a factor of a thousand, with a corresponding difference in the abundance of messenger RNA. A strong promoter favored by genetic engineers is the promoter for the *lac* operon of the common intestinal bacterium *Escherichia coli*. The *lac* operon is the set of genes that includes β-galactosidase, the enzyme that splits the sugar lactose into the simpler sugars glucose and galactose.

The second factor, the efficiency of translation of messenger RNA, has been approached empirically because fewer of the fundamental details of translation efficiency are known. Most bacterial messenger RNA's contain a very short nucleotide sequence, located between the beginning of the RNA molecule and the first nucleotide triplet which codes for the amino acid that starts the protein chain. This sequence is complementary to a nucleotide sequence at the end of the RNA molecule in the small ribosomal subunit. (Ribosomes are intracellular particles made of protein and RNA that serve as workbenches for the translation of the genetic code. The site for initiation of protein synthesis is on the smaller of the two ribosomal subunits that participate in protein synthesis.) This complementary relationship between one end of messenger RNA and the other end of ribosomal RNA was first noticed by John Shine and L. Dalgarno of the Australian National University, Canberra, in 1975.

UPI

Microscopic crystals of synthetic human insulin are the result of a gene-splicing method developed by Genentech for the pharmaceutical company Eli Lilly. Clinical trials of the protein began in London in 1980.

The so-called Shine-Dalgarno sequences of bacterial messenger RNA's are highly variable, and until 1980 it was not even known whether they were quantitatively significant in terms of protein synthesis. However, recombinant DNA methods, involving fusion of the Shine-Dalgarno sequence for one gene onto a different structural gene, led to several clear results: not only is the precise sequence of nucleotides important, but also the distance between the Shine-Dalgarno sequence and the nucleotide triplet in the message that codes for the first amino acid is crucial in determining the efficiency of translation of the message into protein.

The third factor seems to be less amenable to simple solution. Bacteria contain proteases, which are enzymes that split the peptide bonds of proteins, essentially converting the product of protein synthesis back to the starting material, the amino-acid building blocks. The physiological importance of proteases to bacteria is not completely clear. Under conditions favorable to bacterial growth there is very little breakdown of normal bacterial protein. When bacteria are starved for nitrogen, without which they cannot make new amino acids, they will use proteases to convert some of their existing proteins back to amino acids. In the case of certain bacterial species—those that are capable of fixing nitrogen (converting atmospheric nitrogen to ammonia)—the sacrifice of a few proteins to make amino acids is reasonable because the amino acids will be used to synthesize enzymes for nitrogen fixation. Soon the bacteria will resume growth, using new amino acids produced as a consequence of atmospheric nitrogen fixation. Other bacterial species may use proteases to "digest" foreign protein, converting it into amino acids that can be transported into the bacterial cell and used for the bacteria's own protein synthesis.

Proteases appear to serve yet another function in bacteria, and it is this process that plagues genetic engineers. By mechanisms that are not understood, bacteria recognize intracellular proteins that are "wrong" and destroy them with proteases. Wrongness is still being defined experimentally. Replacement of one amino acid by another, by mutation at a particular location in a protein, may or may not lead to destruction of that protein. Mutations that cause premature termination of synthesis of a protein (nonsense mutations) with the resultant formation of truncated proteins may or may not lead to destruction of the truncated fragment. Finally, there exists a number of chemical compounds called amino-acid analogues that are sufficiently close in structure to amino acids to be taken up by cells and incorporated into protein. Fluorophenylalanine, for example, will occasionally fool a cell and be incorporated into proteins in place of the normal amino acid phenylalanine. But the amino-acid analogues do not fool the proteases. The latter enzymes recognize analogue-containing proteins as wrong and destroy them.

The problem facing a genetic engineer is that some of the proteins he or she wishes to make in bacteria by recombinant DNA methods are perceived as wrong by proteases. So little is understood of the phenomenon that one cannot predict whether a product of a given foreign gene that has been introduced into the bacteria will be destroyed or not. Empirically it may be possible simply to select bacterial strains which lack the proteases that destroy wrong proteins.

Transposable genetic elements. Nature, of course, has been involved in genetic engineering for a rather long time. One of the most exciting developments in molecular biology during the past several years is the recognition of transposable genetic elements in the chromosomes of virtually every organism studied, from bacteria to yeast, to fruit flies, to plants, and probably to man. It is even possible that transposable elements play a crucial role in the transformation of normal cells into cancer cells.

Although such elements are responsible for variegated pigmentation in the kernels of corn, studied at the Cold Spring Harbor (N.Y.) Laboratory by Barbara McClintock more than 30 years ago, their molecular properties were not elucidated until they were recognized in bacteria. The history of transposable genetic elements in bacteria is a wonderful tale of a large number of apparently disparate discoveries that recently have merged to form a unified, although still incomplete, picture.

The story begins with the isolation, in three different laboratories, of bacterial mutants with unusual properties. The particular mutations in question, all in well-studied genes of *E. coli*, were polar; that is, they resulted in the loss of function of nearby genes as well as of the genes in which the mutations occurred. When these mutations were studied by physical means, it turned out that each of them was due to the insertion of a stretch of DNA several thousand nucleotide pairs long into the middle of the mutated gene. Further study revealed that the inserted sequence (called an IS element) could be found at many different locations in the same gene, or in other genes. In other words, it could be transposed from one part of the bacterial chromosome to another.

Up to this point it was difficult to determine when and where an IS element lurked in a chromosome because it could be recognized only in terms of the kind of mutation it created. The situation changed as a result of a second, unrelated line of investigation: the transfer of antibiotic resistance from one bacterium to another. Such resistance was known to be due to the activity of genes for enzymes that inactivate specific antibiotics. Transfer was due in turn to small circular DNA molecules (plasmids) capable of both autonomous replication in bacteria and of transfer from one cell to another. These molecules were named resistance transfer factors (RTF's) or R plasmids.

It had been noticed in Japan and elsewhere that certain antibiotic-resistance genes could be transferred from an RTF to the bacterial chromosome itself or from an RTF to another plasmid carried in the same bacterium. The latter observation was of crucial importance because plasmids can be physically separated from chromosomal DNA and rather easily characterized. It soon became apparent that when a plasmid acquires an antibiotic-resistance gene it acquires a new stretch of DNA inserted into itself. Put another way, the antibiotic-resistance gene behaves like an IS element; *i.e.*, it is transposable from one piece of DNA to another. The great advantage afforded by antibiotic-resistance genes over IS elements is that the former can be selected directly on the basis of the new property which they confer on the cells that carry them.

Physical characterization of newly transposed antibiotic-resistance genes produced several surprises. First, the length of DNA transposed was much more

(Top diagram & photo) Stanley N. Cohen, Stanford University School of Medicine; (bottom diagram) Joany Chou and Malcolm Casadaban, University of Chicago

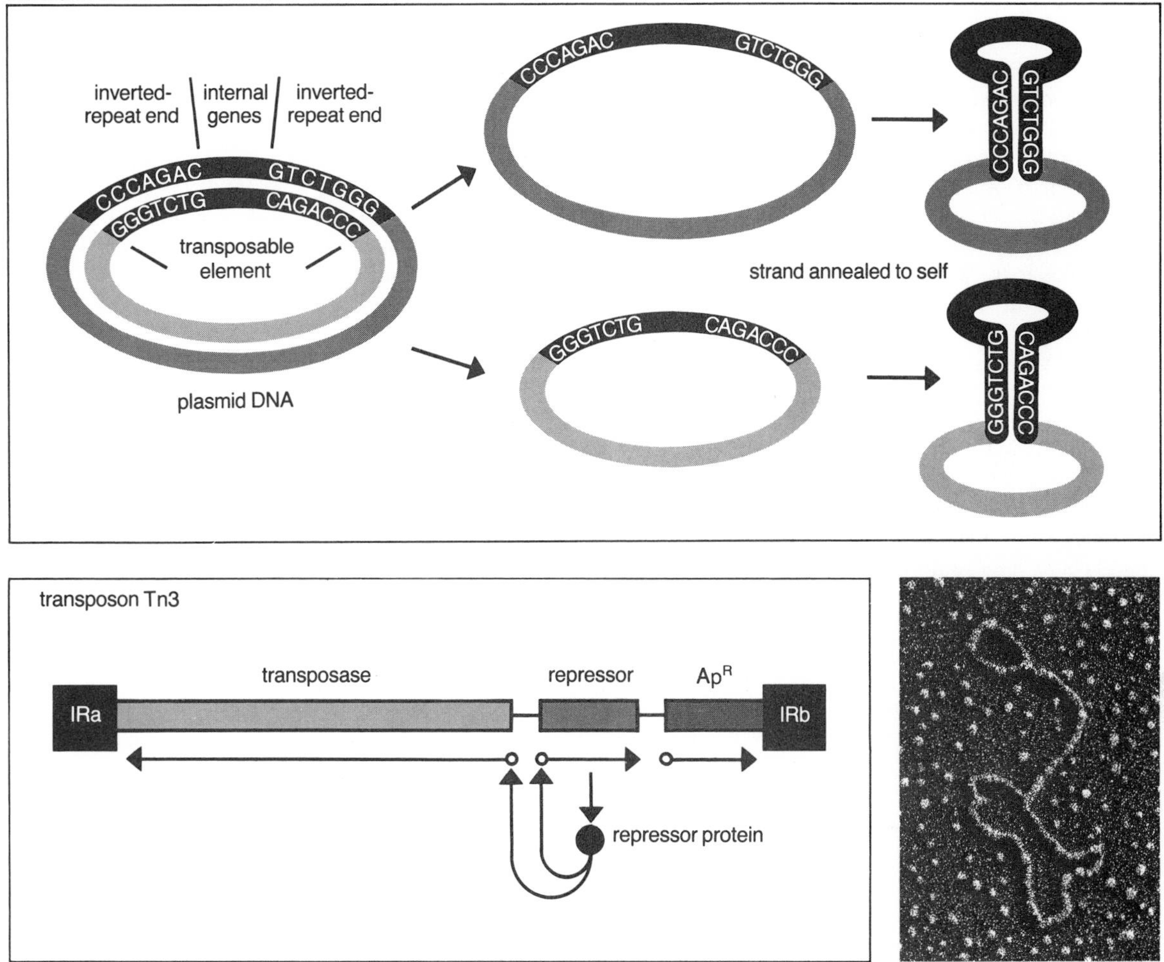

Diagram (top) shows the repeated-but-inverted nucleotide sequence (seven pairs long here) at opposite ends of a plasmid transposon. The plasmid's two DNA strands are linked by their complementary nucleotide building blocks: adenine (A) pairs with thymine (T), and guanine (G) with cytosine (C). When separated experimentally and allowed to anneal to itself, each strand forms a characteristic loop-and-stem structure that is visible microscopically (above right). Transposon Tn3 (lower diagram) has at least four different regions: invert-repeat ends (IRa, IRb), a gene (Ap^R) for ampicillin resistance, and genes for transposase and a repressor protein. Straight arrows show directions of DNA transcription for the various regions.

than needed to code for the drug-inactivating enzyme alone. Second, the nucleotide sequence at one end of the transposed segment was exactly duplicated, but inverted, at the other end of the segment. Finally, and most marvelously, the DNA at each end of the transposed segment was found to be identical to an IS element. These discoveries, taken together, led researchers directly to the definition of a transposon: a transposable genetic element in which genes for some cellular function (they need not be for antibiotic resistance) as well as genes for transposition are flanked by inverted repeated nucleotide sequences that are themselves IS elements.

Significant advances in understanding the process of transposition were made in recent months. First, determination of the nucleotide sequences at the sites where transposons become inserted showed that short lengths of DNA (five, nine, or eleven nucleotide pairs, depending on which IS was studied) were directly duplicated, immediately flanking the inserted material. Second, determination of the entire nucleotide sequence of a transposon, along with analysis of the function of parts of the transposon by fusing those parts to the gene for β-galactosidase, revealed two genes in addition to the antibiotic-resistance gene. One is called transposase; it codes for an enzyme that initiates the biochemistry required for transposition. The second gene has been called repressor because it regulates transcription of the transposase gene; yet it does more, also participating in the biochemistry of transposition directly.

One of the great possibilities opened up by recombinant DNA technology is the detailed examination of complex chromosomes. Among the early fruits of such labors has been the discovery of transposable genetic elements in yeast and in the fruit fly *(Drosophila)*. Much less is known of these elements than of their bacterial counterparts, but the parallels are striking. An element consists of several thousand nucleotide pairs flanked by two IS elements; wherever they insert, the recipient sequence is duplicated. In *Drosophila* there are as many as 30 different elements, and each can occur at 30 or more locations in the *Drosophila* chromosomes. They can transpose to new locations, and when they do they cause mutations or alter the regulation of genes in the vicinity.

Although there is no direct evidence for IS elements or transposons in man, there is intriguing new information about the nucleotide sequences of retroviruses, the RNA-containing viruses that cause cancer in birds and rodents. When such viruses infect cells, the viral RNA is copied into DNA, called the provirus, which can be inserted in many places in the cellular chromosomal DNA. It turns out that proviral DNA contains repeated IS-like sequences at both ends and that, wherever it inserts, a small segment of host DNA is duplicated on either side of the insert. If it is subsequently shown that transposition of the provirus plays an important part in transformation of normal cells into cancerous ones, the direct relevance of fundamental studies of bacterial genetics to the cancer problem will have been established beyond doubt.

—Robert Haselkorn

For a description of bacterial resistance to antibiotics and its medical implications, see *1979 Yearbook of Science and the Future* Feature Article: THE MICROBES FIGHT BACK.

Zoology

Pioneering achievements in cellular zoology during the past year included the modification of cells in living animals by means of gene transfer and the maintenance of a transferred gene through an animal's developmental stages. A link was found between stress, hormones, and the phenotypic expression of differentiating nerve cells. In physiology, scientists explored a system for the control of muscles based on the opposing effects of pairs of chemicals, noted the occurrence of neurons containing multiple transmitters with differing effects, and identified neurons with common functions through their possession of common molecular surface markers. Research continued on the mechanisms employed by animals for orientation, and semantic communication among monkeys was studied. The opening of a very large marine aquarium facility offered investigators the chance to study coral reef communities under controlled conditions. In evolutionary research significant findings were related to an extraterrestrial cause of mass extinctions, to tooth-making potential in birds, and to primate relationships revealed through fossil evidence.

Cellular and developmental zoology. Martin J. Cline of the University of California at Los Angeles and his colleagues reported the first successful transfer of a selected gene into living animals. Previously DNA had been transferred into animal cells only in culture. Cline's group found that mouse bone marrow cells given a gene for drug resistance and then returned to the mice replicated the foreign gene with each cell division and ultimately conferred drug resistance upon the mice. If genetic engineering is to be used to greatest effect, however, it is necessary that genes be inserted into early embryos and be replicated successfully throughout all the developmental stages of the animal.

Frank H. Ruddle of Yale University and his co-workers took a step in that direction. They injected thousands of copies of a bacterial plasmid, a circular self-replicating DNA molecule that exists independent of the bacterial chromosome, into each of many newly fertilized mouse eggs; the plasmid had first been altered by insertion of a viral gene coding for a specific enzyme. The eggs were then implanted into foster mothers. Later, DNA was extracted from the newborn mice, and, in a small number of cases, a DNA sequence was found that matched that of the foreign gene. As this technique is refined, researchers may be able to determine if foreign genes can be passed to an animal's offspring.

Findings of fundamental importance in understanding nerve cell development and function were reported by G. Miller Jonakait and her colleagues at Cornell University Medical College, New York City. They reported that stress alters the phenotypic expression (the expression of an organism's genetic potential) of developing nerve cells in rats. It had been known that the drug reserpine, when used as a stressor (stress-producing agent) in pregnant rats, causes behavioral and neurological disorders in the offspring. In the fetus Jonakait found that reserpine changed the manner in which the embryonic nerve cells expressed their genetic potential for production of the chemical neurotransmitter noradrenaline; the period of phenotypic expression was increased. It is known that reserpine increases blood levels of the glucocorticoids. When these hormones were directly implanted in pregnant rats, nerve cells in the fetus expressed noradrenaline production for longer than usual; however, when a glucocorticoid inhibitor was used in conjunction with reserpine, the period of noradrenaline production was normal. Thus, the glucocorticoids appear to have a direct influence on the expression of the fetal nerve cells. The researchers suggested that other maternal hormones may play intermediary roles in conjunction with other stressors in altering the phenotypic expressions of neurons.

Margaret S. Livingstone, Princeton University

Lobster (upper photos) and crayfish (lower photos) exhibit opposing postures when injected with one of two normally circulating neurohormones. Serotonin activates neurons that excite flexor muscles and neurons that inhibit extensor muscles (left), whereas octopamine produces the opposite effects (right). Both postures normally are used by lobsters and crayfish in antagonistic encounters.

Physiology. Stereotyped and repetitive activity often involves the use of opposing, or antagonistic, muscles. Muscle control during such activity requires integration of centrally and peripherally directed nerve impulses and central programming and appears to involve interneurons called command fibers. The command fibers help to provide sequencing and modulation of the opposing muscles. Current research is providing information on the chemical basis of these control systems. Margaret S. Livingstone and her coworkers at Harvard University Medical School showed that injection of serotonin and octopamine into lobsters and crayfish triggers opposing postures, a hyperextension of the tail and claws with octopamine and a flexed pose with serotonin. The effects were traced to the central nervous system, where the chemicals have opposite effects on the activity of the nerve cells innervating the muscles. Livingstone suggested that an interaction of the chemicals with the command fiber system occurs and raised the intriguing possibility that control of antagonistic muscles by the opposing effects of paired amines may be of general occurrence.

Work with somewhat similar implications was reported by Tomas Hökfelt of the Karolinska Institutet, Sweden, and Benjamin S. Bunney and collaborators at Yale University. They demonstrated that some neurons contain two or more neurotransmitters. For example, certain midbrain cells release dopamine and, in addition, cholecystokinin (CCK). Dopamine is inhibitory, while CCK activates nerve cells. Other examples are known, and it is becoming apparent that paired transmitters with opposing effects may interact to provide a precise yet flexible mechanism for the regulation of cell and organ function.

Progress was made in identifying networks of neurons that make up functional units. It has been suspected that neurons participating in a common function (for example, processing information from pressure sensors) are "marked" by specific molecules on their surfaces. Birgit Zipser and Ronald McKay of the Cold Spring Harbor (N.Y.) Laboratory reported that they had been successful in detecting the molecular markers in leeches. They made monoclonal antibodies to the whole isolated nervous system of the leech; of the specific antibodies produced, about 300 kinds bound to neuron surfaces. More important, 41 kinds bound to restricted sets of neurons. The network of neurons that makes up such a set and that is tagged by a given antibody is apparently active in a given function or group of related functions.

Animal behavior. Recent work adds to earlier reports (see *1981 Yearbook of Science and the Future* Year in Review: LIFE SCIENCES: *Zoology*) that pigeons and honeybees possess crystals of magnetite (magnetic iron oxide). The resulting sensitivity to the Earth's magnetic field is thought to be the basis for some kinds of directional behavior, such as homing by pigeons. In bees it influences informational dances and comb-building orientation. James L. Gould and Joseph L. Kirschvink of Princeton University and their collaborators showed that the magnetic sensitivity of a bee is probably due to the presence of some 100 million crystals of magnetite. Perhaps movement of the crystals as they maintain alignment with the external field stretches or compresses the structures holding them, providing a basis for a detector system.

In the pigeon a similar number of larger magnetite crystals act as stable magnets. David Presti and John Pettigrew of the California Institute of Technology recently suggested that the detector system of the pigeon might involve a coupling of the tiny magnets to sensitive spindle fibers of the complexus muscle in the neck; presumably, torque from the magnets would be communicated to the spindles. The concept of magnetic guidance systems was extended to dolphins with the report, by John Zoeger of Los Angeles Harbor College in Wilmington, Calif., and colleagues from the University of California, of a strong magnetic moment (20 times stronger than that of the Earth's field) in tissues between the roof of the skull and the brain, presumably based upon magnetite.

In most animals a part of the environment more limited than the magnetic field of the Earth is used in direction finding. Berthold Hölldobler of Harvard University found that African stink ants use "canopy orientation" in returning to the nest, after foraging. As it leaves the nest, the insect senses and "fixes" the pattern of the forest canopy against the sky. Then, after foraging as far as 5 m (about 16 ft), it uses the canopy pattern to relocate the same exit from which it departed. Canopy orientation is particularly well-suited to the reduced lighting typical of tropical forests.

In primate communication research attention for some years has focused on the ability of tutored chimpanzees and gorillas to communicate abstractly. Semantic communication, in which specific objects are referred to by systematic use of signals, has been neglected despite its importance in nature. Recently Robert M. Seyfarth and co-workers of the Rockefeller University Field Research Center in Millbrook, N.Y., reported its use in nature among vervet monkeys in Amboseli National Park, Kenya. These monkeys make alarm calls virtually only when they see their chief predators—eagles, pythons, and leopards—even though they commonly see many species of birds, reptiles, and mammals. The signals for the predators are acoustically distinct and produce responses that are appropriate to the hunting behavior of the particular predator. For example, the staccato grunts of the eagle alarm make the monkeys look up, while the high-pitched python alarm makes them look down. Interestingly, infant monkeys give the eagle alarm for any bird and the python alarm for any snake or long, thin object.

Environmental zoology. Concern for endangered animals during the past year turned again to the African elephant. Between 50,000 and 150,000 are killed each year, primarily by poachers responding to the inflated price of ivory. A recently completed census found only 1.3 million elephants left, and it is possible that the death rate has exceeded the rate of reproduction. The World Wildlife Fund launched a campaign to finance several programs to stem the slaughter.

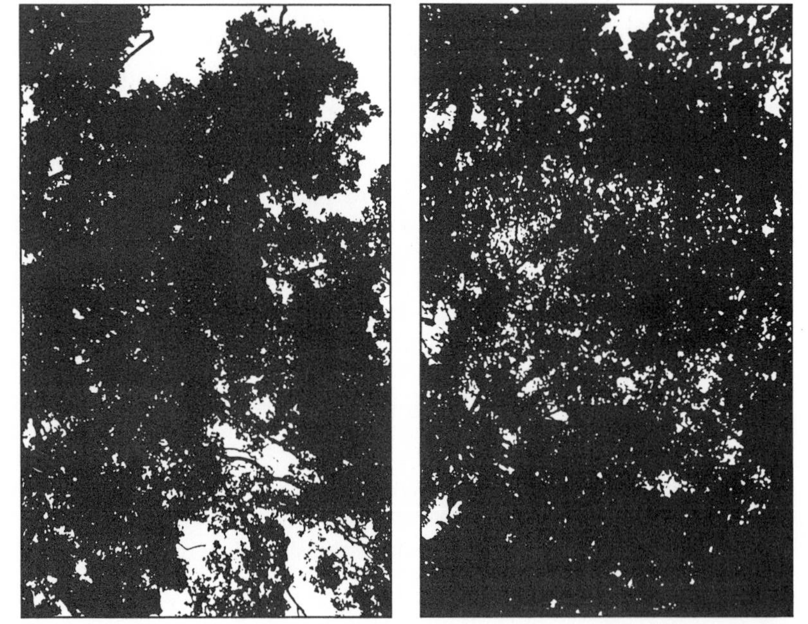

Wide-angle photographs of differing forest-canopy patterns as seen from the ground aided a study of the African stink ant, which was found to use these patterns to orient itself in its home range. More conspicuous pattern at left seemed to help the ant much more than the pattern at right.

Photos, Bert Hölldobler, Museum of Comparative Zoology, Harvard University

Norman Myers—Bruce Coleman Inc.

Devastating slaughter of the rhinoceros by poachers is the primary cause of its fast declining numbers. The animal is sought for its fibrous horn, which is made into dagger handles and is highly valued in many Asian countries as an aphrodisiac and a medicine.

Although humans are the primary threat to such animals as the elephant, rhinoceros, and whale, natural disasters take their toll, too. J. D. Woodley of the Discovery Bay Laboratory in Jamaica reported that when Hurricane Allen skirted the coast of Jamaica in August 1980 it caused heavy damage to the island's beautiful fringing reefs. Enormous numbers of echinoids, corals, sponges, and fish were destroyed as reef structures were smashed or toppled; in the shallowest regions the destruction was almost total. This event, though tragic, provided an opportunity for initiation of research into reef community formation and survival.

The dangers and frustrations of reef research were eliminated for some marine biologists when in October the Smithsonian Museum of Natural History in Washington, D.C., opened for public viewing a 11,400-l (3,000-gal) tank system housing a coral reef community. As of early 1981 the community was the largest to be kept alive in isolation from the sea. It includes 20 species of corals and totals about 200 species of plants and animals. Wave action, lighting that simulates natural intensities and photoperiods, and scrubbers to maintain chemical balance are features of the system. A multifaceted program of research into reef community ecology under controlled conditions is now well established at the Smithsonian.

Evolutionary zoology. About 65 million years ago a relatively sudden extinction of plant and animal groups, including the dinosaurs, pterosaurs, and many marine reptiles, occurred during a period defined as the boundary between the Cretaceous and Tertiary periods. An interesting controversy has developed between those who interpret these extinctions in terms of a slow-acting, Earth-based mechanism and those who support a catastrophic hypothesis based upon extraterrestrial events. In recent months several researchers presented evidence and arguments in strong support of the latter. Their arguments are based upon studies of the occurrence and pattern of abundance of the noble metals in Cretaceous-Tertiary boundary sediments, in meteorites, and in typical terrestrial basalts. Walter Alvarez of the University of California at Berkeley and his co-workers reported anomalously high levels of iridium in the Cretaceous-Tertiary boundary layer in New Zealand, Denmark, and Italy. Others obtained similar results in Spain. Most recently R. Ganapathy of the J. T. Baker Chemical Co. in Phillipsburg, N.J., reported that the pattern of abundance of most of the noble metals in the boundary layer is very similar to that in samples from meteorites.

These scientists believe that their observations can be explained in terms of the impact of an asteroid or comet upon the Earth. Presumably, meteoric material and dust ejected from the impact crater and carried into the stratosphere were eventually deposited evenly over the Earth, forming the boundary layer in question. According to the "catastrophic hypothesis," the dust layer in the stratosphere shaded the Earth for several years so that photosynthetic activity virtually stopped. The resulting disruption of food chains produced the mass extinctions that marked the end of the Cretaceous Period. (*See* Feature Article: OF DINOSAURS AND ASTEROIDS.)

Birds first appeared in the Jurassic Period. Probably evolving from a line of early saurischian dinosaurs, they possessed reptilelike teeth. Toothed birds persisted into the Cretaceous Period, but suffered extinction as modern, toothless birds appeared. Edward J. Kollar

The Times, London

Mother puma Betsy shows new cub Bonny around their enclosure at the London Zoo. Bonny is the world's first large cat to be born as a result of artificial insemination. Scientists expected to use the technique in the future to preserve such endangered species as the cheetah, snow leopard, and clouded leopard.

and Christopher Fisher of the University of Connecticut School of Dental Medicine used a complex system of embryonic tissue grafts to show that "tooth-making genes" were not lost during avian evolution but, rather, became inoperable. Kollar and Fisher found that chick tissue in mouse-chick tissue grafts produced teeth and that the teeth were more reptilian than mammalian. They suggest that an evolutionary change in developmental pattern, perhaps associated with the evolution of the bird skull to accommodate a beak, came to prevent the necessary inductive tissue interactions that would lead to tooth formation.

Well-formed tooth resulted from an experimental combination of embryonic chick and mouse tissues. Its enamel is birdlike, suggesting that the genes for tooth formation were not lost during avian evolution.

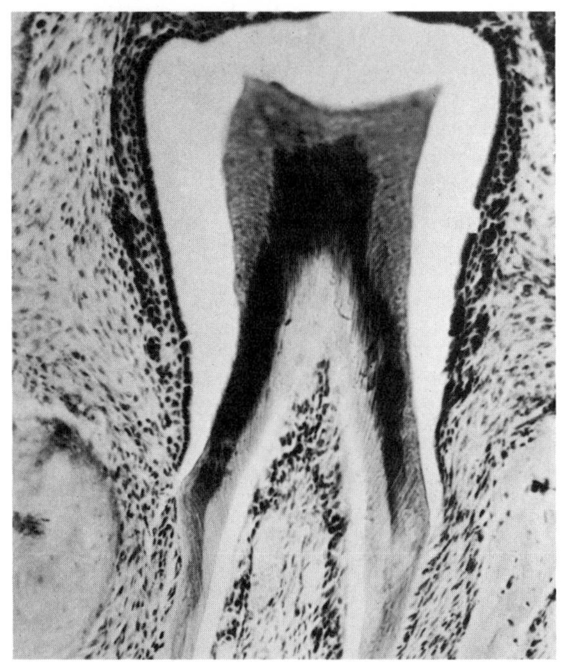

Edward J. Kollar, University of Connecticut School of Dental Medicine

In the continuing study of human ancestry John Fleagle of the State University of New York at Stony Brook and Richard Kay and Elwyn Simons of Duke University, Durham, N.C., strengthened the argument that *Aegyptopithecus* was a common ancestor of humans and apes. *Aegyptopithecus* was a small ape, weighing some 4.5 kg (10 lb), that lived about 30 million years ago. Study of jaw fossils uncovered since 1977 in the Fayum Depression of Egypt enabled Fleagle and his colleagues to infer that *Aegyptopithecus* had a social structure in which males competed for group dominance, that males were somewhat larger and heavier than females, and that they were diurnal vegetarians. *Aegyptopithecus* was anatomically more like *Dryopithecus* (a generally accepted precursor of primitive humans) than was *Propliopithecus*, a contemporary of *Aegyptopithecus*. This, together with the social advancement of *Aegyptopithecus*, supports the view that it was a forerunner of human beings whereas *Propliopithecus* was not.

The search for clues to human origins and primate relationships proceeded at the subcellular level as well. When chromosomes are stained, they exhibit definite patterns of light and dark horizontal bands. In the 1970s Jorge Yunis of the University of Minnesota Medical School refined the banding technique such that by 1981 he could visualize more than 2,000 bands on 23 pairs of human chromosomes. Yunis's application of his high-resolution banding technique to comparative studies of the chromosomes of the great apes and humans revealed that virtually every band seen in human chromosomes is matched by a counterpart in the chimpanzee. This is consistent with the results of gene-mapping studies, in which more than 30 specific genes have been pinpointed on 18 corresponding chromosomes in the human, gorilla, and chimpanzee. Of course, differences exist. Yunis argued, however, that the observable differences at the chromosome level are too slight to provide an explanation of the large phenotypic differences between primate species. Such large phenotypic differences may have to be explained in terms of small gene changes.

—John A. Mutchmor

See also Feature Articles: SYMBIOSIS AND THE EVOLUTION OF THE CELL; PLANT BREEDING TO FEED A HUNGRY WORLD.

Materials sciences

Conservation of energy during manufacturing processes was a major concern of materials scientists in recent months. Ceramicists worked to develop solutions of organometallic compounds that could be densified for fabricating ceramics with much less heating than the powders normally used. Silicon nitride and silicon carbide underwent experimentation in diesel and turbine engines as materials that could facilitate improved fuel efficiency. Metallurgists attempted to improve the design and operations of blast furnaces.

Ceramics

Many of the year's developments in ceramics were focused on environmental protection and energy conservation. The search continued for an acceptable ceramic substitute for asbestos. Because of its incombustibility asbestos had been used for centuries for such purposes as lampwicks and cremation cloths, but its widespread employment as thermal insulation began with applications in steam engines in the 19th century. By the 1960s worldwide asbestos production had reached about three million tons per year.

Unfortunately, there is now convincing evidence that asbestos is harmful. Studies have shown that asbestos workers tend to develop asbestosis, a form of chronic pneumonia associated with the permanent deposition of asbestos fibers in their lungs, and also have a higher-than-average incidence of lung cancer. Even low-level exposures can represent a health hazard, and U.S. government air-quality standards now allow not more than two asbestos fibers per cubic centimeter in work areas. As a result manufacturers and consumers are increasingly reluctant to use asbestos products, and asbestos producers are curtailing output.

It has been difficult, however, to find effective, inexpensive asbestos substitutes. Asbestos fibers can be woven into tough, wear-resistant fabrics for continuous thermal insulation to 540° C (1,000° F) and intermittent use to 980° C (1,800° F). Asbestos is also fabricated into fire-resistant construction materials and products for the chemical industry.

Several types of glass and ceramic fibers, often inter-

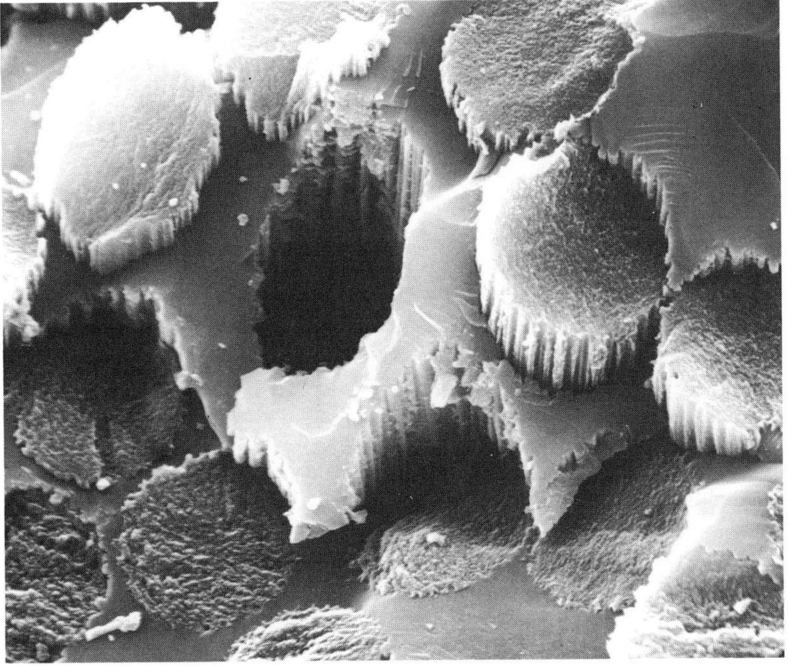

Courtesy, United Technologies Research Center

Structure of a glass-fiber composite material is shown magnified 3,000 times in an electron micrograph. Rough, column-shaped graphite fibers are embedded in a smooth glass matrix. Stability, low density, and good mechanical properties at high temperatures make such composites desirable for aerospace applications.

woven with polymer fibers for improved wear and strength characteristics, have been developed for use in thermal protection fabrics. Until recently, however, it was difficult to replace asbestos-cement building boards that were used in fire walls and heat shields where a rigid structural material was needed. Machinable calcium silicate boards have now been shown to be an excellent substitute in many of those applications. Asbestos millboard and rollboard have been even more difficult to replace, but silicate fiberboards with some flexibility and compressibility have recently been introduced for those applications. These new fiberboards have added advantages in that they can be used to 1,000° C, which is somewhat higher than the use temperature for asbestos millboard.

Because the high-temperature reactions and densification processes required for ceramics to develop their useful properties consume large amounts of energy, researchers have begun working to develop processes that would use less energy. The sol-gel process has particular promise in this regard and received much attention in recent months. Solutions of organometallic compounds, such as metal alkoxides and alkoxysilanes, or of organometallics and metal salts are polymerized to form amorphous (noncrystalline) gels. These gels are then dehydrated to form very fine, reactive, amorphous powders or monolithic porous structures. These reactive preforms can be densified with much less heating than the conventional powders normally used in the fabrication of ceramics. The homogeneity and fineness of the starting materials inherent in the sol-gel process may also lead to fewer and/or smaller defects and, therefore, to improved properties. For example, U.S. Naval Research Laboratory scientists studied this process as a means of forming high-strength single-phase and composite ceramics with exceptionally uniform, fine microstructures.

The sol-gel process was also studied as a low-energy and very pure method of forming glasses. Westinghouse Research and Development Center scientists found a way to form glassy networks by bringing alkyl and hydroxyl groups together in polymer solutions. These groups react to form the bridging oxygen bonds characteristic of oxide glasses. The active polymerizing species are reacted at room temperature to form a gel, which is then dried. In this case, however, the gels do not yield loose powders or porous structures when they are dried and heated. Instead, relatively large pieces of transparent glass can be obtained by heating them to temperatures of the order of 300°–500° C (570°–930° F). Formation of the same glasses by melting techniques would require much higher temperatures. Westinghouse researchers suggested a variety of applications for their novel glass-forming process, including high-purity optical fibers, nuclear waste disposal glasses, electronic ceramics, and antireflection coatings for solar cells.

Battelle-Columbus Laboratories and the U.S. National Aeronautics and Space Administration (NASA) were also studying the sol-gel process as a first step in the preparation in space of ultrahigh-purity optical glasses. The purity of these glasses now is normally limited by contamination from the container in which the glass is melted. Under the very low gravity and high vacuum conditions of space, acoustic positioning (the holding of an object in one place by means of sound waves) could be used to permit containerless melting. Battelle researchers were using sol-gel processing to produce starting materials that contained impurity levels below ten parts per billion. These gels would then undergo containerless melting in space to retain their purity. After the material is brought back to Earth, ultrahigh-purity fibers would be drawn from it for testing as low-loss optical waveguides for communications systems.

Containerless melting in space could also be used for compositions that cannot be made on Earth. Magnetic glasses, such as iron borate containing a large atomic percentage of ferromagnetic iron, might be produced in space for applications in magneto-optic devices. Similarly, containerless melting might permit the making of neodymium laser glasses with a high calcium content. An Owens-Illinois Co. study suggested that improved laser efficiency in these glasses might justify the additional cost of space processing.

The use of silicon nitride and silicon carbide in diesel and turbine engines for high-temperature, fuel-efficient operation has been a goal of much recent research. During recent months several turbine engine-automobile company teams have been working on this problem for the U.S. Department of Energy. For example, Ford Motor Co. and the AiResearch division of Garrett Corp. have been trying to develop and demonstrate by 1985 an advanced gas turbine engine that would power a car weighing about 3,000 lb (1,360 kg), achieve a fuel economy of about 36 mpg, meet mid-1980 pollution standards, and operate on a variety of fuels. To meet these goals the engine would operate at a turbine inlet temperature of about 1,400° C (2,500° F), some 225°–280° C (400°–500° F) hotter than current small turbine engines, and would probably use a variety of uncooled ceramic parts.

The producibility of high-quality ceramic turbine parts at an affordable cost may be a determining factor in the success of these high-temperature turbine engine programs. The High Pressure Laboratory of the ASEA Co. in Sweden announced that it had fabricated a complete one-piece turbine wheel with a solid hub and thin blades with trailing edges only 0.3 mm (0.01 in) thick. This high-density complex silicon nitride shape with close tolerances and good surface finishes was produced economically, marking a major step forward for that firm in its program with United Turbine of Sweden. That program aimed to develop a multi-shaft automotive engine in which the turbine and transmission form an integral unit, with much lower stresses on the ceramic turbine wheels.

In the U.S. the Carborundum Co. worked closely with U.S., European, and Japanese automakers to develop sintered silicon carbide parts for diesel engines. Recent scuff tests of silicon carbide run against steel in an oil spray showed that silicon carbide has a much lower coefficient of friction than alumina, silicon nitride, or chilled cast iron under those conditions. Experimental silicon carbide valve-train components also performed well and demonstrated long life when running against metals under poor lubrication conditions.

Carborundum Co. was also testing sintered silicon carbide precombustion chambers in a variety of test engines in order to evaluate the life of silicon carbide components under severe thermal stress conditions. The firm also announced progress on lightweight silicon carbide turbocharger rotors that could minimize the inertial lag in turbocharger response, reduce wear on bearings, and permit more efficient operation. A turbocharger rotor under joint development with Volkswagen passed its first design milestone of 120,000 rpm.

The use of ceramics as electro-optic image storage devices has interested scientists for many years. Sandia Laboratories researchers worked extensively on the development of transparent lead-lanthanum-zirconate-titanate (PLZT) for these applications. They discovered several years ago that images can be stored in PLZT by an intrinsic photoferroelectric effect, as opposed to the photochemical effects operative in most light-sensitive materials. The photoferroelectric effect is based on the switching of microscopic dipole areas—domains—in the PLZT ceramic. A thin, flat plate of transparent PLZT is sandwiched between two transparent indium-tin oxide electrodes. When a voltage is applied and an image is projected onto the plate with near-ultraviolet (UV) light at the PLZT's energy level band gap of 3.35 eV, domains are selectively reoriented within the PLZT in proportion to the incident light intensity. The domains form the stored image as a pattern of light-scattering centers within the ceramic. Either positive or negative images can be stored in this way, and images can also be formed by scanning the plate with an intensity-modulated UV light source, such as a suitable laser beam. Stored images can be viewed directly or projected onto a screen.

Sandia scientists recently demonstrated that ion implantation greatly increases the photosensitivity and reduces the exposure energy needed for image formation in these devices. Using a positive ion accelerator to implant hydrogen, helium, argon, or mixtures of argon and neon ions to a depth of 0.5 to one micrometer in the surface of the PLZT plates, the scientists reduced the exposure energy by as much as 10,000 times. The effect is apparently associated with an increase in the efficiency of the photoexcitation process and the density of trapping sites due to atomic disorder induced by the ion implantation. These photoferroelectric PLZT image storage devices may have important potential as nonvolatile but erasable image storage and contrast enhancement devices.

Courtesy, Hughes Aircraft Co.

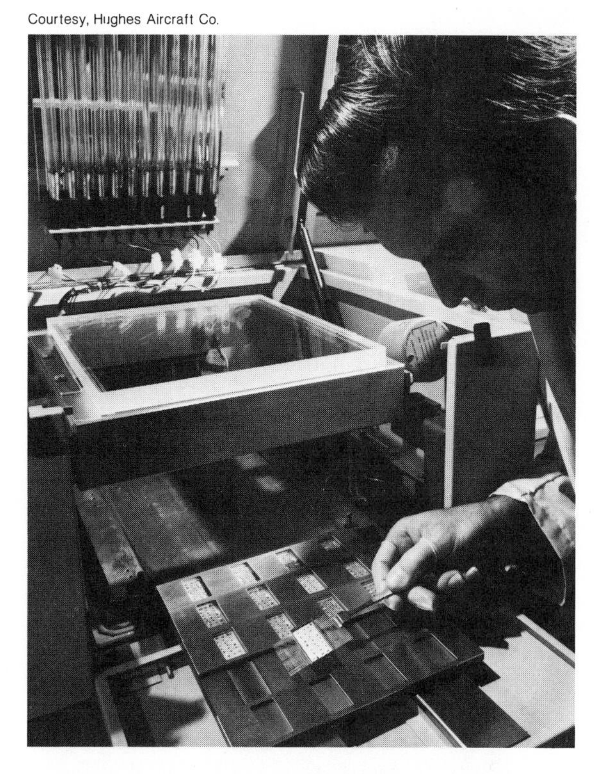

Microcircuits are prepared to undergo photochemical vapor deposition (coating with silicon nitride). This is designed to eliminate wire bond corrosion and failure due to loose conductive particles.

The use of ceramics for the storage of hazardous nuclear wastes received renewed attention in recent months. Some products of uranium fission, such as strontium-90 and cesium-137 with half-lives of about 30 years, will no longer be a problem in nuclear wastes after a few hundred years. Some of the isotopes formed by transmutation in reactor fuel rods, however, such as plutonium-239 and americium-241, have half-lives ranging from approximately 500 to 2,000,000 years. These elements will need safe storage for hundreds of thousands of years. Research on crystalline ceramics that would contain these long-lived radioactive elements over geological time spans was spurred by Pennsylvania State University reports that the borosilicate glasses originally proposed for this application would dissolve slowly if exposed to high-temperature, high-pressure groundwaters.

Many U.S. and European research groups were working on crystalline ceramics, including a variety of titanates and orthophosphates, that would offer considerably more nuclear waste protection than the glasses proposed to date. Meanwhile, the French, who had already begun production-scale incorporation of spent reactor fuel rod wastes into borosilicate glasses, were studying ways of assuring that their materials are not stored underground until safe temperature levels are achieved. Storage temperatures could be kept below critical levels by suitable dilution of the radioactive material, but this would involve additional storage volume and increased cost. As an alternative, radioactive waste-containing glasses could be allowed to cool for 70 or 80 years above ground before being buried for long-term storage.

—Norman M. Tallan

Metallurgy

The consumption of energy in metallurgical extraction and refining processes was a subject of major interest and cause for increasing concern during the past year. Although this concern was being expressed throughout the entire extraction metallurgy industry, it was most keenly felt by steelmakers. Conventional steelmaking in a modern integrated plant involves four distinct batch operations: the chemical reduction of iron ore in a blast furnace to produce "hot metal," which is molten iron at about 1,500° C (2,700° F) containing carbon, silicon, manganese, phosphorus, and sulfur as impurities; the refining of the hot metal in a basic oxygen furnace to produce molten steel; the casting of ingots or continuous casting of the steel; and hot- or cold-forming operations, which involve mechanical working of the cast ingots or continuously-cast strands into the required shapes. Of the total energy consumed in integrated steelmaking plants in the U.S., more than half is used in the blast furnace; therefore, the production of iron in general, and blast furnace technology in particular, were being subjected to close examination.

In principle, the blast furnace is a simple internally heated countercurrent retort in which solid materials (iron ore, coke, and slagmaking material) travel downward, and air, injected near the base, reacts with the coke to produce hot reducing gases that travel upward. Heat is tranferred from the rising gases to the descending solids, and oxygen is transferred in the reverse direction. Hot metal and slag are removed periodically from the hearth at the base of the furnace. In practice, however, the blast furnace is a delicate balance of simultaneous and sequential reactions, and attempts to alter the nature of any one of them must include consideration of the effect on the others.

Although iron has been smelted for some 2,000 years, it is only in the last 30 years that application of the increased understanding of the fundamental aspects of such physical and chemical phenomena as heat and mass transfer, and the development of better furnace design and operating procedures have produced a significant improvement in efficiency. Within this period the fuel requirement per unit quantity of iron produced has been cut by 40%, and furnace productivity, as measured by the mass of iron produced per unit time and unit volume of furnace space, has been increased by a factor of four. During the last decade the size of blast furnaces has increased considerably, and in 1980 there were 15 with an interior volume greater than 4,000 cu m (141,000 cu ft) operating in Japan; the largest of these was capable of producing 13,000 tons of hot metal per day.

Inventors demonstrate a new electroplating system, one that avoids the need for the overlaid "masks" used in conventional photolithographic circuit fabrication. A finely focused laser beam heats a small area on which metallic material is to be deposited. The heating produces convection effects that promote the desired deposition.

Courtesy, IBM

These gains in efficiency and productivity, however, have been seriously offset by the greater increase in the cost of primary energy. Proposed remedies for this problem range from attempts to improve the efficiency of existing blast furnaces, through construction of more efficient furnaces, to the employment of radically different processes using as yet undeveloped technologies. Of these proposals the first is appropriate only for the short term and the second is becoming increasingly impractical for private industry by virtue of the prohibitively high capital costs involved. Therefore, in the long term the third is the only viable alternative.

The ease with which an iron ore can be reduced—purged of all nonmetallic components—is a variable of major importance to the efficiency of ironmaking. In simple terms, the more reducible the ore the less fuel is required for its reduction. Recent laboratory studies have shown that the reducibility of an ore is critically dependent on the temperature at which reduction is started. It has been found that reduction of dense sintered hematite (Fe_2O_3) ore pellets in a highly reducing gas proceeds with the inward advance of distinct product layers of iron (Fe), wustite (FeO), and magnetite (Fe_3O_4). Diffusion of the reducing gases through the fine pores formed in the product layers to the reaction site is slow enough that even when reduction is 90% complete hematite still exists at the center of the pellet. However, with pellets or lump ore of high porosity, which provides easy diffusion paths for the reducing gases, there is rapid "internal reduction" of the hematite to wustite, without the formation of distinct product layers, followed by relatively rapid internal reduction of wustite to iron.

It is thus apparent that formation of distinct product layers at the lower temperatures prevailing in the upper parts of the furnace decreases the reducibility of the ore at the higher temperatures lower down in the stack. Thus the rate of coke consumption of a furnace can be cut by ensuring that the low-temperature reducibility of the ore is such that product layers do not form in the early stages of reduction in the upper regions of the stack. Laboratory studies have also shown that the high-temperature reducibility, at about 1,300° C (2,350° F), is significantly enhanced by adding relatively small quantities of lime (CaO) to the ore during sintering. Iron ore concentrates normally contain 4–5% silica (SiO_2) which, if untreated, reacts with the iron oxide in the ore to form a liquid iron silicate at about 1,300° C. The retention of this liquid by capillary action in the pores of the ore particles eliminates the diffusion paths for the reducing gas and thus retards the reduction process. The addition of lime prevents the formation of a liquid phase and therefore increases the high-temperature reducibility of the ore.

At temperatures greater than 950° C (1,700° F) the carbon dioxide (CO_2) produced by the reduction of wustite by carbon monoxide (CO) reacts with the coke to regenerate CO; therefore, the coke must be reactive enough to produce CO at a rate sufficient for the reduction of the wustite. It has been found that the reaction of CO_2 with the coke to produce CO is significantly catalyzed by the presence of alkali oxides (Na_2O and K_2O) on the surface of the coke. However, within the last few years, identification of what has been referred to as the "alkali problem" has required significant changes to be made in the operation of some blast furnaces. It has been found that in normal practice alkali oxides, which occur at infinitesimal levels as impurities in the ore, accumulate in the furnace. They evaporate at the higher temperatures lower down in the furnace, rise with the ascending gas stream, and condense at the lower temperatures in the upper parts of the furnace. This eventually leads to the development of massive deposits that cause unacceptably irregular descent of the solids in the furnace.

The problem was solved by making significant changes to the slag chemistry in the furnace. In normal practice a mixture of dolomite or limestone and basic oxygen furnace slag serves as a flux for the silica and alumina (Al_2O_3) impurities in the ore and coke. The flux is added in sufficient quantity to produce a slag in which the ratio by weight percentage of CaO to SiO_2 is about 1.3–1.4. This basic slag acts as a chemical sink for the unwanted sulfur and results in sulfur levels by weight percentage in the hot metal of about 0.03. The alkali problem was eliminated by decreasing the basicity of the slag to a level at which the thermodynamic activities of the alkali oxides, dissolved in the slag, are low enough to prevent significant evaporation of the alkalis from the slag. This required decreasing the ratio of CaO to SiO_2 so that it would be between 1.00 and 1.03. The consequence was that the sulfur content of the hot metal increased to 0.065 by weight percentage. Because this sulfur level exceeds that which can be removed in the basic oxygen furnace, external desulfurization of the hot metal prior to the steelmaking process had to be introduced.

One process that has been suggested as an alternative to the blast furnace is direct reduction (DR), which involves producing solid iron directly from the ore. The advantages of such a process include lower capital costs, elimination of the energy required for melting the iron, an increase in the number of gaseous reducing agents that can be used, the ability to use fuels of lower grade than is acceptable in the blast furnace, and the attainment of very low sulfur levels. The introduction of continuous casting has contributed to the potential usefulness of the DR process. It has done so by increasing the yield of solid steel from the 80–85% level obtained when the steel is cast as ingots to 95% and thus has significantly decreased the supply of in-plant scrap steel available for recycling through the basic oxygen furnace. Since the most economical operation of the latter is attained with a feed containing

Courtesy, General Electric Research and Development Center

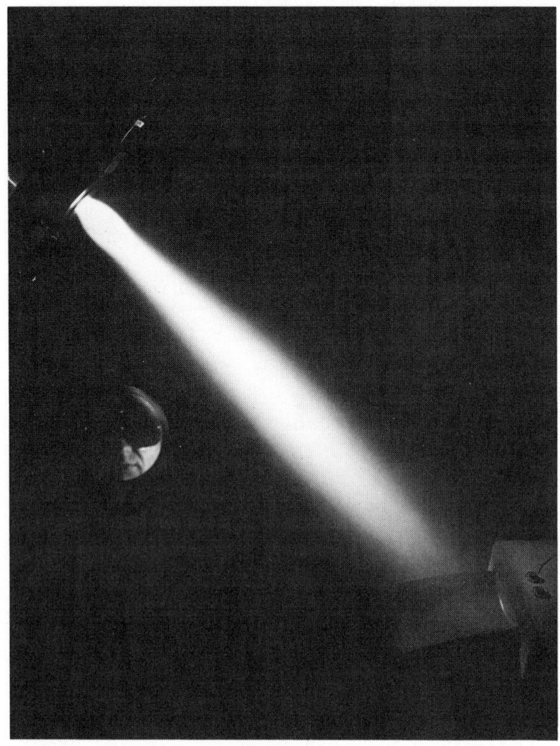

White-hot stream of ionized gases, created by a high-energy spray gun inside a low-pressure chamber, applies molten metal to gas turbine buckets. The stream is passed across the bucket surface several times in order to coat it with the metal thickly enough to prevent "hot corrosion," a problem that has plagued the gas turbine industry for many years.

25–27% scrap, it has been proposed that direct reduced iron (DRI) be used as a supplement to the scrap as a coolant. Alternatively, it has been proposed that briquetted DRI be used as the feed material for electric steelmaking furnaces, in which case the blast furnace/basic oxygen furnace combination could be replaced by a DR process/electric furnace combination. As of 1981 in North America it appeared that the latter had an economic advantage over the former only if the annual production rate of the plant is less than 1.5 million tons of steel per year.

The various radically innovative steelmaking technologies proposed include direct steelmaking, in which the separate ironmaking step would be eliminated. In such a process iron ore, coal, and oxygen would be fed to a vessel containing molten iron. The ore would melt and react with the coal to produce additional molten iron or semisteel, and the necessary heat would be provided by combustion of additional coal. The major practical difficulty to be expected in such a process is the coexistence of both oxidizing and reducing reactions in the same vessel.

Another suggestion is a process in which iron ore and coal are fed to a fluidized bed in which partial reduction of the ore and partial combustion of the coal occur. The gases generated by the reaction would be used to generate power for a plasma-arc furnace in which the iron would be melted and the reduction completed. The gases generated in this process would also be fed to the power station. A further suggestion involves a combination of magnetohydrodynamic (MHD) power generation and direct reduction, in which the gases produced by the MHD power generator would be used for direct reduction of the iron ore and the MHD generator would supply power to an electric furnace in which the reduced ore would be melted. (Magnetohydrodynamics deals with phenomena arising from the motion of electrically conducting fluids in the presence of electric and magnetic fields.) Because these proposed novel technologies exist at present only as ideas or, at best, at the laboratory-scale stage, it is impossible to guess which might eventually be adopted. All that is certain and agreed upon is that timely implementation of any such process requires that immediate attention be paid to the necessary research and development work.

—David R. Gaskell

Mathematics

Research in the mathematical sciences generally proceeds in a steady, uneventful manner: in a subject where important results are spread over a 2,500-year history, major new theorems are not regular events. But in 1980 mathematicians had the good fortune to celebrate a truly major achievement, one that occurs only a few times each century. After a 100-year effort they finally completed the classification of the finite simple groups, the basic building blocks of a major part of modern algebra.

Classification of complex structures into simple constituents is one of the most powerful paradigms of scientific knowledge. The premier example is the periodic table of elements, a classification of primitive components out of which all substances can be formed. The physicists' search for elementary particles is another example, as is the mathematicians' identification of prime numbers as the basic factors of all whole numbers.

The attempt to classify all finite simple groups is perhaps less well known than these other classification programs, but it is surely as compelling and significant. Groups are of profound importance not only in mathematics but also in such diverse fields as quantum theory, crystallography, architecture, and industrial engineering. Enormous effort has been devoted to the complex structural puzzle posed by the finite simple groups. The final steps in this program were completed in 1980. When all details are set down on paper, it will represent one of the century's major mathematical achievements—more than 5,000 pages of detailed proofs and classification arguments.

Groups are abstract representations of symmetry. They were introduced in the early 19th century by a French mathematics student, Évariste Galois, as a device to solve one of the most vexing mathematics problems of his time: to discover a formula for solving polynomial equations of degree greater than four. The quadratic formula taught in high-school algebra solves equations of degree two, and similar but more complex formulas were developed in the late Renaissance to solve equations of degrees three and four. But by 1800, three centuries after these formulas had been discovered, no one had been able to find a formula for polynomial equations of degree five or greater.

Galois showed—in notes scribbled down the night before he died in a duel at the age of 20—that no such formula could exist: the possible symmetries (or permutations) of the roots of fifth-degree polynomial equations exceed in complexity the symmetries that can be represented by algebraic formulas that are based on the four arithmetic operations and extraction of roots. Amazingly, a young Norwegian mathematician, Niels Henrik Abel, developed concurrently yet independently a similar solution to the problem of polynomial equations.

The concepts of Galois and Abel lay fallow for nearly half a century until a Norwegian mathematician, Sophus Lie, used the same strategy to explain why certain elementary differential equations could be solved whereas others could not. These efforts led to a productive theory of what are now called Lie groups, which link the discrete structure of permutations with the continuous variation of differential equations.

The definition of a group incorporates the most basic behavioral features of functions and operations, and so in some sense it is the most fundamental structure in algebra. It also provides an apt idiom for expressing geometric features such as rotation, reflection, and symmetry. Since groups represent a confluence of fundamental patterns from major branches of mathematics, it is not surprising that their structure contains the key to many diverse phenomena.

In this century the role of group theory in both pure and applied mathematics has grown enormously. A major family known as the linear groups, introduced about 1900 by the U.S. mathematician Leonard E. Dickson, has turned out to be of crucial importance in the classification of elementary particles; indeed, these groups provided much of the theoretical basis for the work that led to the 1979 Nobel Prize for Physics. The ability of groups to capture the subtle essence of symmetry has made them no less useful to chemists working in crystallography and spectroscopy. Results from group theory, for example, enabled Rosalind Franklin, James Watson, and Francis Crick to reduce from the infinite to the manageable the number of possible arrangements of molecules in their search for the structure of DNA.

Adapted from original art by John Ellis and John Handwork

An Alternating Group

Let five objects be represented by the letters A, B, C, D, E. An even permutation of these objects occurs when an even number of pairs are interchanged. For example, if A and C are interchanged, as well as D and E, the order becomes C, B, A, E, D. This permutation is called a double transposition; if it is repeated the original order will be restored, and so this permutation is said to have period 2.

The alternating group on 5 elements consists of all even permutations, such as the one illustrated above. There are 60 such permutations:

1 Identity, leaving all letters unchanged, with period 1.

15 Double transpositions (such as A ↔ C, D ↔ E), each of period 2.

20 Cyclic permutations of three letters (such as A → C → D → A), each of period 3.

24 Cyclic permutations of five letters (such as A → C → E → B → D → A), each of period 5.

60 Total even permutations of five letters.

This group also represents the rotational symmetries of a regular icosahedron, the platonic solid with 20 equilateral triangular faces.

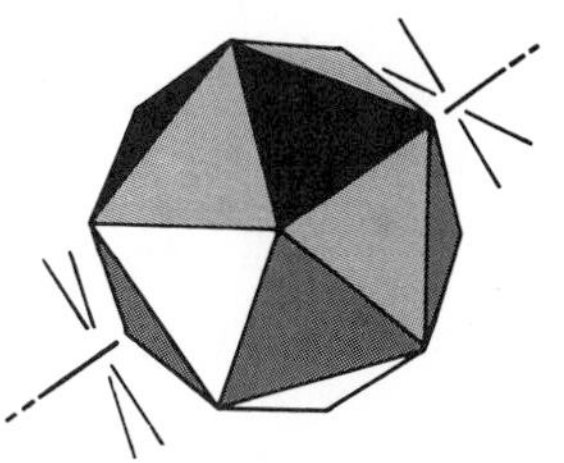

There are 60 rotations of the icosahedron that cause the vertices and faces to shift to new locations:

1 Identity, leaving all vertices unchanged, with period 1.

15 Rotations of 180° about lines joining midpoints of pairs of opposite faces, each of period 2.

20 Rotations of 120° about lines joining centers of opposite faces, each of period 3.

24 Rotations of 72° about lines joining opposite vertices, each of period 5.

60 Total rotational symmetries of the icosahedron.

The alternating group on five objects also represents symmetries of the roots of certain fifth degree polynomial equations. By showing that this group was simple (roughly speaking, that it could not be factored into smaller groups for which algebraic solutions might be possible), Niels Henrik Abel in Norway and Évariste Galois in France showed (independently and concurrently) that it is impossible to solve the general fifth degree polynomial equation.

Finite groups can be built up from combinations of smaller groups by a process analogous to multiplication; just as each whole number can be expressed as a product of prime numbers, so each finite group can be expressed as a combination of certain factors known as simple groups. Simple groups, first identified by Galois, are the ones that cannot be factored; they are the irreducible constituents of all finite groups. The classification problem for finite group theory was simply to find all finite simple groups.

By 1900 the finite simple groups of order (or size) less than 2,000 were all known; by 1963 the classification had been completed up to order 20,000; by 1975 it had been completed through order 1,000,000. Five years later, in 1980, mathematicians finally proved that the classification pattern revealed by those earlier studies held true for all finite simple groups.

Most simple groups belong to one of three major families: the cyclic groups, the alternating groups, and groups of "Lie type." Cyclic groups consist of cyclic permutations of a prime number of objects, such as the rotations by 72° increments of a regular pentagon. Alternating groups consist of even permutations—those permutations that are formed by interchanging the positions of two objects an even number of times. (The 60 total rotational symmetries of a regular icosahedron, a 20-sided solid figure, form an alternating group.) The groups of Lie type include 16 subfamilies, each associated with a particular family of continuous groups introduced by Sophus Lie in his study of the solutions to differential equations.

Unfortunately, some simple groups do not belong to these families, nor, apparently, to any family. These simple groups are called sporadic. Although five sporadic groups were discovered in the mid-19th century by Émile Mathieu, from that time until 1964 every other simple group that was discovered belonged to one of the three major families—cyclic, alternating, or Lie type.

This situation changed in the mid-1960s when Walter Feit of Yale University and John Thompson of the University of Cambridge verified a conjecture first made more than 50 years earlier by the British mathematician William Burnside. It had stated that, apart from the cyclic groups, all finite simple groups have an even number of elements. The proof by Feit and Thompson of this fundamental result was full of new insights into the classification problem and required an unprecedented 250 journal pages. Their work launched an intensive effort to complete the classification program.

Thompson led the assault with a 400-page analysis—published over a seven-year period from 1968 to 1975—showing how the structure of a major class of groups (called solvable groups, a term originating with Galois's use of groups to investigate solutions of polynomial equations) could be used to infer the structure of the finite simple groups. But in 1965 Zvonimir Janko discovered a new sporadic group, of order 175,-560. This was the first exceptional group to be discovered since Mathieu's work in the 19th century, and it presaged a period during which various investigators discovered more sporadic groups, one after the other, in a rush that seemed destined to undermine the entire enterprise with more exceptions than rules.

Some of the most interesting of these new groups were discovered by John Horton Conway of the University of Cambridge using techniques based on geometric considerations dealing with efficient packing of objects (spheres) into boxes in 24-dimensional space. Inexplicably, Conway's largest group contains many of the other sporadic groups as subgroups, suggesting a family structure for sporadic groups that has not yet been discovered.

The symmetries reflected in the sporadic groups found important application in the design of error-correcting codes. Because special patterns in these codes allow information obscured by noise to be reconstructed, they have been widely used in crucial military and space applications. Selecting a good code turns out to be equivalent to picking a collection of spheres that touch a given one but which are as widely spaced as possible. The symmetries of these patterns in 24-dimensional space yield one of Conway's sporadic groups and also provide a particularly efficient error-correcting code.

The most exotic of these new sporadic groups was one predicted by Bernd Fischer and Robert L. Griess Jr. in 1974, nicknamed the Monster. Fischer and Griess (who worked independently) did not actually discover the group; they merely found evidence suggesting that it might exist. Like cosmologists investigating black holes, Fischer and Griess used properties of the known simple groups, together with the massive body of theory that had emerged in the classification effort, to identify a possible new sporadic group of enormous size. Thus was launched the great Monster search, an effort to find the missing sporadic group whose existence was consistent with all known theory.

The difficulty with this search was that the Monster is unimaginably big: it contains 8×10^{53} elements. Yet, early in 1980 Griess discovered it. Armed with knowledge of its properties, he was able to show that it is a group of rotations in a space of dimension 196,883.

Griess's discovery of the Monster confirmed the directions of current research and provided renewed momentum for the final sprint in this extraordinary endeavor. Ultimately 26 sporadic groups were discovered. The final results were confirmed in an exchange of correspondence between Michael Aschbacher at the California Institute of Technology (Pasedena) and Daniel Gorenstein of Rutgers University (New Brunswick, N.J.), two group theorists who had been coordinating the classification effort.

A single theorem with a proof exceeding 5,000 pages is without precedent in mathematical history. Gorenstein admitted that the written proof, when completed, will inevitably contain certain local gaps—mistakes in reasoning, omitted steps—that break the logical chain of proof. But researchers in this field, like those who worked on the four-color problem—another problem with an extraordinarily long proof—know from experience that these short gaps can always be bridged by the routine application of known methods. These refinements will undoubtedly be made as reports of the classification work are prepared for scholarly publication.

The classification effort also spawned numerous intriguing sidelights, clues to structure not yet fully explored. For example, certain numbers describing key properties of the Monster appear as coefficients of important functions studied more than 150 years ago in an entirely different context. No one knows why these connections occur, but their existence points to fundamental structure yet to be discovered and understood.

—Lynn Arthur Steen

Medical sciences

Insulin made by genetically engineered bacteria was injected into a human diabetic for the first time in 1980. Also during the year the first large-scale trial of a vaccine against hepatitis B proved it to be effective, and research scientists made progress in developing a vaccine against tooth decay.

General medicine

The identification of an often severe disease associated with the use of tampons was among the year's significant developments in medicine. A long-term study revealed that people of below average weight have higher mortality rates than those slightly overweight.

Toxic shock syndrome. A disease that dramatically strikes down young, healthy women during the menstrual period was news in 1980 as physicians and epidemiologists tried to track its causes. The disease was associated with tampon use, and consequently one brand of tampon was removed from the market and warnings were included with the other brands.

Toxic shock syndrome (TSS) was considered to be a new disease. Although still rare—affecting fewer than one out of 10,000 menstruating women per year—it appeared to be on the rise. Approximately 800 cases with 69 deaths had been reported to the U.S. Center for Disease Control by the end of 1980, and the vast majority of those were 1980 cases. Approximately 95% of the victims were women, most of whom were stricken during or immediately after a menstrual period. The syndrome's symptoms begin suddenly with a high fever, followed by severe vomiting and diarrhea. Then a bright body rash appears, most prominently on the palms and soles, and the blood pressure drops sharply.

The link of the disease to tampons was proposed by state health departments, and in September the Center for Disease Control confirmed that association. In several studies of those women whose illness began within several days of the beginning of the menstrual period, all or almost all were using tampons. Further investigation by the CDC revealed that an unusually high proportion of TSS victims used the tampon called "Rely." The syndrome, however, has stricken users of all major

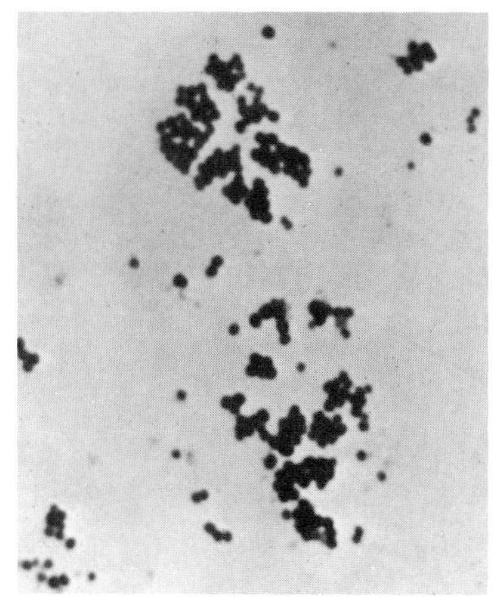

Tampons, especially the Rely brand (left), were found to be associated with cases of toxic shock syndrome. The microbial cause of the illness was the bacterium Staphylococcus aureus *(above).*

(Left) Jim Pozarik—Gamma/Liaison; (right) Center for Disease Control, Atlanta tampon brands. In September Procter and Gamble Co. halted production of the Rely tampon and withdrew the product from stores. Other tampon manufacturers voluntarily included a warning stating that tampons had been associated with a rare disease.

A common bacterium, *Staphylococcus aureus*, was indicated as the microbial cause of TSS. Toxins produced by that bacterium can cause fever and lethal shock. There was speculation that a recent genetic change in the bacterium allowed it to produce a new toxin or to proliferate more effectively. Scientists worked to identify the toxin responsible for TSS and to develop an antidote. Patients were successfully treated with antibiotics such as cephalosporins and penicillinase-resistant penicillins in order to reduce the chance of the disease recurring during subsequent menstrual periods.

Advice was given on how women could avoid TSS, but because researchers did not know how, or even definitely whether, tampons cause the disease, much of the advice was contradictory. Physicians did agree that avoiding tampon use altogether could virtually eliminate risk of the disease. But it seemed unlikely that the estimated 50 million American women who use tampons (approximately 70% of the menstruating population), would heed such advice and revert to sanitary napkin use.

Fat and thin. That fat is unhealthy has been dogma for many years. Overweight people are subject to a variety of risks and complications, including diabetes; high blood pressure; cardiovascular, gall bladder, and degenerative diseases; and increased danger during surgery. But in 1980 scientists discovered that thin may have its disadvantages too. While people weighing in well below the average were still considered fashionably trim, several studies showed that on the average they do not live as long as their more weighty neighbors. As further revenge for the 80 million Americans who were overweight, other experimental results supported the contention that for biological reasons some people gain weight more easily, and lose weight with greater difficulty, than do others.

The finding that thin people have higher mortality rates than more robust people came from a continued analysis of the 5,209 men and women of Framingham, Mass., who participated in a heart disease study from 1948 to 1972. Comparison of death rates with height and weight tables revealed minimum mortality rates for women of average weight, and higher rates for those who were substantially lighter or heavier. For example, among women 63–66 in tall, those weighing less than 115 lb and more than 195 lb had the highest mortality rates. Among the men the lightest group had the highest death rate, and the heaviest group had the lowest. For men 67 to 70 in tall, the lightest group weighed less than 135 lb and the heaviest group weighed more than 215 lb. Neither recognized illness nor smoking habits could explain the increased mortality among the thinnest group. The researchers, Paul Sorlie and Tavia Gordon of the National Heart, Lung and Blood Institute and William Kannel of Boston University Medical Center, suggested that at least part of the explanation for the high death rate among thin people may be subclinical illness.

These results were quite different from the data on height, weight, and mortality used by the insurance companies. The standard actuarial tables indicate that the higher a person's weight the greater the risk of death. That data, reported in 1959, covered five million people insured from 1935 to 1954. In 1980 a reexamination of the more recent portion of that data supported the finding of increased mortality at weights more than 20% below average.

Reubin Andres of the National Institute on Aging reviewed 16 studies on the relationship between obesity and mortality in groups throughout the world. None of the studies linked obesity to early death, and he found data in many of them which indicated that the heaviest group of people in a given population actually had the longest survival times. Andres suggested that there are some poorly understood or entirely unknown benefits of moderate obesity. He concluded that physicians should be cautious in advising weight loss for people who are plump but healthy and who do not have a predisposition to obesity-related disease. Sorlie and colleagues said that their data do not argue against sensible weight reduction for people much heavier than the average; however, they concluded that the findings did raise questions as to health benefits from weight reduction in persons of average or near average weight.

People who wanted to lose weight continued to insist that this goal is not achieved simply by eating less. In 1980 investigators found evidence to support this claim. Jeffrey Flier and colleagues at Beth Israel Hospital in Boston took blood samples from obese and normal people and measured the level of an enzyme called sodium-potassium ATPase. That enzyme uses energy to pump ions across cell membranes. The researchers reported in October that almost all the obese people in their sample had ATPase levels at least 20% below normal. The reduced enzyme activity suggested that in overweight people food calories were being stored instead of being burned. Because two very obese subjects had unusually high instead of low ATPase levels, Flier and colleagues believed that the reduction in enzyme activity observed in the other obese people was not just a result of obesity. In addition, the enzyme level did not change in 12 subjects who lost a significant amount of weight by dieting.

Heat treatment for cancer. Microwaves, radio waves, and ultrasound, as well as hot-water blankets and heated blood, were used in 1980 in experimental attacks on cancer. At least ten major U.S. research

Dan Bernstein, New England Medical Center

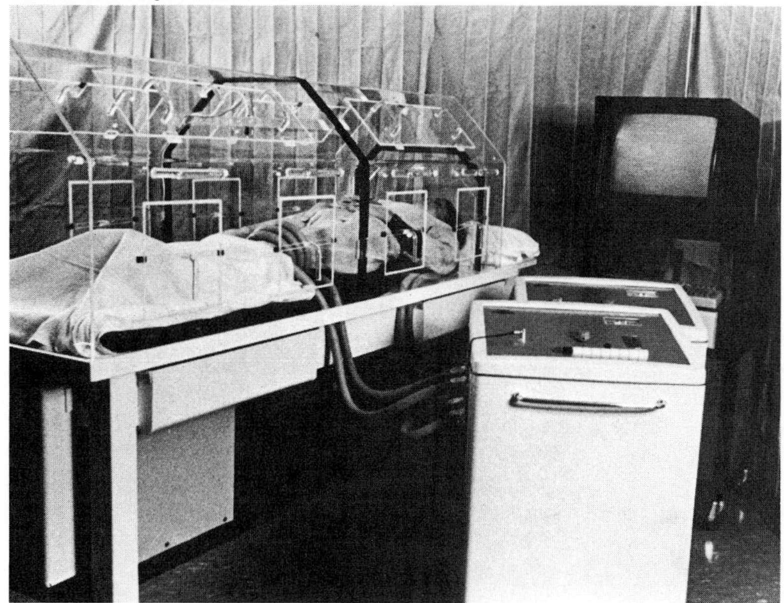

Cancer patient undergoes heat treatment at the New England Medical Center Hospital. Preliminary results indicate that the inducement of high temperatures in patients weakens the structure of their cancer cells and encourages action by antitumor lymphocytes in the blood. The treatment is expected to be most effective in combination with other therapies.

centers offered treatment based on the concept that heat can kill cancer cells. In most cases the heat treatments supplemented radiation or chemotherapy and were used on patients who had not responded to the more traditional methods alone. While scientists remained uncertain as to exactly how such hyperthermia therapy works, the preliminary results indicated that it can be effective.

Major differences in the procedures for administering heat therapy arose from disagreement on whether it is best to apply heat to the tumor alone, to a region such as a limb, or to the whole body. Sophisticated methods of applying heat allowed investigators to raise the temperature extensively in a tumor without damaging the surrounding tissue. On the other hand, some researchers suggested that heating a wider area would fight cancer cells that had spread from the original tumor and prevent recurrence of the disease.

Inserting an antenna into a brain tumor to apply microwaves was one specific heat application undertaken. Neurosurgeon Michael Salcman, using a technique he developed with biophysicist George Samaras at the University of Maryland, implanted an antenna and a tiny thermometer into the tumor of a 28-year-old European businessman and twice in the following 48 hours heated the brain tumor to 45° C (113° F). They reported indications of a decrease in tumor size and observed no ill effects, such as scarring of healthy tissue. More time is required before the success of the operation can be evaluated.

Radio waves were used to heat specific regions. Kristian Storm at the University of California at Los Angeles beamed such waves at tumors, including those of the lung and liver, of 175 patients. He reported some dramatic responses. Because radio waves are less damaging to normal tissue than microwaves, they could be used even when the site of a deeply embedded tumor was not precisely known.

Using an ultrasound transducer strapped over a tumor while cooling the skin with flowing water, Peter Corry at M. D. Anderson Hospital in Houston, Texas, raised tumor temperatures as high as 50° C (122° F). He reported that in half of more than 200 patients whose tumors were irradiated and then treated with ultrasound or by magnetic induction, hyperthermia therapy reduced the tumor size at least 50%.

A new apparatus also was developed for reaching deep-seated tumors with ultrasound without harming overlying tissue. The device converges multiple beams on a precisely located tumor. In earlier ultrasound experiments Stanford University investigators heated tumors to 43° C for 30 minutes. That treatment shrank tumors with no undesirable effects, but it did not cure patients whose cancer had already spread.

A cancer-treatment center specializing in hyperthermia opened in June at the Henry Ford Hospital in Detroit. There the approach was to heat an area slightly larger than the tumor in order to kill cancer cells that had begun to migrate. Haim Bicher reported that in an 18-month period this microwave treatment caused complete remission in 60% of 85 cancer patients and partial regression in another 35%.

Heating an even larger area is the procedure used by John Stehlin at the St. Joseph Hospital in Houston. To treat recurrent melanomas of the arms or legs, he cuts an artery of the affected limb and pumps in heated blood for about two hours. Stehlin has treated hundreds of patients since the procedure was devised in the 1960s. He reported that 77% of them survive five years with drug treatment and regional hyperthermia compared with 22% who survive five years on drug therapy alone.

Finally, selected patients appeared to benefit from heating of the whole body to 41.8° C. Joan Bull of the University of Texas at Houston wrapped patients in plastic hot-water-circulating blankets for four hours. That treatment was used along with chemotherapy to treat melanoma patients. Because heating the whole body puts much stress on the patient, it was expected to be useful only for patients with strong cardiovascular systems.

Why heat can destroy tumors selectively was not completely clear. Some scientists suggested that tumor cells are more sensitive than normal cells to heat. Others said that poor blood circulation in tumors prevents them from dissipating the heat to the rest of the body. Other suggestions were that lack of oxygen, excess acid, or poor nutrition makes tumor cells more vulnerable. The successful combination of hyperthermia and chemotherapy suggested that heat makes cancer cells more permeable to drugs. Whatever the mechanism, the new studies confirmed the concept that heat can fight cancer.

Interferon. A protein produced in minuscule amounts by the human immune system, interferon has been regarded as a potential cure for afflictions ranging from the common cold to cancer. By 1980 new techniques for making interferon and increased funding for research efforts made possible large-scale clinical trials. Early results of trials sponsored by the American Cancer Society were somewhat disappointing, but they did provide evidence of interferon's cancer-fighting ability.

Although interferon was identified in 1957 as a protective chemical produced by animal cells when they are infected with a virus, only a few hundred patients had received a dose as of 1980. Researchers were thwarted because cells make such small amounts of the material and because interferon extracted from animals is ineffective in people.

In May 1980 the American Cancer Society announced the first progress reports on a series of clinical trials of interferon begun in 1978. Small amounts of interferon purified from blood cells were used to treat four kinds of cancer—breast cancer, non-Hodgkin's lymphoma, multiple myeloma, and the skin cancer melanoma. The first results for breast cancer and multiple myeloma indicated anticancer activity less than that reported in earlier studies. Regression of the disease followed treatment in only 5 out of 16 breast cancer cases and 4 out of 14 multiple myeloma cases. The society said that impaired potency of some freeze-dried interferon used may have contributed to the disappointing results.

Future studies were expected to use more abundant, less expensive, and purer interferon as a result of research developments in 1980. Scientists at the British Medical Research Council announced a technique that purifies interferon 5,000-fold in a single step without destroying the activity. They made a monoclonal antibody (*see* below) that binds only interferon from white blood cells. When that antibody was linked to a support column, it bound interferon molecules from an impure preparation flowing by.

Scientists in 1980 also reported the order of the approximately 150 amino acids that make up each of two varieties of human interferon. Although laboratory synthesis of so long a molecule seemed impractical, chemical synthesis of only a segment might be sufficient to mimic interferon's activity.

Genetic engineering seemed the most promising source of interferon. The Geneva-based biotechnology

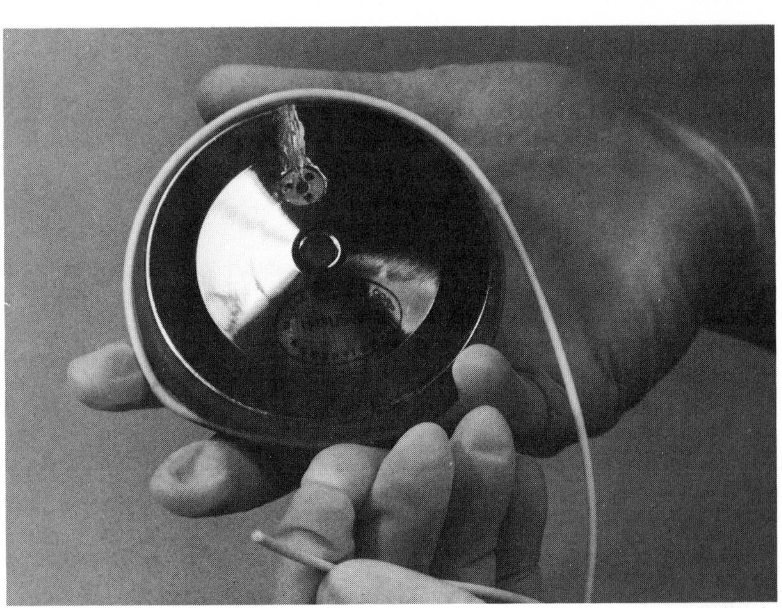

Insulin pump that can be implanted within the body of a diabetic was devised to provide a steady flow of insulin and thereby prevent the fluctuations in blood-sugar levels caused by once-a-day injections. The pump consists of two chambers separated by a flexible membrane. Compressed fluorocarbon gas in one chamber forces a trickle of insulin out the other and into the patient's bloodstream. Every two weeks an injection through the patient's skin refills the insulin chamber.

University of Minnesota

company, Biogen S.A., announced in January the first production of human interferon by genetically engineered bacteria. By the end of the year other groups of researchers had also transferred human interferon genes into bacteria. The bacterially produced material was active in protecting laboratory animals from viruses and also protecting human cells growing in tissue culture.

The somewhat frenzied interest in interferon was reflected in the allocations of research money from a wide variety of organizations. The American Cancer Society, the Interferon Foundation in Houston, and the National Institutes of Health budgeted more than $20 million for experiments with interferon treatments, which cost as much as $30,000 per patient. The increased abundance of human interferon was expected to allow physicians in the near future to work out the best dosages and treatment schedules and to compare the three known types of human interferon to determine whether any of them will be as effective, and less distressing, than cancer treatments already in use.

Bacterially produced insulin. A Wichita, Kan., homemaker in December 1980 became the first diabetic to be injected with insulin made by genetically engineered bacteria. Such insulin is identical to the hormone normally produced in the human body and differs in one of its 51 amino-acid components from the pig insulin many diabetics receive. Scientists expected that the human-style hormone would reduce the incidence of allergic reaction to insulin among diabetics and prevent a shortage of insulin should the supply of animal pancreas glands become low.

Insulin was the first product of genetic engineering technology to reach clinical trials. Moving the human genes for insulin's two chains into bacteria in 1978 was one of the earliest gene-splicing feats. In July 1980 healthy volunteers in the U.K. (followed by others in the U.S., Greece, and West Germany) tested the bacterial insulin for safety and effectiveness. The preliminary tests showed the insulin to be approximately as effective in lowering blood sugar as naturally produced insulin. The drug firm Eli Lilly and Co. then began construction of two facilities to produce insulin with bacterial techniques.

Pure human antibodies. Two teams of scientists in 1980 made laboratory lines of human antibody-producing cells by combining them with cells of human tumors. The scientists obtained cells that can grow in the laboratory indefinitely and that produce a limited amount of pure human antibody. A hybrid cell and all its descendants churn out the exact same antibody, which is thus called a monoclonal antibody.

The prospect of abundant pure human antibodies allowed physicians to envision a wide variety of applications. Human monoclonal antibodies were expected to serve as especially specific and safe vaccines. They were also expected to detect cancers and to deliver drugs directly and exclusively to tumors.

Jim Pozarik—Gamma/Liaison

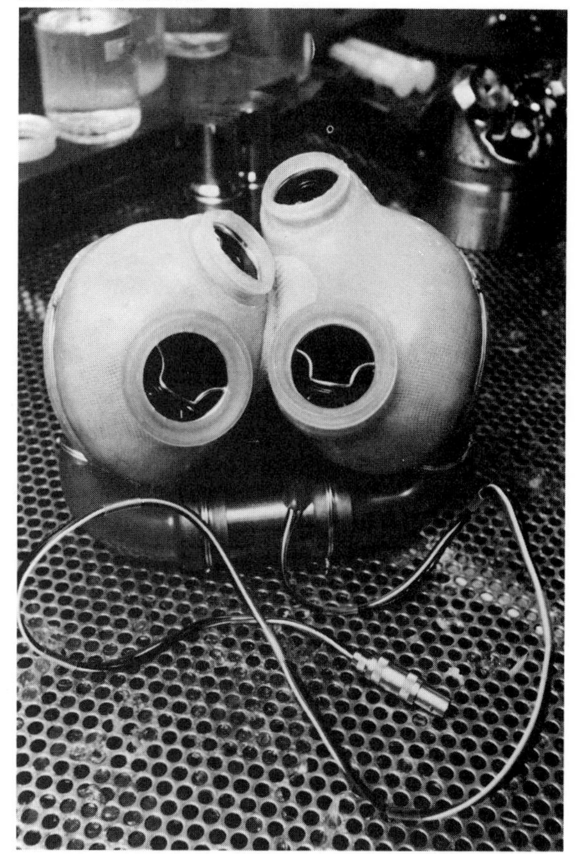

Artificial heart, developed at the University of Utah, is made of polyurethane and aluminum. It is powered by compressed air that passes from a compressor through six-foot-long plastic hoses in the patient's abdomen.

A 19-year-old woman in a Philadelphia hospital was the source of one team's antibody-producing cells. She was comatose with a disease called subacute sclerosing panencephalitis, and her blood contained an extraordinarily high concentration of antibody to measles virus. Carlo Croce and colleagues at the Wistar Institute of Anatomy and Biology combined white blood cells from the woman and cells of a bone marrow tumor. They obtained hybrid cells that produce a single antibody which binds to the protein coat of the measles virus. Croce predicted that the antibody could be applied against any measles infection and that the general approach—making human cell hybrids—theoretically could produce any type of human antibody.

In California Stanford University scientists independently created human antibody-producing hybrid cells. Henry S. Kaplan and Lennart Olsson used as the antibody-making member of the cell partnership spleen cells taken from patients as part of the clinical evaluation of Hodgkin's disease. The spleen cells were combined with cells derived from malignant bone marrow.

Massachusetts Institute of Technology; photo, Calvin Campbell

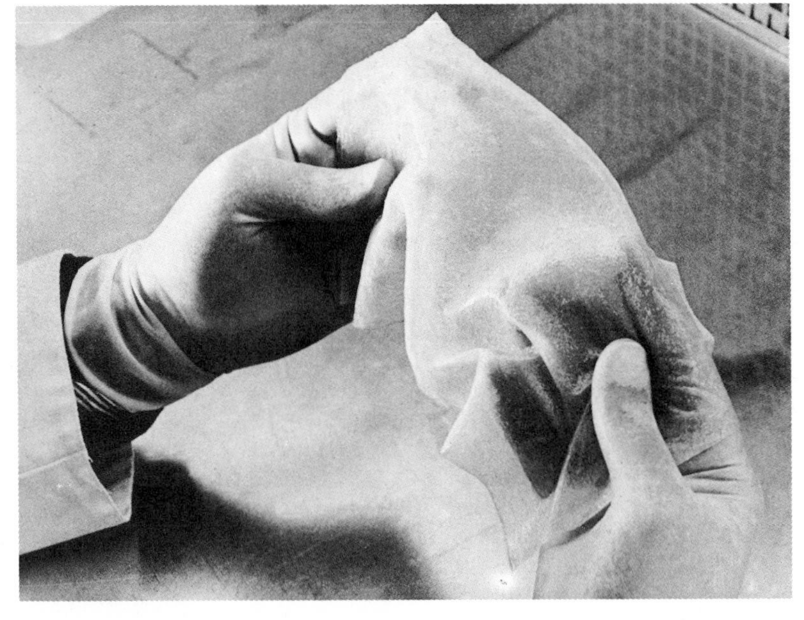

Artificial skin has been used successfully to treat human burn victims. Developed at the Massachusetts Institute of Technology and Harvard University Medical School, it consists of a bilayer of transparent silicone elastomer bonded to a bottom layer of a porous crosslinked collagen-glycosaminoglycan polymer material.

The result was hybrid cells that grow indefinitely in the laboratory and produce a specific antibody.

Human monoclonal antibodies showed potential as weapons against autoimmune diseases. In some forms of diabetes, for example, the patient makes antibodies to his own insulin. Croce and colleagues generated hybrid cells that produce an antibody that binds this antibody to insulin. The scientists also undertook a similar approach to counter myasthenia gravis, a disease in which a person's antibodies attack the receptors on muscles that receive signals from nerves.

The technique for creating antibody-producing hybrid cells was first worked out with mice in 1975 by César Milstein and Georges Köhler at the British Medical Research Council Laboratory of Molecular Biology in Cambridge. While antibodies produced by mouse cells are not considered as valuable therapeutically as human antibodies, the mouse cell products have been useful in immunological research and as guides to therapy. For example, in 1980 the U.S. Food and Drug Administration approved a trial of mouse monoclonal antibodies to the rabies virus. Those antibodies were expected to be useful both in detecting the virus in tissue samples and in boosting the resistance of people bitten by rabid animals. Monoclonal antibodies also began to be used to screen donated blood more effectively for hepatitis. Creation of a monoclonal antibody which binds to the protein that coats the malaria parasite in its infective stage offered hope for an effective vaccine against malaria. This antibody was shown to protect mice from malaria transmitted by mosquito bites.

Surgical correction for vision. Novel eye surgery during the year was found to be effective in correcting both nearsighted and farsighted conditions, but most ophthalmologists remained wary of the risks of such operations. The most sensational technique was developed in the Soviet Union by Svyatoslav Fyodorov of the Moscow Clinical Eye Institute for Surgery. With a procedure taking only ten minutes for each eye, he claimed to have treated 2,000 nearsighted patients in the last five years. He said that 90% of patients whose vision was 20/400 before surgery had vision of 20/40 or better after the operation.

Fyodorov's procedure involves making 16 radial incisions on the outside of the cornea. By using a computer, Fyodorov and his followers calculate from a patient's visual acuity and eye shape the length and depth of incision necessary to flatten the cornea sufficiently to bring the image onto the retina.

Despite reservations about its safety and effectiveness, the procedure grew in popularity in the United States. Surgery on the cornea can lead to scarring and infection, and if the cornea is accidentally perforated, blindness might result. In addition, some ophthalmologists believed that the improvement in vision achieved by the surgery was only temporary.

In May the National Eye Institute's advisory council expressed "grave concern about potential widespread adoption" of the technique, which is called keratotomy. The council called for carefully controlled animal and human trials. The American Academy of Ophthalmology also expressed concern about possible abuse of the procedure by untrained surgeons.

Farsighted as well as nearsighted patients obtained benefit from another type of eye surgery called keratomileusis (or, literally translated, "cornea-carving"). José Barraquer of Bogotá, Colombia, pioneered this technique, in which the front of the cornea is sliced off and the slice is quickly frozen and then reshaped with a lathe according to computer calculations. The sculpted slice is finally stitched back onto the eye. To correct farsightedness, pieces are trimmed off the outer edges of the slice to increase the cornea's curvature. Nearsightedness is corrected by trimming at the center of the slice to decrease the curvature. Each operation takes about an hour.

Barraquer reported good results in the approximately 4,000 operations he has performed over 15 years. Recently ophthalmologists at a few medical centers began performing such operations in the United States. In general, ophthalmologists held that the surgical procedures remained too uncertain for people whose vision is adequately corrected with glasses or contact lenses. The operations remained in demand, however, by such people as airline pilots and police officers, who are required to pass a vision test.

Bone marrow treatment. The first direct application of genetic engineering to a human hereditary disease provoked controversy in several arenas. Martin Cline of the University of California at Los Angeles revealed in October that he had transferred genes into bone marrow cells of two patients, one in Israel and one in Italy. Some people accused Cline of bypassing his own university's committee on human experimentation, a committee that deliberated a year over his proposal to do a similar experiment at UCLA and then, after he had already performed the gene transfer overseas, rejected his proposal. Other people were shocked that any genetic engineering in humans was begun without further ethical consideration. Among medical scientists, however, the objection most prominently voiced was that the attempt was premature and should have been delayed for further animal experimentation.

In April Cline published the first report that a selected gene could be successfully inserted into a living animal. He and colleagues transferred the gene for a drug-resistance factor into bone marrow cells of mice. Other experiments with a viral gene confirmed that genes could be transferred successfully with this method.

When the results of the animal studies were announced, Cline was quoted as saying human applications of the technique were at least three years in the future. Surprisingly just months later he used a similar method to transfer into two patients a gene for a blood protein—called beta globin—that they were unable to make. The patients had no functional hemoglobin and were kept alive by regular blood transfusions. Each had only a one- to two-year life expectancy.

In Cline's experimental therapy he took bone marrow cells from each patient and incubated them with the human gene for beta globin, which had been reproduced in bacterial cells. He injected the cells back into the patient's blood in the hope that they would return to the bone marrow and proliferate there. Whether the experiment was successful remained unknown as of early 1981.

Other scientists challenged the value of the experiment. They charged that in animal experiments there was no evidence that transplanted globin genes could produce useful amounts of protein. Cline and his colleagues argued, however, that the production of even low levels of beta globin could be considered progress. They expected to obtain important information from the experiment even if the patients did not benefit therapeutically. Most scientists who expressed opinions did not think there were practical risks from the experiment.

The UCLA committee on protection of human subjects and the National Institutes of Health began an investigation as to whether Cline broke their rules. Meanwhile, Cline resigned as chief of the Division of Hematology-Oncology but remained a professor on the UCLA medical faculty.

The procedures for human experimentation also caught Cline in another argument in 1980. He was one of a team of six researchers rebuked for performing bone marrow transplants (in this case without any genetic engineering) on terminally ill patients. Such transplants, considered an experimental procedure, required approval of the UCLA Human Subject Protection Committee. In some cases the transplants had been done after approval lapsed, or the procedure used differed from the method approved by the committee. Robert Gale, who headed the transplant team, argued that bone marrow transplantation had been sufficiently demonstrated as beneficial and was no longer an entirely experimental procedure. The dean of the medical school stressed that the university's objections were with the scientists' failure to adhere strictly to approval procedures; it was not a case of patients being subjected to unnecessary risks.

New vaccines and drugs. The first large-scale trial of a vaccine against hepatitis B showed it to be almost totally effective. The study conducted at the New York (City) Blood Center enlisted more than 1,000 homosexual men, chosen because their risk of contracting the disease is ten times that of the general population. Among more than 500 men who received the three-shot program, 96% responded with antibodies against the virus. None of those who had the antibodies developed hepatitis B. Among a group who received a placebo instead of the vaccine, 73 cases of hepatitis B were reported in 21 months of follow-up investigations.

A new rabies vaccine was approved by the U.S. Food and Drug Administration in June. It was produced from viruses grown in laboratory cultures of human fetal lung cells. The virus is inactivated before use. Because the technique does not use duck eggs, as did earlier rabies vaccine preparations, it did not provoke allergic reactions in people sensitive to eggs. The new vaccine was less painful to receive and caused fewer local reactions than did the earlier duck-egg vaccine.

In November the Food and Drug Administration approved Zomax (zomepirac sodium) for prescription use. The drug inhibits the body's production of prostaglandins, hormonelike substances implicated in many aches and pains. Aspirin and acetaminophen (Tylenol and others) reduce prostaglandin production at the site of injury, but Zomax acts both there and in the central nervous system, according to the drug's producer, McNeil Pharmaceutical. Tests on more than 4,000 people showed Zomax having side effects similar to aspirin and being effective on lower back pain, headache, chronic joint and muscle pain, orthopedic pain from sprains and fractures, and pain following surgery.

—Julie Ann Miller

See also Year in Review: LIFE SCIENCES: *Microbiology; Molecular biology.*

Holistic medicine

A belief that the body, mind, and spirit form a single unit is the essence of holistic medicine. When these three elements are in a state of harmony a human being is "whole," or healthy and free of disease, according to holistic theory. Holistic medicine, holism, holistic healing, or "The New Medicine," as this growing phenomenon has been variously called, uses a wide-ranging variety of techniques borrowed or adapted from ancient healing practices, religious concepts of both East and West, yoga, transcendental meditation, acupuncture, biofeedback, folk medicine, faith healing, and other methods, including aspects of the occult.

Emphasizing the importance of the individual as self-healer, holistic medicine attempts to cure with as little use of drugs, surgery, and technology as possible. Holism is also concerned with the prevention of disease and the maintenance of good health through sound living habits, the reduction of stress, and an awareness of the mind-body unity. But holistic medicine is not regarded by its responsible practitioners as a substitute for conventional medicine nor as a panacea for all human afflictions.

Although not officially recognized as a medical specialty by the American Medical Association, holistic medicine and health care was recently the subject of AMA meetings. These seminars were held to inform doctors of developments in holistic medicine and to enable them to answer their patients' questions on the subject. Among the speakers was C. Norman Shealy, clinical associate professor of neurosurgery at the Universities of Wisconsin and Minnesota, and director of the Pain & Health Rehabilitation Center in La Crosse, Wis. Shealy, an eminent pain specialist and a respected practitioner of holistic medicine, proposed that courses in holism become part of the curricula of the nation's medical schools.

The medical profession in general has adopted a cautious attitude toward holistic healing. But in recent years a number of respected and reputable members of the medical profession have begun using holistic techniques. Physicians who practice holism are usually more experimental than others in the profession; many will treat patients with "whatever works" no matter how unorthodox, providing the treatment does no harm. No ethical or conscientious practitioner of holistic medicine, however, would suggest it as a replacement for conventional medical care. A competent physician should be consulted for any symptom of disease, for injury, or other condition requiring medical treatment.

Psychosomatic diseases. Conventional and holistic medicine are perhaps most in agreement in the area of psychosomatic disease. Both recognize the influence of mind over body. But while conventional medicine prescribes drugs and a special diet for high blood pressure patients, for example, the holistic approach might al-

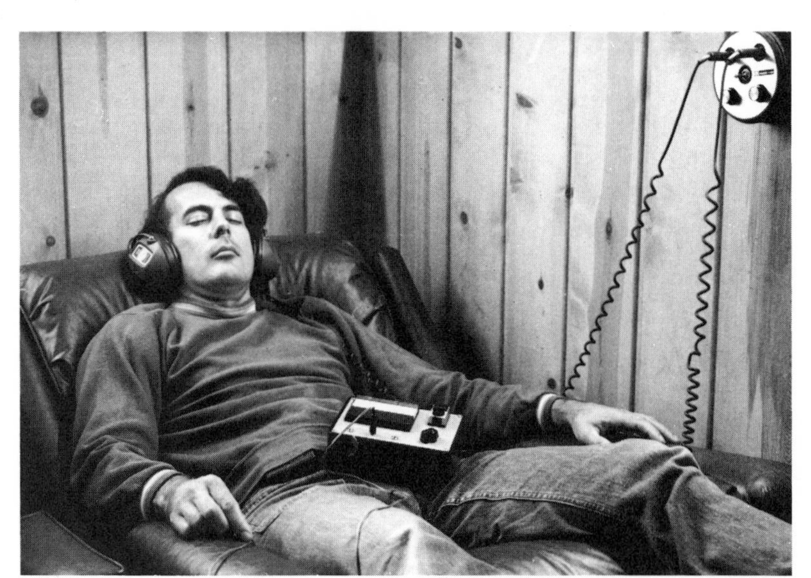

Patient undergoes biofeedback training in the control of body sensation. On his lap and attached to his right finger is a temperature-recording device. Through the earphones relaxation exercises and suggested visualizations are directed at the patient. After a few minutes he opens his eyes to check on his progress in raising his temperature.

C. Norman Shealy, The Pain & Health Rehabilitation Center, La Crosse, Wis.

so recommend meditation, biofeedback treatment, and yoga exercises.

Some estimates claim that as much as 85% of a physician's practice is the treatment of psychosomatic complaints, symptoms created by emotional or psychological factors. Such diseases are real enough, and clinical diagnosis confirms that the body has indeed been damaged or is malfunctioning. Yet in many of these patients a placebo may be just as effective as standard pharmaceutical therapy or sometimes even surgery. Studies have shown that the pain of terminal cancer in many patients can be controlled or eliminated through the use of placebos. Since a placebo, or so-called sugar pill, has no direct chemical effect on the body, the healing, if and when it occurs, is assumed to have come from some process within the patient. Some physicians would extend the definition of placebo to include any therapy producing the desired results through psychological means. Using that definition, some holistic practitioners insist that even faith healing may be a form of placebo and may be, therefore, a useful alternative treatment when traditional medicine fails to cure.

The placebo effect demonstrates an apparent self-healing ability that all people possess in varying degrees, according to holistic thinking. Certain self-healing skills may be learned or enhanced and successfully employed through such methods as biofeedback and yoga. Common to these techniques is the assumption that increased awareness of the mind-body unity accompanied by a consciousness of body functions is the key to good health.

Until the mid-1960s it was believed that the autonomic nervous system, which regulates such "involuntary" actions as breathing, heart rate, blood pressure, and body temperature, could not be voluntarily controlled. Biofeedback, yoga, and meditation, among other methods, have disproved that theory. Under strict laboratory controls patients have repeatedly demonstrated an ability to regulate their own heartbeat, blood pressure, brain waves, muscle fiber tension, body temperature, and other physical phenomena once thought to be beyond voluntary regulation. Elmer Green, a biopsychologist at the Menninger Foundation, pioneered a successful treatment for migraine headache using a simple electronic biofeedback device by means of which a patient monitors and controls his own blood flow to reduce vasoconstriction in the scalp and cranial arteries and thus reduce or eliminate pain.

Green and his wife conducted extensive research into the potential of similar techniques in the treatment of high blood pressure, asthma, gastrointestinal problems, neuromuscular disorders, anxiety tension states, epilepsy, and cerebral palsy. All of these conditions were treated with some degree of success using variations on the biofeedback method and/or through yoga meditations that "quiet" the body and mind, reducing stress. Hypertension is being effectively treated by a number of physicians using conventional therapies combined with meditation or biofeedback.

Yoga is among the methods that have been discovered effective in allowing a person to control his or her autonomic nervous system, which regulates such "involuntary" actions as breathing and heart rate.

Faith healing. Among the most controversial aspects of holistic medicine is faith healing. No comprehensive and authenticated statistics concerning its effectiveness are available. But genuine "miracle cures" and spontaneous remissions have occurred in a small fraction of patients with diseases diagnosed as incurable before they were treated by faith healers.

The healers have a variety of explanations for their abilities. Some claim to act as a medium through which the "universal love-energy" is channeled and transmitted to the patient. Others claim to collect "the power and love of God" or an all-pervasive "cosmic energy" and pass it on in high concentrations to the afflicted.

Another explanation was proposed by Shealy after extensive studies of faith-healing phenomena. His theory holds that the faith healer sets up a strong expectation of cure in the patient, and the patient's mind and body then combine efforts to produce the anticipated results. The patient participates in his own cure by balancing his autonomic nervous system, which in turn produces the biochemical changes in the body required for the restoration of health. Many physicians

The Bettmann Archive

A forerunner of today's holistic medical practitioner was the American Indian medicine man. When faith in him was strong, he probably achieved a high cure rate for stress-related diseases.

realize that faith, which has no measurable physical properties, is nevertheless part of the healing process. The will to live, the patient's faith in his physician, and the empathy that the physician communicates to the patient are all believed by many to be important factors in effecting a cure.

Holistic healers see in the ministrations of the faith healer a direct relationship to the medicine man, witch doctor, or combination physician-priest of ancient or primitive societies. In early American Indian cultures, for example, where faith in the medicine man was probably universal, the cure rate for stress-related disease must have been remarkably high. Herbs and medications with potent chemical properties were also used effectively by so-called primitives, and both holistic healers and modern pharmacologists have employed these remedies.

Similar to some aspects of yoga meditation and biofeedback training is a visualization technique known as guided imagery developed by O. Carl Simonton of Fort Worth, Texas. Patients using this technique mentally visualize their tumors being attacked and reduced by the body's disease-fighting mechanisms. The exercises are repeated three times daily over a period of weeks, and are eventually expanded in scope to include the patients' visualizing themselves completely restored to good health. Simultaneously, the patients are treated with radiation (and/or chemotherapy) and psychotherapy. Working with his wife, Simonton found that cancer patients frequently have a poor self-image and certain other psychological problems, which he believes constitute a "cancer personality." The Simontons design their treatment to combat this combination of physical disease and psychological distress. A study of 50 patients treated with the Simontons' visualization and psychotherapy techniques, along with conventional cancer treatment, showed 37 (74%) responding with "excellent" or "good" results. The Simontons have been challenged by other physicians, however, for failing to provide conclusive scientific data connecting cancer to psychological factors.

Acupuncture. A Chinese method of healing about 5,000 years old, acupuncture involves the use of long, slender needles to puncture the body at specific points. According to Chinese thought this restores a balance of yin and yang, a concept of opposing elements that are found throughout nature; when out of balance in the human body, they cause disease. Although known to a few Westerners over the years, acupuncture was not widely publicized in the U.S. until 1971 when Chinese physicians used it to relieve the post-surgical pain of a U.S. newspaper correspondent who underwent an emergency appendectomy while on assignment in Beijing (Peking). Since then acupuncture has been used with some success to relieve acute pain but has not been as effective in the treatment of chronic pain. Acupuncture has also been used in the West with varying degrees of success for other medical problems.

How and why acupuncture works is not fully understood. The Chinese theory contends that the needles permit surplus yin and yang to be released from the body until a perfect balance of these factors is achieved and good health is restored. Invisible connections are thought to link various acupuncture points on the body, and certain combinations of these points influence specific organs. When improperly used, acupuncture can cause severe injury or be fatal.

The list of holistic therapies extends to Zen Buddhist meditation techniques, folk remedies, fasting, special diets, massage, and other increasingly esoteric concepts. Beyond the fringes of science holistic medicine may include astrology, psychic diagnosis, clairvoyance, and other aspects of the occult. But while holistic medicine may be all-encompassing and open to "what-

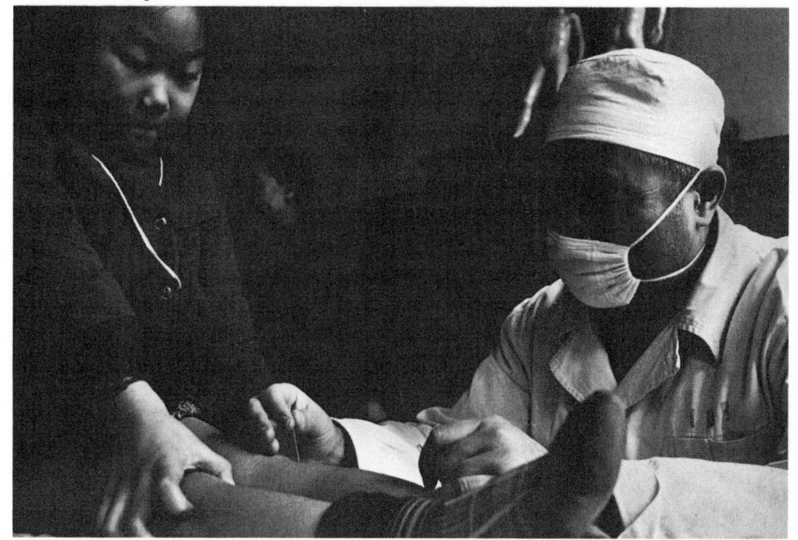

Acupuncture treatment is given to a child at the Children's Hospital in Beijing (Peking), China. This ancient Chinese procedure has achieved success in relieving acute pain.

ever works," the individual faith healer or acupuncturist, for example, may not necessarily subscribe to general holistic theory that extends beyond his or her particular specialty.

Future prospects. Behind the expanding interest in holistic medicine is an increasing disenchantment with the impersonal nature of modern medicine despite its miraculous advances. A small number of physicians seem to share this attitude with a growing number of laymen. Other factors contributing to the growth of holistic medicine include the ever-climbing cost of health care, the need for alternative methods of treating the poor and elderly, and a general frustration with conventional medicine's treatment of stress-related diseases. The West's renewed interest in the cultures of the East—India and China with their yogis and Zen masters, their altered states of consciousness and their ability to control "involuntary" body functions—is another element behind the growth of holism. Holism may also express modern man's reawakening desire for things natural even as his civilization becomes more complex and therefore more unavoidably reliant on technology. Finally, holism suggests a simultaneous turning inward of human consciousness and a turning outward as well, a conception of man as neither divisible from himself nor separate from the universe.

Holistic medicine seems likely to grow in respectability in the years to come. An increasing number of physicians will probably study and acquire holistic skills, especially in the areas of biofeedback and meditation techniques. Proposals have already been made to develop licensing standards for professional holistic paramedics. Shealy has suggested a plan for the establishment of holistic health and healing centers run by physicians and fully staffed by medical personnel but with holistic sub-specialists available for consultation and assistance. Such centers would protect the public from quackery and offer the full range of both conventional and holistic medical treatment. —Marc Davis

Holistic medicine is not officially approved nor endorsed by the American Medical Association. People are urged to seek competent medical care promptly when the need arises. The most ethical and responsible practitioners of holism advise their patients to consult physicians before undergoing holistic treatment and also while being treated by holistic methods.

The public should be on guard against quacks and charlatans who have entered the field of holistic medicine. Legitimate practitioners of holism do not regard it as a substitute for conventional medical care.

Dentistry

Innovative and constructive concepts for the delivery of dental care to previously underserved segments of the population were among the significant developments in dentistry during the past year. Also of note were some startling research findings heading the list of developments during 1980.

Making dental care more accessible for the underserved became the overall goal of the American Dental Association (ADA) for the 1980s. The ADA efforts were directed at target populations consisting of the elderly, handicapped, institutionalized, homebound, poor, working poor, uninsured worker, and remote area residents. Noting that there are 33 million handicapped persons in the United States, the ADA announced that it was working closely with the National Foundation of Dentistry for the Handicapped (NFDH). The NFDH initiated programs in a number of states in which dental hygienists will instruct homebound handicapped persons in oral hygiene practices and, if needed, arrange for referral to local dentists.

Photos, courtesy, National Foundation of Dentistry for the Handicapped

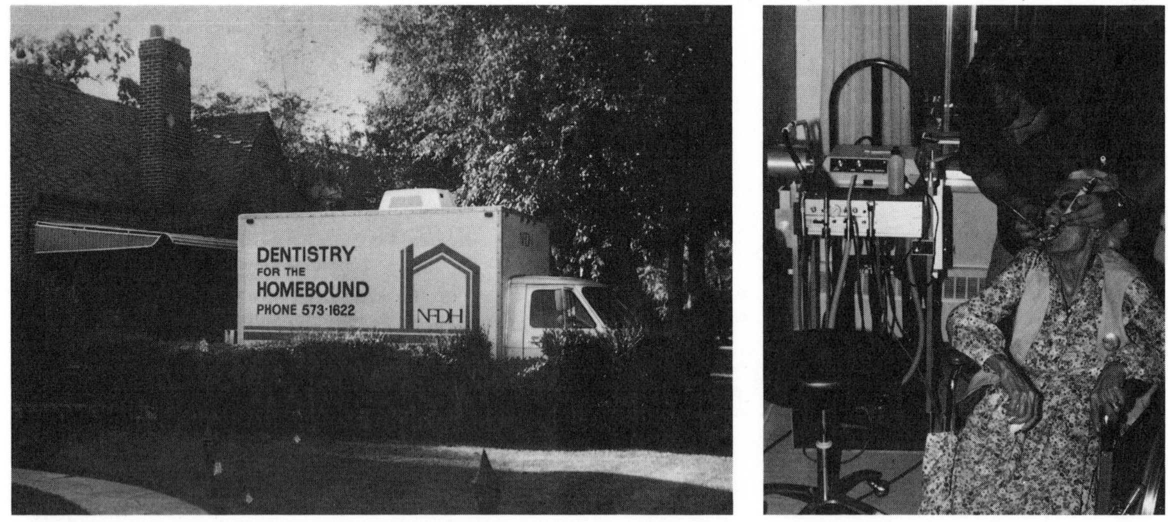

Customized van (left) allows dentists in metropolitan Denver to make house calls to homebound people and residents of nursing homes. Equipment modules can be taken from the van into a home and organized into a complete dental office in order to treat a patient (right).

At its 1980 annual session in New Orleans the ADA called on all of its state societies to develop reduced-fee comprehensive dental care programs for indigent elderly residents. As of the end of the year the ADA reported 157 reduced-fee dental care programs in 43 states plus the District of Columbia and Puerto Rico sponsored by dental societies or dental schools.

Vaccine research. Intensified efforts to develop a vaccine against tooth decay drew probably the greatest attention among the year's research developments. According to the National Institute of Dental Research four separate research teams were working on producing a vaccine that would be both effective and safe for human use. All four teams based their work on the same concept, that of stimulating natural antibodies in human saliva. Antibodies mix with the saliva and prevent the bacteria, *Streptococcus mutans,* from sticking to the teeth.

William H. Bowen of Bethesda, Md., chief of the caries (decay) prevention and research branch of NIDR, said that adults generally develop fewer cavities because they have built up a natural immunity against *Streptococcus mutans* by swallowing tiny amounts of the bacteria over the years. But acquiring immunity naturally is a slow process and does not protect children who would benefit from a vaccine by strengthening their natural immunity against cavities. Bowen, one of the first scientists to begin vaccine studies at the Royal College of Surgeons of England in London, demonstrated that a vaccine injected into five monkeys was effective in preventing decay.

In addition to vaccine research at NIDR, similar studies were being conducted at the University of Alabama in Birmingham, the State University of New York at Buffalo, and the Forsyth Dental Center in Boston. One of the vaccines tested with a small group of human volunteers was contained in capsules. Other forms of vaccine application under study included slow-release capsules and a nasal inhaler.

Decay inhibitors. In related research Bowen announced some puzzling findings in the area of nutrition. One of these indicated that cheddar cheese may possibly inhibit the tooth decay process. When rats were fed semi-processed cheddar cheese immediately after they had eaten sugar, a known culprit in the development of tooth decay, they developed less decay than did rats who were fed the sugar but not the cheese. Thus far no explanation why cheddar cheese should be an antidecay weapon has been found. One theory is that it might interfere with the acid that leads to decay or with the bacteria that produce the acid.

Another surprising discovery reported during the year by NIDR researchers was that sweetened cereals proved about equally potent in causing cavities whether they contained 8% sugar or almost eight times that much. The Institute's scientists devised a technique for testing the cariogenicity of foods by tube-feeding rats to provide nutrition while giving them test foods by mouth in a feeding machine that insured a predetermined content over a specific period of time. All test foods were then compared with sugar, which was selected as the standard food item causing tooth decay. Breakfast cereals containing 8%, 14%, and 60% sugar were found almost equally decay-producing in the rats. Foods that contained sugar and dicalcium phosphate, an additive sometimes used to improve a product's physical consistency, had lower decay potential than the same foods with sugar alone.

"Our results show that the relative cariogenicity of a food cannot be assessed simply on the basis of concentration of sugar," Bowen explained, "although clearly all foods tested that contained sugar were to some degree cariogenic." The researchers found that peanuts do not produce decay but that potato chips had a higher risk index than caramel or a chocolate bar and were almost as high as sugar-coated chocolate candy. In all cases, the more frequently teeth were exposed to the food, the greater the risk.

Washington's wooden teeth. A UCLA dental scientist during the year disputed the myth of George Washington's wooden dentures. "He had teeth of elephant ivory, walrus tusks, teeth from cows and a hippopotamus, even human teeth including one of his own," said Reidar Sognnaes of Los Angeles, dean emeritus of the UCLA School of Dentistry. "But he never had teeth made of wood."

Sognnaes said that he duplicated some of Washington's teeth in the same way they originally were made. Speaking at the annual meeting of the American Association for Dental Research, he noted that hippo teeth were popular at the time because one could be carved into several human teeth. He noted that hippo teeth were used for dentures at least as far back as 1730 and were described in the literature at the time. The teeth were mounted on a base of ivory and held in place with a sticky wax. John Greenwood, Washington's favorite dentist, made "a beautiful set with hippo teeth carved into segments and fitted into ivory with gold plate in the upper jaw, and the two parts were connected with gold springs in the back," Sognnaes said. The springs lost tensile strength over a period of time and pushed the plates forward, which may have given Washington the outthrust-jaw appearance revealed in many portraits of him.

Preventive orthodontic therapy. Early treatment by the family dentist may sometimes correct minor orthodontic problems in children, thus avoiding the need for future extensive orthodontic therapy, a Chapel Hill, N.C., dentist suggested. "Unfortunately, parents too often think of orthodontic treatment only as a full mouth of braces," said Henry W. Fields. "They don't realize that there are numerous other orthodontic appliances that can help correct or avoid problems in the developing dentition."

Speaking at the annual session of the ADA in New Orleans, Fields described several frequently encountered occurrences that may benefit from minor orthodontic intervention. "A child may prematurely lose a primary tooth during a playground accident. The space for a permanent premolar which has not yet erupted may be crowded by adjacent teeth. Or a permanent tooth may erupt in an aberrant direction, crowding and jeopardizing an adjacent tooth. Some of these conditions could be resolved by careful diagnosis and minor orthodontic treatment," he stated.

Dental care for the elderly. Every time an older person loses a tooth, he feels he's one step closer to dying, according to a San Francisco dentist. Sidney Epstein pointed out that tooth loss often is viewed by the elderly as evidence of the disintegration of the body and loss of independence. "If we casually remove teeth as an oral expedient," he said, "this represents our failure as dentists. Complete dentures should be viewed as evidence of indifference, lack of availability of appropriate dental care, fear, and economic circumstances." In many cases, he maintained, extractions can be avoided in favor of modified treatment of remaining teeth.

But patients who must use dentures should not feel inadequate or guilty, Epstein stressed. "Moreover, the patient must realize that the insertion of full dentures is not the end of the line for his oral health problems. Regular home care and periodic examinations by the dentist are most important. Dentures must be checked, corrected, modified, and even replaced." He noted, however, that patients should not necessarily discard ill-fitting dentures to purchase new appliances. "It is now possible to modify existing prosthetic devices to make them more comfortable. Money is not the primary factor here; the difference is in the patient not having to adapt to something totally new, which may be beyond his capacity. Patients who have dentures should be comforted and counseled so they don't develop feelings of loss, guilt, or inadequacy."

Dietary fluoride supplements. Whether or not dietary fluoride supplements given during pregnancy may benefit the offspring came under close scrutiny at the ADA annual session. Anders Thylstrup of Copenhagen and William S. Driscoll of Bethesda, Md., said that there is no sufficient scientific evidence to assume any benefits of prenatal fluorides to the offspring. "Although it has been hypothesized that the developing bones of the fetus may serve as a repository for fluoride, which could be mobilized postnatally to benefit both deciduous (baby) and permanent first molars, the actual existence of such phenomenon has yet to be demonstrated," Driscoll said.

The ability of the fluoride to pass through the placenta to the fetus was also widely investigated. And although there appeared to be little doubt that at least part of the fluoride ingested during pregnancy crosses the placenta and is incorporated in the developing bones and teeth of the fetus, there "is some disagreement as to whether or not the placenta partially bars the passage of fluoride," Driscoll explained.

—Lou Joseph

Veterinary medicine

The role of companion animals in human well-being received increasing attention by the veterinary profession of the United States and Great Britain during

Washington State University College of Veterinary Medicine, Pullman; photo, Ken Porter

Student in a therapeutic riding class leads out his horse. Such classes help mentally or physically handicapped people develop self-confidence and the ability to concentrate on a task.

1980. A major development was the implementation of a "People-Pet-Partnership Program" and a program of animal-facilitated therapy by the College of Veterinary Medicine at Washington State University. As stated by Leo K. Bustad, dean of the college, this broadly based program had as its objectives: (1) education of children about the responsibilities of pet ownership and the potential of pets for enriching their lives; (2) utilization of animals in therapy for the mentally disturbed; (3) promotion of pets and companionship programs for the elderly, the lonely, students in dormitories, and others; (4) establishment of a system to provide information on pet programs and to enable persons to obtain such specially trained pets as hearing dogs for the deaf; and (5) expansion of the veterinary neuroscience program to include instruction in animal behavior and the animal-human interface together with development of animal profiles to determine the types of animals best suited for persons with specific needs.

A project to provide social work support at the University of Pennsylvania's Small Animal Hospital resulted from a perceived need of veterinary clinicians to know more about human behavior in order to deal with problems they encountered with clients. According to Eleanor Ryder of the University of Pennsylvania School of Social Work pets are useful as a social bridge between people, and they may be especially good for children with developmental problems and for isolated elderly people. In one study persons with serious heart disease lived longer if they owned a pet. At first it was thought that the regular mild exercise a person gets from walking a dog might explain this benefit, but persons having only a bird or goldfish also had significantly longer survival rates. Other studies focused on the emotional involvement of owners whose pets were ill or had died and the problems of communication between veterinarians and pet owners.

The Mt. St. Helens volcanic eruption in May 1980 killed an estimated 11,000 deer and elk, 200 bears, 100 mountain goats, 500,000 salmon and trout, and 1.5 million small mammals and birds. The total dying from starvation as a result of the destruction of vegetation was expected to go much higher, however, and veterinarians surmised that the life span of many domestic animals might be shortened because of fibrosis of the lungs due to inhalation of volcanic dust. About 200 cattle, horses, dogs, cats, and pigs were rescued after being abandoned in the evacuated area, and another 1,000 or more were gathered at animal shelters, where veterinarians treated large numbers for eye irritation. Immediate and long-term losses to livestock owners in the affected area were estimated at $3–5 million.

As in the 1979 nuclear accident at Three Mile Island in Pennsylvania the need for planning evacuation programs for both people and animals in advance of potential emergency situations in relatively remote as well as populated areas became evident. As an aftermath of the Three Mile Island evacuation, veterinarians had drawn up recommendations for the appropriate handling of animals in emergencies, but lack of time and facilities at Mt. St. Helens prevented implementation of organized efforts.

The "scare story" of the year with regard to animal disease was the widespread outbreak of a highly contagious and fatal canine parvovirus infection, commonly dubbed "parvo." The disease had been recognized in Australia in 1978 and may have been present in Texas about that time. In 1980 veterinarians in nearly every state reported outbreaks, particularly among puppies. In Corpus Christi, Texas, more than 1,000 dogs out of a total canine population of about 50,000 were said to have died within a few months, and one veterinarian in Pennsylvania saw some 450 cases during one 20-day period.

The virus is closely related to or may be the same as that of feline panleukopenia ("cat distemper") and mink enteritis. At the time the disease was recognized as a serious threat, only one vaccine approved for use in dogs was available. The manufacturer stepped up production to one million doses a month and had a backlog of orders for another two million. Other vaccines were being produced by the fall of 1980, but meanwhile many veterinarians were using feline vaccine for dogs. The feline vaccine was later approved for use in dogs, and 45 million doses were produced by early 1981. As in feline panleukopenia the virus attacks and destroys white blood cells (leukocytes) and cells lining the small intestine, causing diarrhea and vomiting. Even with treatment the mortality in some areas was 50–100%, although as veterinarians came to understand the disease better some reported nearly 100% success using large doses of antibiotics and corticosteroids with supportive fluid therapy. In older puppies and dogs the virus causes severe enteritis, but in young puppies it attacks the heart and causes myocarditis with virtually 100% mortality despite treatment.

The sudden appearance of a virus that affects only dogs but is similar to that of feline panleukopenia caused considerable speculation regarding its origin. One theory postulated that canine parvovirus was a mutant form of feline panleukopenia virus, and a number of veterinarians recalled having seen many dogs 15 or 20 years ago affected with what appeared to be "cat distemper," which was widespread in cats at that time. Since then a large proportion of cats have been routinely vaccinated against feline panleukopenia, and this may have accounted for the disappearance of the "look-alike" disease in dogs. Because the virus could mutate again, perhaps causing an even more deadly disease, some veterinary virologists cautioned against using live-virus vaccines and suggested serious consideration be given to use of only killed vaccines.

"People-Pet-Partnership Program," developed at Washington State University, has as one objective the use of companion animals to provide a feeling of well-being among such groups of people as the elderly.

Tom Wurm—Black Star

An outbreak of African swine fever (ASF) in the Caribbean area had devastating effects on the swine industry of several countries in that region and posed a major threat to the U.S. An outbreak in Cuba in 1971, thought to have been introduced in pork scraps in garbage from an airliner, was contained largely within the province of Havana and eradicated by the destruction of about 500,000 pigs there. The disease appeared again in January 1980, possibly by way of refugees from Haiti, although the Pan American Health Organization expressed concern that Cuban troops returning from Angola may have been responsible.

Herds in the Dominican Republic became infected in 1979, and by early 1980 ASF had reduced the swine population there by two-thirds. It was hoped that some isolated herds could be saved for breeding, but later the remainder of the nation's entire 1.2-million swine population was destroyed as a preliminary step toward repopulation with clean animals. Because of the proximity of the U.S. to Cuba and the influx into the U.S. of Cuban refugees, veterinarians of the U.S. Department of Agriculture feared it would be impossible to avoid the introduction of ASF into the U.S., where all swine would be susceptible.

Although ASF resembles hog cholera (swine fever) in many respects, the differing and complex immune mechanisms in ASF have prevented development of a vaccine. The immediate threat posed by ASF caused it to be moved from sixth to second place (foot-and-mouth disease has been first for many years) on the U.S. Department of Agriculture (USDA) priority list of 45 diseases potentially dangerous to U.S. livestock. This list is used to determine how much should be spent on research and surveillance of each disease. In third and fourth place are Rift Valley fever (confined to Africa) and Newcastle disease of birds and poultry (recently introduced into 45 states), both of which can also infect man.

After an alarming increase in canine rabies during 1979 in Texas, where three persons died of the disease, there was a further 50% increase in cases during early 1980. The disease was probably much more prevalent than the number of confirmed cases indicated, since public health officials issued orders to kill suspected rabid animals on sight rather than having them confined for observation. As in many other areas increased hunting of foxes and raccoons for their pelts had greatly reduced the numbers of those predators and had thereby allowed a population explosion among skunks, which are a primary reservoir for the rabies virus.

Bernard Gotfryd–Newsweek

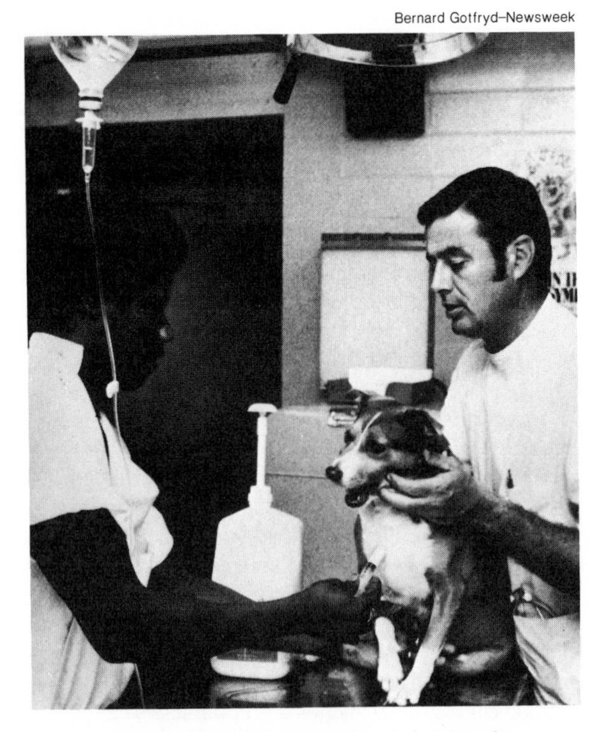

Dog is vaccinated against parvovirus infection that spread throughout the U.S. in 1980. In young puppies the highly contagious disease attacks the heart, causing a mortality rate of over 70%.

The cost involved in dealing with even a single rabid animal may be large. In South Carolina a raccoon "adopted" from the wild bit or scratched 12 people and was proved to be rabid, and six other persons were exposed to a second raccoon from another litter. The total cost of rabies vaccine and immune globulin ($5,538) for these 18 persons, together with administrative and other costs, exceeded $10,000. In Laredo, Texas, a three-and-a-half-month rabies outbreak was controlled by mass immunization of owned dogs, intensified stray animal control, and increased surveillance at a direct cost of about $137,000; this total, together with an estimated loss of nearly $2 million in tourist revenues, resulted in a cost of about $25 per inhabitant of this city.

In March 1980 the U.S. Air Force Veterinary Corps was abolished by act of Congress, and its 280 veterinary officers were reassigned to the AF Biomedical Sciences Corps or permitted to transfer into the Army Veterinary Corps. This was a reorganizational move designed to increase administrative efficiency, and in either case the veterinarians would continue in the same types of jobs that they had previously held. Duties of the Veterinary Corps have included food inspection and sanitation of food-handling facilities, military dog care, human and animal disease research, and control of diseases that are common to animals and man.

New schools of veterinary medicine at Tufts University (Boston) and Virginia Polytechnic admitted their first classes in 1979 and 1980. Total enrollment at the 28 veterinary medicine schools in the U.S. and Canada in 1979–1980 was 8,714, up 390 from the previous year, but applications for admission had dropped from 6.2 per opening in 1974 to 3.2 in 1980. Annual tuition for in-state students was in the $3,000 to $6,000 range at a number of schools, rising to about $16,500 for out-of-state or noncontract students.

—J. F. Smithcors

Optical engineering

The year 1980 was marked by evolutionary changes in technical optics and also by some rather remarkable innovations in optics as applied to everyday life. One revolution that occurred was the introduction of electronics and automatic control circuitry for cameras. Most of the principal manufacturers of 35-mm cameras introduced automated models. In several cases notable advances were made in the optics (lenses and related parts) associated with such cameras. Traditionally, 35-mm cameras have been relatively heavy, partly due to the size of the lenses that were used. During 1980, however, several compact fixed-focus conventional lenses were introduced, as was the direct mounting of optical elements in plastic lens mounts. Significant savings in cost resulted, and there was no appreciable degradation in the image quality produced by the cameras.

In addition, two techniques for the automatic focusing of cameras were in common use by the year's end. One was optical, using a combined detector and computer chip to collect light from a pair of offset images. A correlation or comparison is performed between the two images, and the electronic signal so obtained is then used to drive the lens to an appropriate focal position. Thus an optical focus adjustment can be made in the instant before exposure. When automatic focusing is added to automatic control of exposure and selection of flash, the amateur photographer is relieved from the concerns normally associated with the setting and adjustment of cameras.

The technique described above is complementary to another in which automatic focus is determined by using sound waves. The Polaroid SX-70 instant camera system added an ultrasonic radar which is able to sense the distance between the camera and the subject. The time-out-and-return of an ultrasonic pulse that is reflected from the object provides the distance measurement and adjusts focus prior to an exposure being made.

The design of lenses in the new 35-mm cameras was tending toward the replacement of standard 50-mm focal-length lenses by variable magnification or zoom lenses that cover the range of 30–90-mm focal length. These zoom lenses are compact, generally about two-thirds the size and one-half the weight of previous designs. As yet most cameras still have a relatively conventional look, because designers have not used the advantages of full automation to invent new forms of miniature cameras.

One innovation of potentially great importance was the announcement by Ilford, Ltd., of the U.K. that it was marketing a black-and-white film that recovered more than 95% of the silver used in the manufacturing process. This was accomplished by producing a final dye-coupled image rather than a silver image in the negative. Normal silver halide photosensitive material is used in taking the exposure, and in the processing of the film the silver is recovered from the emulsion and dye couplers to obtain a dye image. Although the use of this material is not yet widespread, it offers significant savings in the future in conserving the use of relatively expensive silver in photographic films.

In somewhat more technological areas several innovations were made in systems for the production of microcircuit chips. The principal development along this line was the introduction of reflective systems by Perkin-Elmer Corp. in its "Micralign" series of circuit mask printers. This unit magnification system enabled circuits of extremely reliable quality to be produced on wafers of larger size than ever before. Two spherical mirrors are employed as the principal imaging components, rendering complicated refractive elements for microcircuit lenses virtually obsolete.

New optical devices for the measurement of vision were produced in recent months. Automated instruments for measuring the refraction of the eye use an electro-optical system to detect the power and cylindric errors in an individual's eyes. These instruments can substitute for many of the routine measurements made by an optometrist and allow automatic rapid screening for visual defects. In addition, these devices may aid in obtaining more precise eyeglass prescriptions for patients.

The influence of electro-optics in all areas of measurement has been strongly felt. Several firms marketed light-measuring devices and color-measuring instruments based on detectors that are controlled by microprocessors. These devices can provide calibration and measurement in any selected set of units.

An interesting general-purpose automated spectrophotometer was introduced by Hewlett-Packard Co. Conventional instruments of this sort disperse the light from a source into a spectrum and then measure the transmission of light through a sample at individual wavelengths. The new spectrophotometer passes white light through the sample and then disperses it using a diffraction grating. Detector arrays, along with microprocessors, are used to provide a parallel readout of the spectrum transmitted through the sample over a selected wavelength range. Calibration and comparison of samples can be done rapidly because the spectrum is read out in parallel on the detector array, and fast changes in absorption of the material can be measured. For example, the course of action of a photochemical reaction over short time periods—of the order of fractions of a second—can be read, sampled, and evaluated by this device.

Lawrence Livermore National Laboratory; photo, Dave Proffitt

Laser light shining through the glass tip at the left stimulates the substance in the vial to glow. Optical fibers collect the glowing light and can transmit it to instruments a mile away to analyze the substance.

New applications of optics to space exploration were at hand, as the 2.4-m-diameter primary mirror for the Space Telescope neared completion. Two primary mirrors, a flight model and a backup, were constructed of lightweight, rectangular-cell, ultralow-expansion fused silica produced by Corning Glass Works. One mirror was being polished by Perkin-Elmer Corp. and the other by Eastman Kodak Co.

The measurement and testing of precision optics reached a high state of development with the completion of several improved interferometers for studying the characteristics of optical surfaces. The most recent of these uses an array of light detectors and a computer to view interference fringes produced by the test surface and thereby provide a virtually instantaneous readout of the manufacturing errors present on an optical surface.

Authenticated News International

Engineers at Perkin-Elmer Corp. check the radii of curvature of the 2.4-meter- (7.9-foot-) diameter primary mirror for the Space Telescope of the U.S. National Aeronautics and Space Administration. The cell structure of the mirror's interior can be seen through the concave face plate. The Space Telescope is scheduled to be launched by the space shuttle in 1983.

A future application of optics in space may include a proposed coherent lidar. This is a form of optical engineering. This device will sense the motion of scattering particles in the atmosphere relative to the motion of the Earth and to the motion of the satellite carrying the instrument. The measurements are to be made by measuring the Doppler shift of the ten-micrometer-wavelength laser radiation that is transmitted from the satellite and scattered by aerosol particles. The small change in the frequency of the return radiation as a result of the scattering can be used to measure the direction and magnitude of the wind at points located over the entire surface of the Earth. Since real-time measurements of winds over isolated portions of the Earth and the oceans are not now available, the use of this instrument should allow significant improvements in the ability to forecast weather patterns on both a short- and long-term basis.

Finally, progress continued on projects begun in earlier years. Work on laser-induced nuclear fusion proceeded to the point of demonstrating that short-wavelength or blue light couples best with the fusionable pellet. Current proposals indicated that short wavelengths would be required for all future experiments in laser fusion. Plans were announced for massive improvements in facilities to provide higher levels of laser radiation for future fusion experiments.

The production of lenses and mirrors by using direct-machining diamond-turning techniques continued to be developed so that equipment as large as one meter in diameter could be produced. Techniques using high temperatures that approach the softening point of glass during diamond turning were shown to be promising for the production of aspheric lenses for laser television recorders.

Future developments in optical engineering are almost certain to be related to the marriage of electronics and optics. New devices providing direct interaction between electrons and photons to permit the development of optical computers are beginning to appear. Such techniques have been developed and demonstrated in limited form in laboratories. One can look forward in the future to integrated detector and processing systems that should provide sensing systems with many of the capabilities of an intelligent observer.

—Robert R. Shannon

Physics

During the past year experiment and theory sent tremors of doubt through two long-accepted beliefs of high-energy and nuclear physics: the indefinite stability of the proton and the zero rest mass of the neutrino. Satellite photos of the ocean surface revealed curious bands of rough and smooth water that proved to be related to mathematical entities called solitons, and several groups of physicists participated in an intriguing hunt that ultimately linked an ancient technique for polishing gem diamonds with a solution to a critical problem in the microelectronics industry.

General developments

The discovery of giant solitary waves in the Earth's oceans and resurrection of a concept for a nuclear fusion reactor, abandoned in the 1960s, highlighted research in physics during the past year. Investigators also succeeded in coaxing the fourth state of matter, ionized plasma, to emit laser light.

Giant solitons. Only very rarely does a discovery on the planetary scale relate to microscopic physics, but it happened in 1980. The U.S. Landsat Earth resources satellite and the Apollo-Soyuz manned space mission set the process in motion in the mid-1970s with photos of the Andaman and Sulu seas in the Far East. The pictures showed traveling groups of striations more than 100 km (60 mi) long. These were interpreted to be surface manifestations of "internal solitons," solitary ripples of the boundary between the warm upper layer of the ocean and the cold lower depths.

Two scientists who studied this phenomenon in the Andaman Sea, Alfred Osborne of the Exxon Production Research Co., Houston, Texas, and Terrence Burch of EG&G Environmental Consultants, Waltham, Mass., found that the striations are kilometer-wide bands of very choppy water moving several kilometers per hour. Behind each band follows about two kilometers of unusually smooth, undisturbed water. Osborne and Burch were able to backtrack to the origin of the bands. Apparently, tidal currents washing over shallow undersea ridges between islands create deep disturbances along the boundary between warm and cold ocean waters. These internal ripples in turn serve to launch the alternating bands of rough and smooth surface water. The rough sea in the bands is the signature of a wedge of warm water extending hundreds of meters down past the thermal boundary. The extraordinary nature of the bands was brought out by observations that they traveled more than 400 kilometers (250 miles), remaining intact for two days. Maintained by the internal ripples below them, they did not disperse and fade out, as would "packets" created by adding together many ordinary sea waves.

This ability to survive in the turbulent sea suggested a very coherent phenomenon, much like the solitons that have been studied as mathematical entities for a century. Unlike ordinary waves, solitons have crests but no troughs. They can arise on the surface of channels of water and were reported, though not studied, as early as 1834 when a naval architect in Scotland saw "a well-defined heap of water" moving up a channel at "some eight or nine miles an hour."

The surface bands seen in the Andaman and Sulu seas have a cause akin to such heaps of water, though in this case the heaps are internal. A wedge of warm, upper water intrudes massively into cold water below. The wedge maintains its identity because of deep, rapidly circulating currents. A sign of its relation to a soliton, even though the intrusion cannot be seen directly on the surface, is that the amplitude of the subsurface warm-water intrusion affects its velocity. Measurements using deep probes showed that velocity increased for the larger amplitude wedges. This is a distinctive sign of all solitonlike phenomena.

At the leading edge of an oceanic soliton, currents descend, pass under the entire soliton, and reemerge at the tail. The downward current produces the chop at the surface and serves to "sweep the sea clean" of small waves thereafter, thus producing a calm zone behind the soliton. These currents produce strong horizontal forces as the soliton passes. Exxon oil-drilling rigs have suffered from such strong surges; this was a prime motivation to begin the study. Future deep-sea drilling will have to take account of such phenomena. Solitons and their strong currents are even suspected of causing the mysterious sinkings of several submarines.

Satellite image of the Sulu Sea (left) records groups of striations interpreted to be surface effects of internal ripples in the boundary between cold and warm layers of ocean. Interpretative map (right) locates underwater measuring stations set up by John Apel and James Holbrook of the Pacific Marine Environmental Laboratory, Seattle, Washington, in studies of the evolution and structure of these ripples.

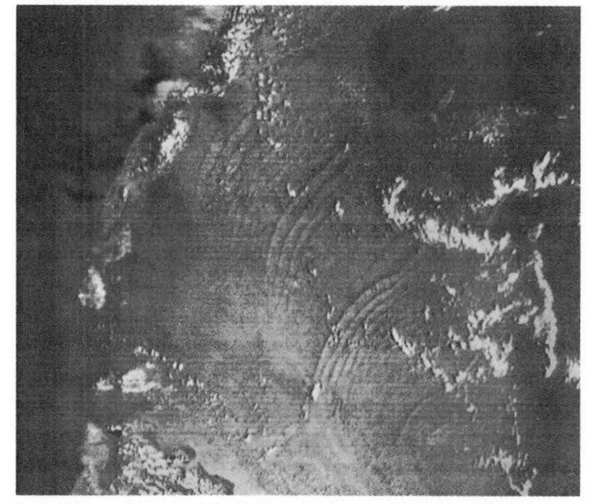

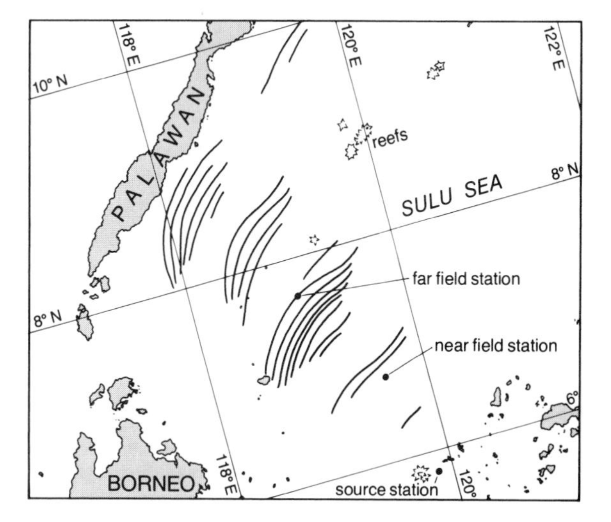

Photo and map, John R. Apel, Pacific Marine Environmental Laboratory, NOAA

Courtesy, IBM Corp.

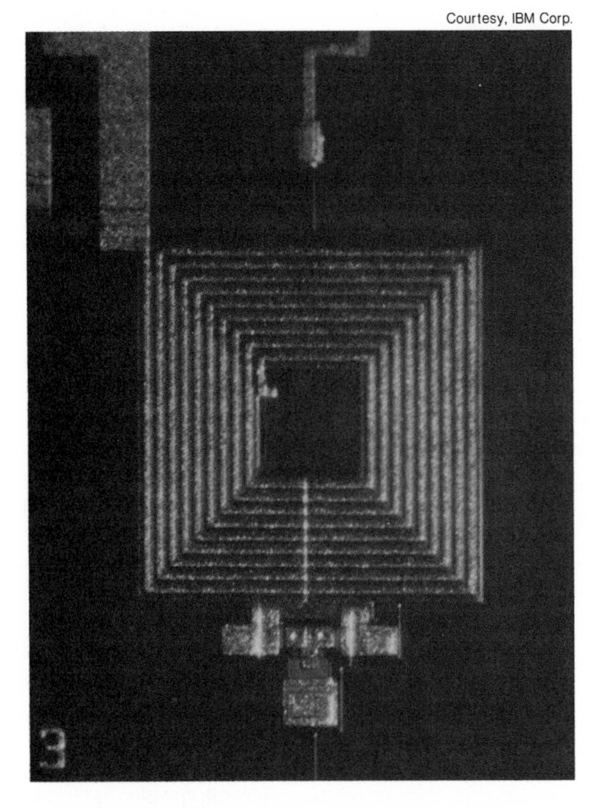

Superconducting circuitry developed at IBM has the potential for sensing changes in magnetic fields as small as one trillionth (10^{-12}) the field of the Earth. The square spiral input coil is about 175 micrometers wide.

Although solitons have been explored mathematically, their production and study in the laboratory has been limited. The most convincing laboratory measurements involved regions of compressed electric waves only millimeters (a few hundredths of an inch) in size. Solitons have application to such wide-ranging fields as organic conductors, plasma turbulence, coherent radiation from the Sun, superconductivity, and even the Great Red Spot of Jupiter. They maintain a constant shape while propagating, even when they pass through each other. Calculations show that solitons may form at the quantum level, entering directly into field theory. If so, they are a common feature of nature, spanning 22 orders of magnitude in size.

Fusion futures. During the 1970s progress toward controlled thermonuclear fusion had focused on attaining higher plasma densities, temperatures, and confinement times. In 1980 the international fusion program took the time to stand back from the breakneck pace of the last few years and considered the longer view.

The present leading candidate for a future fusion reactor is the tokamak, a design that generates a doughnut-shaped, or toroidal, magnetic field for plasma confinement. Until the late 1960s the U.S. fusion program had invested its largest effort in another toroidal-field device, the stellarator. Disappointing trials of the machine came in the same years that the Soviet tokamak began showing good confinement times, leading to abandonment of the stellarator. In 1980, however, a team at the Max Planck Institute for Plasma Physics in Garching, West Germany, revived the stellarator by demonstrating that it can provide the basic equilibrium and confinement envisioned by its inventor, Lyman Spitzer of Princeton University.

In all toroidal reactors there is a basic magnetic field, the toroidal field, running around the axis of the machine. Retaining the plasma for any appreciable time, however, requires another field, called a poloidal field, to add a helical twist. The tokamak provides this second field by running a circulating current in the plasma itself. The current heats the plasma as well, helping to boost the temperature in the reactor to the level needed to ignite the fusion of a mixture of heavy isotopes of hydrogen (deuterium and tritium). By contrast, in the stellarator helical coils outside the main jacket provide the poloidal field.

The Garching successes proved what some theorists had come to suspect: that the early stellarators failed not because of some fundamental flaw in the concept but because they were not big enough. External field coils were needed to provide the helical twist, and these encumbered the machine enough to limit its cross section. The magnetic doughnut of the stellarator was so "skinny" that the plasma it confined could not be heated enough. Like tokamaks, early stellarators used plasma currents to heat, and they could not drive enough current through their thin cross sections to heat their plasmas without losing the ability to confine. The obvious solution would be to greatly expand stellarator size, but this is prohibitively expensive. The Garching group sidestepped that problem by not using plasma currents to heat. Instead they injected into the stellarator geometry an intense, uncharged particle beam. This collided with the plasma, heating it. Confinement was good, at least five times as high as that attained by tokamaks of comparable size.

This result aroused enthusiasm among workers on stellarators in the U.K., Japan, and the Soviet Union, because the long-range prospects of a stellarator reactor promises efficient running characteristics. The helical stellarator field appears to stop the rapid plasma breakup and consequent vessel-wall damage that can occur in tokamaks. Such instabilities often occur during the pulsed tokamak cycle, as the plasma current drops under the influence of changing plasma conditions.

The most attractive aspect of stellarators, however, is the possibility of steady operation. Tokamaks make their internal currents by pulsing external coils, which drive the currents by means of a transformerlike magnetic coupling. Tokamak optimists believe that a working reactor could achieve a fusion "burn" for as long as an hour. Still, any interruption presents huge engineering problems. The varying heat and radiation loads within the machine will strain the materials near the fusion fire; cracking and materials failure can result. Some planners have proposed that the "up time" in tokamaks could be extended if the plasma was heated steadily by yet another external agency—electromagnetic waves injected at radio frequencies. Although this idea was being tested in 1980, it represents yet another complication of the already complex tokamak design.

The stellarator, in contrast, is the only scheme yet proposed that would require no further external heating to keep the burn going once the fuel was ignited. Only the external field coils need current to run them, while the heat generated by the burn itself keeps the plasma in a steady, energy-yielding condition. The simplicity of the stellarator is thus its greatest asset.

In 1980 long-range planning began for a giant tokamak beyond the present generation of four machines under construction. This machine is foreseen to have seven times the working volume of the Princeton Tokamak Fusion Test Reactor, which when built will attempt to achieve break-even fusion (generated power exceeds power applied to the plasma) with a deuterium-tritium fuel. Nevertheless, the consensus is that options for such resurgent ideas as the stellarator—a dark horse coming up fast on the outside—should be kept open.

Plasma laser. Since the invention of the laser in 1960, many kinds of solids, liquids, and gases have been used to produce its narrow, powerful beams of light. In many cases it was only necessary to put mirrors on the ends of a tube filled with the substance, pipe in a stimulating glow of light, and the substance would lase. In 1980 the fourth state of matter, ionized plasma, was made to lase for the first time. Workers at Bell Laboratories, Murray Hill, N.J., used an ionized metal vapor, which emits light when its ions recombine with the electrons they have lost. As the electrons are recaptured, they lose the energy that they gained when they left their parent atoms by dropping in steps through discrete energy levels. In practice this well-defined cascade is easily disrupted by collisions within the fast-moving ion-electron cloud. Making a coherent laser beam requires that the collisions be avoided so that the stepwise recombination can proceed. Nearby mirrors then reflect the emitted light back into the plasma cloud, egging on other electrons to yield up their energy and rejoin their ions.

The difficult trick is stopping the interfering collisions while encouraging the essential recombination. This can be accomplished if the plasma is allowed to expand against a surrounding gas, which cools it rapidly. One way to make the plasma in such surroundings involves a kind of miniature lightning effect. Small patches of metal film are laid down on a nonconducting base material, and then a large electric current is applied. Sparks jump from patch to patch, pulling off metal atoms and forming tiny clouds of plasma between the patches. The hot clouds of high-density metal plasma expand, cool, and recombine before many collisions can occur.

It helps to use a metal such as indium, which has a large gap between two of its energy states and which thus requires large energy transfers from the recombining electrons. This makes it difficult for a random collision to provide just the right energy to stimulate recombination; hence, the chances of collision-induced recombination are lowered. Instead, recombination must wait for stimulation by precisely tuned incoming radiation, which is simply the light from other, earlier recombinations reflected from nearby mirrors. The plasma laser has sophisticated physics behind it but is simple in operation. The sparking metal patches are stuck to glass strips lying inside a tube full of gas. Quick pulsing of the driving current can yield thousands of laser bursts per second. The metal gets used up, ending the process after about 100,000 pulses. Continued development promises longer operating times. Whereas the indium-based device lases in the near infrared, use of different metals may push the emitted light into the high ultraviolet, opening a new region of the electromagnetic spectrum to lasers.

—Gregory Benford

High-energy physics

During the 1970s an enormous development of elementary-particle physics occurred. At last scientists have a notion of what are the most elementary building blocks of matter and believe that they understand the nature of fundamental forces by which those particles interact. Three different forces are known to underlie the behavior of the microworld. Electromagnetism keeps atomic electrons in orbit about the nucleus. The strong nuclear interaction keeps protons and neutrons bound within the nucleus. The weak nuclear force allows neutrons to turn into protons in the process termed beta decay. Presently all three forces are described by a gauge theory wherein the basic particles are of two kinds, called force particles and matter particles. Practically all complex physical phenomena result from iteration of a primal act of becoming: the emission or absorption of a force particle by a matter particle.

Matter particles. How many different species of matter particles are there, and what are their properties? Matter particles that partake in the strong interactions are known as quarks. Those that do not are known as leptons. The neutrons and protons that make up atomic nuclei were once regarded as elementary particles. Today they are widely believed to be made of

Courtesy, CERN

Workers assemble a large streamer chamber—a particle-track recording device—that will be used in proton-antiproton colliding beam experiments at CERN's Super Proton Synchrotron in Geneva. The chamber, which can make visible particle tracks several meters long, will aid high-energy physicists in the search for exotic events.

quarks. Each proton (p) contains two up quarks (*u*) and one down quark (*d*) and may be designated p = (*uud*). Each neutron (n) has the configuration n = (*ddu*). It follows that quarks must carry fractional amounts of the proton's electric charge. This fraction, Q, is $^2/_3$ for the up quark and $-\,^1/_3$ for the down quark.

The strange quark (*s*) has $Q = -\,^1/_3$ like the down quark, but it is heavier. Strange particles are particles containing one or more strange quarks. Cousins of neutrons and protons with such quark configurations as (*sud*), (*ssu*), and (*sss*) are well-studied particles. Their weights are 20–80% more than that of the proton, and they are all unstable particles with lifetimes of about 10^{-10} seconds. They can be produced naturally by cosmic rays or with the help of high-energy accelerators.

Confirmation of the existence of charmed particles in 1976 verified the theoretical prediction of a fourth kind of quark, the charmed quark (*c*), which carries electric charge $Q = ^2/_3$. Charmed particles are very short-lived. Their mean lifetimes are only about 10^{-13} seconds, a thousand times shorter than strange particles. Between the time of their production and their subsequent decay, they travel only a fraction of a millimeter. Nonetheless, experimenters in 1979 and 1980 succeeded in directly viewing a record of this process. High-energy neutrinos are directed at a target consisting of many liters of photographic emulsion. Occasionally a charmed particle is produced, lives its brief life span, and decays into other particles. These processes are recorded in the emulsion. Electronic detectors tell when this happens and direct the experimenter to the exact location of the event, which then can be studied in detail. Several such experiments were performed by European collaborations at the Super Proton Synchrotron at the European Organization for Nuclear Research (CERN) in Switzerland. The most accurate information available as of early 1981 was accumulated by a U.S.-Canadian-Japanese-Korean team working at the Fermi National Accelerator Laboratory (Fermilab) in Batavia, Ill.

The first experimental indication of the existence of charmed quarks was the simultaneous discovery in 1974 of the J/psi particle at Brookhaven (N.Y.) National Laboratory and at the Stanford Linear Accelerator Center (SLAC) in California. For this discovery Samuel C. C. Ting and Burton Richter shared the Nobel Prize for Physics in 1976. A similar experiment to that at Brookhaven was later done at Fermilab, but at considerably higher energy. It resulted in the discovery of the upsilon particle in 1977 by Leon Lederman and his collaborators. This particle was the first indication of the existence of yet a fifth species of quark, the bottom (or beauty) quark (*b*). The upsilon particle and related particles are currently under study at CESR, an electron-positron colliding-beam machine at New York's Cornell University that began operation in 1979. As of early 1981 four different particles were found that contain bottom quarks. Their behavior indicates that the strong, weak, and electromagnetic interactions of bottom quarks accord with theoretical expectations.

From study of the quarks so far identified it appears that quarks come in pairs, with electric charges $Q = ^2/_3$ and $Q = -\,^1/_3$. Thus, *u* and *d* form a pair, as do *c* and *s*. The mate of the bottom quark has not yet been discovered. It is called the top (or truth) quark (*t*). Sensitive searches for the top quark were performed during 1979 and 1980 at PETRA, the electron-positron machine in Hamburg, West Germany. The J/psi, which signals the existence of the charmed quark, shows up clearly in electron-positron collisions at a total energy of about three GeV (one GeV equals one billion, or 10^9, electron volts). The upsilon, sign of the bottom quark, shows itself in similar experiments at about ten GeV. Experiments at PETRA demonstrated that the characteristic sign of the existence of the top quark does not lie at any value of energy less than 36 GeV. This means that the top quark must be heavier than 18 GeV. Perhaps it will be discovered by the next generation of high-energy accelerators.

The most familiar lepton is the electron, with a mass about $^1/_{2,000}$ that of the proton and opposite electric charge, $Q = -1$. Two heavier charged leptons are known to exist. The muon, discovered in cosmic rays in 1936, is about 200 times heavier than the electron. The tau lepton was discovered at SLAC in 1975 by Martin Perl and his collaborators; it is about 16 times heavier than the muon. Corresponding to each of the three charged leptons there is believed to be a neutral ($Q = 0$) lepton called a neutrino. Neutrinos are much lighter than their electrically charged partners. In the past it was generally thought that neutrinos were truly massless particles. Experiments done in the Soviet Union and at the Savannah River reactor in South Carolina in 1980 suggest that they may have small but observable masses. All told, there is evidence for the existence of six species of leptons, three with one charge and three with another. The parallel with the quarks is striking: there are six species of quarks, three with $Q = ^2/_3$ and three with $Q = -^1/_3$.

Why are there so many different kinds of matter particles? Familiar matter is made of up and down quarks and electrons. One kind of neutrino is needed to explain the mechanisms that power the Sun. But as for the other particles, one may borrow from Nobel laureate I. I. Rabi, who said upon the discovery of the muon, "Who ordered that?"

"Left bend," a row of energy-saving superconducting magnets more than 140 meters long, went into routine service at Fermilab in 1981 (below). The magnets direct high-energy protons from the main ring to the Meson Laboratory, one of three major experimental sites. Upper end of the left bend, where protons for the meson and neutrino areas diverge, is at right.

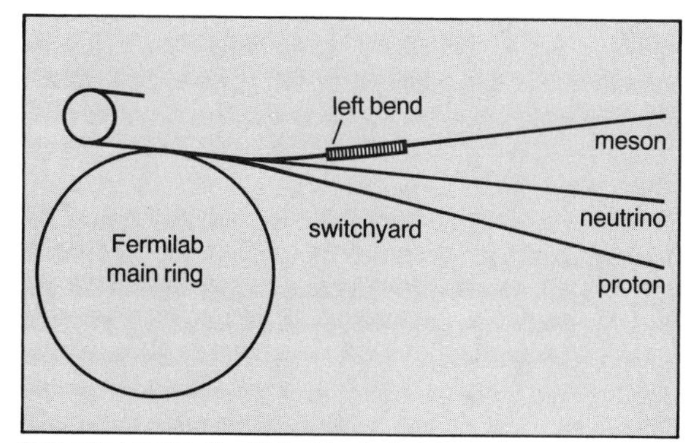

Photo and art, courtesy, Fermi National Accelerator Laboratory

Force particles. Two matter particles may "exchange" a force particle, rather in the way two soccer players exchange the ball. Such an exchange produces a force, which is why the force particle is called as it is. Any two particles that carry electric charge (*i.e.*, any of the matter particles but neutrinos) may exchange a photon, a packet of electromagnetic radiation. If the particles are oppositely charged, the resulting force is an attraction. Thus it is that an electron, by exchanging photons with the quarks in a proton, is attracted to form a hydrogen atom. The photon is said to mediate the electromagnetic force. Other kinds of force particles mediate the strong and weak interactions.

The force particles that mediate strong interactions are known as gluons, and there are supposed to be eight distinct varieties of them. Quarks are the only matter particles that may exchange gluons. It is the exchange of gluons among three quarks that keeps them confined as a proton or neutron. A less direct consequence of this force is the nuclear interaction that causes protons and neutrons to bind together to form atomic nuclei.

Although the electromagnetic and strong forces are analogous, they are not identical. The attraction between an electron and proton decreases with increasing separation. Thus, it is possible to detach the electron and to ionize the hydrogen atom. By contrast, as far as can be determined, the strong force between quarks does not decrease with distance. A quark apparently cannot be dislodged from the particle of which it forms a part. Quarks and gluons are permanently confined and cannot be produced in isolation. Therefore, it is difficult to acquire direct evidence for the existence of quarks and gluons. Recent experiments at PETRA, however, provide convincing indirect evidence. When an electron and its antimatter counterpart, a positron, collide, they often produce a quark-antiquark

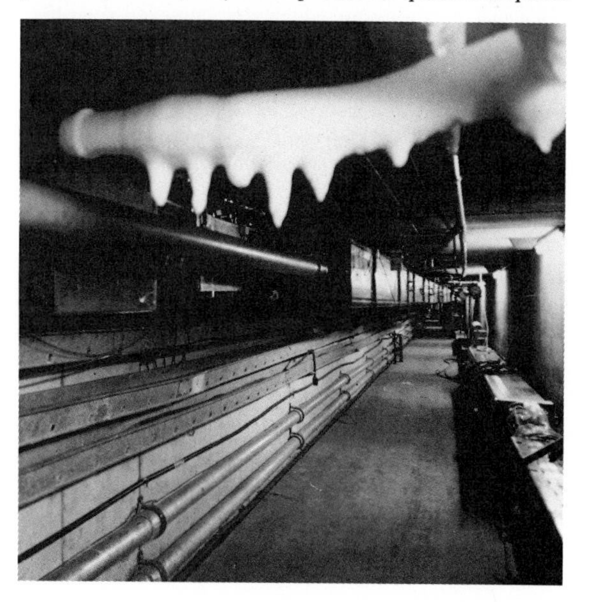

pair. These particles move off in opposite directions in a doomed attempt to separate from one another. The end result is the production of two narrow jets of particles in opposing directions—a two-jet event. Upon occasion one of the quarks emits an energetic gluon. This produces a characteristic three-jet event. Both two-jet and three-jet events have been observed at PETRA, and their properties are just as expected from the theory of the strong force.

Weak interactions are mediated by particles called $W^{\pm}$ and Z^0. Unlike photons and gluons, these force particles are very heavy, so much so that none of the accelerators that have been built are capable of producing them. The $W^{\pm}$ is responsible for familiar weak interactions like nuclear beta decay. The Z^0 particle mediates a class of weak phenomena that was first seen only in 1973—neutral currents.

A unified theory of weak and electromagnetic interactions, which was formulated in the 1960s, successfully predicted the existence of the neutral currents and earned the 1979 Nobel Prize for Physics for Abdus Salam, Steven Weinberg, and Sheldon Glashow. The most recent test of this "electroweak" theory took place at SLAC in 1978, where experimental results provided strong evidence of a mechanism mediated by the predicted Z^0 (see *1980 Yearbook of Science and the Future* Year in Review: PHYSICS: HIGH-ENERGY PHYSICS). The ultimate test of the electroweak theory is the existence of $W^{\pm}$ and Z^0 with predicted properties. In 1980, particle accelerators with enough energy to produce these particles were under construction both in Europe and in the U.S. The first of these machines to be built, an apparatus at CERN for colliding beams of protons and antiprotons that are made to circulate in opposite directions, began operation in early 1981. Confidence runs very high that $W^{\pm}$ and Z^0 will be found in the 1980s.

Grand unified theories. Having achieved the effective synthesis of weak and electromagnetic interactions, scientists are encouraged to pursue the further unification of strong, weak, and electromagnetic forces. Such theories have been invented, and they seem to be in agreement with experimental data. They predict the existence of a new force, one much weaker than the weak force. The new force is responsible for the decay of the proton, long thought to be eternal, and hence of all familiar matter. In these theories the half-life of a proton is about 10^{31} years. Despite such a long lifetime the search for proton decay is feasible. In 1980 large underground experiments were being mounted in Ohio and in Utah. Should a proton decay in a very large tank of highly purified water, it would produce a small flash of light. This signal of proton decay would be picked up by hundreds of sensitive electronic eyes within the tank. Work during the next few years should determine whether the prediction that protons decay is fulfilled.

The dream of Einstein was to formulate a truly unified theory that included gravity, the dominant force in the macroworld, as well as the forces of the microworld. The greatest challenge to theoretical physics remains the fulfillment of this dream.

—Sheldon L. Glashow

See also *Nuclear physics,* below.

Nuclear physics

In nuclear physics, as in all of fundamental physics, the past year was one of bold new theories and conjectures, new measurements, and new insight into the atomic nucleus, one of the basic quantum systems in nature.

Neutrinos and protons. Nuclear beta radioactivity is the process wherein a neutron inside a nucleus transforms into a proton, or vice versa, accompanied by the emission of a negative or positive electron. To explain it Wolfgang Pauli in 1931 had to postulate the existence of an elusive new entity, the neutrino. Just how elusive is indicated by the fact that a neutrino with one million electron volts of energy can pass through a thickness of lead equal to the distance that light, traveling at 300,000 km (186,000 mi) per second, traverses in a year. The neutrino was supposed to have zero mass and thus, like the proton, to travel always at the velocity of light.

During the past year a group at the Institute of Theoretical and Experimental Physics in Moscow very carefully examined the beta radioactivity of the heaviest hydrogen isotope, tritium, and concluded that the neutrino really does have a mass equivalent to greater than 14 electron volts and most probably in the vicinity of 30 electron volts. If verified, this measurement will have far-reaching consequences. If the neutrino associated with the electron has mass, so also do those neutrinos associated with the electron's heavier relatives, the muon and the tau particle. And if so, these neutrinos probably exist in clouds around large galaxies where they could account for the long-sought missing mass of the universe. Their collective mass could explain what binds together large clusters of galaxies. This mass also may be enough to allow the force of gravitation to slow and eventually reverse the apparent expansion of the universe. David Schramm of the University of Chicago and Gary Steigman of the Bartol Research Foundation, Swarthmore, Pa., won the 1980 Gravity Research Foundation Prize for their essay on this topic.

The proton—the nucleus of the lightest atom of all, hydrogen, and long considered one of the few really permanent parts of the universe—may itself turn out to be transitory. The best measurements reported to date, by Frederick Reines, Friedel Sellschop, and their collaborators working in a special laboratory established several kilometers below the Earth's surface in

Camera Press/Photo Trends

Soviet workers contemplate experimental model of a scintillation telescope located at the Baksan neutrino observatory in the northern Caucasus. The telescope will be used to study the neutrino flux that arises from nuclear reactions occurring during such violent cosmic events as stellar collapse. Information thus gathered should give astronomers and physicists a better understanding of the internal structure of stars and various global processes of the universe.

the Rand Gold Mine in Johannesburg, South Africa, suggest that protons live longer than 10^{30} years. But the grand unified theories for which Steven Weinberg and Sheldon Glashow of Harvard University and Abdus Salam of the International Center for Theoretical Physics, Trieste, Italy, received the 1979 Nobel Prize for Physics suggest that the proton may not be eternal and that it may eventually decay into lighter particles, *e.g.*, a positive electron and a neutral pion. The predicted lifetime is about 10^{31} years, and experimenters have been in hot pursuit of this fundamental decay. One U.S. experiment under construction in 1980 involves the use of 1,000 tons of water closely observed by nuclear detectors in a Utah silver mine, while a second experiment employs a 10,000-ton water sample deep in an Ohio salt mine. (The larger sample includes 10^{33} protons, so that about a hundred decays are predicted to occur each year.) If detected, proton decays will be the most direct evidence yet obtained for the correctness of the grand unified theories of the natural forces —Einstein's dream.

Supersymmetry. While some of the basic building blocks of nuclei are thus under close scrutiny, entirely new and fundamental dynamic symmetries are evolving from study of heavier nuclei. Quantum mechanical systems can be described either by inserting the appropriate interaction into the basic equation of quantum theory, the Schrödinger equation, and solving for the quantum states of the system or, alternatively, by the so-called spectrum-generating approach; *i.e.*, by postulating certain elementary entities that are then assembled, subject to the stringent mathematical laws of group theory, to yield the quantum states. Whereas the quarks are these elementary entities in particle physics (see *High-energy physics*, above), Franco Iachello of Yale University, Igal Talmi of the Weizmann Institute in Israel, and Akito Arima of the University of Tokyo have shown that in nuclear physics they are coupled pairs of nucleons (neutrons or protons).

Physicists have long believed that elementary entities could be separated into two classes: fermions, like the neutron and proton, which have half-integral spins; and bosons, like a coupled nucleon pair, which have integral spins. Moreover, it was a matter of faith that the behavior of these two classes was so different that no single theory could treat both simultaneously; in other words, that bosons always remained bosons and that fermions remained fermions at the most fundamental level.

Bruno Zumino of the European Organization for Nuclear Research (CERN), however, suggested several years ago that in bringing gravitation within the realm of quantum theory (it had previously been a classical picture) it was necessary to consider a "supersymmetry" wherein bosons and fermions can be interconverted and are treated on an entirely equal footing. The same necessity surfaced even more recently in the grand unified theories in bringing the strong and weak nuclear interactions together. But, there was no evidence whatsoever in nature that such supersymmetries actually existed.

During the past year Iachello extended his work on nucleon pairs from nuclei having an even number of neutrons and protons (thus entirely paired and bosonlike) to those having an odd number (thus a bosonlike core with an odd valence neutron or proton that is obviously a fermion). He predicted that supersymmetries in these mixed systems would lead to very special

Photo and art, courtesy, CERN

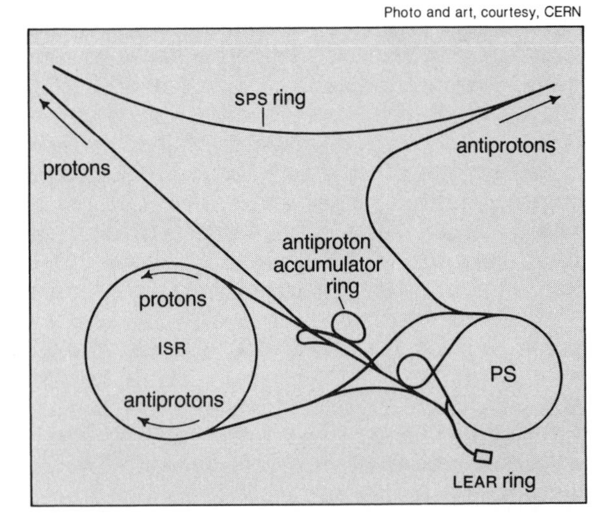

Diagram (above) shows layout of CERN's new proton-antiproton collider. Antiproton beams are stored in the accumulator and routed, by way of the Proton Synchrotron (PS), to the Intersecting Storage Rings (ISR) and the far larger Super Proton Synchrotron (SPS). An aerial view of CERN is at left.

mass and structure relationships between members of nuclear families having both equal and odd numbers of neutrons and protons and that nuclear transfer reactions linking odd and even nuclei would have certain specific properties. Measurements by Jolie Cizewski and her collaborators at Yale and at Los Alamos (N.M.) National Laboratory recently found the predicted relationships with remarkable purity. This combination of theory and experiment thus resulted in identification of the first supersymmetries in nature.

Stretched configurations. New techniques have also continued to uncover unexpected simplicities in nuclei. Richard Lindgren of the University of Massachusetts at Amherst and his collaborators from the Massachusetts Institute of Technology and the University of Indiana showed that, when high-energy electrons and high-energy protons—those having several hundred million electron volts of energy—are scattered off nuclei, they have remarkably high probabilities of moving a single neutron or proton from its original shell-model orbit in the target nucleus into a previously empty higher energy one and, moreover, of doing it in a very particular way that leads to the maximum possible spin or total angular momentum. Discovery of this selective population of so-called stretched configurations promises to provide a very powerful probe for the details of nuclear structure in experiments using high-energy electron and proton beams. Previously, conventional wisdom had held that essentially all possible states would be populated with roughly equal probability so that all simple features would be hopelessly obscured.

Laser use. At the opposite end of the energy regime Curt Bemis and his collaborators at the Oak Ridge (Tenn.) National Laboratory capitalized upon the availability of tunable lasers to pump atoms systematically into a particular (hyperfine) quantum state in which the atomic and the nuclear spins have a precise relationship. In this way it is possible to produce nuclear systems in which all the nuclei have their spins lined up at a particular and completely controllable angle to the laser beam. In his initial measurements Bemis lined up americium nuclei, known to be football-shaped, and found that when those nuclei fissioned the fragments were emitted along the long axis of the football. He also found that although the long axis of the americium nucleus is normally about 40% greater than the short axis, intermediate states along the route to fission that are quasi-stable—the so-called fission isomers—are much more deformed, with a long axis twice the length of the short one.

In 1980 this laser technique was being applied for the first time to the preparation of targets for use with the new 25-million-volt Holifield tandem accelerator at Oak Ridge. When used with polarized projectile beams from that accelerator, it will make possible the first complete study of fundamental spin phenomena in nuclear physics. In one of the earliest measurements, for the first time a direct experimental measurement will be made of the difference in the height of the Coulomb barriers (the electrostatic repulsion) encountered by projectiles approaching a target nucleus along the football axis or perpendicular to it. An entirely new class of nuclear experiments will be opened up.

Research facilities. During the past year the Chinese revealed that construction was proceeding on a major nuclear research facility for heavy-ion physics in Lanzhou in northwestern China. When completed, the installation will be more powerful than any other in the world and will complement a number of additional, less ambitious projects under way at other Chinese research centers. Simultaneously the Soviets, in a collaboration between the I. V. Kurchatov Institute of Atomic Energy in Moscow and the Joint Institute for Nuclear Research in Dubna, announced that they were building what will be by far the world's highest energy nuclear facility, capable of accelerating any nucleus to several billion electron volts per unit mass. There is a hint from work of Erwin Schopper in Frankfurt and Harry Heckman of the University of California at Berkeley and their collaborators, carried out at Dubna and at the Berkeley Bevalac, that nuclear matter of an entirely new sort, first suggested theoretically by T. D. Lee and Gian-Carlo Wick of Columbia University, N.Y., may be produced for the first time when this new Soviet facility becomes operational.

Finally, dealing with even higher energies, William Willis and his collaborators at Brookhaven (N.Y.) National Laboratory and CERN proposed to convert the CERN proton ring to the acceleration of much heavier nuclear species to super-relativistic energies (energies at which the accelerated species closely approach the velocity of light). Thus far, in elementary-particle physics the trend had been to pump ever more energy into ever smaller volumes in the search for new phenomena. In the proposed work, even higher energies will be pumped into much larger volumes—volumes capable of holding many neutrons and protons—with the expectation of observing totally new collective phenomena that involve many nuclear particles simultaneously. The accelerator and the detector systems are already available at CERN. All that is required is the conversion at relatively modest cost to heavy nuclear projectiles. This will almost certainly be attempted during 1981.

—D. Allan Bromley

Solid-state physics

At first glance, certain surface properties of diamond may seem an unusual subject for a review of solid-state physics. Nevertheless, recent discoveries about diamond not only have increased scientific insight but also hold important implications for integrated circuits (IC's) and the large industry that manufactures them. Their account also provides an intriguing story that illustrates the way ancient practical techniques, in this case the procedure for polishing gem-quality diamonds, can contribute to modern science. Moreover, it illustrates the considerations and techniques needed to understand the surface of a material.

As is often true, understanding the surface of a material requires understanding fully the bulk properties of that material. Diamond, which is one form of pure carbon, has the same bulk lattice structure (cubic) as silicon, from which IC's are formed. Each carbon atom is surrounded symmetrically by four neighbors to which they are bonded completely by covalent bonds; *i.e.*, bonds formed by sharing electrons between adjacent atoms. However, whereas the cubic form is stable for silicon, it is not for diamond.

Diamond is nature's hardest known substance. It is not so widely realized, however, that diamond is a metastable form of carbon. Solids can have several different states, but the favored one is always that of least energy. Graphite, a black crystal of unexciting appearance, is the stable form of carbon at the pressures and temperatures that exist on the surface of the Earth. Diamonds are formed only under conditions of very high pressure and temperature. Such conditions, for example, might take place deep in the Earth during a volcanic disturbance. Once diamond is formed, an energy barrier must be overcome (in other words, an

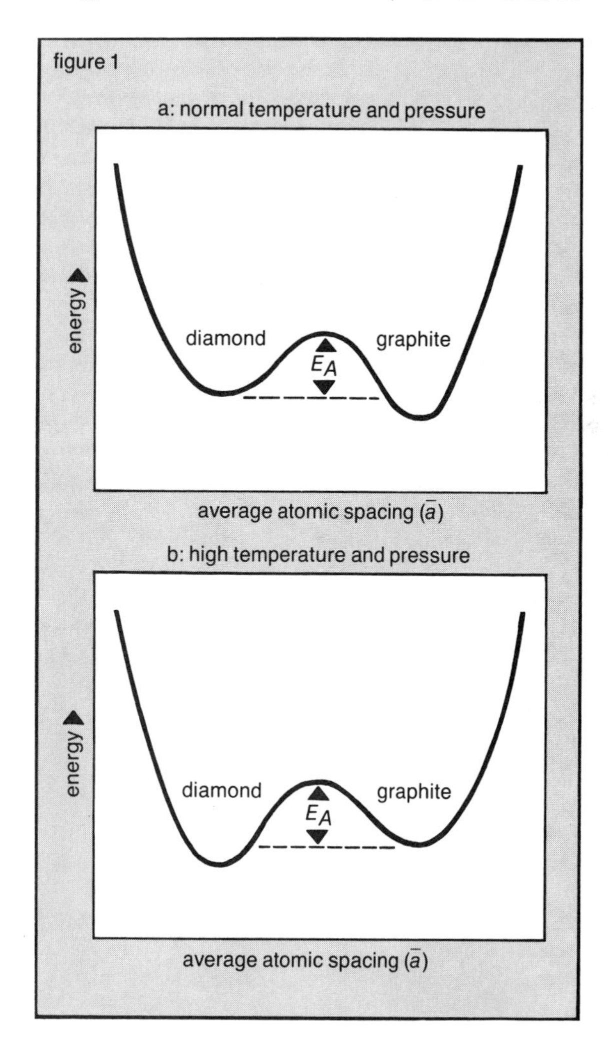

activation energy must be supplied) to transform diamond back to graphite. This is shown graphically in figure 1, in which the energy of solid carbon is plotted against the average spacing of atoms, $\bar{a}$, in the carbon lattice for the two conditions of temperature and pressure described above. The two plots, a and b, treat all of the atoms in the carbon solid in terms of the distances from their actual neighbors and do not distinguish between surface atoms with three neighbors and atoms in the bulk of the solid with four neighbors.

At high pressure and temperature, diamond is the lowest energy state for carbon. As pressure and temperature are reduced (as after the disappearance of volcanic conditions), graphite becomes the lowest energy state. Note the energy "hump" between the diamond and graphite state; the height of this hump, E_A, is the activation energy. If the carbon solid is cooled quickly enough as the pressure is reduced, it is trapped in the metastable state because not enough thermal energy is available to allow the carbon to overcome the energy barrier and to gain its stable state as graphite. Thus can diamond persist on the Earth's surface.

Although this process has been known for decades, it is of current interest because of its application to studies of the surface of diamond. To create the surface, one or two of the four covalent bonds forming diamond is broken. The breaking of this bond greatly reduces the activation energy requirement. Recent experimental and theoretical research at Stanford University, the University of California at Berkeley, and the IBM Thomas J. Watson Research Center, Yorktown Heights, New York, produced evidence that, in fact, the surface of pure diamond easily reverts to graphite. This behavior has strong implications not only for solid-state physics and for IC technology but also for the luster and quality of gem diamonds. Most amazingly, by purely empirical means diamond polishers thousands of years ago learned to stabilize the surface of diamond, preventing or retarding its transformation into graphite.

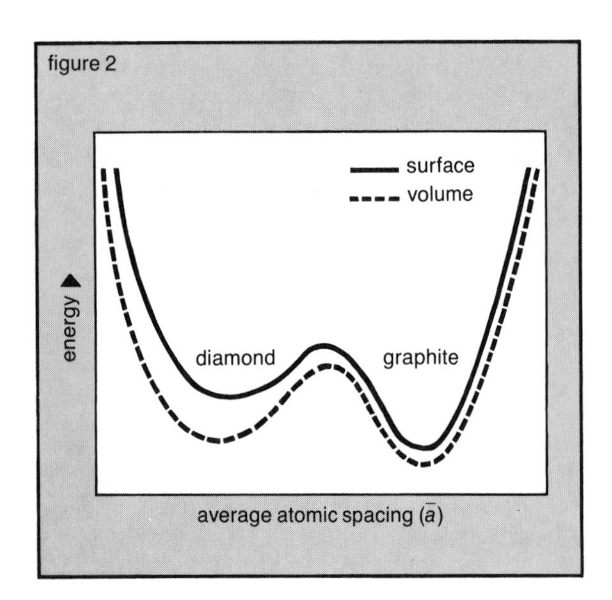

It will help to examine in more detail why diamond tends to revert to graphite at the surface. Diamond, like silicon and germanium, is held together by four covalent bonds with its nearest neighbors. When a crystal is terminated by breaking, one and sometimes two of these bonds are broken. Diamond is a cubic crystal; its bonds with neighboring atoms in every direction are equally strong. By contrast, graphite is a planar crystal; there are strong bonds within one plane with very weak bonding between the planes.

Graphite is an excellent lubricant and is finding increasing use in friction-reducing materials, including graphite-base automobile crankcase oil. The lubricating properties of graphite are due to the weakness of the binding between the structurally strong planes of the graphite crystal; as a result, the crystal planes move easily across one another, providing lubrication. Diamond, because of its cubic structure, is probably one of the world's worst lubricants. Rather than reducing friction, its sharp and hard corners act to tear up materials.

Figure 2 illustrates the difference between the energies for surface and bulk carbon atoms at normal atmospheric pressure. For the graphite state (right-hand side of diagram) this difference is small because of the weak bonds already present between graphite layers within the solid. For the diamond state, however, the energy of its surface atoms is notably higher than that of atoms completely within the cubic diamond structure, where all bonds are equally strong. Thus, the energy barrier inhibiting transformation of diamond to graphite is almost destroyed at the surface, making it very easy for the transition to take place.

When diamond was studied with some of the recently developed tools of surface science, its surface was found to be quite different from that of silicon or germanium despite the similarity in crystal structure and bonding (although silicon and germanium are stable in the cubic form and do not exhibit the planar form). For silicon and germanium the task of creating an ideal surface centers on the problem of removing such foreign impurities as oxygen from the surface. The principal problem with diamond is different; it is the presence of graphite on the surface. Auger electron spectroscopy (AES) has been the tool used to detect foreign impurities on semiconductor surfaces and the presence of a graphite form on diamond. Details of AES are shown in figure 3. As can be seen from the accompanying caption, AES depends on first removing an electron from an inner electronic shell of the atom. Hydrogen has no such inner shell, however, since it has only one electron in its electronic orbitals. Thus, hydrogen cannot be detected by AES. As will be seen, this is a critical consideration in understanding diamond.

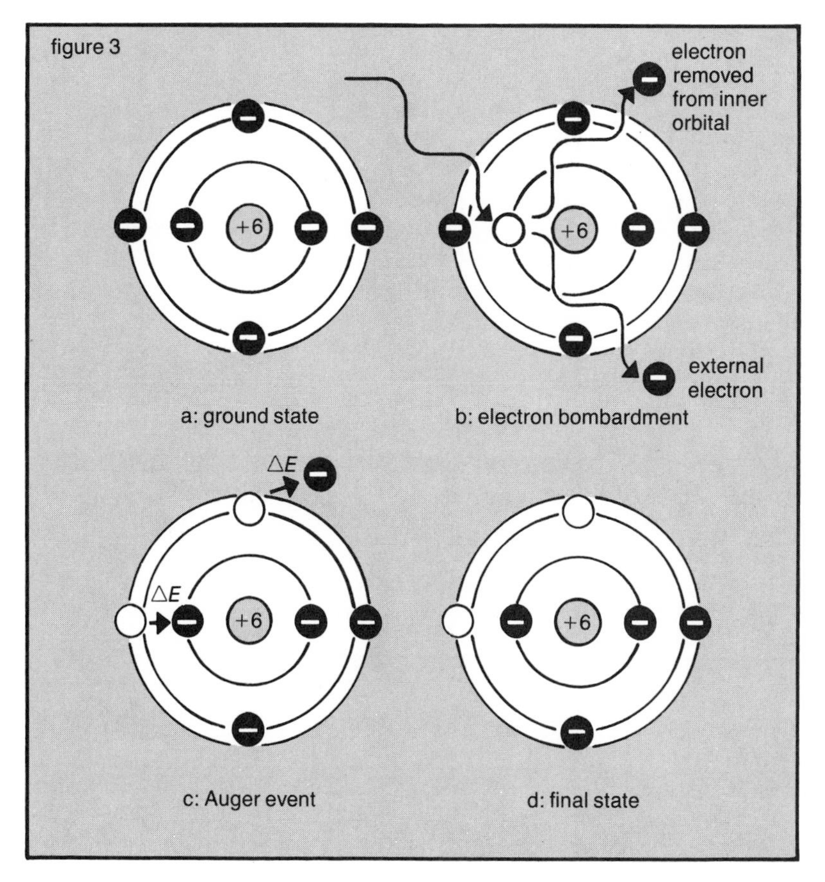

Figure outlines the Auger event in carbon. The carbon atom has six electrons in its normally unexcited, or ground, state (a). An energetic external electron removes an electron from an inner shell (b). The Auger event occurs when an electron falls from an outer shell into the vacancy in the innermost shell (c). To conserve energy, a second electron is ejected from an outer shell. This electron has an energy determined by the energy difference, ΔE, between inner and outer shells. ΔE is a characteristic signature of the element in which the Auger event has occurred. In its final state (d) the carbon atom is left with two electron vacancies in its outer shell.

Once graphite is on the surface, it is extremely difficult to remove. The standard way of removing unwanted impurities is by sputtering. This process involves striking the surface with rare-gas ions (which are not chemically reactive) that have energies in the range of a thousand electron volts. These ions knock off atoms from the surface, including any atoms of impurity. They also disturb the underlying carbon layers, turning diamond into graphite. The net result is that, although the original graphite is removed, more is produced. Other methods including heating in oxygen generally have proved ineffective in removing the graphite.

After studying these failures a graduate student and a Japanese visiting professor involved in day-to-day work at Stanford University decided to see what would happen to the surfaces of diamonds when polished by a diamond polisher who worked for the gem industry. The diamond surfaces provided in this way proved to be much cleaner than had ever before been seen despite the fact that they had been exposed to air for hours before being examined in the laboratory, a circumstance that normally results in a heavy layer of oxide or other undesirable material on most substances (including silicon and germanium). This discovery presented a mystery since the broken bonds on diamond were expected to be just as chemically active as those on silicon or germanium.

Meanwhile, theoretical work at Berkeley combined with experimental work at IBM and Stanford uncovered another mystery. Theory that had been well tested on other semiconductors predicted that surface states should appear in the band gap of diamond. Experiments with commercially polished diamond, however, uncovered no such surface states. The Berkeley group suggested that this absence might be due to foreign atoms—probably atoms of such monovalent elements as hydrogen, fluorine, or chlorine—bonding to the broken bonds on the diamond surface and thus chemically satisfying them. It is well established that such termination removes surface states from the band gap and removes any chemical activity of the surface. Further, the Stanford group pointed out that hydrogen was the only element consistent with all of the experimental data. Any of the halogens would have been detected with AES, but none were.

The lack of chemical activity of the diamond surface would be clearly explained if it were terminated with hydrogen. Likewise, such a termination would make the energy of a surface atom similar to that of a bulk atom; therefore, the energy barrier for atoms at the surface would be similar to that of the bulk. As a result, there would be no tendency for the surface to revert to the graphite form. But then yet another key mystery came out of these considerations. How did the diamond

Courtesy, Lawrence Livermore Laboratory; photo, Dave Proffitt

Courtesy, Lawrence Livermore Laboratory

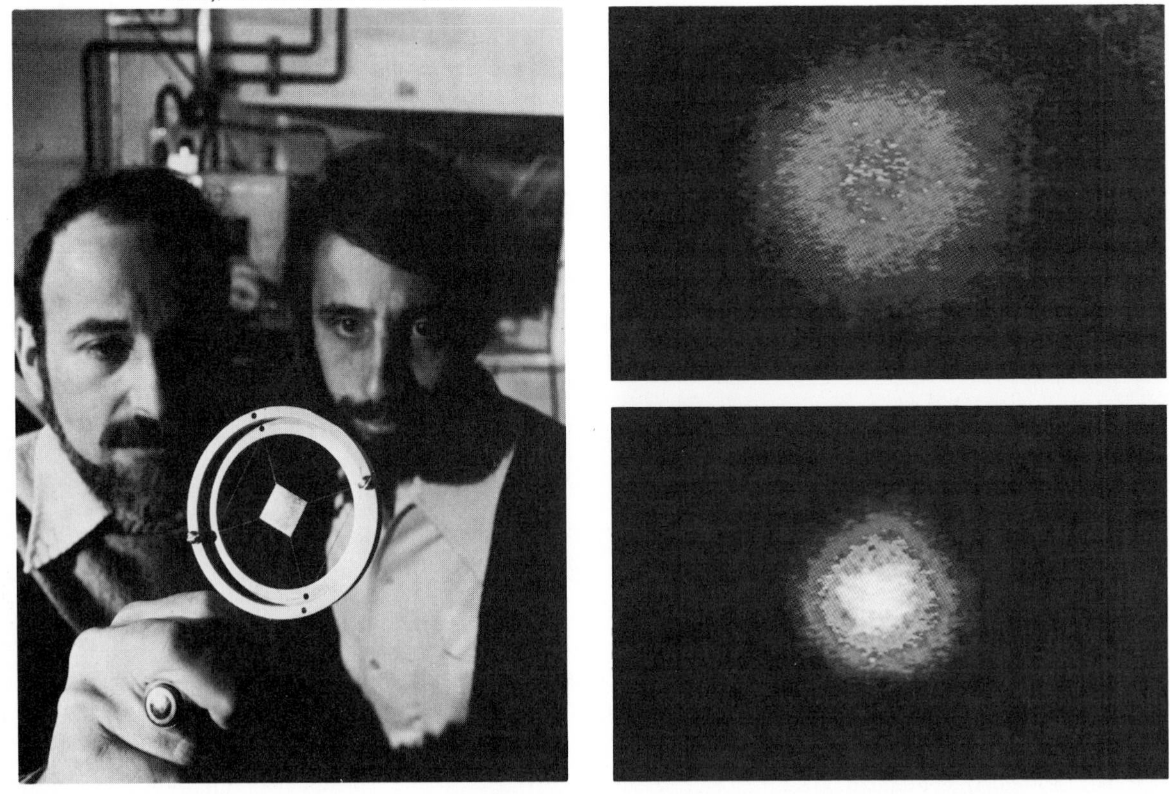

Specimen of single-crystal silicon (left) was used to create a new source of tunable, highly directional X-rays called channeling radiation. The radiation is emitted by charged particles moving near light speed as they are channeled between atomic planes of the crystal. Outputs of a detector receiving unchanneled X-rays and X-rays from a properly oriented crystal are compared (right, top and bottom).

surface come to be terminated by hydrogen in the polishing procedure?

Polishing of diamond is done using a very fine diamond powder suspended in a liquid, which normally is water. The Stanford scientist in charge of the work asked his co-workers to have another diamond polished by the commercial polisher provided that the polisher would allow the graduate student to observe the complete process. The results were quite revealing.

It was found that the liquid used for suspension of the diamond powder was not water but olive oil. Significantly, olive oil contains a large amount of rather weakly bound hydrogen. Thus, as the diamond grit removes surface carbon (including loosely bound graphite), hydrogen is immediately available to bond to and satisfy the broken surface bonds. Thus, all the mysteries appear to be resolved.

The diamond polishing technique is thousands of years old and was probably developed in the region of the Mediterranean Sea, where olive oil would have been a readily available liquid with which to experiment. Realizing that surface film would badly hurt the luster of a diamond gem, ancient diamond polishers probably searched by a slow process of trial and error for the optimum polishing technique. Ultimately some lucky and ingenious artisan made the key breakthrough while trying olive oil.

More recent experiments have added support to the concept of a critical role played by hydrogen in stabilizing the surface of diamond. The definitive experiment, however, will be the direct detection of hydrogen. Toward this end the Stanford group has joined forces with researchers from the U.S. Naval Laboratory at China Lake in California and from Sandia Laboratories in New Mexico. The Navy and Sandia groups have been involved with a new method to detect surface impurities using synchrotron radiation. Preliminary results were encouraging, and definitive experiments were scheduled for 1981.

The topic of diamond surface stabilization is not only of academic or popular interest. A key problem for the solid-state electronics industry is the termination of silicon bonds at the surface of the crystal used to fabricate devices and IC's. This problem is becoming increasingly critical as IC's move into a more mature state in which manufacturing yields (the fraction of devices or IC's started in the manufacturing cycle that produce sellable end products) and the strikingly reduced size of individual devices within IC's are placing strong new demands on surface control.

To date the surface of silicon (Si) has been prepared for IC fabrication by growing a comparatively thick (1000-angstrom) layer of silicon dioxide (SiO_2) on silicon (one angstrom equals a ten-billionth of a meter). Other treatments, however, such as post-growth anneals are also necessary to form "good" interfaces between the Si and SiO_2. In 1965 Peiter Balk at IBM suggested that hydrogen was playing a key role in satisfying the broken bonds of silicon that were missed by the oxygen in SiO_2. That some bonds may be missed is clear because the SiO_2 crystal lattice can not be matched perfectly to the silicon lattice. Recently work at the Xerox Palo Alto Research Center in California and at the U.S. Army Electronics Research and Development Command laboratory in Fort Monmouth, N.J., offered strong evidence for the importance of hydrogen at the Si-SiO_2 interface.

The diamond work will be expanded to include studies of hydrogen and hydrogen and oxygen on silicon, and joint work is foreseen between groups. A prime objective is the use of knowledge gained from study of diamond in developing new preparation methods for the critical Si-SiO_2 interface. These methods should bring the present empirical state of knowledge to a situation in which "scientific engineering" can be applied to the surface in a well-controlled manner. The ultimate aim is to provide much better controlled and flexible treatment of the silicon surface. If this challenge is not successfully met, it will have dire effects on the health of the electronics industry.

—W. E. Spicer

See also Feature Article: A NEW WORLD OF GLASSY SEMICONDUCTORS.

Psychology

The most newsworthy events relating to psychology generally involve matters of intense public interest, and in this respect the past year was no exception. Probably the most significant development was the emergence of governmental interest in the effectiveness of psychotherapy. Stemming from practical problems of payment for psychotherapy under Medicare and Medicaid, a number of separate investigations were initiated. By the beginning of 1980 it had become apparent to governmental policymakers that there was little in the way of hard data to support the assumptions of efficacy for the myriad forms of psychotherapeutic endeavor (estimated at 130 to 250 different varieties, in recent reckonings). Although this fact has been well known to psychologists, the full impact of its implications for public policy were only now beginning to be felt.

Psychology, psychiatry, and government. During the year a number of massive investigative efforts were threatened or actually initiated. The first official examination of the efficacy of psychotherapeutic intervention was started by the U.S. Senate Committee on Finance, which conducted hearings in an attempt to determine an appropriate governmental payment scale for this $2 billion-a-year business (psychiatrists charge from $45 to $75 per hour; psychologists average about $10 per hour less). The lack of professional endorsement of psychotherapy at the early committee meetings apparently surprised government officials and set the stage for later investigations.

Several such major efforts were begun. For example, the National Institute of Mental Health arranged pilot studies, themselves programmed to cost $3.4 million, for a long-term research project involving five therapies for depression. Potentially most far-reaching, a Senate bill was proposed that would establish a federal commission to ask the same efficacy and safety questions of psychotherapy that are routinely asked about new drugs.

The importance of these efforts is indicated by estimates that there may be as many as 50,000 psychologists (practitioners without medical degrees but including many with an M.A. or Ph.D. in clinical or counseling psychology) and 30,000 psychiatrists (with M.D. degrees) treating from 5 to 10% of the adult U.S. population.

Throughout these legislative maneuverings, there was a continuation if not an intensification of the adversarial relationship between the medically trained psychiatrists and the clinically trained psychologists, who are traditional competitors in this lucrative, difficult-to-evaluate, but nonetheless important activity. The most significant events of the year occurred in the state of Virginia, where a group of clinical psychologists had been contesting the reimbursement policies of local Blue Shield organizations which held that in order to be paid for psychotherapy psychologists must be medically (which generally means psychiatrically) supervised. The "Virginia Blues" case, as the legal battle came to be called, centered on the failure of the medically controlled Blue Shields to conform to "freedom of choice" legislation enacted by the state of Virginia; this legislation permits licensed psychologists to engage in independent (not medically supervised) practice.

Some clear legal victories were won by the psychologists. First, the Virginia Supreme Court denied a Blue Shield appeal of a prior ruling requiring conformance with the statute. Second, three federal judges in the U.S. Court of Appeals for the Fourth Circuit granted an appeal by the psychologists of an earlier adverse ruling by a federal judge who had rejected the charge of antitrust violation on the part of the Blue Shields. These two victories offered promise of full professional freedom for clinical psychologists, but there was no reason to anticipate an early end to legal actions in connection with this issue.

© Les Lockey and Robin Baker, NEW SCIENTIST

Student in Britain points toward home after being taken from it blindfolded over a winding route of many miles. When the blindfold was removed, the subject, along with others in the study, pointed in the wrong direction.

Cooperation as well as competition occurs between psychologists and psychiatrists. For example, representatives of the two disciplines joined forces to form a new 800-member organization, the Society of Behavioral Medicine, whose first annual meeting was scheduled in conjunction with meetings of the Association for the Advancement of Behavior Therapy. Another example was an innovative behavioral medicine program at St. Joseph's Hospital in Hamilton, Ont. There a psychologist played the central role in treatment supervision. Joint interviews with clients were conducted by an internist, a psychiatrist, and a behaviorally oriented psychologist.

The year marked the tenth anniversary of a particularly noteworthy success story. One of the most influential and widely copied of all contemporary clinical efforts, William Masters and Virginia Johnson's sex-therapy program, began at Washington University, St. Louis, Mo., in 1970. These two clinicians, one a physician, the other a psychologist, had succeeded in overcoming a host of deep-seated prejudices, both medical and nonmedical. But the dramatic growth of sex therapy, spearheaded by Masters and Johnson, was not without its problems. Mainly, these involved the credentials of the approximately 5,000 persons now claiming competence, only about 1,200 of whom were certified professionals. Also, the Masters and Johnson program itself had become the target of criticism. Two California psychologists recently questioned their claims of success, chiefly on the grounds that failures have been inadequately measured and reported.

Memory. The study of memory continued to be actively pursued, both for its own sake and for its relevance to many practical problems. The latter was well illustrated by two research areas that were attracting steadily increasing attention.

Two recent books (*The Psychology of Eyewitness Testimony* by Daniel Yarmey and *Eyewitness Testimony* by Elizabeth F. Loftus) deal with legal processes as they are affected by memory and related behavioral functions. Both authors had actively engaged in eyewitness research. One experiment reported by Loftus provides some of the flavor of this kind of research. All subjects were shown a film containing views of a collision between two cars. Half the subjects were then asked a question about the cars "bumping" one another, the other half a question about the cars "smashing" into one another. Later, on a recall test, the second group of subjects were more likely to remember seeing broken glass. This kind of result has obvious implications for the way in which memory can be led, deliberately or otherwise, in testimony following apparently simple observations.

The second area of relevant research on memory concerned the forensic (legal) use of hypnosis, largely to improve recall. Here scientific caveats were much more necessary, because the use of hypnosis—for example, in police investigations—had clearly outrun its research support. There were strong criticisms of the quick and easy training of police officers in hypnosis; in the Los Angeles Police Department, for example, a clinical psychologist reportedly trained approximately 800 persons from all parts of the country.

Memory is one of the most crucial problems in the rapidly growing field of gerontological psychology. British psychologist P. M. A. Rabbitt reported some interesting research results relating memory and aging. Addressing the question of why older persons so often seem ill at ease and have trouble in remembering when they engage in group as compared with one-on-one conversations, he had both old and young subjects recall strings of digits. As is usually found, there was little age difference, only a slight decline in the older persons' scores.

Rabbitt then put the same digits in four quadrants, so that the subject had to remember not only the number itself but also its spatial location. All recall scores were depressed, but those for the older subjects were much more severely affected. Rabbitt obtained similar results in an alphabet-recall task involving repetition of letters, apparently because the older subjects lost track of where they were. The general conclusion that could be drawn from these experiments was that older subjects lose their indexing ability more than their memory per se. This accounts for their confusion in group conversations; that is, their inability to identify particular speakers' comments rather than their forgetting the content of the conversation itself.

Gifted children. There was a renewal of interest in gifted children and especially in their social, emotional, and intellectual adjustment in a society that is closely attuned to the "normal" individual. The best known life-span study, following more than 1,500 "genius" (IQ over 135) California children throughout their lives, was started by Lewis Terman at Stanford University in 1921. With most of the original group now past retirement age, it was quite clear that neither monetary success nor quality of life had been precluded by high intelligence, at least in these cases. Furthermore, within this high-IQ population the more successful persons had lived substantially longer than the least successful, a result that is consistent with the newer view that heart attacks and similar symptoms of stress are not necessarily associated with high motivation and achievement.

Research problems concerning today's "severely gifted" children were being identified by means of the first sizable funding—over $6 million in 1980—for the Department of Education's Office of Gifted and Talented. One of the promising new research directions was the expansion of the defining measures for giftedness beyond IQ alone. The Office of Gifted and Talented developed a five-point definition: high IQ, high academic achievement, creativity (identified as divergent thinking), special artistic talent, and "leadership." A somewhat different definition was proposed at the University of Connecticut—excellence in some particular (unspecified) area, motivation (as reflected in high task commitment), and creativity. This new emphasis on intellectual as well as other types of achievement was hailed as a particularly refreshing development in the context of the anti-intellectualism, including conformity to "normal" (often mediocre) standards, that had marked American culture in recent years.

Research on twins. Another kind of searching look at behavioral determinants was initiated by Thomas Bouchard, a developmental psychologist at the University of Minnesota. Nine pairs of identical twins reared apart were located and intensively examined by a team of psychologists, psychiatrists, and physicians. Some of the early results showed remarkable similarities (*e.g.*, the seven rings, three bracelets, and one wristwatch worn by each of two British housewives who apparently first met when they arrived in the U.S. for the testing). The publicity given these early findings resulted in the location of an additional 11 pairs of twins who had been reared apart.

In this program each pair of twins is subjected to six days of highly concentrated tests (including, for example, a total of 1,500 questions). This study, the first of its kind in the U.S. in more than four decades, should provide massive amounts of carefully collected data on many facets of the heredity-environment issue. Although the most interesting early findings were the

Psychologist Thomas Bouchard, in the center of the picture at the left, administers tests to identical twins James Springer (at the left in both photographs) and James Lewis. The twins, who had been raised apart and first met at the age of 39, discovered striking similarities in their backgrounds.

Photos, Thomas S. England

similarities, finer analyses may be expected to uncover some significant differences (smoking was one characteristic that had shown differences thus far). Coming so soon after the debacle of Sir Cyril Burt and his fabricated twin-intelligence data, the Minnesota study should also help to restore credibility to this kind of research.

Verbal reports. The analysis of conscious processes retained its attraction for a group of cognitively oriented researchers and theorists. A major event during the year was the publication by K. A. Ericsson and H. A. Simon of their comprehensive methodological approach to this problem ("Verbal Reports as Data," *Psychological Review,* May 1980). Because the article had been widely distributed as a "working paper," its impact had been spread over the past several years.

In the study verbal reports are incorporated into an information-processing model. The article is in part a rebuttal of R. E. Nisbett and T. D. Wilson's earlier sweeping attack on the validity of verbal reports, especially with respect to the accuracy of retrospective conscious identification of the determinants of behavior. Accelerated progress toward resolution of these key issues should occur soon if interested parties subject them to experimental as well as conceptual attack, a development that may be anticipated now that the issues have been so well articulated.

Reexamining Freud. Freudian interpretations of consciousness and related phenomena, while still predominant in many aspects of psychiatric theory and practice, were being questioned on a number of behavioral fronts. One example was Freud's interpretation of behavioral slips in terms of unconscious motivational factors. Certainly some errors seem to lend themselves to this interpretation (*e.g.,* the reluctant chairman of a meeting who opens it by loudly declaring that it is now "closed"), but many others can be interpreted more simply.

Cognitive researcher Donald Norman made an interesting analysis of behavioral slips. Some entertaining incidents were among the more than 200 that he collected from his students and colleagues. One of his students reported coming home from a track workout and tossing his sweaty T-shirt into the toilet bowl rather than into the laundry bucket in the next room; a colleague, returning home one morning to get his forgotten briefcase, found himself unbuckling his wristwatch instead of the seat belt after turning off the ignition.

Norman classified these errors into a number of categories. One prominent category is that of "capture" errors, so-called because strongly entrenched habits often dominate behavior (as William James noted late in the 19th century). This type of slip is illustrated by the student who found himself counting 1, 2, 3, 4, 5, 6, 7, 8, 9, 10, jack, queen, king while operating a copying machine; he had been playing cards earlier.

Norman's general interpretation is that most of the slips can be accounted for in terms of "accidents" in information processing, produced by the intrusion of such factors as "stray information." These accidents are most likely to occur when conscious control is limited. While this interpretation cannot be expected to account for all types of human error, the slips that it addresses are among the most important as well as the most interesting of behavioral misdirections.

Another area in which orthodox Freudian psychoanalytic theory was being questioned was dream analysis. Standard Freudian dream interpretation has long emphasized hidden motivational factors—the "latent" as contrasted with the "manifest" content. A number of psychologists were looking closely at dream content, and a consensus was developing. The newer and simpler interpretation is that dream materials have a great deal in common with spontaneous (conscious) verbal output and can be interpreted in the same way. Thus, studies have shown that the same themes that dominate an individual's dreams also appear in his stories and in his responses to Rorschach test cards. Moreover, clear evidence was found of a high correlation between somatic states and dream content (*e.g.,* hypertense patients report more hostile dreams; persons who have exercised vigorously before sleep report more physical activity).

The general conclusion emerging from this work was that analysis of the manifest content of dreams is sufficient to explain them, so that there is no need to bring in more complex explanatory mechanisms, such as unconscious wishes. Whether or not a case could be made for unconscious motivation as a determinant of a special class of dreams, there seemed little doubt that the great majority of dreams can be interpreted in a simpler manner. While Freud's historical stature as a great pioneer in psychology seemed secure, this whittling away at so many of his key concepts was progressively reducing the contemporary impact of his general theoretical position.

—Melvin H. Marx

See also Feature Article: THE AMAZING NEWBORN.

Space exploration

The probe of Saturn by the U.S. spacecraft Voyager 1 and the first orbital flight of the U.S. manned space shuttle "Columbia" were the highlights of the year in space exploration. During the Saturn flyby Voyager took hundreds of photographs of the planet and its satellites, providing astronomers and space scientists with a wealth of new information. The much-delayed "Columbia" performed well during its two days in orbit and became the first winged vehicle to leave and reenter the Earth's atmosphere. It was the first manned mission in space for the U.S. since 1975.

(Left) Michael Pugh—UPI; (right) UPI

U.S. space shuttle orbiter "Columbia," with astronauts John Young and Robert Crippen aboard, is launched from Cape Canaveral, Florida, on April 12 (left). Above, accompanied by a "chase" plane, "Columbia" glides in for a landing in California on April 14 after 36 Earth orbits.

In other space developments two Soviet cosmonauts spent six months in the Salyut 6 space station. Pioneer Venus completed two years in orbit around that planet and continued to transmit data to the Earth.

Manned flight

The "Columbia," mated to its external fuel tank and booster rockets, moved to Launch Complex 39A at the National Aeronautics and Space Administration's (NASA's) Kennedy Space Center in Florida during the final week of 1980. Soviet space station Salyut 6 during the year had tenants and house guests aboard for the fourth year, despite having been designed for one year's operation.

Space shuttle. Orbiter "Columbia" made the $3^1/_2$-mi trip to the launch pad with all of its 30,922 silica glass heat shield tiles firmly bonded to its aluminum skin. New adhesive materials for attaching the tiles to an intermediate underlay between tiles and skin and a process for strengthening the inner surfaces of the tiles seemed to have solved the heat shield problems that had plagued "Columbia" since the spacecraft had been ferried piggyback atop a Boeing 747 carrier aircraft in 1979 from California.

An in-orbit tile repair kit was developed during the year for possible use on "Columbia's" first orbital flight. After enough confidence in the reliability of the heat shield tiles was gained, the decision was made not to carry the repair kit. To use it during a mission one of "Columbia's" two-man astronaut crew would have had to "fly" in a space suit on a nitrogen-jet–propelled backpack—caulking gun and trowel in hand—to patch or replace damaged tiles along the spacecraft's 37-m (122-ft) length. Testing of the orbiter's main engine at NASA's National Space Technology Laboratories in Mississippi during 1980 brought about several changes in the design of engine components. "Columbia's" main engines were removed and rebuilt in order to incorporate the improvements.

A 20-second full-throttle burn of the flight engines in February 1981 cleared the way for "Columbia's" first orbital flight on April 12–14. The shakedown flight orbited the Earth 36 times in 54 hours, 22 minutes, with a successful landing on Rogers Dry Lake at Edwards Air Force Base in California. Veteran astronaut John W. Young and Robert L. Crippen served as crew. Originally scheduled to be launched on April 10, "Columbia" remained on the ground for two extra days because of a computer malfunction. During the launch several of the heat shield tiles that covered "Columbia" were lost or damaged, but this posed no threat to the safety of the mission. "Columbia's" second flight, to be piloted by Joe Engle and Dick Truly, was scheduled for August 1981. An Earth resources survey experiment was planned for the mission. NASA's second operational orbiter, the "Challenger," was not scheduled for delivery by the builder until June 1982.

Soviet missions. Cosmonauts Leonid I. Popov and Valery V. Ryumin reactivated space station Salyut 6 for a 185-day stay after their launch April 9 aboard Soyuz 35. Salyut 6 had remained dormant for more than seven months following a 1979 visit of 175 days during which Ryumin was also flight engineer. The unmanned freighter-spacecraft Progress 8 awaited the cosmonauts' arrival at Salyut for unloading of its cargo and the pumping of propellant from it to Salyut's tanks. Progress 8 was then loaded with trash and excess equipment and deorbited on April 25.

During their six-month stay aboard Salyut 6 Popov and Ryumin operated a farm for observing the weightless growth of wheat, pea, and onion seedlings, and they also grew orchids in a greenhouse. Materials technology research begun by earlier Salyut inhabitants and reminiscent of U.S. Skylab experiments in 1973–74 was continued by Popov and Ryumin. The Kristall furnace nurtured the growth of indium antimonide, gallium arsenide, gallium antimonide, and germanium semiconductor crystals. Electron-beam crucibles vacuum-deposited layers of gold, silver, aluminum, and copper on metal, glass, and plastic surfaces. Semiconductor alloys of cadmium, mercury, and tellurium were formed in another experiment. Earth resources experiments included Biosphere, for observation and photography of weather formations and ocean currents, and the photography of Soviet agriculture. In the Lotos experiment polyurethane foam structural members were formed under weightless conditions. Salyut's small gamma-ray telescope again measured the near-Earth electron flux.

Arnaldo Tamayo Méndez of Cuba (left) and Yuri Romanenko of the Soviet Union (right) piloted the Soviet spacecraft Soyuz 38 to a rendezvous with the Salyut 6 space station in September 1980.

UPI

Unmanned Progress supply vessels and manned Soyuz spacecraft shuttled routinely to Salyut 6 while Popov and Ryumin were aboard. Progress 9, launched April 27, 1980, took up more scientific instruments and propellants; Popov and Ryumin controlled the docking of the unmanned craft with the station. After an orbit adjustment boost to an apogee of 369 km (229.3 mi), Progress 9 was jettisoned to burn up in the atmosphere on May 22. In Soyuz 36, launched on May 26, Valery Kubasov (a member of the 1975 Apollo-Soyuz U.S.-Soviet joint mission) and Bertalan Farkas of Hungary brought with them a joint Soviet-Hungarian experiment for weightless synthesizing of human interferon. After a seven-day stay the pair returned to the Earth on June 3 aboard Soyuz 35, following the Soviet practice of leaving the newest spacecraft for the space station crew.

Soyuz T-2 was launched on June 5 with Yuri Malyshev and Vladimir Aksenov as the crew in the first manned flight of the improved Soyuz T spacecraft. The cosmonauts wore a newly designed lightweight space suit that provided better mobility and helmet visibility. An unmanned Soyuz T had docked with Salyut 6 in December 1979 for a 100-day test. In the Soyuz T both the main propulsion and the attitude and maneuvering thrusters were fed from a single propellant system, and the craft also featured an improved environmental control system, solar panel charging of chemical batteries, an onboard digital computer, and a star tracker attitude-control system. Malyshev and Aksenov returned to the Earth on June 9 aboard Soyuz T-2. Between the Progress and Soyuz visits Popov and Ryumin flew the remaining Soyuz around to the forward docking port in order to vacate the aft port for the next resupply vessel or visiting crew.

Progress 10, launched on June 29, docked with Salyut 6 on July 1; it was jettisoned over the Pacific on July 19. Viktor Gorbatko and Pham Tuan of Vietnam brought up a Vietnamese crystal growth experiment in Soyuz 37, launched on July 23. They returned to Earth on July 31 aboard Soyuz 36.

In Soyuz 38, launched on September 18, Yuri Romanenko and Arnaldo Tamayo Méndez of Cuba took to the Salyut the Cuban "Caribe" germanium-indium crystal growth experiment; they returned to the Earth on September 26 without exchanging spacecraft. Progress 11 was launched September 28; Popov and Ryumin unloaded the freighter while preparing Salyut for dormant, unmanned flight. Progress 11 was jettisoned on December 9 after boosting Salyut to a 290 × 374 km (180 × 232 mi) parking orbit. Then Popov and Ryumin left the Salyut and returned to the Earth on October 11 in Soyuz 37. As a result of the mission, Ryumin gained the individual time-in-space record with 362 days.

Soyuz T-3 was launched on November 27 and docked with Salyut 6 the next day. On board were

Leonid Kizim, Oleg Makarov, and Gennadiy Strekalov, the first Soviet three-man crew since the fatal Soyuz 11 flight in 1971. The mission was mainly a further test of the Soyuz T, but at the Salyut the trio replaced equipment and ran a maintenance checkout of the space station's systems. Also, experiments concerning plant growth and semiconductor crystals were undertaken before Soyuz T-3 returned to the Earth on December 10.

The director of the Soviet Union's cosmonaut training program, Lt. Gen. Vladimir Shatalov, hinted at expanded Soviet space station efforts when he said after the Salyut 6 crew return, "Imagine Salyut as a basis to which modules are linked up. One is for meteorological observations of the atmosphere; another for geological studies; and a third for astronomy and space physics." Shatalov also said that Mongolian, Romanian, Indian, and French cosmonauts would soon take part in Soyuz-Salyut flights.

Astronauts. During 1980 NASA selected 19 astronaut candidates (8 pilots and 11 mission specialists), who began a year's training at the Johnson Space Center near Houston in July. The Aviation Hall of Fame in Dayton, Ohio, inducted former astronaut Charles "Pete" Conrad, Jr., who flew on the Gemini 5, Gemini 11, Apollo 12, and Skylab 2 missions.

—Terry White

Space probes

Clearly the most spectacular performance by a space probe in 1980 was that of Voyager 1. The enormous amount of scientific data it gathered in passing Saturn on a trajectory that would send it out of the solar system in 1990 would keep scientists busy analyzing it for many years.

Probing Saturn. Launched in September 1977, Voyager 1 first visited Jupiter, making its closest approach to that planet on March 5, 1979. After coming within 277,500 km (172,475 mi) of the planet, Voyager used the tremendous pull of Jupiter's gravitational field to bend its trajectory toward Saturn. Photographs taken by the probe on its flyby revealed the existence of Jupiter's fifteenth and sixteenth satellites. The former orbits the planet some 151,000 km (94,000 mi) above its cloud tops and between the orbits of its sister moons Amalthea and Io. It is approximately 80 km (50 mi) in diameter. The discovery was made by Stephen P. Synnott of the Jet Propulsion Laboratory in Pasadena, Calif., who later found the sixteenth satellite, about 40 km (24 mi) in diameter, orbiting Jupiter about 56,000 km (35,000 mi) above its atmosphere.

In October 1980, Voyager 1 was approaching Saturn at a velocity of 74,000 km/hr (46,000 mph), relative to the Sun. On October 10 a slight course correction was made to ensure that the probe would assume the desired trajectory to intercept the planet. It entered its "Far Encounter Phase 2" on November 2, a period of observation that would continue until eight hours before the point of closest approach to the planet on November 12. At that time it would be about 124,200 km (77,000 mi) above the cloud tops of Saturn after having passed a day earlier within 4,000 km (2,500 mi) of the surface of the moon Titan.

Bursts of radio noise from the planet received by Voyager 1 and its companion, Voyager 2, as they approached Saturn permitted scientists to refine their estimates of the length of the Saturnian day. The Voyagers measured 10 hours and 39.9 minutes.

As Voyager 1 drew nearer to Saturn, attention focused on the system of its rings, which had first been seen from the Earth by the crude telescope of Galileo in 1610. Astronomers using Earth-bound telescopes of increasingly greater resolving power learned that the rings were clearly distinct and divided and began giving them letters as they were identified. Thus, Pioneer 11 discovered the F-ring, which is well beyond the A-ring. As Voyager 1 flew closer to Saturn, it detected a faint G-ring, and it became obvious that the rings were far greater in number and more complex in structure than the best telescopes on Earth could discriminate.

At the time Bradford A. Smith, head of the Voyager imaging team, said, "It's getting harder to tell where the rings begin and end. In an area between the A and B rings we thought was relatively clear—called the Cassini Division—we now see a band in it with a line in that band." Objects as large as three meters (nine feet) in diameter were detected in it. Later pictures,

Saturn, three of its rings (A, B, and C), and two of its moons (Tethys, above, and Dione) appear in a photograph taken from Voyager 1 at a distance of 13 million kilometers (8 million miles).

Courtesy, Jet Propulsion Laboratory/NASA

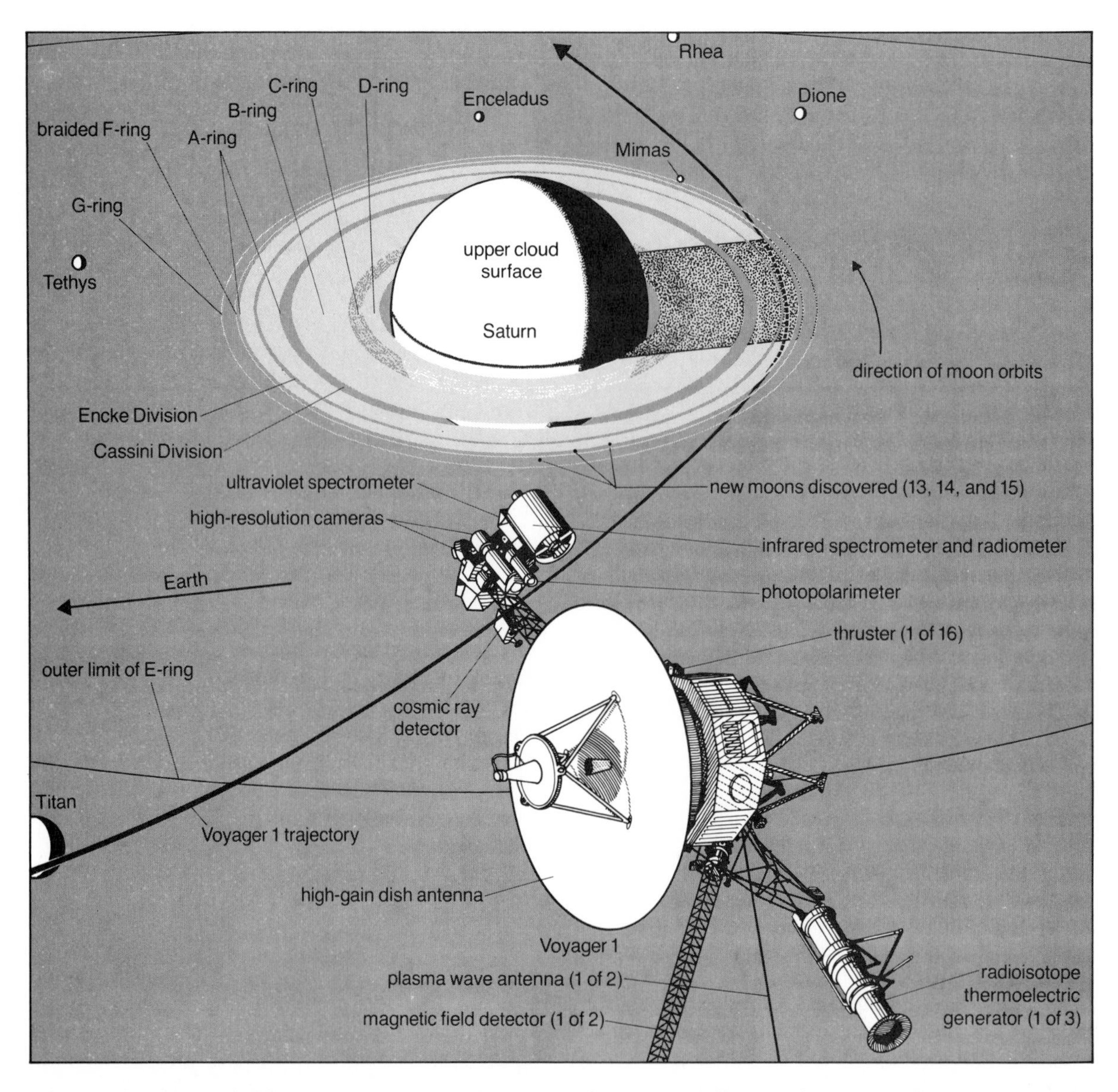

when analyzed, revealed from 500 to 1,000 ringlets in the complete ring system.

Inside the B-ring appeared features that looked like spokes, oriented perpendicularly to the rings and the planet's surface. Scientists were at first unable to explain the mechanism by which these "spokes" could retain their structure or shape as they rotated around the planet. Because of their great length their bases were traveling at much faster velocities than their tips. Logically, then, they should lose shape as they circled the planet. However, they did not. Scientists eventually concluded that the spokes were composed of small charged particles lifted above the ring plane and moving in step with Saturn's rotating magnetic field. Other pictures indicated that the F-ring might consist of three rings, two of which are interwoven or braided with one another. Said Smith, "As with a lot of other things, we really don't have an explanation; certainly, it was most unexpected." He added, "It defies the laws of pure orbital mechanics, as I understand them.... We have continued to watch the rings as resolution improves, and with the discovery of the two eccentric rings we thought that perhaps we had seen all that there was to see. But in this strange world of Saturn's rings the bizarre has become commonplace."

Voyager 1 discovered three new moons of Saturn, bringing the total number of satellites to 16. Two of them are located on either side of the F-ring, orbiting 141,700 km (88,000 mi) and 139,400 km (86,580 mi) above the planet. They are each about 200 km (125 mi) in diameter. The sixteenth moon is 50–100 km (30–60 mi) in diameter and orbits the planet at a distance of 138,200 km (85,800 mi), just outside the edge of the A-ring.

Studies of the atmosphere of Saturn by Voyager 1 indicate that it has a turbulence almost like that of Jupiter. Indeed, Saturn has its own Red Spot similar to that of Jupiter. It is about 11,200 km (7,000 mi) long and is located at 55° S latitude. Other features include brown spots at least 9,600 km (6,000 mi) long. Various cloud bands observed by the probe aroused the interest of scientists because of their number, which is even greater than on Jupiter.

The winds on Saturn in the equatorial region have velocities up to 1,440 km/hr (900 mph) and travel in the same direction rather than in alternating easterly and westerly bands as on Jupiter. The planet is also surrounded by a flattened doughnut-shaped region, a million kilometers thick, of neutral hydrogen gas.

Voyager 1 also made images of several of Saturn's 16 satellites. Mimas, about 384 km (240 mi) in diameter, has a pockmarked surface similar to that of Mercury and the Earth's Moon. It also has a huge crater approximately 128 km (80 mi) in diameter with crater walls 9 km (5.4 mi) high and a central peak of 5 km (3 mi). Tethys, about a third the size of the Earth's Moon, is also heavily cratered and has a trench some 800 km (500 mi) long and 64 km (40 mi) wide. Rhea, also heavily cratered, has such features as large as 296 km (185 mi) in diameter. Some of the craters have white patches in them, possibly indicating freshly exposed ice on their slopes. Dione, about 1,120 km (700 mi) in diameter, has impact craters as large as 96 km (60 mi) in diameter and wispy features forming complex patterns on the surface of the satellite. It also has valleys that are probably formed by geological faults. Iapetus, as does Mimas, has a large circular feature, some 192 km (120 mi) in diameter, probably an impact crater.

In regard to the satellites, Voyager 1 devoted the most time to studying Titan, the largest of them. Scientists had not expected the probe's camera to penetrate the dense atmosphere of Titan and, indeed, it did not. However, they learned many interesting facts about the satellite. It had been known from Earth observations that the atmosphere of Titan contained methane, ethane, and acetylene. It was widely believed that the atmosphere was rich in methane. Voyager 1's instruments indicated, however, that only 1% of the atmosphere is composed of that gas, with almost all of the other 99% being nitrogen. Edward Stone, a Voyager project scientist, said, "We have found a very important compound called hydrogen cyanide.... It is important in terms of the possible evolutionary processes that continued here on Earth but which presumably did not continue on Titan because of the very low temperatures." Other information on the atmosphere of Titan included the fact that its pressure at the surface is 1.5 times that of the Earth, that it extends some ten times farther into space than does that of the Earth, and that it is approximately five times as dense at the surface as that of Earth's.

With an estimated surface temperature of −146° C (−295° F), the surface of Titan would scarcely be like that of Earth, Venus, Mars, or even the Earth's Moon. "If you land on Titan, you won't be looking at rocks and you probably won't be looking at impact craters, because all those features will be under accumulated aerosols—if the layer of accumulated aerosols is as deep as it might be—and modified, conceivably, by a warm history in which liquid ammonia rather than liquid water was flowing across the surface," said Tobias Owen, a project scientist. Other features of the atmosphere included a "hood," or closed layer, over the northern pole but the lack of a similar one over the southern pole.

After considering the success of Voyager 1 and its mission, Smith said, "Practically everything that we are seeing now on Saturn is brand new. That's what makes it so different from where we were at this time in the Voyager encounter with Jupiter—where a lot of what we saw was confirming our ground-based suspicions or theories." Scientists hoped that new information would be supplied by Voyager 2, due to rendezvous with Saturn in August 1981.

Mission to Venus. On Dec. 4, 1980, the Pioneer Venus orbiter completed two Earth years circling the planet. Since entering the orbit on Dec. 4, 1978, the probe had made 730 orbits of Venus. It mapped 93% of the planet's surface and returned more than 1,000 pictures of its clouds taken in the ultraviolet spectrum.

The maps, made from information transmitted by the probe's radar, revealed a startling planetary geography. Huge continents and mountains on them as high as Mt. Everest together with deep rift valleys make an awesome topography. The major features are the two continents.

The largest of the continents was tentatively named Aphrodite Terra, pending official approval by the Working Group on Planetary Nomenclature of the International Astronomical Union. It is approximately half the size of Africa and may be the remnant of a plate tectonic collision before crust formation choked off tectonic action completely. However, it could as well be the result of local lifting forces such as those that produced the Sierra Nevada range in California. Aphrodite Terra has two mountainous regions separated by a lower plain and lies almost on the equator of the planet. It is approximately 9,600 km (6,000 mi) long and extends in an east-west direction. The western mountainous region rises about 7,900 m (23,000 ft) above the surrounding terrain and some 9,000 m (27,000 ft) above Venusian mean sea level. To the east the mountains are about 3,300 m (10,000 ft) above the neighboring terrain and about 4,300 m (13,000 ft) above mean sea level. The lack of uplifted plateaus or volcanic mountains indicate that Aphrodite Terra may be older than its sister continent, Ishtar Terra.

About the size of Australia, Ishtar Terra is a very

Courtesy, NASA

Artist's conception of Ishtar Terra, the highest continent on Venus, is based on measurements by the Pioneer Venus space probe. At the left is the smooth plateau Lakshmi Planum, about 3 kilometers (10,000 feet) above "sea level." At the right is Maxwell Montes, the highest region yet found on Venus. It rises to a peak of 10.6 kilometers (34,980 feet) above "sea level."

high plateau dominated by several mountain ranges. Its western part is a relatively smooth plateau called Lakshmi Planum and is about 3,300 m (10,000 ft) above mean sea level. The area is enclosed on the west and north by mountains varying between 2,300 m (7,000 ft) and 3,300 m (10,000 ft). The western chain has been tentatively named Akna Montes, and those to the north have been named Freyja Montes.

The entire eastern end of Ishtar Terra is dominated by the highest terrain on Venus. Maxwell Montes is 10,600 m (34,980 ft) at its highest point above mean sea level and 9,000 m (29,700 ft) above the Lakshmi Planum. It extends generally in a northwesterly and southeasterly direction. Its eastern flank contains a feature that is 100 km (62 mi) in diameter and some 1,000 m (3,300 ft) deep. Scientists theorize that it may be a volcanic crater.

Two other significant topographical features were revealed by the radar mapping of Pioneer Venus. Beta Regio, centered about 30° N latitude, appears to consist of two shield-shaped volcanoes. The region is larger than the Hawaii-Midway chain of islands in the Pacific Ocean. It may be located on a very long fault line. The two mountains, Theia Mons in the south and Rhea Mons in the north, adjoin each other and cover a span of some 2,100 km (1,300 mi). Each is about 4,000 m (13,000 ft) above mean sea level. That they are of volcanic origin was supported by the data returned to the Earth from the Soviet probes Venera 9 and 10, which earlier had landed just east of Beta Regio.

Alpha Regio is about 25° south of the equator and 6,400 km (4,000 mi) west of Aphrodite Terra. It rises about 1,800 m (6,000 ft) above the central plain and is extremely rough, with parallel features extending across its surface. In some ways it resembles the range and basin structure of the western U.S.

Some 60% of Venus is covered by rolling plains marked by many features that appear to be impact craters. The craters typically have diameters of 400 km (250 mi) to 600 km (370 mi) but depths of only 200 m (600 ft) to 700 m (2,300 ft). Bright features seen in the plain are probably central peaks associated with such craters. The mapping also revealed that Venus, like the Earth, has a Grand Canyon. In the eastern part of Aphrodite Terra is a trench deeper than the Earth's Dead Sea, though only one-fourth the depth of the Marianas Trench in the Pacific Ocean. It is approximately 2,900 m (9,500 ft) below mean sea level.

Early in 1981 Pioneer Venus took up a new mode of operation. Instead of maintaining a stationary position over the planet it began drifting, subject to the forces of the solar wind and the gravity of Venus and the Sun. By 1986 it was expected to be over the planet's equator. Scientists estimated that the probe would enter the Venusian atmosphere in 1992 and burn up. The long period of drifting and floating would permit instruments aboard the probe to make much more accurate measurements of the planet's gravity, long-term changes in cloud circulation, and changes in the planet's upper atmosphere during a period of decreasing solar activity.

The Soviet Venus probe Venera 12, its mission completed, was called upon in March to serve yet another scientific task while it was more than 190 million km (114 million mi) from the Earth. It was commanded to orient its ultraviolet spectrometer toward Comet Bradfield. Since the comet would not return to the Earth for some 300 years, Soviet scientists were taking every opportunity to study it. Data returned from Venera 12, when fully analyzed by scientists, would permit quantitative estimates of such elements in it as hydrogen, helium, argon, and oxygen.

Vikings on Mars. The Viking mission to Mars was largely complete by the end of 1980. Originally conceived as a 90-day mission, it far exceeded that goal. On April 12 the Viking 2 lander was commanded to shut down after $3^1/_2$ years of transmitting data from the surface of Mars after a power failure occurred aboard the craft. (Its orbiter had ceased operation in mid-1978). Since touching down on the planet on Sept. 3, 1976, the probe had taken more than 1,800 pictures of the surface, while other instruments aboard it analyzed the Martian atmosphere. Indeed, for budgetary reasons NASA had been using the lander mainly as a weather station for some time previous to shutting it down altogether. One of its most important discoveries was that a thin layer of water frost covers the ground in the far northern latitudes each winter.

The Viking 1 orbiter literally ran out of gas and was commanded to shut down on August 7. The supply of nitrogen gas used to maintain proper attitude of the probe with respect to the Martian surface, the Sun, and the Earth was exhausted. In orbit since June 19, 1976, the probe obeyed the command on its 1,489th revolution around Mars. However, before this action took place the orbiter made a significant discovery. On February 22, an exceptionally clear day on the planet, the craft photographed several meteorological features rarely seen on Mars. One consisted of a sharp, dark line curving north and east from the volcano Arsia Mons in the Tharsis Ridge. It was interpreted as being either a weather front or an atmospheric shock wave. In some ways it appeared to be similar to bow waves often seen in the oceans on Earth. The other feature consisted of a group of four clouds that cast distinct shadows on the surface. The cluster was just north of the crater Lowell. The largest cloud was some 32 km (20 mi) long, and the four were at an altitude of nearly 28 km (17 mi) above the surface. Such shadows are extremely rare in Viking pictures.

After studying surface photographs made by both Viking orbiters since 1976, Dag Nummedal, associate professor of geology at Louisiana State University, concluded that Mars once had large underground bodies of water that were released by volcanoes some 100 million to 3 billion years ago. The water was derived from snow and was stored as permafrost, as on parts of the Earth. It was released randomly because of volcanic activity and the resultant melting of ice pockets.

The demise of Viking Orbiter 1 prompted Alan Wood, a spokesman for the Jet Propulsion Laboratory, to say, "It marks the end of an era. This is the formal end of the Viking project; and, really, it's the end of the '70s for space exploration by the U.S. We're looking at a long, dry period now." Tumbling slowly, Viking Orbiter 1 was expected to take at least 75 years to spiral down through the Martian atmosphere to the surface of the planet.

At the end of 1980 only Viking Lander 1 remained active on Mars. It was programmed to make weekly reports to the Earth for the next 14 years. Unless there are failures or malfunctions in the probe's electronic components or scientific instruments, its lifetime is limited only by its decreasing power supply, a radioisotopic thermoelectric generator.

Studying the Sun. As 1980 ended, the U.S. probe Pioneer 6 had been in orbit around the Sun and studying it for more than 15 years. It set an operational record for an interplanetary probe. Amazingly, original specifications for the probe had called for it to function for only six months. Its longevity encouraged Richard Frimmel, mission manager, to state, "Pioneer 6 is such a good spacecraft that we may get another ten years out of it."

Also still circling the Sun were Pioneers 7, 8, and 9. Data supplied by the four probes, whose positions relative to the Sun change constantly, are used by the Solar Disturbance Forecast Center of the U.S. National Oceanographic and Atmospheric Administration to predict solar storms for some 100 primary users, including commercial airlines, power companies, communications companies, the U.S. military services, and the Federal Aviation Agency.

Future probes. With the end of 1980 prospects for future interplanetary probes before the 1990s appeared bleak in the U.S. Ever-tightening budgets meant there were to be few "new starts" in an area of technology that may require as much as a decade to proceed from concept to data acquisition from the target planet. Work continued on the U.S. Galileo probe to be launched aboard the space shuttle in 1984 for a rendezvous with Jupiter in 1987. By November 1980 contract negotiations were almost concluded for the design and development of the spacecraft "bus" that would carry the probe to the planet.

The Soviet Union planned to launch two probes in November 1981 that would land on Venus and take samples of its soil for compositional analysis, as did the Viking landers on Mars. The Soviet Union also revised its proposed 1984 Venera mission to Venus. While specific details were not available by the end of the year, it was known that the probe will now incorporate an encounter mission with Halley's Comet in 1986. The encounter distance would be approximately 10,000–50,000 km (6,200–31,000 mi). The Venus portion of the mission was expected to include a lander containing a French-built balloon or balloons to be released when the probe landed on the surface. They would carry French-developed instruments into the Venusian atmosphere for analysis of its chemical components and pressure.

The European Space Agency (ESA) announced that it would go ahead with Project Giotto, a probe to investigate Halley's Comet. The probe would fly at a speed of 70 km per second (43 mi per second) through the comet at a distance of less than 1,000 km (600 mi) from the coma, or head, of the comet. Giotto was to be based on the GEOS satellite, developed by ESA, and was to have as principal instruments a camera and mass spectrometers for measuring the atomic composition of the comet. NASA's own Comet Science Working Group preferred the development, by NASA, of a separate and more sophisticated U.S. probe; however, funding for such a probe seemed improbable. The only alternative was for NASA to supply a Delta launch vehicle for Giotto and become a junior partner in the venture, but ESA rejected U.S. offers of participation on such a basis. At year's end the prospects for any U.S. space effort aimed at the comet were practically nil.

—Mitchell R. Sharpe

See also Year in Review: ASTRONOMY.

Transportation

The predominant trend in transportation technology continued to be the great attention given to the conservation of petroleum-based fuels, on which the transportation industry is almost fully dependent. Several major programs were also being continued to find other sources of energy for transportation. Much of the effort was focused on coal, both for direct use and indirectly as fuel to produce electrical energy. The adverse effects of continued inflation and economic recession resulted in lower priority for many non-energy–related transport research and development (R and D) programs.

The switch from petroleum to coal as an energy source was stimulating R and D efforts in that direction. One development, announced by Babcock & Wilcox Co. was the successful burning of a new coal-water slurry without the expensive dewatering required with most conventional water-based slurries. Named Co-Al, the slurry consisted of 71% eastern bituminous coal along with water and a small amount of chemical additives. Not only could the slurry be stored in existing liquid fuel tanks, thus eliminating the need to build special storage facilities, but it also could be transported by pipeline, tankers, tank trucks, or rail tank cars. Babcock & Wilcox was funding preliminary combustion tests and was discussing construction of a pilot plant to produce 1,000–3,000 bbl of Co-Al per day.

Air transport. Delta Air Lines, citing a promise of higher fuel savings and greater technological advances than competing engines, selected Pratt & Whitney's new PW2037 (formerly JT10D) engine for its 60-plane order of Boeing 757s. The total cost of the engines, including spares and test equipment, was estimated at about $600 million. The B-757 is one of several new-generation air transports featuring extensive use of aluminum and composite minerals to reduce weight and having redesigned wings expected to achieve a 30–35% improvement in fuel efficiency. The twin-engine plane was designed in varying configurations to seat from 174 to 196 passengers and is a narrow-bodied jet, as compared with the new-generation B-767, which is also twin-engine but is wide-bodied and seats from 211 to 255 passengers.

Previous orders for the B-757 called for the use of General Electric or Rolls-Royce engines that are smaller-thrust derivatives of the engines by those manufacturers used in the jumbo wide-bodied B-747, L-1011, and DC-10. The PW2037, on the other hand, is a new engine developed by Pratt & Whitney at a cost of about $2 billion specifically for use by an aircraft the size of a B-757. Thus, the latter builder was competing strongly for a major share of the engines that will be needed to equip the predicted total of 1,500 B-757s.

First prototype of the eight-passenger Lear Fan 2100 was shown late in 1980. The plane is made of a graphite-epoxy composite considered to be as tough as titanium but so light that the craft weighs only 1,800 kilograms (4,000 pounds). Two turbine engines power a single pusher propeller located at the rear.

Courtesy, Lear Fan Corp.

Courtesy, Airship Industries Ltd.

Skyship R-40 is one of four such airships scheduled for cargo service by Redcoat Cargo Airlines of Great Britain. Its overall cargo capacity of 1,069 cubic meters (37,775 cubic feet) is more than 50% greater than that of a jumbo jet.

A new aircraft designed especially for general aviation reached the prototype-testing stage. By early 1981 LearAvia Corp. had 180 orders for its Lear Fan 2100, even though delivery was about two years away. The company was being aided by the British government, which extended $50 million in grants and guaranteed loans to help pay for the conversion of a former Royal Air Force base at Aldergrove near Belfast, Northern Ireland, into production facilities. LearAvia also raised $30 million on its own to help finance construction of the eight-passenger Lear Fan, which was to have an airframe made of a graphite-epoxy composite material said to be as tough as titanium but so light that the plane will weigh only 1,800 kg (4,000 lb). It will cruise 3,700 km (2,300 mi) at 563 km/hr (350 mph) on only one-fifth the fuel used by present executive jets. Two Pratt & Whitney of Canada 650-hp turbine engines were to power the single pusher propeller, a four-blade design located at the rear of the aircraft.

A high-altitude wind shear warning system to complement the low-level wind shear alert system already being used successfully at several major U.S. airports was being developed by the U.S. Federal Aviation Administration (FAA). The new system was designed to increase the recording of wind shear from the 18-m (60-ft) maximum height of the existing system to elevations up to 490 m (1,600 ft) along aircraft approach and departure paths. In operation since 1978, the test system consists of a 4.5-m- (15-ft-) diameter parabolic radar antenna and a wind shear measuring system developed by the National Oceanic and Atmospheric Administration for the FAA. The system uses the Doppler effect to detect wind speed and direction at preselected altitudes so that pilots can be forewarned of wind shears and possible loss of airspeed or control during takeoffs and landings.

A report by the Office of Technology Assessment (OTA) of the U.S. Congress recommended that research be continued on an advanced supersonic transport (ASST). The OTA noted that technological advances had been made since the design of the British-French Concorde SST. These include advances in areas of aerodynamics, structures, propulsion systems, and noise reduction that could make the ASST economically viable if fuel costs do not become too high.

McDonnell Douglas Corp. began preliminary work on an ASST that it claimed would fly 50% farther without refueling and almost 10% faster than the Concorde, and carry about 250 passengers, or two and a half times that of the British-French SST. While the proposed new plane would reduce the takeoff and landing noise level about 7%, it would still have the sonic boom problem that has to date forced the U.S. to restrict the SST to subsonic speeds over land. One possible solution, according to the U.S. National Aeronautics and Space Administration (NASA), is the use of a pivoting wing, which, when swung oblique to the fuselage while in supersonic flight, eliminates the sonic boom.

British Airways reported that it has made a number of changes to help save fuel and improve the performance of its operationally successful but economically unsuccessful Concorde. These included increasing the size of engine air intakes, adjusting aft limits of the center of gravity to increase fuel-carrying capacity, increasing the speed limit of tires to boost takeoff weight, using tank expansion areas for fuel consumed in taxiing before takeoff, and adopting new approach procedures to permit faster, more fuel-efficient speeds.

Predicting major savings in operating costs, fuel consumption, and turnaround time, Redcoat Cargo Airlines, a small British cargo airline headquartered near London, ordered four lighter-than-air cargo vessels for delivery by 1984 at a cost of about $9.5 million each. To be built by Airship Industries Ltd., the 183-m × 36-m (600-ft × 120-ft) airships are claimed to provide over 50% more cargo capacity than a jumbo jet and to be able to operate at freight rates about 30% lower than those for narrow-bodied air transports. A 50-m-(164-ft-) long prototype was scheduled to enter service in mid-1981; if so, it would be the first such airship in commercial service since the German Zeppelins last flew in 1937.

The four large airships, designated Skyship R-40s, were designed to have an interior cargo deck capable of handling containers, an overall cargo capacity of 1,069 cu m (37,775 cu ft), and a short-range payload of 58 metric tons. Operating at a cruising speed of 68 knots, they would be able to make an Atlantic crossing in about one-third the time required by a modern surface cargo vessel. Lift would be provided by helium, a nonflammable gas, and forward propulsion by four 1,-120-hp turboprop engines.

Airship Industries announced that it planned to build by 1986 a 300-m × 60-m (1,000-ft × 200-ft) vessel able to lift 100 metric tons of freight, although actual commercial operations with either this airship or the R-40s must await licensing by the British Civil Aviation Authority. Also, problems relating to the sophisticated docking facilities required for these craft must be resolved.

Highway transport. A zinc/chlorine system to power electric motor vehicles was unveiled and demonstrated by Gulf & Western Industries, Inc. An outgrowth of research into an energy storage system to help electric utilities deal with periods of peak demand, the new system was developed over eight years in cooperation with the Electric Power Research Institute and what became the U.S. Department of Energy at a cost of $33 million. The chief advantages of the zinc/chlorine battery are its long electrode life (a prototype system was put through 1,400 charges with no signs of deterioration) and its ability to deliver peak power during 95% of its discharge cycle. A Volkswagen Rabbit, fitted with the Gulf & Western system, went 240 km (150 mi) at 89 km/hr (55 mph) between recharges. The manufacturer claimed that operating costs are one-third those of gasoline-powered cars getting 20 miles per gallon (mpg), but this does not include the higher initial cost of the new power system or any eventual battery replacement costs.

Though the electricity-generating potential of a zinc/chlorine battery has long been known, several breakthroughs were necessary before a practical system could be developed. First was the discovery that inexpensive graphite could be substituted for costly titanium electrodes. Second was the development of an inexpensive plastic that could contain the highly corrosive chlorine, and third was the discovery that the chlorine could be stored safely if it is mixed with chilled water to form slushy chlorine hydrate.

The power system consists of a chilled storage unit containing the chlorine hydrate and a "stack" with positive and negative electrodes in a zinc chloride solution. To create electricity zinc chloride is pumped to the storage unit, where it reacts with the chlorine hydrate to produce chlorine gas. The chlorine is then pumped to the stack, where it reacts with the zinc on the cathode to produce zinc chloride again and electricity. Recharging reverses the process, redepositing zinc on the cathode and chlorine hydrate in the stack.

Not long after the introduction of the zinc/chlorine battery, the U.S. Department of Energy (DOE) reported problems with it. These included a power output only 65% of that expected, which, the department said, was caused by freeze-ups in the heat exchanger part of the power pack that required defrosting for a few minutes every hour during the complex recharging process. The DOE also said that recharging the battery was so difficult that it could be done only by highly trained personnel. Both Gulf & Western and DOE, however, expressed confidence that most of the problems could be resolved.

General Motors Corp., the only major U.S. automaker working to develop electric cars, announced that it was delaying from the original date of 1984 until 1985 its announced plans to begin mass production of electric-powered commuter cars. While not giving details for the delay, GM said that it needed more time to test untried new technology before committing itself to large equipment and tooling costs. It indicated that the delay was related to the power plant that GM itself was developing. GM indicated that it favored a nickel/zinc battery, which it said would power a car about 160 km (100 mi) at 70 km/hr (45 mph) between charges and run for 48,300 km (30,000 mi). Much simpler than the zinc/chlorine battery, it resembles the common lead/acid battery with alternating plates of nickel and zinc suspended in an electrolyte.

Daihatsu Motors in Japan reported that it was testing a subscription electric car service. Subscribers would have a magnetically encoded credit card, which, when inserted in a control box at any of several strategically spaced subscription lots, would record driver information and direct the driver to his assigned car. Inserting the card in a slot in the side of the car would unlock the door. The car could be returned to any lot within its 60-km (37-mi) range, at which time the driver would insert the card in a control box that would calculate the fee and issue orders for the vehicle's recharging and servicing. The system is being tested using company employees without assessment of charges.

Automatic computerized engine controls in 1981-model General Motors cars are expected to provide better fuel mileage and to meet antipollution regulations. This was to be done through use of an electronic control module about the size of a cigar box. Linked to a system of sensors and controls, the unit can pick up early warnings of upcoming trouble and automatically take corrective action such as advancing or retarding the spark time, adjusting the air-fuel mixture solenoid, adjusting the flow of fuel into the engine, and managing the pumping of air to the manifold and catalytic converter. Most adjustments would be needed to meet 1981 auto-emission standards. The cost of these controls was estimated to average about $725 per car, although GM said that it would absorb $240 of this.

General Motors also began offering an automatically adjustable V-8-6-4 engine in its new Cadillacs to reduce fuel consumption and avoid payment of the 1981 U.S. gas-guzzler tax of $200–650 on models averaging less than 17 mpg for overall driving. As constant driving speed increases, the engine cuts off two and then four cylinders, with the car then achieving about 30 mpg at steady highway speeds of 89 km/hr (55 mph) according to GM.

Pipelines. Further improvements were being sought in the use of transient pressure wave monitors that help to detect, confirm, and locate the occurrence of leaks or ruptures in liquid pipelines in time to permit rapid response. The use of such monitors was made possible by the fact that a leak of any liquid in a line under pressure of 40 psi or more will create a pressure wave at acoustic velocity through the fluid in both directions from the point of origin. Such waves can be detected, depending on the size of the leak and the pressure loss, at distances of 320 km (200 mi) or more in liquid lines. The monitors, installed approximately 60 m (200 ft) from the station, consist of two dynamic transducers. One of these is located about 60–90 m (200–300 ft) from the other and is designed to receive and adjust the wave shock coming from the leak so that the closer unit can detect the proper amplitude. This, in turn, permits use of a standard distance/signal discriminator module to locate the exact position of the leak, as measured by its distance from the station. While such monitors have been widely used for a number of years in natural-gas transmission lines and to a lesser extent in propane and butane lines, recent tests were conducted by CRC Bethany International, Inc., and Shell Pipe Line Corp. to evaluate their effectiveness and expanded use.

Pipelines serving the Louisiana Offshore Oil Port (LOOP) deepwater oil port under construction in the Gulf of Mexico off the Louisiana coast, were to consist of 160 km (100 mi) of pipe 120 cm (48 in) in diameter and larger. Each of the three mooring buoys, capable of accommodating tankers as large as 700,000 tons deadweight, would be connected to the pumping platform by 2,440 m (8,000 ft) of 142-cm (56-in) underwater pipeline, the largest-diameter line ever laid in the U.S. A 120-cm underwater line will transport the oil from the platform, 31 km (19 mi) offshore, to the Clovelly salt dome for storage. From the dome another pipeline will deliver the oil to the Capline system for delivery to local refineries and for transportation to other regions.

In addition to the crude-oil pipeline a 45-km-(28-mi-) long, 76-cm (30-in) brine line was also being laid through swampland from the salt dome to the Gulf of Mexico. To force oil out of the salt dome cavities, of which there were eight for storing different grades of crude, brine was to be pumped in. When oil from LOOP is pumped into the cavities, the brine will be displaced.

The 120-cm and 132-cm pipelines laid offshore were coated with concrete to provide negative buoyancy and were buried from one to three meters (three to ten feet) below the floor of the Gulf. Initial capacity of the $640 million LOOP project, scheduled to be completed in 1981, was designed to be 1.4 million bbl per day.

Rail transport. Alternatives to internal-combustion locomotives were being considered as the price of diesel fuel continued to climb and its availability was no longer assured. A bid to return the reciprocating steam-engine locomotive to the rails was advanced by American Coal Enterprises (ACE). The ACE 3000 locomotive would consist of a power unit and a permanently coupled tender called a "service module" to carry coal and water. Coal would be loaded into the service module in three 11-ton "coal-pack" containers and moved by conveyer to the firebox in the power unit. Combustion would be a two-stage process, with hot air under the coal bed driving gases up for combustion above the bed. The drive system would consist of two engines, each with two cylinders driving two axles through conventional valve gear and rods. Axles two and three would be internally rod-connected in 180° opposition, thus stabilizing movement of the locomotive and making speeds as high as 130 km/hr (80 mph) practical.

Though most major appliances and functions would be similar to those of steam locomotives of 30 years ago, the stoker, combustion and exhaust air, boiler water feed, and valve setting operations would all be precisely and automatically balanced by microprocessors responsive to a single throttle. To make the ACE 3000 environmentally acceptable, ash would be separated out from exhaust gases and conveyed to a five-ton "ash pack" module for removal during servicing.

Besides being less costly to fuel, the steam locomotive would be competitive with the diesel in other areas, according to ACE. It could run 800 km (500 mi) between fuel stops and 1,600 km (1,000 mi) between water stops. It would be bi-directional, with cabs at both ends of the two-unit locomotive and with a capability for multiple-unit operations with other similar units or with diesels. It would have the tractive power of a modern 3,000-hp diesel unit but more high-speed horsepower. ACE estimated that two prototypes could be rolling in two years for an outlay of $25 million.

Photos, Keystone

The world's first oceangoing tanker with sails, the "Shin-Aitoku Maru," was built by the Japanese firm Aitoku. A computer makes automatic adjustments to achieve the best combination of sail and diesel power, and the sails fold against steel masts automatically when not in use (left).

Electrification of U.S. railroads was drawing increased attention as a means of conserving petroleum-based diesel fuel and of assuring an adequate supply of nonpetroleum fuel for essential railroad operations in the future. The U.S. Federal Railroad Administration (FRA) submitted, for consideration by the nation's railroads, utilities, and the Congress, a proposal to electrify the 14% of the nation's rail lines—about 42,000 km (26,000 mi)—that produce more than half the gross ton-miles. The FRA stated that the 20-year program would cost $20 billion, with funding to be shared by government, railroads, and the utilities. Savings to the rail industry in fuel operations and maintenance, according to the FRA, would be $44 billion over 29 years.

Advances in high-speed intercity rail passenger service continued, highlighted by British Rail's program to place its Advanced Passenger Train (APT) in service between London and Glasgow. The APT, capable of speeds of 250 km/hr (155 mph), would make the London-Glasgow run at 200 km/hr (125 mph), cutting time for the 645-km (400-mi) trip from five hours to four hours. Both the APT and the high-speed train (HST), which in 1981 was already in service on several routes, can operate on conventional mainline tracks, unlike Japan's "bullet trains," which have their own specially constructed rights-of-way. The APT is able to take curves at much greater speeds than other trains because of a design feature that electronically tilts coaches up to 9° in order to counter centrifugal forces. Extensive use of plastics and aluminum help keep both weight and fuel consumption down. British Rail planned to build 60 APT's at a cost of $360 million.

An Americanized version of the British Leyland railbus went into revenue testing service between Lowell, Mass., and Concord, N.H. The hybrid vehicle consisted of a bus shell mounted on a two-axle underframe. It was modified to increase passenger capacity from 40 to 56 and was equipped with a new and tougher suspension system designed to handle rides on the rough U.S. trackage. Sponsored by FRA under a $3.2-million experimental program, the railbus was expected to be easy to maintain and inexpensive to operate. Its cost of $400,000 per unit was compared to nearly $1 million per unit for the Budd Company's SPV-2000, the only self-propelled rail passenger car made in the U.S.

Another innovative dual-purpose vehicle, one designed for combination truck-rail movements of freight without interchange, took another step toward full production when the North American Car Corp. placed an order for 250 units with the Budd Co. The RoadRailer is a 14-m (45-ft) truck trailer equipped with a dual-purpose, steel/rubber-tired wheel system that can be adjusted to permit movements over either railroad tracks or highways. During tests RoadRailers have been operated at speeds of 170 km/hr (105 mph) on FRA's Pueblo, Colo., testing facility and at sustained speeds of 105 km/hr (65 mph) on the Seaboard Coast Line main line between Richmond, Va., and Jacksonville, Fla. In rail service they are coupled together behind a locomotive to form a special train, or they can be individually coupled to a truck power unit for over-the-road movements. FRA cleared the way for commercial production by waiving certain regulations pertaining to brakes, handholds, and coupler height.

Proposals for a rail tunnel under the English Channel were again being given serious consideration. The last Channel tunnel project, which was stopped in 1975, would have featured a two-track railway and cost an estimated $4 billion. The latest proposals—one by the British and French national rail systems and the other by a consortium of British, French, Dutch, and West German engineering companies—would have a single-track railway and cost considerably less, about $1.5 billion. The "Chunnel" would be about 35 km (21 mi) long and connect Dover, England, and Calais, France. Trains would take about 30 minutes to cross, and, to minimize traffic delays on the single-track line, blocks of five or six trains moving in the same direction could alternate. Motorists could make use of the Chunnel by having their cars placed on the trains.

Water transport. Ship operators during the year tried both old and new technologies as ways to conserve fuel. The new was represented by the low-speed marine diesel engine, which was about 30% more fuel-efficient than the oil-fired steam turbines used in most U.S. ships and which could burn very low-grade fuel oil. American President Lines Ltd. received from Allis-Chalmers Corp. the first large low-speed diesel engine ever built in the U.S. It was one of three 43,200-hp units to be used to propel three 49,360-ton container ships being built by Avondale Shipyards, Inc. for delivery starting in 1982. The 12-cylinder engine, weighing 1,160 tons, was designed to power the huge ships at speeds as high as 26 knots, although they would operate more economically at lower speeds.

Significant savings in fuel oil for ocean ships through use of sails and wind power was demonstrated by the world's first oceangoing oil tanker with sails, the "Shin-Aitoku Maru." It was powered by a 1,600-hp diesel engine and two 12 m × 8 m (39 ft × 26 ft) plastic sails. The builder, the Japanese firm Aitoku, launched the vessel at the Imamura Shipbuilding Co. dockyard in Kure, Japan. A computer makes automatic adjustments to achieve the optimum combination of sail and diesel power, even changing engine output according to wind velocity. Sails fold against steel masts automatically when not in use. The sails and slim hull give the ship a range of 14,500 km (9,000 mi) between refuelings, compared with 4,800–6,450 km (3,000–4,000 mi) for similar size all-diesel ships. The fuel savings of up to 50% were expected to more than offset the ship's 20% greater construction cost.

A return to coal-fired steam turbine power was being tried by the Australian firm Bulkships Ltd. which signed a contract with Italcantieri of Italy for construction of two 75,000-deadweight tons coal-fired bulk carriers for use in the Australian coastal bauxite trade. The government-owned Australian National Line also contracted with Japan's Mitsubishi Heavy Industries, Ltd. for construction of two similar ships.

—Frank A. Smith

U.S. science policy

The year found the U.S. still struggling with the problem of how to employ its historic facility for technological advances to improve its floundering economy. The administration of Pres. Jimmy Carter had placed a great deal of emphasis on technological productivity, and in early 1980 the budget planners allocated considerable public funds to several proposed cooperative research programs in universities and industry. By the end of the year, however, budgetary cutbacks in response to inflationary pressures had effectively eliminated any opportunity to test the merits of the various programs. The incoming administration of Ronald Reagan made it clear that it had no intention of funding such ventures.

An article appearing in the June 24, 1980, issue of the *Wall Street Journal,* however, suggested that industry had already recognized the value of collaboration with academic research groups and was not waiting for federal leadership. The *Journal* cited a Battelle Memorial Institute study reporting a 12% increase in funding of academic research by industry over 1979, to $179 million. The chemical firm E. I. Du Pont de Nemours & Co. was supporting research at the California Institute of Technology looking into the therapeutic potential of the naturally occurring protein interferon. Westinghouse Electric Corp. had given $1 million to Carnegie-Mellon University to study industrial robots. Exxon Corp. had committed up to $8 million over a ten-year period to support research in combustion technology.

Edward E. David, Jr., president of Exxon Research and Engineering Co. and former science adviser to Pres. Richard Nixon, told the *Journal:* "Powerful forces are pushing industry and universities closer together.... With a long-term commitment, they [university researchers] don't have to worry about a proposal every year. Furthermore, we're much more concerned with the quality of their work, not the bookkeeping, which is more to their liking." To scientists on university campuses, the freedom from bookkeeping may well have been the most attractive factor in the arrangement with Exxon, for this was a year in which the financial relationship between the academic community and the federal agencies reached a new level of exasperation.

Another possible spur to the U.S. technological research effort was seen in the appointment near the end of 1980 of John B. Slaughter, academic vice-president and provost of Washington State University and also an engineer, as new director of the National Science Foundation (NSF). Shortly before taking the post, Slaughter announced that the agency would place more emphasis on applied engineering and technology over the next few years. To concerned scientists, he promised that he would attempt to maintain current levels of support for fundamental research, especially in the life sciences, but he added that "some strong effort has to be made to provide support for the engineering community."

He also stressed the need "to improve the interactions between university and industrial researchers" and to do something about the "stagnation that appears to have occurred in science education in our country." Another matter of special concern to him, as the first black director of NSF, was "the sorry state of the involvement of minorities in science and engineering."

Shortly after the Reagan administration took office, however, it became clear that allocation of the resources available to NSF would not depend entirely on the desires of the director. Before the new president had made his first official statement on the fiscal year 1982 budget, word came from the administration that not only was NSF's budget to be cut severely as part of the overall budget trimming but also that several areas had already been singled out for near or total elimination. They were funding for the social, behavioral, and economic sciences, funding for new instrumentation, science education, and international activities.

Birth of a growth industry. While efforts to improve productivity in the medium-technology heavy industries appeared to be floundering, research in recombinant DNA, loosely called genetic engineering, began to take off. New technologies applied to a mature medium-technology industry can rarely have more than a slight incremental effect, whereas new techniques applied to a young, high-technology industry can be explosive. Such was the case in 1980 with the successful application of recombinant DNA research to the pharmaceutical industry.

Perhaps the most remarkable feature of the entry of recombinant DNA research into the marketplace was the speed with which it took place. If one were to choose a series of milestones in the industrialization of genetic engineering, the first would probably be the 1973 publication in the *Proceedings of the National Academy of Sciences* of a paper reporting the invention by Stanley N. Cohen of Stanford University, Herbert W. Boyer of the University of California at San Francisco, and others of a technique for the construction of biologically functional DNA molecules that combined genes from different sources.

Second would be the filing in 1974 by Stanford University of an application for a patent covering the technique for gene splicing and cloning and for the products resulting from use of the technique. Third would be the founding in 1976 of a corporation called Genentech, Inc., by Boyer and Robert A. Swanson, a venture capitalist who knew a good thing. Fourth was the issuance in October 1980 of the first stock offering by a genetic engineering firm; a million shares of stock in Genentech went on the market at $35. Within hours the price shot up to $89, and within days the phenome-

Dan McCoy—Rainbow

Recombinant DNA research continued to expand rapidly in the U.S. during the past year. Among its most notable products are synthetic analogues of human insulin, interferon, and growth hormone.

non had been brought to the attention of readers of almost every major metropolitan newspaper in the country. But then the hysteria subsided, as did the price of the stock. By December 1980 Genentech was selling for $45.75.

Nevertheless, there was much in the promise of recombinant DNA research to attract venture capital. There was the reasonable assurance that the technique could be used to develop synthetic analogues of naturally occurring substances difficult to accumulate (such as insulin or human growth hormone). Furthermore, some scientists were speculating that the new techniques could be used to produce compounds that would extend to some viruses the same level of control that antibiotics now offer in the prevention and cure of bacterial disease.

One such substance was interferon, which occurs naturally in the human body. It had been isolated in such small quantities there was not even enough to supply the leading medical research laboratories, much less clinics for the treatment of the ill. But the hope it offered was tantalizing—perhaps a cure for some forms of cancer, but more likely an active antiviral drug. The optimists were calling it "a penicillin for viruses," a group of pathogens that had previously evaded almost every kind of therapy.

Thus the field attracted not only high-rolling investors but also such established companies as Monsanto

Co., Schering-Plough Corp., Dow Chemical Co., Lubrizol Corp., Koppers Co., Inc., and Emerson Electric Co. The president of one of the new genetic engineering firms predicted worldwide sales of related products amounting to $40 billion in 15 industries (including energy and chemicals) by the year 2000.

DNA and Harvard. Stanford University had already moved to collect some of the financial rewards of its faculty's efforts through successful application for a major patent. It was following a well-established precedent. Indiana University holds a lucrative patent for stannous fluoride, the University of Wisconsin for a rat poison, and the University of Florida for Gatorade. In 1980, however, Harvard University decided to go all the rest one better. In doing so, it not only established a precedent for a major university, it almost set off a civil war among its own faculty.

On Oct. 9, 1980, Harvard revealed that it was considering a business venture in the field of recombinant DNA. According to *Harvard Magazine:* "The planned company would be jointly owned by a group of scientists including at least one Harvard professor, by the company's top management, by Harvard 'as a minority owner,' and by a number of venture-capital firms." The university would put up no money, according to the proposal, but it would have a "stake in a promising new industry and better control over the demands that outside work made on its faculty."

The response to the proposal, both from Harvard's faculty and from the outside, was quick and angry. Richard C. Lewontin, Agassiz professor of zoology at Harvard, described the proposal as "the gravest threat to the integrity of the University in its history." The *Boston Globe* followed up with a critical editorial; a few days later the *New York Times* weighed in with a softer reproach.

It was clear that the ensuing debate grew out of a larger issue. The economic well-being of the nation has come to depend more and more on high-technology industries whose growth, in turn, is heavily dependent on the products of academic research. The universities are being squeezed by the higher costs of undertaking that research. And in the search for new sources of funding, research universities are finding themselves in unfamiliar relationships with both government and industry.

As these relationships have developed, university science faculties have endeavored to ensure that their traditional freedom to determine their own course of investigation is not eroded by external factors. It appeared to many influential and vocal members of the Harvard faculty that this independence might not survive the new arrangement. The stakes were too high. As Harvard's own magazine heard the difficult questions: "Could Harvard guarantee free research to a professor who was also a major business partner? Would such a scholar be allowed to pursue a new and commercially unprofitable line of research, no matter how big the University's investment in his work? And how could the University avoid favoring subjects promising large profits?" Lewontin asked, "What about the rest of us who are so foolish as to study unprofitable things like poetry...?"

In November Harvard's president, Derek C. Bok, announced that the university had decided to drop the idea, for the time being, at least. "The preservation of academic values is a matter of paramount importance to the university," he declared, "and owning shares in such a company would create a number of potential conflicts with these values. After consulting with the faculty, I have concluded that Harvard should not take such a step, even on a limited, experimental basis, unless we are assured that we can proceed without the risk of compromising the quality of our education and research."

But, he added: "It is possible...that the participation of Harvard and its professors in commercial ventures can be structured in ways that are wholly consistent with its academic values.... I therefore believe that the University would continue to consider various means of participating in ventures of this kind...."

Science and the bureaucrats. Over the past decade stresses have also developed between the scientific community and the agencies of the federal government that have supported their research activity. Here the conflict is between the independence of the investigator and concern on the part of the government that public funds are being spent frugally and to some stated purpose.

Drawing by Whitney Darrow, Jr.; © 1980 The New Yorker Magazine, Inc.

There never has been a commonly accepted compact between the scientific community and the federal agencies as to the nature of their relationship. It began gradually—in effect, with the publication of *Science, the Endless Frontier* by Vannevar Bush, based on his experience as director of the highly successful Office of Scientific Research and Development in World War II. That seminal report, looking into the distant future, saw the possibility of a federal research budget approaching $50 million. In the 35 years that have elapsed since that prediction was made, the annual federal budget for the support of research and development has grown to a figure a hundred times larger—$5 billion.

This tremendous growth has been incremental in nature—and so has the growth of its fiscal regulation. In recent years the latter process has speeded up considerably. University administrators were the first to be affected. Regulations were issued that limited the amount of money universities could collect for the indirect costs of administering research programs, such as the support of research libraries and the maintenance of physical structures. At the same time a number of other regulations, reflecting the concern of the government for fair employment practices and more precise accounting methods, have had the effect of increasing the actual costs to the university of conducting business.

Thus, the first cries of outrage came from university administrators, who felt that they were now being obliged to devote an increasingly larger fraction of their income to doing the government's work. In 1980, however, the circulation of a new directive from the Office of Management and Budget (OMB) not only set off unprecedented cries of anguish from the scientists but also set them against their own campus colleagues in administration. The anguish came from a provision in Circular A-21 that called upon all scientists whose work was supported by federal funds to report on the number of hours each spent on the actual performance of research.

As the scientists saw it, there were two major difficulties with this dictum. The first was that it is almost impossible for scientists also employed as university professors to designate when they are conducting research and when they are teaching their research assistants. In addition, those scientists engaged in abstract or theoretical studies found it very difficult to identify when they were working and when they were not. It is not uncommon for a scientist to go to sleep with a nagging problem and awake with a promising solution.

The second problem was that the OMB wanted all the scientists' working time accounted for, while the prevailing custom in the academic scientific community is to allow the academic scientist to devote a portion of his working week—one day a week, perhaps—to related activities, such as participation in advisory committees or, in the case of the more fortunate, paid consultation. In the view of some scientists 100% time accounting, because it could involve activities not supported by government funds, would present an opportunity for possible government intrusion into the academic laboratory. Several compromises were proposed by the end of 1980, by both academic scientists and the OMB, but no solution emerged that appeared to be satisfactory to all parties.

Wide World

Harrison Schmitt, a lunar geologist who went to the Moon on an Apollo mission, is one of the few scientists in the U.S. Congress.

The quarrel between scientists and administrators stemmed from the more complicated question of reimbursement for the indirect costs of doing research. Although the question is complex, one of its stickier aspects is that funds turned over to a university as reimbursement for such indirect costs are—in many universities—put back into the general funds and thereby used for the benefit of poets and philosophers as well as scientists. Some heads of scientific departments have argued, so far without success, that the money should remain with them—in order to rebuild laboratories and purchase more sophisticated instruments rather than to pay the costs of staging Greek drama.

Scientist in the Senate. In early 1981 the scientific community found itself facing a glum prospect: not only the likelihood of greatly reduced funding of current research and a high probability of no new starts in large-scale research projects but also the disappearance of several old friends in Congress. On the other hand, one of the few real scientists in Congress had gained new power. Sen. Harrison H. ("Jack") Schmitt (Rep., N.M.) appeared to emerge from the realignment of authority with a unique breadth and intensity of interest in the affairs of the scientific community. As one of the few human beings who had stood on the surface of the Moon, he was especially interested in the program of the National Aeronautics and Space Administration, but he also had a great deal to say about other major research programs.

—Howard J. Lewis

Scientists of the Year

Honors and awards

The following article discusses recent awards and prizes in science and technology. In the first section the Nobel Prizes for 1980 are described in detail, while the second is a selective list of other honors.

Nobel prize for chemistry

How do three molecular biologists react when they find that the popular press is calling them genetic engineers while the Royal Swedish Academy of Sciences has decided that they are the chemists who, during the preceding year, "have conferred the greatest benefit on mankind"? They surely could be forgiven for smiling, serenely but modestly, all the way to Stockholm. Paul Berg of Stanford University, Frederick Sanger of the University of Cambridge, and Walter Gilbert of Harvard University, whose citation by the Academy brought them the Nobel Prize for Chemistry in 1980, have concerned themselves lately with the study of nucleic acids. These substances are the essential components of the genes and chromosomes, which are directly involved in the hereditary transmission of the characteristics of the cells of all living things.

Genetic engineering is a recently coined term that, in the words of a technical review, "usually implies deliberate manipulation of genes of various organisms . . . to achieve useful products of metabolism or to cause a permanent hereditary change. . . ." Its objectives differ from those of Luther Burbank and other breeders of plants and animals, who have brought forth the Shasta daisy and champion thoroughbred racehorses. Rather than study the whole organism formed by the summed expression of all the genes it contains, genetic engineers concentrate on the constitution of the individual genes and their methods of expression.

The youth of the profession is suggested in the review quoted above, which refers to 33 earlier publications, none more than ten years old. Like a teenaged child, genetic engineering has not yet shown what it will be when it grows up. Its prospects, however—which include turning bacteria into factories for scarce proteins or modifying corn so that it will, like peas and beans, harbor bacteria that add nitrogen to the soil—have attracted a great deal of attention, some of it from investors hopeful of eventual profits. Its darker implications—such as the possible creation of new organisms that could cause intractable infections—have drawn their share of hostile criticism.

Berg was awarded half the prize in recognition of "his fundamental studies of the biochemistry of nucleic acids, with particular regard to recombinant DNA [deoxyribonucleic acid]" a citation that he prefers to interpret as acknowledging the merits of an entire body of work rather than recognizing a single achievement. The other half of the prize was shared by Gilbert and Sanger for their independent development of methods for determining the order of the individual links in the chainlike molecules of the nucleic acids. Sanger, who won the Nobel Prize for Chemistry in 1958, became the fourth person—following Marie Curie, Linus Pauling, and John Bardeen—to win a second.

Berg was born in New York City on June 30, 1926. He interrupted his studies at Pennsylvania State College (now University) and joined the U.S. Navy during World War II, after which he returned and graduated in 1948. He received a Ph.D. in biochemistry from Western Reserve (now Case Western Reserve) University, in Cleveland, Ohio, in 1952. He held research fellowships at the Institute of Cytophysiology in Copenhagen, studying fatty acid metabolism with Hermann Kalckar, and at Washington University in St. Louis, where his adviser was Arthur Kornberg. Kornberg, who shared the Nobel Prize for Physiology or Medicine in 1959, moved to Stanford in that year, and Berg, who had joined the faculty at Washington in 1955, went with him.

Paul Berg

UPI

Frederick Sanger

UPI

Walter Gilbert

Keystone

During the 1960s Berg investigated the stages in the process by which genes are expressed in simple organisms. The most clearly recognized action of genes, in bacteria just as in more complex living things, is the control of the structure of proteins formed in the cell. Berg studied the copying, in the cell nucleus, of the segment of DNA that makes up an individual gene onto a molecule of RNA (ribonucleic acid), which then migrates to another location in the cell. There, outside the nucleus, the RNA acts as a template during the assembly of amino acids into the protein coded by the gene. Berg also studied the conversion of the amino acids into the reactive and specific forms that line up with the proper sites on the RNA to be joined together in the new protein molecule.

About 1968 Berg turned to the investigation of gene expression in organisms more intricate than bacteria. He and other biochemists had noted that the induction of tumors by certain viruses is an instance of the expression of the viral genes in the cells of the host animal, and they envisioned the analysis of this phenomenon by extensions of techniques that had been developed for chemically manipulating nucleic acids. There were long-range prospects of applying the findings to cancer research and more immediate hopes of identifying the biological activity of isolated genes. It appeared feasible to dissect chromosomes and to install selected pieces—single genes—in new chromosomal environments where their expression could be examined in the absence of the complicating effects of the other genes that had accompanied them in their native organisms.

Berg's leadership and ingenuity in devising ways of taking DNA molecules apart and putting the pieces back together, and of keeping track of the progress of the events, established him among the founders of the new discipline variously called molecular genetics, genetic engineering, or recombinant DNA technology. He pioneered in executing the feat now generally called gene splicing: patching together, or recombining, fragments of DNA from different organisms. Commonly, the gene to be studied (the target fragment) is inserted in the DNA of a virus or a plasmid, entities that can penetrate the cells of animals or bacteria (acting as vehicles or vectors), where the genes in the new medley can express themselves.

This strategy has been used in bringing about the expression of bacterial genes in mammals and of mammalian genes in different mammals and in bacteria. For example, human insulin and interferon have been produced by bacteria that have been genetically reprogrammed according to the procedures developed by Berg and others adept in this novel art.

Not long after Berg showed that DNA segments could be transplanted between the genetic materials of different species, he and other life scientists became concerned that organisms containing modified DNA's might become the agents of dangerous infections. They organized a committee that called for the deferment of these experiments and for the convocation of an international meeting to consider the problem. At the International Conference on Recombinant DNA Molecules—usually called the Asilomar Conference because it was held at the Asilomar state conference center at Pacific Grove, Calif.—in February 1975, about 140 scientists from 17 countries agreed upon a set of guidelines that classified experiments on recombinant DNA and specified safeguards for each category. Since then the restrictions have been relaxed, as continuing research has shown that the hazards are less serious than had been feared.

Sanger was born on Aug. 13, 1918, in Rendcombe, Gloucestershire, England. He received his bachelor's degree from St. John's College of the University of Cambridge in 1939 and his doctorate in biochemistry from that university in 1943. He remained at Cambridge, working first as a research fellow in the biochemistry department of the university until 1951 and then as a member of the staff of the Medical Research Council. From 1943 he and his collaborators worked out techniques for determining the order in which amino acids are linked in the chainlike molecules of proteins. Concentrating on the hormone insulin obtained from cattle, they found the location of each of the 51 amino acids present in the molecule. They also showed that the insulins from humans, sheep, pigs, and whales are not quite identical to that from cattle; although the overall structure of the molecules is the same, a few of the locations in each one are occupied by different amino acids. The methods devised by Sanger were quickly adopted by researchers throughout the world and have remained the standard procedure for finding the amino-acid sequence of proteins.

During the years that Sanger spent in solving the structure of insulin, other investigators proved that nucleic acids—DNA's and RNA's—are the essential components of the materials responsible for the transmission of hereditary traits. This discovery was announced by Oswald Avery of the Rockefeller Institute for Medical Research, New York City, in 1944, but confirmation and general acceptance of the findings were delayed until 1952. By 1958, when Sanger was rewarded with the Nobel Prize for his success with insulin, he had already switched his attention to the even more difficult problem of unraveling the structure of the nucleic acids.

Earlier investigators of these substances had provided essential pieces of background information. It was known, for example, that nucleic acids are long-chain molecules, even longer than proteins, but most of them are composed of only 4 kinds of links, called nucleotides, instead of the 20 kinds of amino acids found in proteins. Each nucleotide is itself a subassembly composed of a sugar group, a phosphate group, and a nitrogen-containing group, called a base, that belongs to either the purine or the pyrimidine family. The sugar and phosphate groups alternate along the backbone of the nucleic acid chain; one of the nitrogen bases is attached to each sugar group. In DNA's the sugar is deoxyribose and the purine bases are adenine and guanine; the nucleotides containing these two bases are named deoxyadenylic acid (abbreviated dA or just A) and deoxyguanylic acid (dG or G). The pyrimidine bases cytosine and thymine are present in the nucleotides deoxycytidylic acid (dC or C) and deoxythymidylic acid (dT or T). In RNA's the sugar is ribose; the purine bases adenine and guanine are present in the nucleotides adenylic acid (A) and guanylic acid (G). The pyrimidine nucleotides of RNA's are cytidylic acid (C) and uridylic acid (U, containing the base uracil), instead of dC and dT of the DNA's.

In 1953 James Watson and Francis Crick had announced the double-helix structure of DNA. (They and Maurice Wilkins shared a Nobel Prize for this discovery in 1962.) According to their model two nonidentical strands of DNA, differing in the order in which the units A, C, G, and T are connected, twine around one another in such a way that each nucleotide in one strand is adjacent to its complementary nucleotide in the other: A always pairs with T, and C always pairs with G. The double-helix concept meshes with the genetic process by which the paired DNA strands generate duplicates of themselves; the strands disengage, each acts as a template to which individual nucleotides associate themselves in the precise order needed for formation of a new complementary strand, and finally all the new nucleotides become chemically bonded into a new DNA strand twined around the old one as its new partner. In 1955 Severo Ochoa had reported the isolation of the enzyme that connects the nucleotides in forming new strands of RNA; in 1956 Arthur Kornberg had announced the discovery of the enzyme that performs the same function in the case of DNA's. (A Nobel Prize was awarded to Ochoa and Kornberg in 1959.)

Two other Nobel-prizewinning pieces of work that Sanger incorporated in his research were the discovery of the restriction enzymes, which recognize short sequences of nucleotides in DNA's and cut the molecules at those sites (for which Werner Arber, Daniel Nathans, and Hamilton Smith won a Nobel Prize in 1978), and the technique of electrophoresis, which can be used to separate complex mixtures of electrically charged substances (such as nucleotides and chains of them) because they migrate at different speeds along a strip of paper or through a slab of gelatin in an electric field (the discovery of Arne Tiselius, a Nobel laureate in 1948).

One of Sanger's methods of finding the sequence of nucleotides in a nucleic acid involves the following steps. First, the nucleic acid is cut—by restriction enzymes, ordinarily—into segments of manageable size (a few hundred nucleotides), and the segments are separated. Next, each of these segments is used as a template for the synthesis of short complementary polynucleotides from four different mixtures of nucleotides. The four mixtures differ in that, in each of them, one of the nucleotides is partially replaced by a closely similar compound that can pair up with its normal partner in the DNA segment but can undergo chain formation (in the presence of Kornberg's enzyme) with only one of its neighbors rather than both. Wherever one of these modified nucleotides enters the sequence, chain growth stops, and the last link of the aborted chain must be the particular modified nucleotide. Finally, electrophoresis is used to sort out all four mixtures according to the lengths of the chains present in them. If the nucleotide mixture containing modified G, for example, produces chains of 4, 9, and 15 nucleotides, then the nucleic acid segment used as a template must have contained C, the complement of G, at its 4th, 9th, and 15th positions.

Sanger's method can be used to decipher stretches of DNA containing more than 100 nucleotides in a single day. (Protein sequence determinations, by contrast, cannot be performed nearly as fast.) Its effectiveness was demonstrated in an article published in 1977, in which Sanger and eight collaborators reported its use to find the sequence of almost 5,400 nucleotides that make up the DNA of a virus that infects bacteria. From the results they identified nine regions in the nucleic acid that are the genes responsible for the structure of proteins that serve known functions in the virus.

Gilbert was born in Boston on March 21, 1932, and earned bachelor's and master's degrees at Harvard in 1953 and 1954. He then continued his education at the University of Cambridge, where he became friendly with James Watson (of the double helix), concentrated in mathematics and physics (as a student of Abdus Salam, who shared the Nobel Prize for Physics in 1979), and obtained his doctorate in 1957. He returned to Harvard as a theoretical physicist but within a few years redirected his efforts into the problems of molecular genetics, changing his affiliation from physics to biophysics in 1964 and to molecular biology in 1969. He was named American Cancer Society professor of molecular biology in 1972.

During a study of an interaction between proteins and nucleic acids Gilbert observed that the reagent dimethyl sulfate reacts with the purine nucleotides A and G in DNA and breaks the molecule at those positions. In collaboration with Allan Maxam he pursued this finding, discovering one set of conditions in which only A is attacked and another set in which the chain is broken exclusively at the G sites. Seeing in these results the elements of a method for determining DNA sequences, Gilbert and Maxam sought a way of cutting DNA molecules at the pyrimidine nucleotides C and T. Their requirements were met by the substance hydrazine, which also can be made specific for either target by selection of the reaction conditions. The DNA fragments produced by Gilbert's methods, like those built up by Sanger's, have end-members of known identity and are separable by the same electrophoretic technique. From the lengths of the segments formed under the four sets of experimental conditions, the nucleotide sequence of the original DNA molecule can be deduced.

Nobel prize for physics

Wm. Franklin McMahon

James W. Cronin

Clem Fiori—Keystone

Val. L. Fitch

The Nobel Prize for Physics was divided equally between two U.S. scientists who have devoted themselves more to experimental than to theoretical studies of the behavior of subatomic particles. James W. Cronin of the University of Chicago and Val L. Fitch of Princeton University were cited for an experiment they jointly planned and conducted in 1964; its outcome directly implies that the interactions of certain particles would not proceed in the same way if the direction of time were reversed. This finding overthrew the last survivor of a triad of fundamental symmetries formerly thought to apply separately to all the laws of physics. These three principles are time-reversal invariance, which is designated *T* and states that particle interactions should be indifferent to the direction of time; charge conjugation (*C*), which provides that sets of electrically charged particles should interact the same way if all the charges are interchanged; and parity conservation (*P*), which holds that natural processes make no distinction between right- and left-handed behavior.

The first challenge of the validity of *C*, *P*, and *T* brought a Nobel Prize to the physicists T. D. Lee and C. N. Yang in 1957. They had suggested in 1956 that parity might not be conserved in a class of processes governed by forces designated by the physicists as the weak interactions. One of these processes is the decay of the neutron into a proton, an electron, and an antineutrino; this process can be studied by observing neutrons in flight as they emerge from a nuclear reactor (they survive for about 15 minutes), but it is more convenient to make measurements of the decay of neutrons that are at rest. Such measurements are possible because in the nuclei of certain radioactive isotopes neutrons undergo this transformation; the proton produced remains in the nucleus while the electron and the antineutrino escape. This technique was used by C. S. Wu and her collaborators, who made the measurements (of the decay of the isotope cobalt-60) that confirmed the prediction made by Lee and Yang. Experimental verification that parity is indeed violated forced physicists to reconsider the integrity of *C*, *P*, and *T;* they abandoned the position that each is independently maintained and fell back on the view that the violation of *P* requires a simultaneous, offsetting violation of *C*, but the combinations *CP* and *CPT* remain valid.

Cronin and Fitch saw an opportunity to refine some earlier measurements of the behavior of neutral K-mesons, or kaons, which are uncharged particles formed when a beam of accelerated protons strikes a beryllium target. The force that governs their production is the strong force, which also is involved in the cohesion of protons and neutrons in atomic nuclei. The decay of the neutral kaons, however, is subject to the weak force, a phenomenon regarded as strange by its first observers; the term strangeness has since then taken on this special connotation for particle physicists. The weak decay of neutral kaons manifests itself in their relatively long lifetime (about 10^{-10}, or one ten-billionth, of a second, not long except when compared to the 10^{-24}-second lifetime of particles that decay by way of the strong interaction). Two varieties of neutral kaons are observable; the short-lived variety decays to two pi mesons, or pions, and the long-lived to three pions. Both of these decays conform to combined *CP* symmetry.

Cronin and Fitch, with James H. Christenson (then a graduate student at Princeton working under Cronin's supervision) and René Turlay (a French postdoctoral student at Princeton), set up a very sensitive detector to observe these decays and, among about 23,000 photographic records of the paths followed by the pions in magnetic fields, identified 45 events in which long-lived kaons produced two, instead of three, pions. Their demonstration of the violation of *CP* symmetry necessitated another revision of the relationship of *C*, *P*, and *T*. To retain the overall credibility of the concept, it had to be concluded that when *CP* symmetry is violated, *T* symmetry also breaks down.

Commentators on the Nobel prizes speculated that Cronin and Fitch, whose experiment was recognized as a classic as soon as it was reported, were not honored sooner because their finding has only recently been incorporated in attempts to explain the apparent dominance of matter over antimatter in the universe. Earlier interpretations of the big bang theory had assumed that time-invariance was inviolable and foundered on the consequence that the presumably equal amounts of matter and antimatter formed by a symmetrical big bang should have completely annihilated one another. The Cronin-Fitch experiment indicates that particles and antiparticles can indeed decay at different rates. The Soviet physicist Andrey Sakharov was one of the first to formulate a theory invoking *CP* violation to account for the existence of a universe composed of matter to the apparent exclusion of antimatter.

Another source of current interest in *CP* violation has been the emergence of several physical theories that postulate the existence of still undetected particles, such as the fifth and sixth quarks or the Higgs boson. In these concepts *CP* violation appears to be a natural consequence of the underlying mathematical structures.

Cronin was born in Chicago on Sept. 29, 1931. He graduated from Southern Methodist University in 1951 and entered the University of Chicago for postgraduate training, receiving his Ph.D. in 1955. He then joined the staff of the Brookhaven National Laboratory and was appointed to the Princeton faculty in 1958. In 1971 he moved to the University of Chicago. Fitch was born in Merriman, Neb., on March 10, 1923, and switched his interest from chemistry to physics, when, as a member of the U.S. Army, he was sent to Los Alamos, N.M., to work on the Manhattan Project to develop the atomic bomb. He graduated from McGill University in Montreal with a bachelor's degree in electrical engineering in 1948 and was awarded a Ph.D. in physics by Columbia University in 1954. He then joined the faculty of Princeton, where he was named Cyrus Fogg Brackett professor of physics in 1976.

Nobel prize for physiology or medicine

Second-guessing the assemblies who choose the winners of the Nobel prizes each October is not one of the world's better known sports, but a perennial undercurrent of speculation always precedes the announcements and a flurry of criticism invariably follows. There have been recurring complaints that Nobel did not endow prizes in mathematics or the Earth sciences; the lack of a prize in economics has been remedied by the National Bank of Sweden, which since 1969 has funded the Nobel Memorial Prize in Economic Science, awarded—as are the prizes in physics and chemistry—by the Royal Swedish Academy of Sciences. The community of astronomers and cosmologists has been encouraged by seeing its members receive two Nobels, in physics, during the 1970s. When the prize for physiology or medicine was announced in 1979, many observers were surprised that the award went to the developers of computed tomography, a diagnostic technique, instead of to the leaders in the field of immunogenetics; there were rumors of strife

Keystone

George D. Snell

and last-minute compromise in the meeting of the Caroline Institute.

The partisans of the immunogeneticists clearly persisted, however, because in 1980 the prize was divided equally among George D. Snell, now retired from the Jackson Laboratory, Bar Harbor, Maine; Jean Dausset of the Lariboisière-St. Louis, St. Louis Hospital, and the University of Paris; and Baruj Benacerraf of the Harvard Medical School, Boston. The three conducted research in a field that harks back to a member of the Nobel class of 1930, Karl Landsteiner, who in 1901 announced his discovery of the ABO system of human blood groups. (The inheritance of the blood types was demonstrated around 1910.) The hallmarks of immunogenetics are the application of immunological techniques to the study of inherited traits and, conversely, the use of genetic methods to elucidate the properties of the immune system. The biochemical connection between immunology and genetics has been growing stronger since 1944, when the function of the nucleic acids in controlling hereditary traits was discovered. The contributions of the 1980 prizewinners underlie much of present knowledge of the inherited qualities that determine whether tissue can be successfully transplanted from one individual to another and those that affect the susceptibility of persons to an important group of diseases.

Snell, who was born in Bradford (now part of Haverhill), Mass., on Dec. 19, 1903, graduated from Dartmouth College in 1926 and received a doctorate of science in genetics from Harvard in 1930. As a postdoctoral fellow at the University of Texas he studied with Hermann J. Muller, who later (1946) won a Nobel Prize for his discovery (in 1927) that X-rays induce hereditary mutations in fruit flies. Snell extended Muller's finding from insects to mammals by showing that the same effect occurs in mice. In 1935 he joined the staff of the Jackson Laboratory, a private institution set up in 1929 for the study of mammalian genetics. Throughout its existence the laboratory has devoted itself to the breeding and study of mice with impeccable pedigree; it now derives a steady and considerable income from the sale of about two million mice, belonging to about 60 inbred strains, per year.

At the laboratory Snell continued to work on the mutation of mice for several years, but in 1944 he turned his attention to the genetic and immunological elements that affect the success or failure of tissue transplantation. Before Snell had arrived, Clarence Cook Little, the founder of the laboratory, knew that the viability of grafted tumors was strongly influenced by heredity. Little's position was based on several years of experiments in which he had observed that tumors transplanted from one mouse to another usually survive, and the tissues of the donor and the recipient are said to be compatible, if the mice belong to the same genetic strain. Conversely, if the mice are of different strains, their tissues are incompatible and the transplants are rejected. It appeared that several genes were involved in these effects, but Little's results did not reveal the relative strength of these genes or their locations on the chromosomes of the mice.

Snell also was aware of the research conducted by Peter A. Gorer at Guy's Hospital and the Lister Institute of Preventive Medicine in London. In 1937 Gorer had reported his discovery that after a mouse had received an incompatible transplant, its blood serum would cause the red blood cells of the donor to clump together, or agglutinate. He concluded that the immune system of the recipient, when stimulated by the antigens of the graft, develops antibodies that not only attack the tumor but also evoke a specific and easily observable response to tests like those used in blood typing. Even though the fate of a graft might remain uncertain for weeks or months, Gorer's tests gave reliable results within three days after transplantation. Gorer also showed that grafts of normal skin were equally effective in provoking the formation of the antibodies, and he identified a genetic locus responsible for them.

Snell confirmed and extended Little's results by systematically breeding dozens of generations of mice and testing each individual for tissue compatibility. From an initial mating of mice from two distinct strains he selected the offspring that would accept skin grafts from one of the parent strains and mated them with members of the other strain. The next generation was similarly tested, and the selected mice again crossed with tissue-compatible parent strain. The outcome, after about 15 generations, was a new strain genetically identical to one of the parent strains in all respects except that it would accept skin grafts only from the other original strain. At this stage the new strain could be perpetuated by inbreeding; that is, by mating its members with one another. Over the years Snell developed 69 of these inbred strains and detected in them 11 genetic loci associated with tissue compatibility.

Keystone

Jean Dausset

After World War II Gorer spent a year at Snell's laboratory, where the two scientists showed that the most important of the loci found by Snell coincided with the one that determines the structure of the antigens responsive to the serum tests developed by Gorer. The site was first regarded as a single gene, but further breeding experiments revealed that it is an assemblage —now called the major histocompatibility complex— of several genes, each of which can exist in 20 to 30 alternative forms, called alleles. Each allele consists of a unique stretch of DNA that, during cell growth, governs the structure of a protein whose distinctive antigenic properties are expressed equally in the response of the skin to Snell's grafts and in the response of the blood cells to Gorer's specific antiserums.

Dausset, who was born on Oct. 19, 1916, in Toulouse, France, received his medical degree from the University of Paris in 1943. While serving with the Free French forces during World War II, he gained first-hand experience with the reactions caused in recipients of transfusions of incompatible blood. After the war he undertook further training in immunology and immunohematology at the Harvard Medical School and returned to France as director of the laboratories of the national blood transfusion center. In 1952, during a study of blood samples taken from patients who had received many transfusions, Dausset observed that many of these blood serums contained antibodies that reacted with antigens of the leukocytes and platelets present in the blood of other individuals, causing those particles, but not the red cells, to agglutinate. The human leukocyte antigens, or HLA's, were found to be controlled by a genetic locus on one chromosome, like the major histocompatibility complex found in mice by Gorer and Snell. As soon as it was found that these antigens are not confined to the blood cells, but are detectable in most of the tissues of the body, it became possible to use serological tests to predict the compatibility of various tissues before using them as grafts.

A series of International Histocompatibility Workshops, initiated in 1964 by D. Bernard Amos (a former student of Peter Gorer), brought increasing order to a confusing situation that had evolved as independent research teams developed different techniques for identifying human compatibility antigens. Further clarification was provided in 1965 by Dausset, who had by then been appointed associate professor of medicine at the University of Paris, and two scientists working in Czechoslovakia. They proposed that the human gene complex contains several subsites, each of which determines the structure of several antigens. No one doubts that the human histocompatibility complex is just as complicated as that of the mouse, and although systematic genetic experiments cannot be used to unravel it, tissue matching has become an indispensable step in bringing together suitable donors and recipients for transplants of skin, kidneys, and other organs.

More than 40 distinct HLA's have been identified in human tissue samples, though only a few of them can be present in any one individual. The number of different combinations is, therefore, so large that the possibility of a perfect match between unrelated persons is extremely small. The specificity of HLA typing has proved of great value in forensic medicine, particularly in settling cases of disputed paternity, in which the usefulness of blood group determination is limited.

Benacerraf was born in Caracas, Venezuela, on Oct. 29, 1920, and shortly thereafter was taken by his parents to Paris, where he received his early education. He graduated from Columbia University in New York City in 1942 and received his medical degree from the Medical College of Virginia, at Richmond, in 1945. After returning to New York, first as an intern and then as a research fellow at Columbia, he went back to France in 1949 to serve as director of research at the National Center for Scientific Research at the Broussais Hospital in Paris. In 1956 he became professor of pathology at the New York University Medical School and collaborated for a time with Gerald Edelman of Rockefeller University in analyzing the structure of antibodies. (Edelman shared a Nobel Prize in 1972 for his work on this topic.)

Baruj Benacerraf

Joe Wrinn—Keystone

Benacerraf, troubled by the inhomogeneity of the antibodies available, quit the structure problem and began experiments directed toward the preparation of more uniform antibodies by immunizing the simple synthetic antigens of animals. He soon found that antibody production depended strongly upon both the molecular structure of the synthetic antigens and the genetic strain of the experimental animal. Guinea pigs of one strain, for example, would generate antibodies when exposed to one of the antigens but not when exposed to another, while those of a second strain would respond to the second antigen but not to the first. Benacerraf and scientists at other laboratories discovered that the genes (called Ir, for immune response) that control the formation of these antibodies are located within the major histocompatibility complexes of mice, guinea pigs, and other species.

In 1968 Benacerraf was appointed chief of the laboratory of immunology at the National Institute of Allergy and Infectious Diseases (part of the National Institutes of Health at Bethesda, Md.), and in 1970 he was named Fabyan professor of comparative pathology at the Harvard Medical School in Boston, where he is also president of the Sidney Farber Cancer Institute. Since the time he participated in Edelman's research project, his own interests have remained focused upon the details of the way in which the immune system deals with foreign antigens. He has ascertained that the Ir genes affect the production of only one of two major classes of antibodies, those that require the cooperation of so-called helper cells in their function. The genetic loci mapped with the aid of Benacerraf's synthetic antigens were shown to be associated with several diseases that afflict humans; in many of these, the immune system produces antibodies that attack the body's own tissues. Susceptibility to some of these diseases is inherited, and Benacerraf's findings have been essential in guiding studies of the origins and mechanisms of these conditions, which include psoriasis (a chronic skin lesion marked by red, scaly patches), Graves' disease (also called exophthalmic goiter), and ankylosing spondylitis (an inflammation commonest in young males, sometimes leading to fusion of the lower vertebrae).

—John V. Killheffer

AWARD	WINNER	AFFILIATION
ARCHITECTURE		
Architectural Firm Award	Hardy Holzman Pfeiffer Associates, New York, N. Y.	
Gold Medal of the American Institute of Architects	Joseph L. Sert	Sert, Jackson and Associates
International Pritzker Architecture Prize	Luis Barragán	
ASTRONOMY		
Annie J. Cannon Award in Astronomy	Lee Ann Wilson	Iowa State University of Science and Technology, Ames
Dannie Heineman Prize for Astrophysics	Joseph H. Taylor, Jr.	University of Massachusetts at Amherst
Elliott Cresson Medal	Riccardo Giacconi	Harvard/Smithsonian Center for Astrophysics, Cambridge, Mass.
Franklin Medal	Lyman Spitzer, Jr.	Princeton University, N.J.
Henry Draper Medal	William W. Morgan (Ret.)	Yerkes Observatory, University of Chicago, Ill.
Herschel Medal	Gerard de Vaucouleurs	University of Texas at Austin
John Price Wetherill Medal	Ralph A. Alpher	General Electric Research and Development Center, Schenectady, N.Y.
John Price Wetherill Medal	Robert Herman	University of Texas at Austin
CHEMISTRY		
Bingham Medal	Howard Brenner	University of Rochester, N.Y.
Charles Frederick Chandler Medal	Frank H. Westheimer	Harvard University, Cambridge, Mass.
Copley Medal	Sir Derek Barton	L'Institut de Chimie des Substances Naturelles, Gif-sur-Yvette, France
Franklin Medal	Avram Goldstein	Stanford University, Calif.; Addiction Research Foundation, Palo Alto, Calif.
Garvan Medal	Elizabeth K. Weisburger	National Cancer Institute, Bethesda, Md.
Gold Medal of the American Institute of Chemists	Arthur M. Bueche	General Electric Co.
Gregory and Freda Halpern Award	Anthony M. Trozzolo	University of Notre Dame, South Bend, Ind.
Guthrie Medal and Prize	Michael E. Fisher	Cornell University, Ithaca, N.Y.
Howard N. Potts Medal	Stanley G. Mason	McGill University, Montreal, Canada
National Academy of Sciences Award in Chemical Sciences	Frank H. Westheimer	Harvard University, Cambridge, Mass.
Perkin Medal	Ralph Landau	Halcon International Inc.
Priestley Medal	Herbert C. Brown (Ret.)	Purdue University, Lafayette, Ind.
Robert A. Welch Award in Chemistry	Sune Bergström	Karolinska Institutet, Stockholm, Sweden

AWARD	WINNER	AFFILIATION
Victor K. LaMer Award	Wilson Ho	Cornell University, Ithaca, N.Y.
Willard Gibbs Medal	F. Albert Cotton	Texas A & M University, College Station
Wolf Prize in Chemistry	Henry Eyring	University of Utah, Salt Lake City
EARTH SCIENCES		
Arthur L. Day Medal	Henry G. Thode	McMaster University, Hamilton, Ont., Canada
Carl-Gustaf Rossby Research Medal	Roscoe R. Braham, Jr.	Cloud Physics Laboratory, University of Chicago, Ill.
Clarence Leroy Meisinger Award	J. Michael Fritsch	National Oceanic and Atmospheric Administration, Boulder, Colo.
Cleveland Abbe Award for Distinguished Service to Atmospheric Sciences	Stanley A. Changnon, Jr.	Illinois State Water Survey, Urbana; University of Illinois, Urbana
Einar Naumann-August Thienemann Medal	David Frey	Indiana University, Bloomington
George P. Merrill Award	Robert N. Clayton	Enrico Fermi Institute, University of Chicago, Ill.
James B. Macelwane Award	Lawrence Grossman	Enrico Fermi Institute, University of Chicago, Ill.
James B. Macelwane Award	Thomas W. Hill	Rice University, Houston, Texas
James B. Macelwane Award	Norman Sleep	Stanford University, Calif.
Maurice Ewing Medal	J. Tuzo Wilson	Ontario Science Centre, Don Mills, Canada
Penrose Medal	Hollis Dow Hedberg (Ret.)	Princeton University, N.J.
Robert E. Horton Award	William C. Ackermann	University of Illinois, Urbana
Second Half Century Award	Thomas H. Vonder Haar	Colorado State University, Fort Collins
Second Half Century Award	Charles D. Keeling	Scripps Institution of Oceanography, University of California at San Diego
Stacie Prize	Gordon Rostoker	University of Alberta, Edmonton, Canada
Sverdrup Gold Medal	Jerome Namias	Scripps Institution of Oceanography, University of California at San Diego
William Bowie Medal	Charles Whitten (Ret.)	U.S. National Ocean Survey
ELECTRONICS AND INFORMATION SCIENCES		
Alexander Graham Bell Medal	Richard R. Hough	AT&T
Delmer S. Fahrney Medal	Jerome B. Wiesner (Ret.)	Massachusetts Institute of Technology, Cambridge
Distinguished Service Award of the Association for Computing Machinery	Bernard A. Galler	University of Michigan, Ann Arbor
Edison Medal	Robert Adler	Extel Corp.

AWARD	WINNER	AFFILIATION
Founders Medal of the Institute of Electrical and Electronics Engineers	Simon Remo	TRW Inc.
Glazebrook Medal and Prize	Michael C. Crowley-Milling	CERN, Geneva, Switz.
Grace Murray Hopper Award	Robert M. Metcalfe	3Com Corp.
Jack A. Morton Award	Nick Holonyak, Jr.	University of Illinois, Urbana
Lamme Medal	Eugene C. Starr	Bonneville Power Administration, Portland, Ore.
Medal of Honor of the Electronic Industries Association	Arthur A. Collins	Arthur A. Collins Inc.
Morris E. Leeds Award	Wallace H. Coutler	Coutler Electronics, Inc.
Scientist of the Year	Jacob Rabinow	National Bureau of Standards, Washington, D.C.
Turing Award	Charles A. R. Hoare	University of Oxford, U.K.
ENERGY		
Enrico Fermi Award	Sir Rudolf E. Peierls	University of Oxford, U.K.
Enrico Fermi Award	Alvin M. Weinberg	Institute for Energy Analysis, Oak Ridge (Tenn.) Associated Universities
ENVIRONMENT		
Edward W. Browning Achievement Award for Conserving the Environment	A. Starker Leopold (Ret.)	University of California at Berkeley
John C. Phillips Medal	Sir Peter Scott	International Union for Conservation of Nature and Natural Resources, Morges, Switzerland; World Wildlife Fund, Morges, Switzerland
J. Paul Getty Wildlife Conservation Prize	Harold J. Coolidge	Former president of the International Union for Conservation of Nature and Natural Resources
National Academy of Sciences Award for Environmental Quality	Gilbert F. White	University of Colorado, Boulder
FOOD AND AGRICULTURE		
Babcock-Hart Award	Steven R. Tannenbaum	Massachusetts Institute of Technology, Cambridge
Borden Award in Nutrition	Roslyn B. Alfin-Slater	University of California at Los Angeles
Conrad A. Elvehjem Award for Public Service in Nutrition	Sanford Miller	U.S. Department of Health and Human Services
Distinguished Service Award of the U.S. Department of Agriculture	Glenn A. Bennett	U.S. Department of Agriculture
Distinguished Service Award of the U.S. Department of Agriculture	Marion L. Goulden	U.S. Department of Agriculture

AWARD	WINNER	AFFILIATION
Distinguished Service Award of the U.S. Department of Agriculture	Gail M. Shannon	U.S. Department of Agriculture
Distinguished Service Award of the U.S. Department of Agriculture	Odette L. Shotwell	U.S. Department of Agriculture
Distinguished Service Award of the U.S. Department of Agriculture	Robert D. Stubblefield	U.S. Department of Agriculture
International Award of the Institute of Food Technologists	John C. Ayres	University of Georgia, Athens
Lederle Award in Human Nutrition	Jack W. Coburn	University of California at Los Angeles; Wadsworth Medical Center, Los Angeles Veterans Administration Center
Mead Johnson Award for Research in Nutrition	Dale R. Romsos	Michigan State University, East Lansing
Osborne and Mendel Award of the Nutrition Foundation	Lucille S. Hurley	University of California at Davis
Samuel Cate Prescott Award	John W. Erdman, Jr.	University of Illinois, Urbana
William V. Cruess Award	Roy G. Arnold	University of Nebraska, Lincoln
Wolf Prize in Agriculture	Karl Maramorosch	Rutgers, the State University of New Jersey, New Brunswick
LIFE SCIENCES		
A. I. Oparin Medal	Cyril Ponnamperuma	University of Maryland, College Park
American Association for the Advancement of Science—Rosenstiel Award in Oceanographic Science	David H. Cushing	Ministry of Agriculture, Fisheries, and Food, U.K.
Charles Thom Award	Clifford W. Hesseltine	U.S. Department of Agriculture
Common Wealth Award of Distinguished Service	Lewis Sarett	Merck & Co., Inc.
Dobzhansky Memorial Award	Benson E. Ginsburg	University of Connecticut, Storrs
Edgar D. Tillyer Award	Fergus W. Campbell	University of Cambridge, U.K.
Eli Lilly Award in Biological Chemistry	Roger D. Kornberg	Stanford University School of Medicine, Calif.
Gairdner Foundation International Award	Paul Berg	Stanford University School of Medicine, Calif.
Gairdner Foundation International Award	Irving B. Fritz	University of Toronto, Canada
Gairdner Foundation International Award	H. Gobind Khorana	Massachusetts Institute of Technology, Cambridge
Gairdner Foundation International Award	Efraim Racker	Cornell University, Ithaca, N.Y.
Gairdner Foundation International Award	Jesse Roth	National Institutes of Health, Bethesda, Md.
Gairdner Foundation International Award	Michael Sela	Weizmann Institute of Science, Rehovot, Israel
Lawrence Memorial Award	James M. Affolter	University of Michigan, Ann Arbor

AWARD	WINNER	AFFILIATION
Louisa Gross Horwitz Prize	César Milstein	Medical Research Council, London, U.K.
U.S. Steel Foundation Award in Molecular Biology	Phillip A. Sharp	Center for Cancer Research, Massachusetts Institute of Technology, Cambridge
Waksman Award in Microbiology	Julius Adler	University of Wisconsin, Madison
Wilson S. Stone Memorial Award	Michael R. Lerner	University of Pennsylvania, Philadelphia
MATERIALS SCIENCES		
Acta Metallurgica Gold Medal	Johannes Weertman	Northwestern University, Evanston, Ill.
Albert Victor Bleininger Award	Gilbert C. Robinson	Clemson University, S.C.
Arthur Frederick Greaves-Walker Award	William W. Coffeen	Clemson University, S.C.
Berkeley Citation	Joseph A. Pask (Ret.)	University of California at Berkeley
Copeland Award	Peter J. Sereda	National Research Council, Ottawa, Canada
John E. Marquis Memorial Award	Russell K. Wood	American Standard, Inc.
Karl Schwartzwalder-Professional Achievement in Ceramic Engineering Award	J. Richard Schorr	Battelle Memorial Institute, Columbus, Ohio
MATHEMATICS		
Frank Nelson Cole Prize in Algebra	Michael Aschbacher	California Institute of Technology, Pasadena
Frank Nelson Cole Prize in Algebra	Melvin Hochster	University of Michigan, Ann Arbor
Leroy P. Steele Prize	Harold M. Edwards	Courant Institute of Mathematical Sciences, New York University
Leroy P. Steele Prize	Gerhard P. Hochschild	University of California at Berkeley
Leroy P. Steele Prize	André Weil (Ret.)	Institute for Advanced Study, Princeton, N.J.
National Academy of Sciences Award in Applied Mathematics and Numerical Analysis	George F. Carrier	Harvard University, Cambridge, Mass.
Norbert Wiener Prize	Tosio Kato	University of California at Berkeley
Norbert Wiener Prize	Gerald B. Whitham	California Institute of Technology, Pasadena
Oswald Veblen Prize in Geometry	Mikhael Gromov	State University of New York at Stony Brook
Oswald Veblen Prize in Geometry	Shing-Tung Yau	Institute for Advanced Study, Princeton, N.J.
Wolf Prize in Mathematics	Henri Cartan (Ret.)	University of Paris, France
Wolf Prize in Mathematics	Andrey N. Kolmogorov	Moscow M. V. Lomonosov State University, U.S.S.R.

AWARD	WINNER	AFFILIATION
MEDICAL SCIENCES		
Albert Lasker Basic Medical Research Award	Paul Berg	Stanford University School of Medicine, Calif.
Albert Lasker Basic Medical Research Award	Stanley N. Cohen	Stanford University School of Medicine, Calif.
Albert Lasker Basic Medical Research Award	A. Dale Kaiser	Stanford University School of Medicine, Calif.
Albert Lasker Basic Medical Research Award	Herbert W. Boyer	University of California at San Francisco
Albert Lasker Clinical Medical Research Award	Sir Cyril A. Clarke	University of Liverpool, U.K.
Albert Lasker Clinical Medical Research Award	Ronald Finn	Royal Liverpool Hospital, U.K.
Albert Lasker Clinical Medical Research Award	Vincent J. Freda	College of Physicians and Surgeons, Columbia University, New York; Presbyterian Hospital, New York
Albert Lasker Clinical Medical Research Award	John Gorman	College of Physicians and Surgeons, Columbia University, New York; Presbyterian Hospital, New York
Albert Lasker Clinical Medical Research Award	William Pollack	College of Physicians and Surgeons, Columbia University, New York; Ortho Diagnostic Systems Inc.
Bernard B. Brodie Award in Drug Metabolism	Minor J. Coon	University of Michigan, Ann Arbor
Bolton L. Corson Medal	Bruce N. Ames	University of California at Berkeley
Charles F. Kettering Award	Elwood V. Jensen	Ben May Laboratory for Cancer Research, University of Chicago, Ill.
Charles S. Mott Award	Elizabeth C. Miller	McArdle Laboratory for Cancer Research, University of Wisconsin, Madison
Charles S. Mott Award	James A. Miller	McArdle Laboratory for Cancer Research, University of Wisconsin, Madison
CIBA Award for Hypertension Research	Bjorn Folkow	University of Göteborg, Sweden
CIBA Award for Hypertension Research	Arthur C. Guyton	University of Mississippi, Jackson
Distinguished Service Award for Outstanding Contributions in Medicine	Leon Jacobson (Ret.)	University of Chicago, Ill.
Distinguished Service Award of the American Medical Association	Frank H. Mayfield	
Dr. Rodman E. Sheen and Thomas G. Sheen Award	Jonathan E. Rhoads	University of Pennsylvania School of Medicine, Philadelphia
Dr. William Beaumont Award in Medicine	John B. McCraw	Eastern Virginia Medical School, Norfolk

AWARD	WINNER	AFFILIATION
Edward Longstreth Medal	Leonard T. Skeggs, Jr.	Case Western Reserve University, Cleveland, Ohio
Flemming Award	Anthony S. Fauci	National Institute of Allergy and Infectious Diseases, Bethesda, Md.
Founders' Prize of the Texas Instruments Foundation	William E. Paul	National Institute of Allergy and Infectious Diseases, Bethesda, Md.
Heath Memorial Award	Phil Gold	Montreal General Hospital, Canada
Jeffrey A. Gottlieb Memorial Award	Joseph H. Burchenal	Memorial Sloan-Kettering Cancer Center, New York
Jonas Friedenwald Research Prize	Arnall Patz	Johns Hopkins Wilmer Eye Institute, Baltimore, Md.
Joseph B. Goldberger Award in Clinical Nutrition	Russell B. Scobie	St. Luke's Hospital, Newburgh, N.Y.
Meyer and Anna Prentis Award	Henry S. Kaplan	Stanford University School of Medicine, Calif.
Queen's Gold Medal	Henry Harris	University of Oxford, U.K.
Radiation Research Society Award	Martin J. Brown	Stanford University School of Medicine, Calif.
Scientific Achievement Award of the American Medical Association	Harold E. Kleinert	University of Louisville School of Medicine, Ky.
Vestermark Award	Daniel X. Freedman	University of Chicago, Ill.
William D. Coolidge Award	John R. Cameron	University of Wisconsin, Madison
Wolf Prize in Medicine	James L. Gowans	Medical Research Council, London, U.K.
Wolf Prize in Medicine	César Milstein	Medical Research Council, London, U.K.
Wolf Prize in Medicine	Leo Sachs	Weizmann Institute of Science, Rehovot, Israel
OPTICAL ENGINEERING		
Adolph Lomb Medal	David M. Bloom	Hewlett-Packard Co.
Albert A. Michelson Medal	Emil Wolf	University of Rochester, N.Y.
David Richardson Medal	William T. Plummer	Polaroid Corp.
David Richardson Medal	Richard F. Weeks	Polaroid Corp.
Frederic Ives Medal	Aden B. Meinel	University of Arizona, Tucson
R. W. Wood Prize	Anthony E. Siegman	Stanford University, Calif.
PHYSICS		
Alan T. Waterman Award	Roy F. Schwitters	Harvard University, Cambridge, Mass.
American Physical Society High Polymer Physics Prize	Robert Simha	Case Western Reserve University, Cleveland, Ohio
American Physical Society Prize in Fluid Dynamics	Philip S. Klebanoff	U.S. National Bureau of Standards
Apker Award	Richard P. Binzel	Macalester College, St. Paul, Minn.
Biennial Award of the Acoustical Society of America	Peter H. Rogers	U.S. Naval Research Laboratory
Boltzmann Medal	Rodney J. Baxter	Australian National University, Canberra
Charles Vernon Boys Prize	Alan E. Costley	National Physical Laboratory, Teddington, U.K.

AWARD	WINNER	AFFILIATION
Charles Vernon Boys Prize	P. N. Pusey	Royal Signals and Radar Establishment, Malvern, U.K.
Common Wealth Award of Distinguished Service	James Hillier (Ret.)	RCA David Sarnoff Research Center, Princeton, N.J.
Davisson-Germer Prize in Surface Physics	Robert Gomer	James Franck Institute; University of Chicago, Ill.
Earl K. Plyler Prize	Richard N. Zare	Stanford University, Calif.
Ernest Orlando Lawrence Memorial Award	Donald W. Barr	Los Alamos Scientific Laboratory, N.M.
Ernest Orlando Lawrence Memorial Award	B. Grant Logan	Lawrence Livermore National Laboratory, Calif.
Ernest Orlando Lawrence Memorial Award	Nicholas P. Samios	Brookhaven National Laboratory, Upton, N.Y.
Ernest Orlando Lawrence Memorial Award	Benno P. Schoenborn	Brookhaven National Laboratory, Upton, N.Y.
Ernest Orlando Lawrence Memorial Award	Charles D. Scott	Oak Ridge National Laboratory, Tenn.
Fluid Dynamics Prize	Hans Wolfgang Liepmann	Graduate Aeronautics Laboratories, California Institute of Technology, Pasadena
Hewlett-Packard Europhysics Prize	O. Krogh Andersen	Max Planck Institute for Solid State Research, Stuttgart, West Germany
Hewlett-Packard Europhysics Prize	Andries R. Miedema	Philips Research Laboratories, Eindhoven, The Netherlands
Humboldt Senior U.S. Scientist Award	Walter G. Mayer	Georgetown University, Washington, D.C.
Irving Langmuir Prize in Chemical Physics	Willis H. Flygare	University of Illinois, Urbana
James Clerk Maxwell Prize in Plasma Physics	Thomas H. Stix	Plasma Physics Laboratory, Princeton University, N.J.
J. Robert Oppenheimer Memorial Prize	Richard Henry Dalitz	University of Oxford, U.K.
Max Born Medal and Prize	Helmut Faissner	Rhenish-Westphalian Technical University, Aachen, West Germany
Maxwell Medal and Prize	J. M. Kosterlitz	University of Birmingham, U.K.
Maxwell Medal and Prize	David J. Wallace	University of Edinburgh, Scotland
Medal of Honor of the Institute of Electrical and Electronics Engineers	William Shockley	Stanford University, Calif.
New York Academy of Sciences Award in Physical and Mathematical Sciences	Nicholas P. Samios	Brookhaven National Laboratory, Upton, N.Y.
Oliver E. Buckley Solid State Physics Prize	Robert C. Lee	Cornell University, Ithaca, N.Y.
Oliver E. Buckley Solid State Physics Prize	Douglas D. Osheroff	Bell Laboratories, Murray Hill, N.J.
Oliver E. Buckley Solid State Physics Prize	David M. Richardson	Cornell University, Ithaca, N.Y.
Otto Klung Award	Theodor W. Hänsch	Stanford University, Calif.
Pioneers of Underwater Acoustics Medal	Claude W. Horton	Applied Research Laboratories, University of Texas, Austin

AWARD	WINNER	AFFILIATION
Queen's Gold Medal	Sir Denys H. Wilkinson	University of Sussex, U.K.
Rumford Medal	William F. Vinen	University of Birmingham, U.K.
Rutherford Medal and Prize	Paul G. Murphy	University of Manchester, U.K.
Rutherford Medal and Prize	John J. Thresher	Rutherford Laboratory, Chilton, U.K.
Tom W. Bonner Prize in Nuclear Physics	Bernard L. Cohen	Sarah Mellon Scaife Nuclear Physics Laboratory, University of Pittsburgh, Pa.
Trent-Crede Medal	John C. Snowdon (Ret.)	Pennsylvania State University, University Park
Walter Schottky Prize	Klaus Funke	Technical University of Hannover, West Germany
Washington Academy of Sciences Award in Physical Sciences	E. Joseph Friebele	U.S. Naval Research Laboratory
William F. Meggers Award	John G. Conway	Lawrence Berkeley Laboratory, Calif.
Wolf Prize in Physics	Michael E. Fisher	Cornell University, Ithaca, N.Y.
Wolf Prize in Physics	Leo P. Kadanoff	University of Chicago, Ill.
Wolf Prize in Physics	Kenneth G. Wilson	Cornell University, Ithaca, N.Y.
PSYCHOLOGY		
American Association for the Advancement of Science Socio-Psychological Prize	Stephen G. Harkins	Northeastern University, Boston, Mass.
American Association for the Advancement of Science Socio-Psychological Prize	Bibb Latané	Ohio State University, Columbus
American Association for the Advancement of Science Socio-Psychological Prize	Kipling D. Williams	Drake University, Des Moines, Iowa
Distinguished Contribution Award for Applications in Psychology	Edwin A. Fleishman	Advanced Research Resources Organization, Washington, D.C.
Distinguished Professional Contribution Award	Douglas W. Bray	AT&T
Distinguished Professional Contribution Award	Leonard D. Eron	University of Illinois, Chicago
Distinguished Professional Contribution Award	Nicholas Hobbs	Vanderbilt University, Nashville, Tenn.
Distinguished Professional Contribution Award	Zygmunt A. Piotrowski	Temple University, Philadelphia, Pa.
Distinguished Scientific Contribution Award	Albert Bandura	Stanford University, Calif.
Distinguished Scientific Contribution Award	Alvin M. Liberman	Yale University, New Haven, Conn.; University of Connecticut, Storrs
Distinguished Scientific Contribution Award	Michael S. Posner	University of Oregon, Eugene

AWARD	WINNER	AFFILIATION
SPACE EXPLORATION		
Distinguished Public Service Medal of NASA	John M. Bozajian	Hughes Aircraft Co.
Distinguished Public Service Medal of NASA	Steve D. Dorfman	Hughes Aircraft Co.
Distinguished Public Service Medal of NASA	C. Malcolm Meridith	Hughes Aircraft Co.
Distinguished Service Medal of NASA	Lawrence Colin	Ames Research Center, Moffett Field, Calif.
Distinguished Service Medal of NASA	Charles Hall (Ret.)	Ames Research Center, Moffett Field, Calif.
Outstanding Leadership Medal of NASA	Robert U. Hofstetter	Ames Research Center, Moffett Field, Calif.
Outstanding Leadership Medal of NASA	Ralph W. Holtzclaw	Ames Research Center, Moffett Field, Calif.
Outstanding Leadership Medal of NASA	Joel Sperans	Ames Research Center, Moffett Field, Calif.
TRANSPORTATION		
Collier Trophy	Paul B. MacCready	AeroVironment, Inc., Pasadena, Calif.
Henderson Medal	Fitz Eugene Dixon	Delaware River Port Authority, Philadelphia, Pa.
Henderson Medal	Robert Johnston	Port Authority Transit Corporation, Camden, N.J.
Inventor of the Year	Paul B. MacCready	AeroVironment, Inc., Pasadena, Calif.
National Academy of Sciences Award in Aeronautical Engineering	James S. McDonnell	McDonnell-Douglas Corp.
Queen's Gold Medal	J. Paul Wild	Commonwealth Scientific and Industrial Research Organisation, Australia
SCIENCE JOURNALISM		
American Association for the Advancement of Science-Westinghouse Science Writing Award	Mark Bowden	*Inquirer,* Philadelphia, Pa.
American Association for the Advancement of Science-Westinghouse Science Writing Award	David Crisp	*Herald-Press,* Palestine, Texas
American Association for the Advancement of Science-Westinghouse Science Writing Award	C. P. Gilmore	*Popular Science*
American Institute of Physics-U.S. Steel Foundation Science Writing Award in Physics and Astronomy	Dennis Overbye	*Sky and Telescope*

AWARD	WINNER	AFFILIATION
Bradford Washburn Award	Kenneth F. Weaver	*National Geographic*
James T. Grady Award for Interpreting Chemistry for the Public	Robert W. Cooke	*Boston Globe*
National Academy of Sciences Public Welfare Medal	Walter Sullivan	*New York Times*
MISCELLANEOUS		
American Physical Society International Prize for New Materials	LeGrand G. Van Uitert	Bell Laboratories, Murray Hill, N.J.
Coleman Memorial Award	William G. Hyzer	Consulting engineer, Janesville, Wis.
James Murray Luck Award	W. Conyers Herring	Stanford University, Calif.
Leo Szilard Award for Physics in the Public Interest	Sidney D. Drell	Stanford Linear Accelerator Center, Calif.
National Academy of Sciences Award for Distinguished Service	William C. Kelly	Commission on Human Resources, National Research Council, Washington, D.C.
Robert A. Millikan Award	Thomas D. Miner	*The Physics Teacher*
Westinghouse Science Talent Search	1. Amy S. Reichel	Hunter College High School, New York, N.Y.
	2. Douglas A. Simons	Vero Beach High School, Vero Beach, Fla.
	3. Michael M. Dowling	Newington High School, Newington, Conn.
	4. Song Tan	Southwest Miami High School, Fla.
	5. Joel M. Wein	Stuyvesant High School, New York, N.Y.
	6. Terence D. Sanger	Dalton School, New York, N.Y.
	7. Lori E. Kaplowitz	George W. Hewlett High School, Hewlett, N.Y.
	8. Seth S. Finkelstein	Bronx High School of Science, New York, N.Y.
	9. Mark L. Movsesian	Forest Hills High School, New York, N.Y.
	10. William I-Wei Chang	Bronx High School of Science, New York, N.Y.

Obituaries

The following persons, all of whom died in recent months, were widely recognized for their scientific achievements.

Bateson, Gregory (May 9, 1904—July 4, 1980), British anthropologist and philosopher, cultivated an abiding interest in the patterns that unify organisms, and in his search for these became absorbed in such disciplines as anthropology, psychology, biology, and philosophy. After earning a master's degree in anthropology at St. John's College, Cambridge, Bateson went to New Guinea where he conducted field work. In 1936 he published *Naven* and married anthropologist Margaret Mead. The two made classic studies (1936–38) on the people of Bali, which were published as *Balinese Character* (1942). The couple were divorced in 1950. From 1950 to 1962 Bateson studied alcoholism and schizophrenia at the Veterans Administration Hospital in Palo Alto, Calif. He also broke new ground in communication theory with his studies of families with schizophrenic children and with his most important contribution to psychiatry; the double bind theory. This hypothesis asserted that individuals who were sent contradictory messages of love and hate by their parents develop schizophrenia. His interest in communication theory also extended to animals, especially dolphins. In 1976 Bateson was appointed to the Board of Regents of the University of California, and in 1978 he became scholar-in-residence at the Esalen Institute at Big Sur, Calif. His other writings include *Communication, the Social Matrix of Psychiatry* (1951), *Steps to an Ecology of Mind* (1972), and *Mind and Nature: A Necessary Unity* (1979).

Bullard, Sir Edward Crisp (Sept. 21, 1907—April 3, 1980), British scientist, pursued an outstanding academic career as a geophysicist before becoming an adviser to the U.S. government on nuclear power. His work in geophysics was fundamental to the development of the science. He pioneered the theory of continental drift, a concept now widely accepted, that the continents were once joined as a single supercontinent but have since undergone large-scale horizontal movements relative to one another and to the ocean basins. Later scientific investigations provided evidence that continental drift and the tectonic plates on which land masses rest have a profound effect on the occurrence of earthquakes. Bullard also made original contributions in such areas as the study of the Earth's magnetic field, the dating of rocks, and the use of computers in geophysics. During World War II his team developed techniques to protect naval vessels from magnetic mines. He served as professor of physics at the University of Toronto (1948–49) and director of Britain's National Physical Laboratory (1950–55) before returning to Cambridge, where he was professor of geophysics from 1964 until his retirement in 1974. In the U.S. he served (1963–74) as professor at the University of California at San Diego. Bullard was knighted in 1953.

Keystone

Sir Edward Bullard

Carpenter, R. Rhys (Aug. 5, 1889—Jan. 2, 1980), U.S. archaeologist, participated in excavations (1923–24) in Spain and announced in 1925 the discovery of the first Greek settlement there. The port town was named Hemeroskopeion ("lookout post" or "watchtower") and was probably built by Greek mariners and merchants before 600 BC. Carpenter, who founded (1913) the classical archaeology department at Bryn Mawr College in Pennsylvania, headed the department until his retirement in 1955. A Rhodes scholar with a wide range of interests, he also conducted original research in the transmission of alphabets and the cultural geography of the Mediterranean region. His many writings included *The Tragedy of Etarre* (1912), *The Plainsman, and other Poems* (1920), *The Land Beyond Mexico* (1920), *The Esthetic Basis of Greek Art of the Fifth and Fourth Centuries B.C.* (1921), and *The Humanistic Value of Archaeology* (1933). Taken together, they reflect his studies in Greek art, history, archaeology, and literature, and also his flair for writing poetry. During the 1960s Carpenter occasionally returned to Bryn Mawr to teach, and in 1970 he published his last book, *The Architects of the Parthenon.*

Dornberger, Walter Robert (Sept. 6, 1895—June 26, 1980), German-born guided-missile and space-vehicle engineer, managed the development of the German V-2 rocket and Wasserfall missiles during World War

II. A professional army officer, he graduated from the School of Technology in Charlottenberg (now Technical University of Berlin) with a B.A. in 1927 and an M.A. in 1930. In the same year he returned to duty as a captain with the German Army in the ballistics branch of its ordnance department, where he became interested in developing the rocket, a weapon not specifically forbidden by the Versailles Treaty. Dornberger was instrumental in establishing a rocket-testing facility in Kummersdorf, near Berlin, and began recruiting a team of civilian engineers and scientists to staff it. He principally drew on the ranks of the Deutsche Verein für Raumschiffahrt (German Space Exploration Club), a group of young space enthusiasts who were developing liquid-propelled rockets in Berlin. Among the most notable of these were Wernher von Braun, Arthur Rudolph, and Klaus Riedel. As the scope of their activities increased, they established a larger rocket development and test facility, the famous center at Peenemünde (now in East Germany) on the Baltic coast. There, the Dornberger-Braun team developed the V-2 rocket that bombed Belgium, France, and Great Britain during World War II. At the end of the war Dornberger, by then a major general, was imprisoned in Britain for two years. After his release he moved to the U.S. and was a consultant for the U.S. Air Force at Wright-Patterson Field, in Dayton, Ohio; he later served as technical assistant to the president, as a vice-president, and chief scientist of the Bell Aircraft Corp. in Buffalo, N.Y. At Bell he was involved in the design of the Rascal air-to-surface missile and Project Dyna-Soar before retiring in 1965.

Drew, Richard G. (1899?—Dec. 14, 1980), U.S. chemical engineer, was a laboratory assistant for the Minnesota Mining and Manufacturing Co. (3M Co.) when he created Scotch brand cellophane tape by combining a glue and glycerin stickum with a strip of transparent tape. Drew began experimenting with various gummed tapes after automobile factory employees complained that painting two-toned cars was nearly impossible because the paint would often adhere to the masking tape when it was removed. The invention made Drew, who had had only three semesters of college engineering, a millionaire. The product ended his company's search for a clear and easily removable tape to fasten cellophane-wrapped packaging. His original discovery eventually led to the development of some 600 different products. In 1944 Drew was named director of the company's Products Fabrication Laboratory.

Erickson, Milton Hyland (Dec. 5, 1901—March 25, 1980), U.S. psychiatrist, was instrumental in establishing hypnosis as a vital tool in psychotherapy. After obtaining his medical degree from the University of Wisconsin in 1928, he practiced psychiatry in Worcester, Mass., and Eloise, Mich., before opening (1950) a private office in Phoenix, Ariz., and attracting patients from throughout the world. Erickson, who hypnotized some 30,000 people during his career, was also coauthor of a multitude of books and learned articles on hypnosis including: *Successful Treatment of a Case of Acute Hysterical Depression, Time Distortion in Hypnosis, Hypnotic Investigation of Psychosomatic Phenomena, Hypnotic Realities,* and *Practical Application of Medical and Dental Hypnosis.* Several of his theories and techniques were explored by Jay Haley, a former student and colleague for many years, in such books as *Advanced Techniques of Hypnosis and Therapy* and *Uncommon Therapy.* In addition to his studies, Erickson was the founding president (1957) of the American Society of Clinical Hypnosis, editor (1958–68) of its publication, *American Journal of Clinical Hypnosis,* and associate editor (1940–55) of the journal *Diseases of the Nervous System* (now the *Journal of Clinical Psychiatry*).

Farb, Peter (July 25, 1929—April 8, 1980), U.S. anthropologist, conducted research in such widely diverse fields as insect life, linguistics, the evolution of man, and culture in North America. A prolific and entertaining author, who gathered firsthand information during his extensive travels, Farb wrote or coauthored some 50 major articles and nearly 20 books. Some of his most popular works include: *The Insect World, Man's Rise to Civilization as Shown by the Indians of North America from Primeval Times to the Coming of the Industrial State, Humankind, Face of North America: The Natural History of a Continent,* and *Word Play: What Happens When People Talk.* Besides writing free-lance articles for such publications as *Reader's Digest, Saturday Review,* and *Audubon,* Farb was editor-in-chief (1960–61) of the Panorama science series produced by CBS, curator (1964–71) of American Indian cultures at Riverside Museum in New York City, and consultant (1966–71) in scientific education at the Smithsonian Institution. Shortly before his death he had completed *Consuming Passions: The Anthropology of Eating,* and was compiling along with Irven DeVore *The Human Experience: A Textbook of Anthropology.*

Fromm, Erich (March 23, 1900—March 18, 1980), German-born psychoanalyst and social philosopher, discovered that economic and social factors have a profound effect on human behavior and incorporated this new awareness into his concept of Freudian psychoanalysis. Fromm, who was trained in psychoanalysis at the University of Munich (1923–24) and at the Psychoanalytic Institute in Berlin, was an orthodox Freudian until he introduced a confrontational technique that is in direct opposition to the Freudian ideal of the analyst's role as passive and noncommittal. This practice was one of Fromm's major contributions to psychoanalysis and is now widely used by practitioners. Fromm became a giant in the field of psychology through his 20 books, which reflected his broad inter-

Michigan State University

Erich Fromm

ests in such subjects as the nature of man, religion, ethics, and love. After leaving Nazi Germany in 1934, he lectured (1934–41) at Columbia University in New York City, where his views became increasingly controversial. In *Escape from Freedom* (1941) he recounted his theories on the alienation of human beings in a technological society; in *The Sane Society* (1955) he called for international harmony in the nuclear age; and in *The Art of Loving* (1956), a classic among college students in the 1960s, he suggested that "love is the only sane and satisfactory answer to the problem of human existence." Other notable works include *Man for Himself* (1947), *Zen Buddhism and Psychoanalysis* (1960), and *The Revolution of Hope* (1968). He also served on the faculties of Bennington (Vt.) College; Universidad Nacional Autónoma de México (National Autonomous University), Mexico City; Michigan State University, East Lansing; and New York University, New York City.

Gantt, William Horsley (Oct. 24, 1892—Feb. 26, 1980), U.S. behavioral psychiatrist, specialized in the causes of mental illness and repudiated Freud's concept that neurotic and psychotic behavior stem only from childhood traumas and sexual conflicts. By inducing asthma, manic activity, and nervous breakdowns in dogs, Gantt concluded that mental illness could be affected by environment. Gantt was a proponent of Ivan Pavlov, who conditioned dogs to associate food with the ringing of a bell and to salivate even when food was not presented. Gantt extended these studies by teaching the dogs to distinguish the pitches of two different bells and to associate only one with food. He also made pioneering studies of the cardiovascular system and sparked an exercise craze by endorsing exercise to strengthen heart muscles. After Gantt received (1920) his M.D. from the University of Virginia, he studied in the U.S.S.R. for three years and then founded (1929) the Pavlovian Laboratory in the Phipps Psychiatric Clinic at Johns Hopkins University School of Medicine in Baltimore, Md. For many years he was a professor at the university and served as director of the Pavlovian Laboratory from 1929 to 1967. In 1946 Gantt was awarded a Lasker Award for his research into the causes of mental illness.

Ingelfinger, Franz Joseph (Aug. 20, 1910—March 26, 1980), German-born physician, was the editor (1967–77) of the prestigious *New England Journal of Medicine* and an expert in the field of gastroenterology, the study of diseases and disorders of the stomach and intestines. He graduated from Harvard University Medical School in 1936 and then devoted more than 25 years to gastroenterology. He undertook original research into the motions and contractions of the esophagus and intestines and was one of the first to maneuver tubes through the nose and mouth in order to study the physiology of the esophagus, stomach, and intestines. Ingelfinger also studied the role of the liver in various diseases. As the witty editor of the *New England Journal of Medicine* he broadened the scope of the journal by including articles on ethical and social problems in medicine. During his tenure circulation increased more than 60%, and the journal was established as the nation's main medical forum. Ingelfinger's students at Boston University took special pride in being singled out as "Fingerlings."

Johnson, Harold L(ester) (April 17, 1921—April 2, 1980), U.S. astronomer, devised the UBV photometric system of stellar magnitudes (with William W. Morgan), a system based on photoelectric photometry in three broad bands: the ultraviolet (U), blue (B), and visual (V) spectral regions. The method is based on the comparison of stars' magnitudes with a standard sequence of about 400 stars. Many more stars have been measured in the UBV than in any other photoelectric system. Johnson, who also expanded this system into the near-infrared region, graduated from the University of Denver in 1942 and served (1942–45) as a staff member at the Massachusetts Institute of Technology radiation laboratory. In 1948 he earned his Ph.D. from the University of California at Berkeley. During his many years as a teacher Johnson taught at the universities of Wisconsin, Chicago, Texas, and Arizona, and at the time of his death was professor of astronomy at the Instituto de Astronomía, Universidad Nacional Autónoma de México. There he helped found the university's Baja Observatory at San Pedro Mártir in Mexico.

Keeton, William Tinsley (Feb. 3, 1933—Aug. 17, 1980), U.S. ornithologist, helped to explain and identify the various sensory systems used by birds to orient themselves and navigate while flying. While conducting other research, he discovered that pigeons could navigate just as well under heavily overcast skies as they could in sunshine, thus disproving the tenet that birds use the Sun as their only navigational aid. By fastening magnets to pigeons' heads, he proved that the Earth's magnetic field also played a role in the birds' orientation and course selection. Other factors shown to affect the birds' navigational course were ultraviolet and polarized light, barometric pressure changes, and low-frequency sounds, inaudible to the human ear. Keeton, who earned a Ph.D. in entomology at Cornell University, taught that subject at Radford College in Virginia before moving to Cornell University and becoming (1969) Liberty Hyde Bailey Professor of Biology. He was also the author of a widely used textbook, *Biological Science.*

Libby, Willard Frank (Dec. 17, 1908—Sept. 8, 1980), U.S. chemist, was awarded the 1960 Nobel Prize for Chemistry for developing a radioactive carbon technique that dated archaeological artifacts tens of thousands of years old within an accuracy of 120 years. His "atomic clock" measured small amounts of radioactivity in organic or carbon-containing materials and identified older objects as those with less radioactivity. This system proved to be an invaluable tool for archaeologists, anthropologists, and Earth scientists; the latter were able to pinpoint the final period of the North American Ice Age to 10,000 years ago (rather than the previously believed 25,000 years ago). After receiving his Ph.D. (1933) from the University of California at Berkeley, Libby taught chemistry at his alma mater. When World War II broke out, he joined the Manhattan Project at Columbia University, where (1941–45) his laboratory team was responsible for developing a technique to separate uranium isotopes, a process vital to the creation of the atomic bomb. From 1945 to 1959 he worked at the Institute for Nuclear Studies at the University of Chicago, where he conducted his prize-winning carbon-14 research and also proved that tritium, hydrogen's heaviest isotope, was produced by cosmic radiation. As the first chemist to serve on the Atomic Energy Commission (now the Nuclear Regulatory Commission), Libby headed Pres. Dwight D. Eisenhower's "Atoms for Peace" project and conducted studies on the effects of radioactive fallout. In 1959 he joined the faculty of the University of California at Los Angeles, where he remained until his death.

Willard Libby

UPI

Marshall, John Leahy (1936—Feb. 12, 1980), U.S. orthopedic surgeon, specialized in reconstructive knee surgery and became nationally acclaimed for his expertise in sports medicine after treating such stars as tennis player Billie Jean King, basketball forward Julius Erving, and football fullback Larry Csonka. Marshall's popularity was due to both his innovative preventive and rehabilitative exercise programs, and to his ability to instill confidence in athletes undergoing rehabilitation. After receiving (1965) his M.D. from Albany Medical College he worked at the Hospital for Special Surgery in New York City, where he helped establish its Clinic for Sports Medicine. The clinic catered to the 45,000 competitors in the city's Public Schools Athletic League, offering them a weekly three-hour clinic. Marshall, a consultant to the U.S. Olympic ski team, was killed when the plane in which he was traveling to the Winter Olympics crashed.

Matthias, Bernd T. (June 8, 1918—Oct. 27, 1980), German-born physicist, was instrumental in the 1954 discovery of the superconductive niobium-tin alloy (three atoms of niobium and one atom of tin) and was credited with discovering more elements and compounds with superconducting properties than any other scientist. When he entered the field, there were only about 30 known superconducting materials; there are now more than 1,000, many of which were found with his help. His niobium-tin alloy proved to be the "workhorse" material of superconducting generators, magnets, and transmission systems. Superconducting magnets were viewed as a key element in the multibillion-dollar effort to use the fusion of hydrogen atoms to generate power. Matthias, who obtained a Ph.D. in physics from the Swiss Federal Institute of Technology in Zurich, taught at the University of California at San Diego and worked for both Bell Laboratories in Murray Hill, N.J., and the Los Alamos Scientific Laboratory in New Mexico.

Mauchly, John W(illiam) (Aug. 30, 1907—Jan. 8, 1980), U.S. physicist, with John Presper Eckert invented the Electronic Numerical Integrator and Computer (ENIAC), the first electronic digital computer capable of modifying a stored program. With U.S. Army financing, the two men constructed the mammoth computer, which covered 15,000 sq ft in the basement of the Moore School of Electrical Engineering at the University of Pennsylvania in Philadelphia. This invention replaced such time-consuming devices as hand-operated desk calculators, punched-card accounting machines, and differential analyzers, which the Army had been using to recompute its artillery firing tables used in North Africa during World War II. The ENIAC was first used by the U.S. Army in 1947 for ballistics tests in Maryland. In the same year Mauchly and Eckert established the Eckert-Mauchly Computer Corp., and in 1949 they announced their Binary Automatic Computer (BINAC), the first machine to use self-checking devices. Although Mauchly and Eckert sold their company to Remington Rand, Inc., in 1950, the two were instrumental in the development of the Universal Automatic Computer (UNIVAC-I), which was first used (1951) by the U.S. Bureau of the Census and proved invaluable because it could handle both numerical and alphabetical information with equal ease. After serving (1959–65) as president of Mauchly Associates, Inc., and founding (1967) Dynatrend, a computer consulting company, Mauchly returned (1973) to Sperry Rand Corp. as a consultant.

Parrot, André (Feb. 15, 1901—Aug. 24, 1980), French archaeologist, excavated and identified in Syria the site of the ancient Semitic city of Mari, previously known only from references in Babylonian texts. Parrot, a Protestant theologian, began excavations in 1933 at Tell Hariri and, from a temple dedication, was able to identify it as Mari. The site revealed buildings dating from around 3500 BC and thousands of tablets with cuneiform inscriptions of the 19th and 18th centuries BC. Parrot also worked on sites in Lebanon and Iraq. In 1946 he was appointed chief curator of French national museums and undertook a major reorganization of Near Eastern antiquities in the Louvre. He became general inspector of museums in 1965 and from 1968 to 1972 served as the first director of the Louvre. A member of the Académie des Inscriptions et Belles-Lettres, he wrote several books and edited *Cahiers d'archéologie biblique.*

Peshkin, M. Murray (May 23, 1892—Aug. 17, 1980), U.S. physician, was a pediatric allergist who diagnosed and then pioneered a treatment for children with intractable asthma (a term he coined to define severe asthma that does not respond to traditional treatment so long as the child remained at home). Peshkin found that children suffering from anxiety and fear experienced rapid asthmatic relief as soon as they were temporarily separated from their parents and transferred to residential treatment centers with new mother and father figures. In 1953 he speculated that as many as 200,000 children in the U.S. were afflicted with persistent attacks of asthma because of an emotionally troubled environment. After graduating (1914) from the Fordham University School of Medicine in New York City, Peshkin joined (1915) the staff at Mount Sinai Hospital, New York City, where he studied childhood allergies. After his first paper was published in 1922, his work came to the attention of Viennese physician Bela Schick. When Schick became director of the hospital's pediatric department in 1923, he helped Peshkin establish a new clinic specializing in juvenile asthma. After serving until 1952 as chief of the children's allergy service at Mount Sinai Hospital, he was (1953–60) clinical professor of medicine and clinical professor of pediatrics for allergy at the Albert Einstein College of Medicine of Yeshiva University in New York City. Besides writing more than 100 scholarly papers on allergy and asthma, he founded the Asthma Care Association Inc., the Asthmatic Children's Foundation of New York, and the association of Convalescent Homes and Hospitals (now the Association for the Care of Asthma, Inc.), all in New York.

Piaget, Jean (Aug. 9, 1896—Sept. 16, 1980), Swiss psychologist, was thought by many to have been the major figure in 20th-century developmental psychology and was the first to make a systematic study of the acquisition of understanding in children. His massive

Jean Piaget

A.S.L./Pictorial Parade

output, including *Le Langage et la pensée chez l'enfant* (1923; *The Language and Thought of the Child,* 1926), *Le Jugement et le raisonnement chez l'enfant* (1924; *Judgment and Reasoning in the Child,* 1928), *La Représentation du monde chez l'enfant* (1926; *The Child's Conception of the World,* 1929), and *La Construction du réel chez l'enfant* (1937; *The Construction of Reality in the Child,* 1954), influenced generations of teachers, educators, and child psychologists throughout the world, offering them a framework, based on observation, to categorize the different and successive stages in the development of human intelligence. He created a new awareness of the preadolescent years as crucial in intellectual life and revolutionized classroom techniques, notably in the teaching of mathematics and the perceptions teachers had of their roles in the classroom. Piaget, who published his first scientific article at the age of 11 and was an accomplished zoologist by 15, received a doctorate in natural sciences at the University of Neuchâtel in Switzerland in 1918. His interest soon turned toward psychology and, in particular, the intelligence tests devised by Alfred Binet. From 1921 to 1925 he worked at the J. J. Rousseau Institute in Geneva and then was professor of philosophy at Neuchâtel (1926–29). He then served as professor of child psychology at the University of Geneva. His work with children was sometimes criticized because it was based on small samples, in particular his own children, rather than on broader observation. It was Piaget's close study of individuals, however, that allowed him to enter the child's conceptual universe and postulate its evolution through precise stages. In 1955 he founded the International Center of Genetic Epistemology in Geneva, which he continued to direct after his retirement in 1971; he was also a co-director of the Education Department of UNESCO.

Plaskett, Harry Hemley (July 5, 1893—Jan. 26, 1980), Canadian astronomer, as professor of astrophysics (1928–32) at Harvard University and Savilian professor of astronomy (1932–60) at the University of Oxford, advanced the study of solar physics, especially through his work on absorption line profiles. The son of J. S. Plaskett, discoverer of the double star named after him, H. H. Plaskett worked with his father at the Dominion Astrophysical Observatory, Victoria, B.C., from 1919 until his Harvard appointment. A fellow of the Royal Society from 1936, he was president of the Royal Astronomical Society (1945–47) and received its gold medal in 1963. He was responsible for the erection of two solar telescopes at the Oxford Observatory and for the erection (1967) of the Royal Greenwich Observatory's 2.5-m reflecting telescope (Western Europe's largest) at Herstmonceux, East Sussex.

Price, Dorothy (Nov. 12, 1899—Nov. 17, 1980), U.S. endocrinologist, together with Carl Moore conceptualized the "feedback" mechanism, by means of which sex hormones circulating in the bloodstream regulate secretion of hormones by the pituitary gland; the pituitary hormones, in turn, stimulate the dispersal of sex hormones by the gonads. In the female the production of sex hormones by the ovaries is under the control of gonadotrophins (follicle-stimulating hormones) from the pituitary. This finding later led to the development of the birth control pill, which contains synthetic hormones to prevent ovulation. Price, who earned (1935) a Ph.D. in zoology from the University of Chicago, conducted research there for many years with Moore. After retiring from the university in 1965, she became Boerhoave professor at the University of Leiden in The Netherlands.

Rhine, J(oseph) B(anks) (Sept. 29, 1895—Feb. 20, 1980), U.S. psychologist, was credited with coining the term extrasensory perception (ESP) in the course of researching such phenomena as mental telepathy, precognition, and clairvoyance. Rhine initially studied to be a botanist but became fascinated with "psychic occurrences." With psychologist William McDougall he helped to establish in 1930 the Parapsychology Laboratory at Duke University, Durham, N.C. There Rhine held some 90,000 experiments using a wide variety of human subjects. In 1934 his book *Extra-Sensory Perception* created a sensation with the general public but was greeted with skepticism by the scientific community. His *New Frontiers of the Mind* (1937) further explained his experiments. Rhine left Duke in 1965 and formed his own research center, the Foundation for Research on the Nature of Man.

Roberts, Richard Brooke (Dec. 7, 1910–April 4, 1980), U.S. physicist, with Lawrence Hofstad discovered that about 0.7% of the neutrons generated by the fission of uranium-235 are emitted seconds or even minutes after the moment of fission. This discovery was significant because these so-called "delayed" neutrons allow the control rods in a nuclear reactor several seconds to adjust and respond to the huge quantity of "prompt" neutrons being generated immediately; without this adjustment time the reactor could overheat, leading to the possibility of a dangerous accident. During World War II Roberts devised and tested the circuits under consideration for the proximity fuse, a triggering mechanism that detonates a bomb or artillery shell when its sensor detects a target. He also helped in the production of the fuse, which consists of a tiny miniaturized electronic circuit. After the war he returned to the Department of Terrestrial Magnetism at the Carnegie Institution in Washington, D.C., which he had joined in 1937. There he discovered the processes and major chemical synthetic mechanisms by means of which cell duplication occurs, and published his findings in *Studies of Biosynthesis in Escherichia coli,* a standard text in microbiology. Roberts, who earned a Ph.D. from Princeton University, formally retired in 1975 but continued to pursue independent research. The day before his death he discussed with astronomers his calculations analyzing the density of the universe at the time of the Big Bang.

Ronne, Finn (Dec. 20, 1899—Jan. 12, 1980), U.S. explorer, traversed some 5,800 km (3,600 mi) of the frozen wastes at the South Pole by ski and dogsled and charted vast areas during his nine polar expeditions to Antarctica. Ronne, who accompanied Adm. Richard E. Byrd on his 1933 expedition as a radio operator and dogsled driver, also joined the third exploration as chief of staff and executive officer. It was during this punishing expedition that Ronne and a companion spent 84 days delineating 725 km (450 mi) of coastline on Alexander Island. Ronne also discovered a bay, which he named for his father, Martin, who was with Roald Amundsen when Amundsen was the first to reach the South Pole. In March 1947 Ronne led his own party to the area near the Palmer Peninsula in Antarctica. There the group became the first to see the uncharted Weddell Sea area, and by the following March they had explored and mapped some 260,000 sq km (100,000 sq mi) of land and had completed surveying the Weddell Sea coastline, the last uncharted coastline in the world. Ronne's wife, Edith, participated in the scientific expedition and thus became one of the first women to set foot on the continent of Antarctica. During the journey Ronne was able to ascertain that the continent was definitely one continent and not two, as some believed. In later years he turned to lecturing and writing. At the time of his death he had just completed *Antarctica My Destiny*.

Stein, William H(oward) (June 25, 1911—Feb. 2, 1980), U.S. biochemist, was co-winner with his associate Stanford Moore, and Christian B. Anfinsen of the 1972 Nobel Prize for Chemistry for making fundamental contributions to enzyme chemistry. Together with his colleagues Stein deciphered the molecular structure of the digestive enzyme ribonuclease, a complex protein consisting of a single chain of 124 amino acids. This accomplishment, which was fundamental to the progress of medical research, represented a landmark in the field of chemistry because it raised hopes that damaged or defective enzymes, which can cause mental retardation or early death, might be repaired through chemical means. Stein, who earned (1938) a Ph.D. from Columbia University in New York City, joined the staff of Rockefeller Institute in New York City in 1938 and became a professor there in 1952. He also served as editor (1968–71) of the *Journal of Biological Chemistry*, the leading journal in its field. Since 1969 he had been confined to a wheelchair with polyneuritis, a rare paralyzing disease.

William H. Stein

Eugene H. Kone, Rockefeller University

Stern, Elizabeth (Sept. 19, 1915—Aug. 9, 1980), Canadian pathologist, was a specialist in cytopathology, the study of diseased cells, and the first to link the prolonged use of oral contraceptive pills with cervical cancer. She reported (1973) in the professional journal *Science* that research had established a direct link between contraceptive pills and cervical dysplasia, a condition that often precedes cervical cancer. In her most important contribution in this field Stern scrutinized cells cast off from the lining of the cervix and discovered that a normal cell goes through 250 distinct degrees of cell progression before reaching an advanced stage of cervical cancer. Her research prompted others to develop an automated cytologic screening instrument to detect cancer in its early stages. Earlier (1963) she published a case report in which she linked the herpes simplex virus to cervical cancer. Stern, who earned a degree in medicine from the University of Toronto, studied cytopathology under Herbert Traut in San Francisco, Calif., during the 1940s and later became (1965) professor of epidemiology at the School of Public Health of the University of California at Los Angeles.

Van Vleck, John H(asbrouck) (March 13, 1899—Oct. 27, 1980), U.S. physicist, shared the 1977 Nobel Prize for Physics with Philip W. Anderson and Sir Nevill F. Mott for their independent but closely related contributions to the understanding of the behavior of electrons in magnetic, noncrystalline solid materials. Although the trio's discoveries did not lead directly to practical applications, their research work served as a foundation for the development of such electronic devices as tape recorders, office copying machines, lasers, high-speed computers, and solar energy converters. Van Vleck, who dedicated most of his career to the study of magnetism in the structure of atoms, was informally recognized as "the father of modern magnetism." He published his scholarly *The Theory of Electric and Magnetic Susceptibilities* in 1932, and in 1934 became an associate professor at Harvard University. From 1945 to 1949 he served as chairman of Harvard's

Svenskt Pressfoto/Keystone

John H. Van Vleck

physics department and from 1951 to 1969 he was Hollis professor of mathematics and natural philosophy, the oldest endowed chair in the U.S. Van Vleck retired from Harvard in 1969.

Vodopyanov, Mikhail Vasilevich (1899—Aug. 13, 1980), Soviet Arctic aviator, was an adventurous polar flier who established new air routes and took part in several Soviet expeditions and rescue efforts, In 1934 he became a Hero of the Soviet Union after saving crew members of the icebreaker "Chelyuskin." The ship had become stuck in the ice while trying to navigate along the northern coast of Siberia. He was also credited with originating the idea of establishing a floating base camp on the ice at the North Pole, and commanded (1937–38) the flight detachment that landed Ivan D. Papanin's expedition. The ice research station drifted 274 days along the east coast of Greenland before two Soviet icebreakers removed the party from the collapsing floe. Earlier (1936) Vodopyanov had become the first in aviation history to fly over the Bering Sea when he transported the equipment for O. Y. Shmidt's scientific expedition from Moscow to Franz Josef Land. In 1937 he was the first pilot to land on ice at the North Pole. Vodopyanov also took part in an unsuccessful search for Sigismund Levanevsky, who vanished while trying to fly across the North Pole from Moscow to Fairbanks, Alaska. After serving in the air force during World War II, Vodopyanov retired (1946) with the rank of major general. His exploits were chronicled in *Polar Flier* (1952), *A Flier's Life* (1952), and *The Kireevs,* his first novel (1956).

Warren, Shields (Feb. 26, 1898—July 1, 1980), U.S. pathologist, was a pioneer in the field of radiation biology; during his long career he became an acknowledged expert on the harmful and beneficial effects of radiation. While serving as assistant professor of pathology at Harvard University Medical School, Warren was appointed (1947) director of the division of biology and medicine of the U.S. Atomic Energy Commission. In this capacity he studied the usefulness and effectiveness of radioactive isotopes in treating cancer; he ranked radioactive iodine first, cobalt second, and phosphorus third. As a pathologist with the U.S. naval technical mission to Japan, Warren made the first systematic study of radioactive fallout and discovered that more atomic bomb casualties at Hiroshima and Nagasaki resulted from the release of lethal radiation than from the actual blast itself. He found that the second largest group of fatalities were those stricken by flash burns from the "atom light." Besides his work in radiation Warren conducted research on diabetes and cancer and was credited with tracing the spread of major types of cancer. After earning (1923) his M.D. from Harvard University Medical School, he joined (1925) the faculty there and concurrently served for some 50 years as a pathologist at the New England Deaconess Hospital in Boston. His publications include *Medical Science for Everyday Use* (1927), *The Pathology of Diabetes Mellitus* (1930), and *A Handbook for the Diagnosis of Cancer of the Uterus* (1947; with Olive Gates).

Watt, George W(illard) (Jan. 8, 1911—March 29, 1980), U.S. inorganic chemist, contributed significantly to the Manhattan Project by isolating and purifying the plutonium used in the first atomic bomb. Watt earned a Ph.D. in 1935 from Ohio State University and in the same year joined the Goodyear Tire & Rubber Co. as a research chemist. He held patents for his work on the vulcanization of rubber. In 1937 he became a member of the faculty at the University of Texas at Austin, but from 1943 to 1945 he interrupted his teaching career to work on the Manhattan Project at the University of Chicago and at the Atomic Energy Commission's engineering facility in Richland, Wash. Watt also helped develop the hydrogen bomb and was widely recognized for his work in nonaqueous solvents, specifically ammonia and liquid ammonia. He wrote 17 textbooks and laboratory manuals and some 150 research papers. At the time of his death he was investigating the chemistry of hallucinogenic drugs.

Contributors to the Science Year in Review

C. Melvin Aikens *Archaeology.* Chairman, Department of Anthropology, University of Oregon, Eugene.
D. James Baker, Jr. *Earth sciences: Oceanography.* Professor and Chairman of the Department of Oceanography, University of Washington, Seattle.
Fred Basolo *Chemistry: Inorganic chemistry.* Professor of Chemistry, Northwestern University, Evanston, Ill.
Louis J. Battan *Earth sciences: Atmospheric sciences.* Director, Institute of Atmospheric Physics, University of Arizona, Tucson.
Gregory Benford *Physics: General developments.* Professor of Physics, University of California, Irvine.
Harold Borko *Electronics and information sciences: Information systems and services.* Professor, Graduate School of Library and Information Science, University of California, Los Angeles.
D. Allan Bromley *Physics: Nuclear physics.* Henry Ford II Professor and Director, Wright Nuclear Structure Laboratory, Yale University, New Haven, Conn.
Marjorie C. Caserio *Chemistry: Organic chemistry.* Professor of Chemistry, University of California, Irvine.
Roger H. Clark *Architecture and civil engineering.* Professor of Architecture, North Carolina State University, Raleigh.
Marc Davis *Medical sciences: Holistic medicine.* Free-lance medical writer, Chicago.
F. C. Durant III *Electronics and information sciences: Satellite systems.* Special Assistant to the Director, National Air and Space Museum, Washington, D.C.
Robert G. Eagon *Life sciences: Microbiology.* Professor of Microbiology, University of Georgia, Athens.
Lawrence E. Fisher *Anthropology.* Assistant Professor of Anthropology, University of Illinois at Chicago.
David R. Gaskell *Materials sciences: Metallurgy.* Professor of Metallurgy, University of Pennsylvania, Philadelphia.
Sheldon L. Glashow *Physics: High-energy physics.* Higgins Professor of Physics, Harvard University, Cambridge, Mass.
Robert Haselkorn *Life sciences: Molecular biology.* F. L. Pritzker Professor and Chairman of the Department of Biophysics and Theoretical Biology, University of Chicago.
John Patrick Jordan *Food and agriculture: Agriculture.* Director, Colorado State University Experiment Station, Fort Collins.
Lou Joseph *Medical sciences: Dentistry.* News and Information Chief, Bureau of Communications, American Dental Association, Chicago, Ill.
George B. Kauffman *Chemistry: Applied chemistry.* Professor of Chemistry, California State University, Fresno.
John V. Killheffer *Scientists of the Year: Nobel prizes.* Associate Editor, *Encyclopaedia Britannica.*

David B. Kitts *Earth sciences: Geology and geochemistry*. Professor of Geology and Geophysics and of the History of Science, University of Oklahoma, Norman.
Mina W. Lamb *Food and agriculture: Nutrition*. Professor emeritus, Department of Food and Nutrition, Texas Tech University, Lubbock.
Howard J. Lewis *U.S. science policy*. Director, Office of Information, National Academy of Sciences, Washington, D.C.
Allan G. Lindh *Earth sciences: Geophysics*. Geophysicist, U.S. Geological Survey, Menlo Park, Calif.
Robert W. Lucky *Electronics and information sciences: Communications systems*. Director, Electronic and Computer Systems Research, Bell Laboratories, Holmdel, N.J.
Melvin H. Marx *Psychology*. Research Professor of Psychology, University of Missouri, Columbia.
Julie Ann Miller *Medical sciences: General medicine*. Life Sciences Editor, *Science News* magazine, Washington, D.C.
John A. Mutchmor *Life sciences: Zoology*. Professor of Zoology, Iowa State University, Ames.
W. M. Protheroe *Astronomy*. Professor of Astronomy, Ohio State University, Columbus.
Arthur L. Robinson *Chemistry: Physical chemistry*. Research News Writer, *Science* magazine, Washington, D.C.
Saul Rosen *Electronics and information sciences: Computers*. Director, Computing Center, and Professor of Computer Science, Purdue University, West Lafayette, Ind.
Arthur H. Seidman *Electronics and information sciences: Electronics*. Professor of Electrical Engineering, Pratt Institute, Brooklyn, N.Y.
R. R. Shannon *Optical engineering*. Professor of Optical Sciences, University of Arizona, Tucson.
Mitchell R. Sharpe *Space exploration: Space probes*. Historian, Alabama Space and Rocket Center, Huntsville.
Albert J. Smith *Life sciences: Botany*. Professor and Chairman of the Department of Biology, Wheaton College, Wheaton, Ill.
Frank A. Smith *Transportation*. Senior Vice-President, Transportation Association of America, Washington, D.C.
J. F. Smithcors *Medical sciences: Veterinary medicine*. Editor, American Veterinary Publications, Santa Barbara, Calif.
W. E. Spicer *Physics: Solid-state physics*. Stanley W. Ascherman Professor of Engineering, Stanford University, Stanford, Calif.
Lynn Arthur Steen *Mathematics*. Professor of Mathematics, St. Olaf College, Northfield, Minn.
Norman M. Tallan *Materials sciences: Ceramics*. Chief Scientist, Materials Laboratory, Wright-Patterson Air Force Base, Ohio.
William A. Vogely *Energy*. Professor and Chairman of the Department of Mineral Economics, Pennsylvania State University, University Park.
Kenneth E. F. Watt *Environment*. Professor of Zoology, University of California, Davis.
Terry White *Space exploration: Manned flight*. Public Information Specialist, NASA Johnson Space Center, Houston, Texas.
Eric F. Wood *Earth sciences: Hydrology*. Assistant Professor of Civil Engineering, Princeton University, Princeton, N.J.

Index

This is a three-year cumulative index. Index entries to feature and review articles in this and previous editions of the *Yearbook of Science and the Future* are set in boldface type, *e.g.,* **Astronomy.** Entries to other subjects are set in lightface type, *e.g.,* Radiation. Additional information on any of these subjects is identified with a subheading and indented under the entry heading. The numbers following headings and subheadings indicate the year (boldface) of the edition and the page number (lightface) on which the information appears.

Astronomy 82–254; **81**–254; **80**–255
  archaelogical findings **82**–250
  black hole physics **81**–370
  computer simulated galaxies il. **81**–297
  cosmology **81**–88; **80**–88
  Einstein's theories **81**–79
  extraterrestrial catastrophe as cause of species extinction **82**–124
  honors **82**–404; **81**–401; **80**–406
  laser use in simulation **81**–367
  optical telescope **82**–198

All entry headings, whether consisting of a single word or more, are treated for the purpose of alphabetization as single complete headings and are alphabetized letter by letter up to the punctuation. The abbreviation "il." indicates an illustration.

A

B

C

D

H

N

O

P

Q

R

S

U

V

W

X

Y

Z

Acknowledgments

12 (Top) H. Roger Viollet; (center and bottom, left) Mary Evans Picture Library; (bottom, right) Culver Pictures

13 (Top to bottom, left to right) Culver Pictures; SCIENTIFIC AMERICAN, July 25, 1896; Jean-Loup Charmet; Mary Evans Picture Library; The Bettmann Archive

89 (Top to bottom, left to right) Jeff Albertson—Stock, Boston; Paul Conklin; Owen Franken—Stock, Boston; Harvey Lloyd—Peter Arnold, Inc.; © Paul Fusco—Magnum; Harvey Lloyd—Peter Arnold, Inc.; © Arthur Tress—Magnum; © Richard Howard—Black Star; © Joseph F. Viesti

129 Adapted from "Extraterrestrial Cause for the Cretaceous-Tertiary Extinction," Luis W. Alvarez, *et al.* SCIENCE, vol. 208, pp. 1095–1108, June 6, 1980

215 (Top to bottom, left to right) © Georg Gerster—Photo Researchers; Tom Myers; © Russ Kinne—Photo Researchers; © Porterfield-Chickering—Photo Researchers

216 Illustration by John Zielinski

232–233 Illustrations by Ramon Goas and John Draves

To extend the tradition of excellence of your Encyclopaedia Britannica educational program, you may also avail yourself of other aids for your home reference center.

Described on the next page is a companion product–the Britannica 3 bookcase–that is designed to help you and your family. It will add attractiveness and value to your home library, as it keeps it well organized.

Should you wish to order it, or to obtain further information, please write to us at

Britannica Home Library Service
Attn: Year Book Department
P. O. Box 4928
Chicago, Illinois 60680